RNA Technologies

Volume 16

RNA Technologies provides critical and comprehensive discussions of the most significant areas of RNA research, written by leading international authorities. Each volume in the series reviews both general topics, such as RNA interference, RNA biosynthesis and metabolism, microRNAs and diseases, transcriptome analysis and aptamers, as well as highly topical chapters on very recent findings in the field. Recurring themes of the series include the complexities associated with target recognition, RNA-protein interactions, analysis of genetic networks and the promise of small RNAs as cancer diagnostics and therapeutics in combating cancer.

RNA-based approaches have great potential to revolutionize molecular biology, cell biology, biomedical research and medicine. The nucleic acid-based molecules can be used to regulate the level of gene expression inside the target cells and their potential efficacy against several viruses, including SARS-CoV-2, and other chronic diseases represents a promising research area.

RNA Technologies publishes in both print and electronic format since 2010 and is intended for researchers from academia and industry, as well as graduates that look for a carefully selected collection of high quality review articles on their respective field of expertise.

The series is indexed in SCOPUS.

Jan Barciszewski
Editor

Decoding Aging and Neurodegeneration

The Role of Nucleic Acid Epigenetics

Springer

Editor
Jan Barciszewski
Nanobiomedical Center
Adam Mickiewicz University
Poznań, Poland

ISSN 2197-9731 ISSN 2197-9758 (electronic)
RNA Technologies
ISBN 978-3-031-91071-5 ISBN 978-3-031-91072-2 (eBook)
https://doi.org/10.1007/978-3-031-91072-2

This Springer imprint is published by the registered company Springer Nature Switzerland AG
The registered company address is: Gewerbestrasse 11, 6330 Cham, Switzerland

Introduction

Neurodegeneration and Epigenetics

The term epigenetics was coined 80 years ago by Conrad Hal Waddington. The word epigenetics is derived from the Greek word "epi" and means "over" or "above" the genome. It represents a composition of two different earlier views on organism development that have been competing since the time of Hippocrates and Aristotle. Shortly, the first one proposed that all parts of a mature organism are already present at the embryonic stage, and they grow during development, but the second proposed different suggestion that embryonic development involves the formation of new parts.

Whereas genetic alterations usually reflect permanent changes (mutations) of the DNA nucleotide sequence, epigenetic variations are mediated through processes that are reversible. So epigenetics represents a reversible mechanism in regulating the expression of genes without altering the nucleotide sequence of the genome. The epigenome links genotype to phenotype which plays an important role in modulating of aging process in response to environmental stimulation. These influences potentially alter the phenotype of an organism by interacting with and acting on both the genome and epigenome. The reversible nature of epigenetic changes makes them suitable tool for clinical interventions. Many epigenetic modifications can be inherited through mitosis and some have even found to be transgenerational.

Epigenetic regulation occurs on many levels of gene expression: from modifications of DNA and histones, regulating the mechanism of transcription, to decoding of messenger RNAs, and effects on translation. Epigenetics and chromatin architecture-mediated modulations in gene expression play a key role in the pathological implications of various heritable diseases affecting humans, without inducing mutations in DNA.

The epigenetic machinery includes changes in phenotype induced by reversible DNA and histone methylation, histone acetylation, phosphorylation, and post-transcriptional as well as posttranslational modifications of the genetic code. It also involves the action of non-protein coding RNAs (npcRNAs) in altering the decoding messenger RNAs, facilitating diversity in mRNA editing and alternative splicing.

The only epigenetic mechanism that directly modifies mammalian DNA is methylation of the cytosine carbon ring (m5C) in the dinucleotides of the CpG islands. Most of the genome does not contain the dinucleotide CpG repeats; however, it concentrates on the so-called CpG islands located in regulatory regions at 5′ end of the genes. It is estimated that about 70% of CpG dinucleotides in the genome are methylated in normal somatic cells. DNA methylation has been recognized as a normal mechanism of gene silencing with a crucial role in the transcriptional repression of repeated and centromeric regions. This mechanism efficiently regulates the dynamic process in which methyl groups might be added directly or removed stepwise from both DNA strands. The final result of methylation, and oxidative demethylation, is not permanent, which means that the mechanism of regulation is indeed much more complex. Some DNA methylation patterns are erased, whereas others are maintained in the germline contributing to alter genes expression in a heritable manner without affecting the underlying genomic nucleotide sequences. It should be kept in mind that most constitutively expressed genes and parts of tissue-specific genes and regulatory genes contain CpG islands in their promoter regions. It means that epigenetics is really emerging as the missing link among genetic inheritance, environmental influences, and body and brain health status. Epigenetics is of increasing relevance to neuroscience, with mechanisms being implicated in brain development and neuronal differentiation, as well as in more dynamic processes related to cognition. In the brain, specific changes in nucleic acids or their associated proteins in neurons and glial cells might imprint differential patterns of gene activation that will favor either cognitive enhancement or cognitive loss for more than one generation. Furthermore, deregulation of epigenetic signaling is appearing as a significant risk factor for aging, and neurodegeneration.

Aging is a slow but gradual process that is characterized by a continuous decline in the normal physiological functions of living organisms over their lifespan. With age, the body's resilience decreases, making it more sensitive to aging-related diseases, such as neurodegenerative diseases and cancer, and increasing the risk of death. Aging induces a slow decrease in cerebral function, which leads to cognitive impairment, memory loss, motor disorders, and finally, can reach functional decay and death. Among the main features of aging, we might find genomic instability, telomere shortening, epigenetic alterations, loss of proteostasis, nutrient detection deregulation, mitochondrial dysfunction, cellular senescence, stem-cell depletion, and altered intercellular communication.

Numerous studies have revealed how aging occurs and how it is regulated by complex cellular and molecular mechanisms at different stages of life. Many factors affecting the aging process and longevity have been reported, including telomere shortening, nutrient sensing, mitochondrial dysfunction and oxidative stress, deterioration of DNA repair and accumulation of DNA damage, changes in protein homeostasis leading to the accumulation and aggregation of misfolded proteins, and changes in epigenetic regulation.

These phenomena indicate that aging is a complicated process that may be regulated by a variety of different factors. These aging-associated epigenetic changes include DNA methylation, histone modification, chromatin remodeling, non-coding RNA (ncRNA) regulation, and RNA modification. Therefore, understanding of the epigenetic mechanisms in aging will provide new avenues to develop strategies to delay aging. It seems that aging interventions based on manipulating epigenetic mechanisms have led to the alleviation of aging or the extension of lifespan in animal models. Small molecule-based therapies and reprogramming strategies that enable epigenetic rejuvenation have been developed for ameliorating or reversing aging-related conditions. In addition, adopting health-promoting activities, such as caloric restriction, exercise, and calibrating circadian rhythm, has been demonstrated to delay aging. Currently, various clinical trials for aging intervention are ongoing, providing more evidence of the safety and efficacy of these therapies.

Aging has always been considered as a risk factor for neurodegenerative diseases, but there are individual differences and its mechanism is not yet clear. Epigenetics may unveil the relationship between aging and neurodegenerative diseases. Neuronal function and differentiation are tightly regulated by genome and epigenome, and neurodegeneration is the consequence of dysregulation in both.

Neurodegenerative disorders are among the greatest challenges and among the most serious health problems that will have to be faced by the modern societies across the world, especially in light of increasing population age. They are incurable, progressive conditions resulting from continuous degeneration and death of nerve cells. Most of these disorders become more common with advancing age, including Alzheimer's and Parkinson's diseases. The burden of these neurodegenerative diseases is growing inexorably as the population ages, with incalculable economic and human costs. Epigenetic modifications including chromatin modification, DNA methylation, and changes in regulatory RNAs (miRNA) are having a great impact on neurodegenerative disorders as well as memory. Epigenetics classically serves to propagate acquired non-DNA nucleotide sequence information to progeny cells. Neuroepigenetics involves the same chemical modifications of DNA and histones; however, in neurons, which in fact do not divide, these modifications are not propagated to progeny cells. Neurodegenerative diseases typically involve a progressive loss of neuronal integrity and function, followed by neuronal death. Depending on where in the brain the loss of integrity and neuronal loss occurs, various functional disabilities may arise which gradually worsens as neurodegeneration spreads. The underlying cause and localization of the neurodegenerative processes, however, often vary between different neurodegenerative disorders. The exact etiology of most neurodegenerative diseases is far from clear. Several neurodegenerative as well as age-induced dementia are accelerated by genetic as well as environmental parameters. Since the global human population has attained subtle stability and now aging is growing rapidly, several experiments have been performed to comprehend the molecular basis of these pathologies. It seems that epigenetic dysregulation plays a major role in aging and age-related neurodegenerative disorders, such as Alzheimer's, Parkinson's, and Huntington's diseases, where it may mediate interactions between genetic and environmental risk factors, or directly interact with disease-specific pathological factors. A majority of neurodegenerative diseases are characterized by the aggregation of intracellular proteins, but it is not clear whether this is a primary mechanism or a consequence of another disturbed cell function. The potential mechanisms of neurodegeneration are innumerable, including primary effects of protein homeostasis, disturbed protein degradation, gene expression, transcriptional regulation, mitochondrial dysfunction, etc. There is the urgent need to better understand disease pathophysiology in order to improve early diagnosis and development of disease-modifying treatments. Currently, there is no diagnostic test that can clearly indicate the presence, absence, or category of a neurodegenerative disease. Individual diagnosis is based on clinical evaluation of the symptoms and specific neuroimaging characteristics, and it often takes years to make one.

The central nervous system is highly sensitive to environmental factors, thus epigenetic disturbances are associated with neurological diseases. The CNS has a rich diversity of cell types with a highly specialized structural organization and is tightly regulated by the epigenetic mechanism in response to the environment. In addition, expression patterns are being continuously modulated by brain activity, memory, and learning. Alterations in epigenetic information give rise to deregulation of the expression, resulting in the development of several pathologies. The brain is a high energy-consumption organ that requires about 20% of the body's total basal oxygen to function. Therefore, alterations in the cerebral energy metabolism lead to disease, from subtle modifications in neuronal function to cellular death and neurodegeneration. As aging progresses, the brain is susceptible to its intrinsic and extrinsic cell signaling and to the systemic environment.

As previously mentioned, aging and neurodegeneration are closely related. The increase in overall survival, mitochondrial dysfunction, and the inflammatory response has been linked to accelerated aging and the progression of neurodegenerative diseases. Reactive oxygen species (ROS) and reactive nitrogen species are the main molecules contributing to oxidative stress (OS). The most studied free radicals are superoxide (O_2^-), hydrogen peroxide (H_2O_2), hydroxyl (HO•), nitric peroxide (ONO_2^-), and the nitric oxide-free radical (NO). OS is the resulting redox state from disequilibrium between the generation and detoxification of ROS. Overproduction of ROS by dysfunctional mitochondria, in addition to the imbalance in mitochondrial fission/fusion and mitochondrial DNA mutations, can cause oxidative damage to cells, inducing alterations to their biochemical and physiological functions and modify the genetic expression profiles inducing apoptosis and, thereafter, neurodegenerative pathologies.

In the last 20 years or so, the role of epigenetic alterations in the development of neurodegenerative diseases has been increasingly discussed. Epigenetic constitution presents a landscape where environmental factors may interact with genetic make-up of an individual. Additionally, the development of high-throughput technologies for genomic, transcriptomic, and epigenomic profiling now offers insight into epigenetic alterations in neurodegeneration, as well as opportunity for an integrative view of its interactions with various 'omic' levels. Interpretation of epigenomic profiling results in the context of neurodegeneration, and the methodology for integration of heterogeneous 'omic' data opens an array of novel biological and bioinformatics challenges and requires the development of novel approaches towards analysis of these data.

Accumulative damage by free radicals over a certain time leads to major brain inflammation. Therefore, oxidative stress can also interfere with the homeostatic balance between anti- and proinflammatory cytokines and induce neurodegeneration. Neuroinflammation is a physiological response of the brain to an injury, infection, or disease in order to prevent a potential damage, neuronal destruction, or to dispose of a damaging agent.

This book focuses on different epigenetic mechanisms that, by altering DNA structure through DNA methylation, histone modification or modulating gene expression by ncRNAs, such as microRNAs (miRNAs) or long non-coding RNAs, are involved in several processes associated with neurodegenerative diseases, such as the accumulation of misfolded proteins in the central nervous system, neuroinflammation and microglia activation, oxidative stress, apoptosis, and the formation synaptic plasticity. The book contains review papers of recent work on epigenetic regulation of aging and outlines the advances in intervention strategies for aging and age-associated diseases. A better understanding of the critical roles of epigenetics in the aging process will lead to more clinical advances in the prevention of human aging and therapy of aging-related diseases.

Poznań, Poland Jan Barciszewski

Contents

Deciphering (Epigenetic) Mechanisms Behind Aging and Senescence Using Next-Generation Sequencing Technologies

Dominik Saul, Onur Mert Batmaz, Patricia Raude, Vera Leidig, Xiaodong Wang, and Robyn Laura Kosinsky

Contents

Abstract Aging is the primary risk factor for numerous chronic diseases, driven in part by cellular senescence and associated epigenetic changes. Next-generation sequencing (NGS) technologies have transformed aging research by enabling high-resolution analysis of genetic, transcriptomic, and chromatin landscapes. Single-cell and spatial NGS approaches reveal the heterogeneity of aging tissues and uncover dynamic molecular interactions, such as senescence-associated pathways and chromatin remodeling processes. These advancements provide critical insights into the mechanisms of aging and open new avenues for therapeutic interventions targeting age-related diseases.

Keywords Next-generation sequencing · Aging · Senescence · Single-cell sequencing · Chromatin dynamics · Epigenetics · Aging-associated diseases

D. Saul · O. M. Batmaz · P. Raude · V. Leidig · X. Wang · R. L. Kosinsky (✉)
Robert Bosch Center for Tumor Diseases, Robert Bosch Gesellschaft für Medizinische Forschung mbH, Stuttgart, Germany
e-mail: robyn-laura.kosinsky@bosch-health-campus.com

J. Barciszewski (ed.), *Decoding Aging and Neurodegeneration*,
RNA Technologies 16, https://doi.org/10.1007/978-3-031-91072-2_1

1 Introduction

Aging is a complex biological process characterized by progressive decline in physiological functions and increased vulnerability to diseases such as cancer, neurodegenerative disorders, and cardiovascular conditions. While the causes of aging are multifactorial, epigenetic mechanisms, chromatin dynamics, and cellular senescence have emerged as pivotal factors shaping the aging landscape. Cellular senescence, marked by irreversible cell cycle arrest and the secretion of the senescence-associated secretory phenotype, contributes significantly to aging by disrupting tissue homeostasis and promoting chronic inflammation. This process is tightly regulated by epigenetic changes, including DNA methylation and histone modifications, which alter gene expression and chromatin structure (López-Otín et al. 2023).

To unravel the molecular underpinnings of aging, next-generation sequencing (NGS) technologies have become indispensable tools. These methods enable high-throughput analysis of genetic, transcriptomic, and epigenetic changes, offering unparalleled insights into the cellular and molecular basis of aging. For instance, single-cell RNA sequencing (scRNA-seq) has revolutionized our understanding of cellular heterogeneity in aging tissues by identifying distinct cell populations and dynamic changes in gene expression. Similarly, techniques like Assay for Transposase-Accessible Chromatin using sequencing (ATAC-seq) and chromatin immunoprecipitation sequencing (ChIP-seq) have shed light on chromatin accessibility and transcriptional regulation during aging.

Moreover, advancements in spatial transcriptomics have added a critical dimension by mapping gene expression within the tissue microenvironment, elucidating spatially resolved cellular interactions that influence aging processes. Integrating these approaches with bioinformatics tools has facilitated the creation of comprehensive multiomic datasets, empowering researchers to dissect complex aging mechanisms and identify potential therapeutic targets.

This chapter explores the transformative impact of NGS technologies in aging research, focusing on their applications in profiling chromatin dynamics, identifying epigenetic alterations, and understanding cellular senescence. By highlighting recent innovations and methodological advancements, we aim to provide a detailed overview of how these technologies are shaping the future of aging and senescence research, with implications for developing interventions to enhance healthspan and mitigate aging-associated diseases.

2 Innovations in RNA Sequencing Techniques for Aging Research

In recent years, several methods for transcriptome-wide analyses have emerged, varying in cost, workflow, and resolution as outlined in Table 1. At the gene expression level, three principal methodologies are bulk RNA sequencing (bulk RNA-seq), scRNA-seq, and spatial transcriptomics (Stark et al. 2019).

Bulk RNA-seq measures the entire transcriptome of a potentially heterogeneous population of cells, yielding an average expression level for all cells in the sample. Initially, RNA is extracted from a tissue or cell sample, fragmented, and reverse transcribed into complementary DNA (cDNA). The cDNA is then sequenced to quantify gene expression levels. Bioinformatics pipelines are employed to align reads to reference genomes, quantify gene expression, and conduct differential

Table 1 Commonly used next-generation sequencing methods in aging research, highlighting their resolution, key applications, advantages, and limitations

Method	Resolution	Key applications	Advantages	Limitations
Bulk RNA-seq	Population-level (average across all cells in a sample)	Differential gene expression and pathway analysis	Cost-effective, established workflows, high throughput	Lacks single-cell resolution and masks heterogeneity within cell populations
scRNA-seq	Single-cell resolution	Identifying rare cell types and cell-type-specific transcriptomics	Captures cellular heterogeneity, identifies rare populations, and dynamic processes	Computationally intensive, requires high-quality single-cell suspensions
Spatial transcriptomics	Spatial resolution within tissue sections	Linking gene expression to tissue architecture and cell-cell interactions	Integrates spatial and transcriptomic data, retains tissue context	Limited spatial resolution, currently high cost, requires specialized equipment
ATAC-seq	Genome-wide, chromatin accessibility	Identifying open chromatin regions, regulatory elements	Simple, scalable, reveals transcriptionally active regions	Does not provide protein-level information, lacks single-cell specificity
ChIP-seq	Genome-wide, DNA-protein interactions	Mapping histone modifications, transcription factor binding	Powerful for studying epigenetic regulation and chromatin dynamics	Requires antibodies of high specificity, labor-intensive, does not show accessibility
Multi-omics	Single-cell or bulk resolution	Comprehensive profiling of several processes gene	Can unveil, for example, transcriptional regulation and epigenetic changes at single-cell level	Highly computationally intensive, complex data integration, very high cost

expression analyses (Stark et al. 2019). This method is comparatively cost-effective and requires less computational power, yet it can still provide valuable information regarding gene expression changes in large populations of cells. This is particularly advantageous in research areas with high interindividual heterogeneity, such as aging, where larger group sizes are essential to elucidate transcriptomic changes (Stark et al. 2019). For example, this has been used in the creation of the Tabula Muris Senis, where across 10 different ages, 17 murine organs were taken and analyzed. Using this larger sample size, and multiple time points, specific pathways could be closely monitored across aging, making bulk RNA-seq even favorable for this kind of analysis (Schaum et al. 2020). Bulk RNA-sequencing was likewise used by Papaspyropoulos et al. in a model of replicative senescence in fibroblasts to evaluate translational differences after the onset of cellular senescence. The authors found a deterioration of translation but not a specific deregulation pattern to occur but that this regulation was different in oncogene-induced senescence (Papaspyropoulos et al. 2023).

In contrast to bulk RNA-seq, scRNA-seq enables the profiling of gene expression at the level of individual cells, capturing cell-type-specific changes. Cells are first isolated, and a single-cell suspension is prepared. To achieve single-cell resolution, the mRNA content of each cell is labeled with a cell-specific barcode. The RNA is then reverse transcribed into cDNA, amplified, and sequenced. Computational methods reconstruct gene expression profiles for individual cells, facilitating clustering and identification of different cell types using cell-type-specific marker genes (Stark et al. 2019). This approach allows for the investigation of heterogeneity within tissues and the identification of rare cell types or states that may be obscured in bulk RNA-seq. ScRNA-seq also facilitates the exploration of dynamic cellular processes and cell-cell interactions through RNA velocity and ligand-receptor interactions, making it particularly valuable for aging research, which often encompasses dynamic processes that follow specific trajectories with cell-type-specific contributions (He et al. 2020; Uyar et al. 2020). These specific analyses can be performed based on scRNA-seq data if the cell type of interest is clearly defined within the specific organ. Such has been done by Zou et al. for the human skin, where cell-type-specific networks were detected and trajectory analyses used to determine differentiation patterns and then explore the patterns behind this process, uncovering the downregulation of genes like HES1 and KLF6 to be crucial for the acceleration of cellular senescence (Zou et al. 2021).

Spatial transcriptomics integrates transcriptomic data with spatial information, particularly relevant in tissues exhibiting high heterogeneity, such as the aging bone, liver, and brain (He et al. 2020; Li and Wang 2021; Nikopoulou et al. 2023; Xiao et al. 2023; Kilfeather et al. 2024). This method provides a spatially resolved map of gene expression within a tissue section. The tissue is labeled with two spatial barcodes along the *x*- and *y*-axes. RNA is then sequenced, with each read linked to its corresponding spatial barcode, enabling gene expression to be mapped back to specific locations within the tissue (Stark et al. 2019). This information is crucial for understanding tissue architecture and the cellular interactions that influence aging. It is important to note that current technologies allow for sequencing of 55 μm spots, limiting applicability to larger cell aggregates rather than individual cells (Saul and Kosinsky 2023). In a tissue like the brain, where the spatial architecture is well studied through histology, the addition of spatial transcriptomics can add information on where the aging process is changing neuronal and nonneuronal cells. Allen et al. used the combination of single nucleus RNA-seq (snRNA-seq) and a spatial technology (MERFISH) in young and old mice to discover that white matter was a hotspot for aging-related glial and immune cell interaction and that the activation of microglia near oligodendrocytes in the corpus callosum was age-specific (Allen et al. 2023). Before spatial transcriptomics, these insights would have been difficult to obtain by just repetitive staining methods.

Each of these techniques offers distinct advantages and insights into the biology of aging and senescence. Collectively, they enable researchers to uncover the complex mechanisms that drive aging and may reveal new therapeutic approaches for age-related diseases.

3 Techniques for Mapping Chromatin Dynamics in Aging

Besides studying the transcriptome, deciphering epigenetic regulations in aging tissues, chromatin and its organizational structure are of utmost importance. ChIP-seq is a widely used epigenetic assay for studying DNA-protein interactions, such as mapping transcription factor binding sites (Johnson et al. 2007), histone modifications (Barski et al. 2007), identifying enhancer-promoter interactions (Heintzman et al. 2007), and functional annotation of noncoding regions (ENCODE Project Consortium 2012) to understand disease development. Technically, proteins are crosslinked with DNA in cells or tissues using formaldehyde to preserve their interactions. The chromatin is fragmented and immunoprecipitated with target-specific antibodies to isolate regions of interest within the DNA. After reversing the crosslinks, the associated DNA is purified and subjected to high-throughput sequencing. Due to its versatility, ChIP-seq has been utilized in numerous studies involving aging and senescence. ChIP-seq assays provide insights into how cells alter their histone modification profiles as organisms age, which in turn regulates developmental processes. The transcriptional regulation of genes involved in aging and senescence can also be studied by identifying key protein-DNA interactions through ChIP-seq. Overall, ChIP-seq provides a reliable and robust method for studying the effects of aging on the epigenetic landscape and important DNA-protein interactions in cells, offering mechanistic insights into the regulation of aging.

For example, this has been performed by Song et al., who used ChIP-seq in a model of replicative senescence to decipher that histone modifications like H3K4me1, H3Kme3, and H3K27me3 were reduced in these (replicative) senescent cells, whereas in oncogene-induced senescence, H3K27ac and H3K27me3 were even more downregulated and H3K4me1 and H3K4me3 were enriched. These ChIP-seq insights suggest a preference of repressors over activators in senescent cells, leading to specific senescence-associated heterochromatin foci (Song et al. 2022).

Another crucial epigenetic assay is ATAC-seq, which allows for the study of chromatin accessibility and genome regulation. ATAC-seq provides a simple and scalable method to map open-chromatin regions (Yekelchyk et al. 2024), identify regulatory elements (Yang et al. 2024), and analyze chromatin dynamics during development (Galang et al. 2020), differentiation (Finkbeiner et al. 2022), immune responses (Mogilenko et al. 2021), and disease. ATAC-seq utilizes hyperactive Tn5 transposase to load DNA fragment ends with sequencing adapters in a process known as tagmentation. The tagmented DNA is then purified, transposases are removed, and the isolated DNA fragments are sequenced. Comparative analysis of chromatin accessibility dynamics between young and old cells using ATAC-seq can reveal transcription factor binding patterns and gene activation during aging. Karakaslar and colleagues used ATAC-seq to compare cleavage profiles in young and old mice, determining chromatin accessibility profiles for the transcription factors JUN/FOS. Chromatin accessibility around JUN/FOS target genes, such as those involved in the NF-κB pathway, proinflammatory cytokines, chemokines, and genes expressed by activated myeloid and lymphoid cells, was found to increase in the spleens of older mice. The elevation in JUN/FOS signaling, due to increased chromatin accessibility in immune cells, helps explain how the aging process induces immune cells to adopt a more inflammation-prone state (Karakaslar et al. 2023). Similarly, Nikopoulou et al. used scATAC-profiles of young and old liver samples, finding that Nrf1 was enriched especially in old hepatocytes, potentially due to its involvement in protection against oxidative stress. The authors also found hepatocytes to be epigenetically diverse, depending on the spatial localization either pericentral or periportal (Nikopoulou et al. 2023).

Understanding the purpose, advantages, and limitations of each assay is crucial for experimental design. While assays like RNA-seq, ChIP-seq, ATAC-seq, and Hi-C provide valuable information about transcriptomic and epigenetic landscapes, no single assay can capture all aspects. For instance, RNA-seq provides information on gene expression but does not reveal chromatin accessibility, whereas ATAC-seq shows chromatin accessibility at promoter sites but does not determine the presence of transcription factors. This has led to the rising popularity of multiomics, a combinatorial use of transcriptomic, epigenetic, proteomic, metabolomic, lipidomic, and pharmacogenomic assays, to capture a more comprehensive view. A multiomic approach was used to identify a metabolic mechanism as the key driver of aging in skeletal muscle stem cells. Through a combinatorial analysis of transcriptomic, epigenomic, proteomic, and metabolomic datasets from skeletal muscle stem cells, researchers discovered perturbations in glutathione (GSH) metabolism in old mice, where aged cells exhibited a dichotomy of GSH-high and GSH-low states. These metabolic states were associated with higher levels of NF-κB and NRF2 expression, respectively (Benjamin et al. 2023). This multiomic approach not only revealed this novel mechanism in stem cell aging but also identified a potential target for reversing skeletal muscle aging.

The establishment of multiomic datasets provides a multidimensional analytical approach and increases the scale and resolution of data, leading to the identification of novel cell types or biological mechanisms that might otherwise remain undiscovered. Several databases, such as the Human Aging Genomic Resources (Tacutu et al. 2018), AgeFactDB (Hühne et al. 2014), Digital Aging Atlas (Craig et al. 2015), AGEMAP (Zahn et al. 2007), and Aging Atlas (Aging Atlas Consortium 2021), have already been created to analyze the heterogeneous and complex mechanisms of aging across different human organs and tissues. With advances in bioinformatics tools, the integration of various assay types is becoming more accessible. These databases offer information across multiple organ systems, cell types, and age groups, enabling users, even those with limited bioinformatics expertise, to conduct in-depth aging analyses.

4 Histone and DNA Modifications in Age Dynamics

In recent years, NGS technologies, such as ChIP-seq, have revolutionized our understanding of the epigenetic landscape, particularly in the context of aging research as summarized in Table 2. By enabling high-resolution mapping of histone modifications across the genome, NGS has strengthened previous results and provided critical insights into how these modifications influence gene expression and cellular aging processes. Aging cells experience a loss of histones, which was observed in various eukaryotic models. For instance, in aged yeast, a 50% reduction in nucleosomes was documented, leading to increased global transcription levels (Hu et al. 2014). In human fibroblasts, a reduction in histone synthesis correlates with aging, contributing to the loss of heterochromatin, which alters gene expression (O'Sullivan et al. 2010). In addition, aging is associated with the replacement of canonical histones by variants, such as macroH2A and H3.3, which affect gene transcription. For example, H3.3 accumulates in various tissues with age, replacing canonical histones (Piazzesi et al. 2016; Tvardovskiy et al. 2017), and H3.3 was found to positively correlate with lifespan-extending signaling pathways. This has been found and validated via bulk RNA-sequencing in *C. elegans* where H3.3-deficient nematodes with impaired insulin/IGF-1/DAF-2 signaling showed a reduced lifespan (Piazzesi et al. 2016).

N-terminal tails of core histones undergo various posttranslational modifications, including methylation, acetylation, phosphorylation, and ubiquitination. The respective histone, residue, and type of modification determine whether these changes activate or repress transcription (Iwasaki et al. 2013). An imbalance between activating and repressing histone modifications can alter gene expression patterns, contributing to aging-related changes in the transcriptome. Histone methylation plays a dual role in gene expression, with some modifications promoting and others repressing transcription. Specifically, H3K4, H3K36, and H3K79 methylation tends to activate genes, while H3K9, H3K27, and H4K20 are associated with repression (Greer and Shi 2012). Using ChIP-seq, Signal et al. demonstrated that

Table 2 Key epigenetic changes associated with aging, including histone modifications and DNA methylation patterns, with significant contributions from NGS

Epigenetic modification	Observed changes in aging	Biological effects	References
H3K4me3	Reduced in aging cells; stable in mouse neurons but recruitment to senescence-associated genes increases	Alters transcriptional activation, linked to lifespan and senescence regulation	Greer et al. (2010), and Cruz et al. (2018)
H3K27me3	Declines in *C. elegans* but increases in other species like mice and fish during aging	Represses developmental and longevity-related genes; manipulation extends lifespan	Jin et al. (2011), Siebold et al. (2010), and Baumgart et al. (2014)
DNA methylation	Global hypomethylation, hypermethylation at promoter CpG sites, and age-specific "epigenetic clock" markers identified	Gene silencing at promoters, age prediction, and cellular aging regulation	Horvath (2013), Day et al. (2013), and Jylhävä et al. (2017)
H3K9me3	Reduced levels observed during aging in hematopoietic stem cells; correlates with reduced SUV39H1 expression	Impacts chromatin stability and longevity-associated gene regulation	Djeghloul et al. (2016) and Lorbeck et al. (2010)
H3K36me3	Declines with age in multiple species; associated with transcriptional fidelity and longevity	Affects gene expression, lifespan, and aging-related pathways	Sen et al. (2015), Su et al. (2018), and Pu et al. (2015)
Histone variants (e.g., H3.3)	Increased incorporation of variants such as H3.3 in aging tissues	Alters transcription and contributes to chromatin dynamics and cellular aging	Tvardovskiy et al. (2017)
H3K9ac	Decreased levels in aged tissues; linked to impaired SIRT6 function	Impacts gene expression and chromatin accessibility; associated with aging symptoms	Kawahara et al. (2009) and Michishita et al. (2008)

while H3K4me3 enrichment remained stable in aging mouse neurons, H3K27ac occupancy at developmental gene regulatory regions, particularly enhancers, decreased with age (Signal et al. 2024). In addition, the SAGA transcriptional coactivator component SGF29 was found to be recruited to H3K4me3-occupied regions of senescence-associated genes, such as CDKN1A, to assemble transcription factors and coactivator complexes, key regulatory mechanisms in human mesenchymal progenitor cells and fibroblasts (Yan et al. 2023). In *C. elegans*, trimethylation of H3K4 (H3K4me3) negatively correlates with longevity. Loss of the methyltransferase SET2 and proteins WDR5 and ASH2 enhances survival, whereas depletion of the demethylase RBR2 shortens lifespan (Greer et al. 2010). A similar trend was observed in *Drosophila*, where loss of the RBR2 ortholog Lid decreases lifespan (Li et al. 2010). Notably, around 30% of genes with H3K4me3 show significant age-related deregulation (Pu et al. 2018). In yeast, aging results in reduced H3K4me3 levels linked to the upregulation of aging-associated genes (Cruz et al. 2018). Besides H3K4me3, modifications of H3K36 were associated with aging. A lifespan screen in *S. cerevisiae* revealed that reduced H3K36 methylation shortens lifespan. Deletion of the H3K36 demethylase RPH1 extended survival by increasing H3K36me3 levels (Sen et al. 2015). A decline in H3K36me3 during aging has also been confirmed in *C. elegans*, showing changes in gene expression associated with age. In *Drosophila*, inactivating MET1 resulted in reduced H3K36me3 levels and shortened lifespan (Pu et al. 2015). Recent findings indicate that loss of the H3K36 dimethyltransferase SET18 in *C. elegans* extends lifespan in a DAF16-dependent manner, highlighting the significance of H3K36 di- and trimethylation in longevity (Su et al. 2018). Moreover, H3K27me3, regulated by the polycomb repressive complex 2 (PRC2), typically represses transcription. Its levels decline with age in *C. elegans*, and increasing H3K27me3 by depleting the demethylase UTX1 can extend lifespan (Jin et al. 2011). Conversely, increased H3K27me3 has been noted in other species during aging; for example, depleting PRC2 homologs in flies promotes longevity (Siebold et al. 2010). Elevated H3K27me3 has also been observed in quiescent mouse muscle stem cells and teleost fish brain tissue (Baumgart et al. 2014). H3K9me3's role in longevity was suggested by the deletion of KDM4A, a demethylase, which led to reduced lifespan in male *Drosophila* (Lorbeck et al. 2010). This deletion downregulated several longevity-associated genes. Additionally, reduced SUV39H1 expression during aging was observed in human and murine hematopoietic stem cells (Djeghloul et al. 2016). Furthermore, trimethylation of H4K20 (H4K20me3) has been linked to aging, with significant increases noted in liver and kidney tissues of aged rats (Sarg et al. 2002). Senescent human IMR90 fibroblasts also showed an accumulation of H4K20me3 (Bevanger et al. 1989). In livers from aging rats, H3K9ac levels were found to decrease (Kawakami et al. 2009). SIRT6, an H3K9 deacetylase, impacts chromatin state at telomeres and is linked to premature aging symptoms seen in Werner syndrome (Michishita et al. 2008). Global deletion of SIRT6 leads to severe metabolic issues and short survival (Mostoslavsky et al. 2006). SIRT6 also represses NF-κB target gene expression by deacetylating H3K9ac (Kawahara et al. 2009) and regulates lifespan via the IGF1 pathway (Kanfi et al. 2012).

Histone ubiquitination, involving the attachment of ubiquitin to histones H2A and H2B, influences transcription; H2A ubiquitination generally represses, while H2B enhances transcription. Notably, H2B monoubiquitination is essential for the trimethylation of H3K4 and H3K79 (Lee et al. 2007). Age-dependent ubiquitination of H2A has been noted in *Drosophila* and is conserved in humans (Yang et al. 2019), and H2A.Z has been shown to conjoin proteins to process DNA double-strand break repair and is involved in premature muscle aging, likewise reducing lifespan (Belotti et al. 2024).

Besides histones, DNA can be methylated when a methyl group is added to the fifth position of cytosine, forming 5-methylcytosine (5mC). This modification predominantly happens in CpG-rich regions. NGS offers several powerful techniques for detecting DNA methylation, each with distinct advantages. Bisulfite sequencing involves treating DNA with bisulfite, converting unmethylated cytosines to uracils while preserving methylated cytosines, allowing for sequencing that reveals methylation status. Reduced representation bisulfite sequencing (RRBS) focuses on regions enriched for CpG sites, combining restriction enzyme digestion with bisulfite treatment to provide a cost-effective yet detailed methylation profile. Murach et al. used reduced representation bisulfite sequencing to analyze young and aged skeletal muscle, finding hypermethylated promoters in aged muscle to be overrepresented in the tricarboxylic acid cycle regulation, potentially explaining the reduction of these proteins in aged human muscle (Murach et al. 2021). Bisulfite sequencing was likewise used by Franzen et al. in long-term culture of mesenchymal stem cells revealing that changes in the methylation status were not regulated but rather followed an epigenetic drift (Franzen et al. 2021).

Finally, whole genome bisulfite sequencing (WGBS) offers a comprehensive view by treating the entire genome with bisulfite, yielding high-resolution methylation maps across the genome. Applying WGBS on skeletal muscle, Tharakan et al. detected a peak in methylation difference around the age of 52–62 years, focusing on neuronal pathways, while differences at the protein level become evident slightly later on (Tharakan et al. 2023).

Together, these methods have greatly enhanced our understanding of DNA methylation dynamics and their roles in biological processes, including aging (Rauluseviciute et al. 2019).

While methylation in promoter regions is linked to gene silencing, the role of gene body methylation is still insufficiently understood. Methylation can repress transcription by blocking transcription factor binding or recruiting methyl-CpG-binding proteins. The heritable nature of methylation is regulated by DNA methyltransferases (DNMTs) such as DNMT1, which maintains existing methylation, and DNMT3A and DNMT3B, which establish new methylation (Auclair and Weber 2012). 5mC can be reverted through "active" demethylation, mediated by ten-eleven translocation (TET) proteins that oxidize 5mC to forms like 5-hydroxymethylcytosine (5hmC). These oxidized bases are excised by thymine DNA glycosylase and replaced by unmodified cytosine through base excision repair (He et al. 2011). Recent studies found adenine methylation (6 mA) in *C. elegans*, regulated by DAMT-1 and NMAD-1 (Greer et al. 2015), and variations in 5mC levels in *Drosophila* species (Deshmukh et al. 2018). Aging is typically associated with hypermethylation at promoter CpGs and hypomethylation elsewhere (Day et al. 2013). Two key studies identified 353 and 71 differentially methylated CpG sites during aging, which may serve as reliable age predictors (Horvath 2013). These "epigenetic clocks" show strong correlations with chronological age ($r = 0.96$ and $r = 0.91$) and minor deviations (3.6 and 4.9 years, respectively). In comparisons to biological age predictors, the epigenetic clock has emerged as a promising tool to indicate true biological age, though further validation is needed (Jylhävä et al. 2017). In conclusion, histone and DNA modifications play a crucial role in the aging process, influencing gene expression and cellular function, and their evaluation through next-generation sequencing offers powerful insights into the molecular mechanisms underlying age dynamics and potential interventions for age-related decline in health and vitality.

5 Senescence as a Hallmark of Aging

NGS approaches that allow the evaluation of transcriptome-wide changes at the bulk or single cell level enable researchers to gather deep understanding of aging-related molecular processes, such as senescence. Cellular senescence is a cell fate involving stable cell cycle arrest that occurs in response to various stressors, including exogenous or endogenous DNA damage, oxidative stress, and oncogene activation. Recently, specific guidelines for the minimal information needed for senescence attribution have been published (Ogrodnik et al. 2024). One hallmark of a senescent cell is the increase in cyclin-dependent kinase inhibitors such as $p21^{Cip1/waf}$ (encoded by *CDKN1A*) or $p16^{Ink4a}$ (encoded by *CDKN2A*), a decline in cellular proliferation (Ki67, PCNA), or its highly secretory state, indicated by the secretion of a range of factors, the senescence-associated secretory phenotype (SASP). The SASP involves the secretion of proinflammatory cytokines, chemokines, growth factors, and proteases. The SASP can disrupt tissue homeostasis, promote chronic inflammation, and drive age-related pathologies (Saul et al. 2022; Zhang et al. 2022; Calabrò et al. 2024; Mahmud et al. 2024). Therapeutic strategies targeting senescent cells include senolytics, which induce apoptosis in senescent cells, and senomorphics, which suppress the harmful effects of the SASP without eliminating the cells themselves. These approaches have shown promise in preclinical models and are currently being validated in clinical trials (Tchkonia et al. 2013; Di Micco et al. 2021; Zhang et al. 2022).

Chromatin remodeling, including histone modifications and DNA methylation, plays a critical role in both the onset and maintenance of senescence. Polycomb-group proteins, such as EZH2, repress senescence-associated genes but can be inhibited to induce senescence via induction of key players like p21 and p53 (Saul and Kosinsky 2021; Crouch et al. 2022). In adipose precursors, DNA methylation changes, particularly hypomethylation, contribute to the onset of a senescent phenotype and impaired adipogenesis (Spinelli et al. 2022). Moreover, histone acetylation at superenhancers, regulated by factors such as p300, drives senescence (Sen et al. 2019).

By inducing the expression of Yamanaka factors (Oct-3/4, Sox2, Klf4, and c-Myc, combined as "4F"), a partial reprogramming of adult cells to progenitor cells and increase of cell proliferation and subsequent enhanced regeneration capacity can be achieved (Hishida et al. 2022). Mechanistically, senescence-dependent regulation of nicotinamide *N*-methyltransferase is crucial for controlling *S*-adenosyl-L-methionine levels. Intriguingly, 4F reprogramming in old mice reversed several markers of aging, at least in the liver, while $p16^{high}$ cells still limited cellular plasticity. When these were removed, a rejuvenation could be achieved by 4F reprogramming (Grigorash et al. 2023). A partial reprogramming using *Cdkn2a-OSK* AAVs reduced the expression of inflammation-associated genes, delay an aging phenotype in mice, and extend lifespan (Sahu et al. 2024).

Notably, the so-called bystander effect contributes to cellular aging: senescent cells negatively impact neighboring cells via secretions (mainly the SASP), exacerbating aging features in nonsenescent cells. A gradual decline in the proliferative capacity of cells before reaching the full senescent state indicates that cellular aging is a multistep process, rather than just a binary event of cell cycle arrest (Nassrally et al. 2019; Ogrodnik 2021; Victorelli et al. 2023). The crucial role of the SASP in the mediation of senescent cell burden makes this secretome an ideal candidate to intervene and potentially eliminate their effect. Subsequently, a valid biological marker that can be used with next-generation sequencing datasets is imperative. For this reason, SenMayo has been developed. SenMayo is a gene set specifically designed to identify senescent cells and their SASP and predict senescence-associated pathways across tissues. This gene set has been developed based on experimentally validated factors that were increased in and/or secreted by senescent cells. It has been validated in human bone biopsies and shown to be effective in identifying senescent cells at the single-cell level in both human and murine tissues. Particularly in scRNA-seq datasets, the SenMayo panel can be utilized to monitor senescent cell burden and identify the key cell types responsible for the secretion of the SASP. Furthermore, SenMayo can be utilized in bulk RNA-seq datasets to evaluate the efficacy of senolytic drugs, which are designed to selectively eliminate senescent cells (Saul et al. 2022). Interestingly, the gene panel was also used in retinal glia to reveal that microglia attracted immune cells into the retina of aged mice (Suresh et al. 2024).

The reduction of senescent cell burden is essential since the accumulation of senescent cells and their SASP is implicated in various age-related diseases, including osteoarthritis, atherosclerosis, and neurodegenerative disorders. The SASP contributes to a proinflammatory environment, often referred to as "inflammaging," which exacerbates tissue dysfunction and promotes the progression of diseases (Di Micco et al. 2021; Zhang et al. 2022; Calabrò et al. 2024).

In summary, cellular senescence and the SASP play a significant role in aging and age-related diseases. Gene sets such as SenMayo represent a valuable tool for identifying and studying senescent cells in next-generation sequencing datasets, to monitor therapy response and to identify the key cell types that express the SASP. This has potential applications in monitoring and treating age-related conditions.

6 Inflammaging and Cell Communication

A key contributor of aging is chronic, low-grade inflammation referred to as "inflammaging." Persistent inflammation accelerates tissue and organ deterioration, driving the progression of age-related diseases such as neurodegeneration, cardiovascular disease, and cancer. As the organism ages, immune cells undergo phenotypic changes, senescent cells accumulate, and the secretion of the proinflammatory SASP disrupts systemic homeostasis. Understanding crucial players can partly be achieved by evaluating differentially expressed genes between young and old populations using NGS datasets or by applying, for instance, SenMayo. This has exemplarily been performed by de Man et al. in human lung samples, identifying specifically modules referring to the regulation of extracellular matrix, cell signaling, and damage response. They also highlighted the positive correlation of hypomethylated CpG islands with age on genes like *IL1B*, *IL6R*, *TNF*, and *ETS2* (de Man et al. 2023). However, bioinformatic solutions developed over the last few decades have promoted an unparalleled understanding of these processes within specific cellular niches.

Several cell types within the tissue microenvironment play central roles in the inflammaging process. Due to a failure in senescent cell clearance, these cells accumulate in aging tissues. Metabolically active, they release cytokines, chemokines, and growth factors that perpetuate inflammation and tissue damage, including IL-6, CCL2, and TNF-α. Their presence creates a feedback loop where SASP factors drive the recruitment of immune cells, leading to more inflammation and further tissue degradation. With novel computational tools, the communication between groups of cells (e.g., immune regulatory and senescent cells) can be analyzed regarding their specific cell-cell interactions and sustained inflammaging. Applying popular tools like CellChat, CellPhoneDB, or SingleCellSignalR (Cabello-Aguilar et al. 2020; Efremova et al. 2020; Jin et al. 2021) helps decipher cell-cell interactions in both single-cell and spatial datasets, paving the way to identify key cellular drivers. These tools analyze cell-cell communication by modeling and visualizing ligand-receptor interactions within scRNA-seq data based on the expression and abundance of distinct receptors and ligands. They identify and quantify signaling pathways between different cell types, helping to elucidate how cells influence each other in various biological contexts such as aging. For example, CellChat was used in the aged lung to find that the CCL2-CCR2 signaling network was less active in aged fibroblasts (Wu et al. 2023).

For instance, there are nonimmune players contributing to the inflammatory aging microenvironment: fibroblasts and endothelial cells. They alter the extracellular matrix composition and produce cytokines that attract immune cells to aged tissues. By promoting a shift toward a proinflammatory state, fibroblasts contribute to fibrosis and tissue stiffness (Xu et al. 2022; López-Otín et al. 2023).

Immune cells are strongly involved in the inflammaging process, with age-related changes in their function contributing to chronic inflammation. Macrophages, for instance, release higher levels of cytokines such as TNF-α, IL-6, and IL-1β (Alvarez-Rodríguez et al. 2012). Additionally, antiinflammatory cytokines like IL-10 are released less by macrophages as an organism ages (Zhang et al. 2015). Aged macrophages contribute to the chronic inflammatory state of tissues, amplifying the signals that sustain inflammaging and promoting this inflammatory state.

The aging process also affects T cell homeostasis, characterized by a shift in T cell-type proportions. There is a decline in naïve CD4 and CD8 T cells and a rise in memory CD4 and CD8 T cells (Litjens et al. 2024). Furthermore, decreased CD28 expression on T cells has been observed (Moro-García et al. 2013; Qin et al. 2016), which is associated with T cell exhaustion (Huang et al. 2023). As the numbers of T helper cells and T regulatory cells increase with age, their cytokine release decreases (Elyahu et al. 2019), along with their ability to rescue age-related tissue damage (de

La Fuente et al. 2024). In aged individuals, there is a distinct shift in the cytokine landscape compared to younger individuals. Young tissues generally maintain a balance between pro- and antiinflammatory cytokines, with effective resolution of acute inflammation. In aged tissues, however, proinflammatory cytokines such as IL-6, TNF-α, and IL-1β are elevated, contributing to chronic inflammation. Meanwhile, antiinflammatory cytokines like IL-10 and TGF-β are diminished, impairing the resolution of inflammation. The persistence of these cytokines is a hallmark of inflammaging and correlates with an increased risk of age-related pathologies (Li et al. 2023).

Bioinformatically, ligand-receptor analysis can identify signaling molecules that are most active in aged tissues and how they differ from those in younger tissues. This allows for a better understanding of the cellular cross-talk driving inflammation, which may inform therapeutic strategies to modulate the aging process. Applying trajectory interference to tissues of different ages allows for the modulation of a "time course" (pseudotime) to define developmental processes and monitor expression changes along those trajectories.

In summary, inflammaging represents a complex interplay between immune cells, senescent cells, and their tissue microenvironments. The sustained production of proinflammatory cytokines leads to tissue dysfunction and contributes to the progression of age-related diseases. NGS technologies like scRNA-seq and computational tools that help uncover ligand-receptor interactions are essential in unraveling the molecular underpinnings of inflammaging, offering potential avenues for intervention.

7 Aging-Associated Disease Landscapes Revealed by Sequencing

Besides the decline in physiological functions, aging is characterized by increased vulnerability to chronic diseases such as cancer, cardiovascular illnesses, and neurodegenerative disorders (Kennedy et al. 2014).

As the global population continues to age, the incidence of neurodegenerative diseases such as Alzheimer's disease (AD) and Parkinson's disease (PD) is rapidly increasing, posing significant challenges to healthcare systems worldwide (Alzheimer's Association Report 2023). Evidence suggests that cellular senescence plays a critical role in the development and progression of these diseases, affecting a range of cell types, including neurons, astrocytes, microglia, and oligodendrocytes (Stein et al. 1999; Si et al. 2021; Bigbee 2023).

As various cell types undergo senescence, their deteriorating functions contribute significantly to the neurological decline observed in neurodegenerative diseases. Astrocytes, which normally support neuronal metabolism, become less effective at protecting neurons as they age. This decline originates from an increased secretion of SASP factors and the activation of the NF-κB pathway, ultimately promoting the accumulation of Aβ and tau, further impairing neuronal function (Han et al. 2020; Gao et al. 2021). At the same time, senescent microglia, the brain's immune cells, lose their ability to clear toxic proteins, such as Aβ and tau, resulting in chronic neuroinflammation and exacerbated neuronal damage (Vogels et al. 2019). Cellular senescence in oligodendrocytes, the cells responsible for forming and maintaining myelin sheaths, leads to myelin loss, which is an early marker of cognitive decline in AD (Wang et al. 2018; Vanzulli et al. 2020).

Recent studies utilizing NGS have significantly advanced our understanding of Parkinson's disease. Single-nucleus sequencing (snRNA-seq) was performed on postmortem midbrains from 6 idiopathic PD patients and 5 controls, identifying 11 major cell types, including oligodendrocytes, microglia, and dopaminergic neurons (DaNs). They discovered a unique cluster of *CADPS2*^{high} neuronal cells, likely representing degenerating DaNs, along with decreased oligodendrocytes and increased astrocytes and microglia displaying an activated state and heightened inflammatory gene expression (Smajić et al. 2022). Complementing this, Kamath et al. enriched for DaNs using fluorescence-activated nuclei sorting and identified ten DaN subpopulations. They found a PD-vulnerable population enriched for risk genes like *SNCA* and *MAPT*, along with increased astrocyte subpopulations responding to degeneration (Kamath et al. 2022). An independent study on matched substantia nigra and cortex samples revealed associations between PD risk and both DaNs and oligodendrocytes, highlighting the role of glial cells in PD pathology. Here, the authors generated a single-nuclei transcriptomic atlas for the substantia nigra after sequencing approximately 17,000 nuclei (Agarwal et al. 2020). Additionally, a study integrating genome-wide association study (GWAS) results with single-cell RNA sequencing indicated that PD is linked to both cholinergic neurons and oligodendrocytes, supporting the idea that PD may begin in the gut and spread to the brain (Bryois et al. 2020).

Similarly, sequencing approaches identified disease mechanisms in Alzheimer's disease. For instance, one of the largest studies applied snRNA-seq to prefrontal cortex samples from 48 individuals with varying AD pathology. Transcriptional changes occurring early in the disease were identified, particularly noting upregulation of the AD risk factor APOE in microglia and downregulation in astrocytes. This study highlighted the advantages of scRNA-seq over bulk RNA sequencing, especially regarding microglial representation (Mathys et al. 2019). Another snRNA-seq study further explored the role of transcription factor EB in regulating multiple GWAS loci for AD in disease-specific astrocyte populations (Grubman et al. 2019). Furthermore, research focusing on autosomal dominant AD carriers found dysregulation in various pathways, including ferroptosis in neurons and autophagy in oligodendrocytes (Brase et al. 2023). Studies have also indicated the role of brain vasculature in AD, linking GWAS genes to endothelial and extracellular matrix pathways (Yang et al. 2022). Additionally, scRNA-seq of peripheral blood mononuclear cells showed decreased B cells in AD patients, correlating with clinical dementia ratings, while another study found that T cell dynamics are altered in tauopathy, affecting neurodegeneration (Xiong et al. 2021; Chen et al. 2023).

In addition to neurodegenerative conditions, malignant diseases exhibit a significant incidence among the aging population. Thousands of next-generation sequencing (NGS) datasets have been published and are readily accessible to the research community. In recent years, there has been a notable increase in single-cell sequencing approaches, advancing our understanding of the tumor microenvironment, tumor heterogeneity, subpopulations, and therapy resistance.

In acute myeloid leukemia (AML), using ATAC- and RNA-seq, chromatin accessibility was mapped which revealed regulatory patterns associated with mutation burdens (Corces et al. 2016). In addition, Wang et al. found single-nucleotide polymorphisms in AML cells, uncovering polyclonal resistance mechanisms post-IDH1 mutation therapy (Wang et al. 2019). Besides AML, numerous sequencing approaches at the single-cell level were performed using solid tumor material. For instance, recent studies have demonstrated significant transcriptome heterogeneity in liver cancer, with scRNA-seq revealing variability among different tumor regions and the presence of rare subclones rich in $CD24^{+}CD44^{+}$ markers, indicative of unique carcinogenic traits (Ho et al. 2019). In another study, five liver cancer subclones were identified with distinct gene expression patterns, while copy number variations showed that this disease can be monoclonal or polyclonal, with higher heterogeneity in polyclonal tumors (Dong et al. 2021). In breast cancer, scRNA-seq confirmed transcriptional heterogeneity among different subtypes, and clonotype expansion after PD-1 inhibitor treatment suggested persistent immune responses (Bassez et al. 2021). Lung cancer cells, particularly small-cell lung cancer, exhibit transcriptional diversity and increased regulatory heterogeneity upon metastasis, with different tumor cell subclones identified across various tumor regions (Schaff et al. 2021). In colorectal cancer, increased heterogeneity following tumor transplantation has been reported (Ono et al. 2021). Moreover, single-cell triple omics sequencing has shown consistent methylation patterns within tumor subclones (Bian et al. 2018). These studies underscore the transformative impact of NGS in cancer research, offering critical insights into tumor heterogeneity, clonal evolution, and treatment resistance.

8 Conclusion

In conclusion, NGS has emerged as a revolutionary advancement in aging research, providing unprecedented insights into the complex molecular mechanisms underlying the aging process (Fig. 1). Through advanced techniques such as RNA-seq, researchers can dissect gene expression changes at bulk and single-cell levels, uncovering the intricate transcriptional landscapes that characterize aging. Additionally, methods such as ChIP-seq and ATAC-seq enhance our understanding of chromatin dynamics and regulatory elements, shedding light on epigenetic modifications that influence age-related phenotypes.

The exploration of cellular senescence and its implications for aging has been greatly enriched by NGS, offering a deeper understanding of the molecular drivers

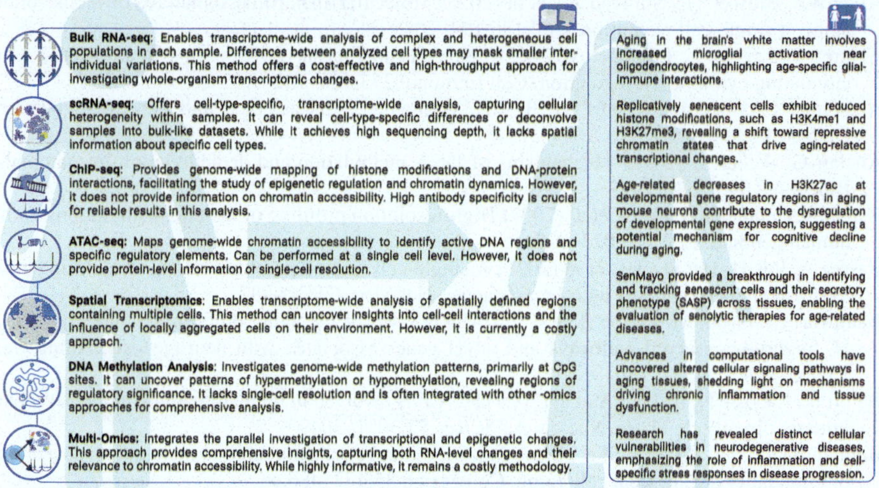

Fig. 1 Integration of NGS technologies in aging research. The left panel illustrates key NGS approaches, such as RNA-seq, ChIP-seq, ATAC-seq, spatial transcriptomics, and DNA methylation profiling, which enable comprehensive exploration of gene expression, chromatin accessibility, and epigenetic regulation. The right panel summarizes major findings from these approaches, including insights into age-related changes in glial-immune interactions, chromatin states, developmental gene regulation, and cellular vulnerabilities contributing to inflammation and neurodegeneration

of this process. Moreover, the study of inflammaging and cell communication has benefited from the integration of NGS data, enabling more nuanced models of age-related inflammation and tissue interactions. As sequencing technologies become increasingly affordable, NGS is poised to serve as a valuable tool for evaluating aging and monitoring the progression, therapy response, and recurrence of aging-associated diseases such as cancer.

Finally, the application of NGS to aging-associated diseases reveals critical genetic and epigenetic factors that contribute to the pathology of these conditions, paving the way for personalized therapeutic strategies. As the field continues to evolve, the potential of NGS to unravel the complexities of aging will undoubtedly lead to new interventions and improved healthspan.

References

Agarwal D, Sandor C, Volpato V et al (2020) A single-cell atlas of the human substantia nigra reveals cell-specific pathways associated with neurological disorders. Nat Commun 11:4183

Aging Atlas Consortium (2021) Aging atlas: a multi-omics database for aging biology. Nucleic Acids Res 49:D825–D830

Allen WE, Blosser TR, Sullivan ZA et al (2023) Molecular and spatial signatures of mouse brain aging at single-cell resolution. Cell 186:194–208.e18

Alvarez-Rodríguez L, López-Hoyos M, Muñoz-Cacho P et al (2012) Aging is associated with circulating cytokine dysregulation. Cell Immunol 273:124–132

Alzheimer's Association Report (2023) 2023 Alzheimer's disease facts and figures. Alzheimers Dement 19:1598–1695

Auclair G, Weber M (2012) Mechanisms of DNA methylation and demethylation in mammals. Biochimie 94:2202–2211

Barski A, Cuddapah S, Cui K et al (2007) High-resolution profiling of histone methylations in the human genome. Cell 129:823–837

Bassez A, Vos H, van Dyck L et al (2021) A single-cell map of intratumoral changes during anti-PD1 treatment of patients with breast cancer. Nat Med 27:820–832

Baumgart M, Groth M, Priebe S et al (2014) RNA-seq of the aging brain in the short-lived fish N. furzeri—conserved pathways and novel genes associated with neurogenesis. Aging Cell 13:965–974

Belotti E, Lacoste N, Iftikhar A et al (2024) H2A.Z is involved in premature aging and DSB repair initiation in muscle fibers. Nucleic Acids Res 52:3031–3049

Benjamin DI, Brett JO, Both P et al (2023) Multiomics reveals glutathione metabolism as a driver of bimodality during stem cell aging. Cell Metab 35:472–486.e6

Bevanger L, Maeland JA, Naess AI (1989) Competitive enzyme immunoassay for antibodies to a 43,000-molecular-weight Francisella tularensis outer membrane protein for the diagnosis of tularemia. J Clin Microbiol 27:922–926

Bian S, Hou Y, Zhou X et al (2018) Single-cell multiomics sequencing and analyses of human colorectal cancer. Science 362:1060–1063

Bigbee JW (2023) Cells of the central nervous system: an overview of their structure and function. Adv Neurobiol 29:41–64. https://doi.org/10.1007/978-3-031-12390-0_2

Brase L, You S-F, D'Oliveira Albanus R et al (2023) Single-nucleus RNA-sequencing of autosomal dominant Alzheimer disease and risk variant carriers. Nat Commun 14:2314

Bryois J, Skene NG, Hansen TF et al (2020) Genetic identification of cell types underlying brain complex traits yields insights into the etiology of Parkinson's disease. Nat Genet 52:482–493

Cabello-Aguilar S, Alame M, Kon-Sun-Tack F et al (2020) SingleCellSignalR: inference of intercellular networks from single-cell transcriptomics. Nucleic Acids Res 48:e55

Calabrò A, Accardi G, Aiello A et al (2024) Senotherapeutics to counteract senescent cells are prominent topics in the context of anti-ageing strategies. Int J Mol Sci 25:1792

Chen X, Firulyova M, Manis M et al (2023) Microglia-mediated T cell infiltration drives neurodegeneration in tauopathy. Nature 615:668–677

Corces MR, Buenrostro JD, Wu B et al (2016) Lineage-specific and single-cell chromatin accessibility charts human hematopoiesis and leukemia evolution. Nat Genet 48:1193–1203

Craig T, Smelick C, Tacutu R et al (2015) The digital ageing atlas: integrating the diversity of age-related changes into a unified resource. Nucleic Acids Res 43:D873–D878

Crouch J, Shvedova M, Thanapaul RJRS et al (2022) Epigenetic regulation of cellular senescence. Cells 11:672

Cruz C, Della Rosa M, Krueger C et al (2018) Tri-methylation of histone H3 lysine 4 facilitates gene expression in ageing cells. elife 7:e34081

Day K, Waite LL, Thalacker-Mercer A et al (2013) Differential DNA methylation with age displays both common and dynamic features across human tissues that are influenced by CpG landscape. Genome Biol 14:R102

de La Fuente AG, Dittmer M, Heesbeen EJ et al (2024) Ageing impairs the regenerative capacity of regulatory T cells in mouse central nervous system remyelination. Nat Commun 15:1870

de Man R, McDonough JE, Adams TS et al (2023) A multi-omic analysis of the human lung reveals distinct cell specific aging and senescence molecular programs. bioRxiv. https://doi.org/10.1101/2023.04.19.536722

Deshmukh S, Ponnaluri VC, Dai N et al (2018) Levels of DNA cytosine methylation in the Drosophila genome. PeerJ 6:e5119

Di Micco R, Krizhanovsky V, Baker D et al (2021) Cellular senescence in ageing: from mechanisms to therapeutic opportunities. Nat Rev Mol Cell Biol 22:75–95

Djeghloul D, Kuranda K, Kuzniak I et al (2016) Age-associated decrease of the histone methyltransferase SUV39H1 in HSC perturbs heterochromatin and B lymphoid differentiation. Stem Cell Rep 6:970–984

Dong X, Wang F, Liu C et al (2021) Single-cell analysis reveals the intra-tumor heterogeneity and identifies MLXIPL as a biomarker in the cellular trajectory of hepatocellular carcinoma. Cell Death Discov 7:14

Efremova M, Vento-Tormo M, Teichmann SA et al (2020) CellPhoneDB: inferring cell-cell communication from combined expression of multi-subunit ligand-receptor complexes. Nat Protoc 15:1484–1506

Elyahu Y, Hekselman I, Eizenberg-Magar I et al (2019) Aging promotes reorganization of the CD4 T cell landscape toward extreme regulatory and effector phenotypes. Sci Adv 5:eaaw8330

ENCODE Project Consortium (2012) An integrated encyclopedia of DNA elements in the human genome. Nature 489:57–74

Finkbeiner C, Ortuño-Lizarán I, Sridhar A et al (2022) Single-cell ATAC-seq of fetal human retina and stem-cell-derived retinal organoids shows changing chromatin landscapes during cell fate acquisition. Cell Rep 38:110294

Franzen J, Georgomanolis T, Selich A et al (2021) DNA methylation changes during long-term in vitro cell culture are caused by epigenetic drift. Commun Biol 4:598

Galang G, Mandla R, Ruan H et al (2020) ATAC-Seq reveals an Isl1 enhancer that regulates sinoatrial node development and function. Circ Res 127:1502–1518

Gao L, Zheng W-G, Wu X-K et al (2021) Baicalein delays H2O2-induced astrocytic senescence through inhibition of senescence-associated secretory phenotype (SASP), suppression of JAK2/STAT1/NF-κB pathway, and regulation of leucine metabolism. ACS Chem Neurosci 12:2320–2335

Greer EL, Shi Y (2012) Histone methylation: a dynamic mark in health, disease and inheritance. Nat Rev Genet 13:343–357

Greer EL, Maures TJ, Hauswirth AG et al (2010) Members of the H3K4 trimethylation complex regulate lifespan in a germline-dependent manner in C. elegans. Nature 466:383–387

Greer EL, Blanco MA, Gu L et al (2015) DNA methylation on N6-adenine in C. elegans. Cell 161:868–878

Grigorash BB, van Essen D, Liang G et al (2023) p16High senescence restricts cellular plasticity during somatic cell reprogramming. Nat Cell Biol 25:1265–1278

Grubman A, Chew G, Ouyang JF et al (2019) A single-cell atlas of entorhinal cortex from individuals with Alzheimer's disease reveals cell-type-specific gene expression regulation. Nat Neurosci 22:2087–2097

Han X, Zhang T, Liu H et al (2020) Astrocyte senescence and Alzheimer's disease: a review. Front Aging Neurosci 12:148

He Y-F, Li B-Z, Li Z et al (2011) Tet-mediated formation of 5-carboxylcytosine and its excision by TDG in mammalian DNA. Science 333:1303–1307

He X, Memczak S, Qu J et al (2020) Single-cell omics in ageing: a young and growing field. Nat Metab 2:293–302

Heintzman ND, Stuart RK, Hon G et al (2007) Distinct and predictive chromatin signatures of transcriptional promoters and enhancers in the human genome. Nat Genet 39:311–318

Hishida T, Yamamoto M, Hishida-Nozaki Y et al (2022) In vivo partial cellular reprogramming enhances liver plasticity and regeneration. Cell Rep 39:110730

Ho DW-H, Tsui Y-M, Sze KM-F et al (2019) Single-cell transcriptomics reveals the landscape of intra-tumoral heterogeneity and stemness-related subpopulations in liver cancer. Cancer Lett 459:176–185

Horvath S (2013) DNA methylation age of human tissues and cell types. Genome Biol 14:R115

Hu Z, Chen K, Xia Z et al (2014) Nucleosome loss leads to global transcriptional up-regulation and genomic instability during yeast aging. Genes Dev 28:396–408
Huang Y, Zheng H, Zhu Y et al (2023) Loss of CD28 expression associates with severe T-cell exhaustion in acute myeloid leukemia. Front Immunol 14:1139517
Hühne R, Thalheim T, Sühnel J (2014) AgeFactDB—the JenAge ageing factor database—towards data integration in ageing research. Nucleic Acids Res 42:D892–D896
Iwasaki W, Miya Y, Horikoshi N et al (2013) Contribution of histone N-terminal tails to the structure and stability of nucleosomes. FEBS Open Bio 3:363–369
Jin C, Li J, Green CD et al (2011) Histone demethylase UTX-1 regulates C. elegans life span by targeting the insulin/IGF-1 signaling pathway. Cell Metab 14:161–172
Jin S, Guerrero-Juarez CF, Zhang L et al (2021) Inference and analysis of cell-cell communication using CellChat. Nat Commun 12:1088
Johnson DS, Mortazavi A, Myers RM et al (2007) Genome-wide mapping of in vivo protein-DNA interactions. Science 316:1497–1502
Jylhävä J, Pedersen NL, Hägg S (2017) Biological age predictors. EBioMedicine 21:29–36
Kamath T, Abdulraouf A, Burris SJ et al (2022) Single-cell genomic profiling of human dopamine neurons identifies a population that selectively degenerates in Parkinson's disease. Nat Neurosci 25:588–595
Kanfi Y, Naiman S, Amir G et al (2012) The sirtuin SIRT6 regulates lifespan in male mice. Nature 483:218–221
Karakaslar EO, Katiyar N, Hasham M et al (2023) Transcriptional activation of Jun and Fos members of the AP-1 complex is a conserved signature of immune aging that contributes to inflammaging. Aging Cell 22:e13792
Kawahara TLA, Michishita E, Adler AS et al (2009) SIRT6 links histone H3 lysine 9 deacetylation to NF-kappaB-dependent gene expression and organismal life span. Cell 136:62–74
Kawakami K, Nakamura A, Ishigami A et al (2009) Age-related difference of site-specific histone modifications in rat liver. Biogerontology 10:415–421
Kennedy BK, Berger SL, Brunet A et al (2014) Geroscience: linking aging to chronic disease. Cell 159:709–713
Kilfeather P, Khoo JH, Wagner K et al (2024) Single-cell spatial transcriptomic and translatomic profiling of dopaminergic neurons in health, aging, and disease. Cell Rep 43:113784
Lee J-S, Shukla A, Schneider J et al (2007) Histone crosstalk between H2B monoubiquitination and H3 methylation mediated by COMPASS. Cell 131:1084–1096
Li X, Wang C-Y (2021) From bulk, single-cell to spatial RNA sequencing. Int J Oral Sci 13:36
Li L, Greer C, Eisenman RN et al (2010) Essential functions of the histone demethylase lid. PLoS Genet 6:e1001221
Li X, Li C, Zhang W et al (2023) Inflammation and aging: signaling pathways and intervention therapies. Signal Transduct Target Ther 8:239
Litjens NHR, van der List ACJ, Klepper M et al (2024) Older age is associated with a distinct and marked reduction of functionality of both alloreactive CD4+ and CD8+ T cells. Front Immunol 15:1406716
López-Otín C, Blasco MA, Partridge L et al (2023) Hallmarks of aging: an expanding universe. Cell 186:243–278
Lorbeck MT, Singh N, Zervos A et al (2010) The histone demethylase Dmel\Kdm4A controls genes required for life span and male-specific sex determination in Drosophila. Gene 450:8–17
Mahmud S, Pitcher LE, Torbenson E et al (2024) Developing transcriptomic signatures as a biomarker of cellular senescence. Ageing Res Rev 99:102403
Mathys H, Davila-Velderrain J, Peng Z et al (2019) Single-cell transcriptomic analysis of Alzheimer's disease. Nature 570:332–337
Michishita E, McCord RA, Berber E et al (2008) SIRT6 is a histone H3 lysine 9 deacetylase that modulates telomeric chromatin. Nature 452:492–496

Mogilenko DA, Shpynov O, Andhey PS et al (2021) Comprehensive profiling of an aging immune system reveals clonal GZMK+ CD8+ T cells as conserved hallmark of inflammaging. Immunity 54:99–115.e12

Moro-García MA, Alonso-Arias R, López-Larrea C (2013) When aging reaches CD4+ T-cells: phenotypic and functional changes. Front Immunol 4:107

Mostoslavsky R, Chua KF, Lombard DB et al (2006) Genomic instability and aging-like phenotype in the absence of mammalian SIRT6. Cell 124:315–329

Murach KA, Dimet-Wiley AL, Wen Y et al (2021) Late-life exercise mitigates skeletal muscle epigenetic aging. Aging Cell 21:e13527

Nassrally MS, Lau A, Wise K et al (2019) Cell cycle arrest in replicative senescence is not an immediate consequence of telomere dysfunction. Mech Ageing Dev 179:11–22

Nikopoulou C, Kleinenkuhnen N, Parekh S et al (2023) Spatial and single-cell profiling of the metabolome, transcriptome and epigenome of the aging mouse liver. Nat Aging 3:1430–1445

O'Sullivan RJ, Kubicek S, Schreiber SL et al (2010) Reduced histone biosynthesis and chromatin changes arising from a damage signal at telomeres. Nat Struct Mol Biol 17:1218–1225

Ogrodnik M (2021) Cellular aging beyond cellular senescence: markers of senescence prior to cell cycle arrest in vitro and in vivo. Aging Cell 20:e13338

Ogrodnik M, Carlos Acosta J, Adams PD et al (2024) Guidelines for minimal information on cellular senescence experimentation in vivo. Cell 187:4150–4175

Ono H, Arai Y, Furukawa E et al (2021) Single-cell DNA and RNA sequencing reveals the dynamics of intra-tumor heterogeneity in a colorectal cancer model. BMC Biol 19:207

Papaspyropoulos A, Hazapis O, Altulea A et al (2023) Decoding of translation-regulating entities reveals heterogeneous translation deficiency patterns in cellular senescence. Aging Cell 22:e13893

Piazzesi A, Papić D, Bertan F et al (2016) Replication-independent histone variant H3.3 controls animal lifespan through the regulation of pro-longevity transcriptional programs. Cell Rep 17:987–996

Pu M, Ni Z, Wang M et al (2015) Trimethylation of Lys36 on H3 restricts gene expression change during aging and impacts life span. Genes Dev 29:718–731

Pu M, Wang M, Wang W et al (2018) Unique patterns of trimethylation of histone H3 lysine 4 are prone to changes during aging in Caenorhabditis elegans somatic cells. PLoS Genet 14:e1007466

Qin L, Jing X, Qiu Z et al (2016) Aging of immune system: immune signature from peripheral blood lymphocyte subsets in 1068 healthy adults. Aging (Albany NY) 8:848–859

Rauluseviciute I, Drabløs F, Rye MB (2019) DNA methylation data by sequencing: experimental approaches and recommendations for tools and pipelines for data analysis. Clin Epigenetics 11:193

Sahu SK, Reddy P, Lu J et al (2024) Targeted partial reprogramming of age-associated cell states improves markers of health in mouse models of aging. Sci Transl Med 16:eadg1777

Sarg B, Koutzamani E, Helliger W et al (2002) Postsynthetic trimethylation of histone H4 at lysine 20 in mammalian tissues is associated with aging. J Biol Chem 277:39195–39201

Saul D, Kosinsky RL (2021) Epigenetics of aging and aging-associated diseases. Int J Mol Sci 22:401

Saul D, Kosinsky RL (2023) Spatial transcriptomics herald a new era of transcriptome research. Clin Transl Med 13:e1264

Saul D, Kosinsky RL, Atkinson EJ et al (2022) A new gene set identifies senescent cells and predicts senescence-associated pathways across tissues. Nat Commun 13:4827

Schaff DL, Singh S, Kim K-B et al (2021) Fragmentation of small-cell lung cancer regulatory states in heterotypic microenvironments. Cancer Res 81:1853–1867

Schaum N, Lehallier B, Hahn O et al (2020) Ageing hallmarks exhibit organ-specific temporal signatures. Nature 583:596–602

Sen P, Dang W, Donahue G et al (2015) H3K36 methylation promotes longevity by enhancing transcriptional fidelity. Genes Dev 29:1362–1376

Sen P, Lan Y, Li CY et al (2019) Histone acetyltransferase p300 induces de novo super-enhancers to drive cellular senescence. Mol Cell 73:684–698.e8

Si Z, Sun L, Wang X (2021) Evidence and perspectives of cell senescence in neurodegenerative diseases. Biomed Pharmacother 137:111327

Siebold AP, Banerjee R, Tie F et al (2010) Polycomb repressive complex 2 and Trithorax modulate Drosophila longevity and stress resistance. Proc Natl Acad Sci USA 107:169–174

Signal B, Phipps AJ, Giles KA et al (2024) Ageing-related changes to H3K4me3, H3K27ac, and H3K27me3 in purified mouse neurons. Cells 13:1393

Smajić S, Prada-Medina CA, Landoulsi Z et al (2022) Single-cell sequencing of human midbrain reveals glial activation and a Parkinson-specific neuronal state. Brain 145:964–978

Song Q, Hou Y, Zhang Y et al (2022) Integrated multi-omics approach revealed cellular senescence landscape. Nucleic Acids Res 50:10947–10963

Spinelli R, Florese P, Parrillo L et al (2022) ZMAT3 hypomethylation contributes to early senescence of preadipocytes from healthy first-degree relatives of type 2 diabetics. Aging Cell 21:e13557

Stark R, Grzelak M, Hadfield J (2019) RNA sequencing: the teenage years. Nat Rev Genet 20:631–656

Stein GH, Drullinger LF, Soulard A et al (1999) Differential roles for cyclin-dependent kinase inhibitors p21 and p16 in the mechanisms of senescence and differentiation in human fibroblasts. Mol Cell Biol 19:2109–2117

Su L, Li H, Huang C et al (2018) Muscle-specific histone H3K36 dimethyltransferase SET-18 shortens lifespan of Caenorhabditis elegans by repressing daf-16a expression. Cell Rep 22:2716–2729

Suresh S, Karthik G, Ouyang JF, Chrysostomou V, Tang SA, Petretto E, Crowston JG, Bell KC et al (2024) SenMayo transcriptomic senescence panel highlights glial cells in the ageing mouse and human retina. NPJ Aging. 2024 Nov 30;10(1):60. https://doi.org/10.1038/s41514-024-00187-9. PMID: 39616175; PMCID: PMC11608257

Tacutu R, Thornton D, Johnson E et al (2018) Human ageing genomic resources: new and updated databases. Nucleic Acids Res 46:D1083–D1090

Tchkonia T, Zhu Y, van Deursen J et al (2013) Cellular senescence and the senescent secretory phenotype: therapeutic opportunities. J Clin Invest 123:966–972

Tharakan R, Ubaida-Mohien C, Dunn C et al (2023) Whole-genome methylation analysis of aging human tissues identifies age-related changes in developmental and neurological pathways. Aging Cell 22:e13847

Tvardovskiy A, Schwämmle V, Kempf SJ et al (2017) Accumulation of histone variant H3.3 with age is associated with profound changes in the histone methylation landscape. Nucleic Acids Res 45:9272–9289

Uyar B, Palmer D, Kowald A et al (2020) Single-cell analyses of aging, inflammation and senescence. Ageing Res Rev 64:101156

Vanzulli I, Papanikolaou M, De-La-Rocha IC et al (2020) Disruption of oligodendrocyte progenitor cells is an early sign of pathology in the triple transgenic mouse model of Alzheimer's disease. Neurobiol Aging 94:130–139

Victorelli S, Salmonowicz H, Chapman J et al (2023) Apoptotic stress causes mtDNA release during senescence and drives the SASP. Nature 622:627–636

Vogels T, Murgoci A-N, Hromádka T (2019) Intersection of pathological tau and microglia at the synapse. Acta Neuropathol Commun 7:109

Wang F, Yang Y-J, Yang N et al (2018) Enhancing oligodendrocyte myelination rescues synaptic loss and improves functional recovery after chronic hypoxia. Neuron 99:689–701.e5

Wang H, Choe S, DiNardo CD et al (2019) Complex polyclonal resistance mechanisms to ivosidenib monotherapy in IDH1-mutant relapsed or refractory acute myeloid leukemia revealed by single cell sequencing analyses. Blood 134:541

Wu R, Zhang X, Zhang X et al (2023) Deciphering the age-dependent changes of pulmonary fibroblasts in mice by single-cell transcriptomics. Front Cell Dev Biol 11:1287133

Xiao X, Juan C, Drennon T et al (2023) Spatial transcriptomic interrogation of the murine bone marrow signaling landscape. Bone Res 11:59

Xiong L-L, Xue L-L, Du R-L et al (2021) Single-cell RNA sequencing reveals B cell-related molecular biomarkers for Alzheimer's disease. Exp Mol Med 53:1888–1901

Xu P, Wang M, Song W-M et al (2022) The landscape of human tissue and cell type specific expression and co-regulation of senescence genes. Mol Neurodegener 17:5

Yan K, Ji Q, Zhao D et al (2023) SGF29 nuclear condensates reinforce cellular aging. Cell Discov 9:110

Yang L, Ma Z, Wang H et al (2019) Ubiquitylome study identifies increased histone 2A ubiquitylation as an evolutionarily conserved aging biomarker. Nat Commun 10:2191

Yang AC, Vest RT, Kern F et al (2022) A human brain vascular atlas reveals diverse mediators of Alzheimer's risk. Nature 603:885–892

Yang P, Corbett R, Daharsh L et al (2024) Definition of regulatory elements and transcription factors controlling porcine immune cell gene expression at single cell resolution using single nucleus ATAC-seq. Genomics 116:110944

Yekelchyk M, Li X, Guenther S et al (2024) Single-nucleus ATAC-seq for mapping chromatin accessibility in individual cells of murine hearts. Methods Mol Biol 2752:245–257

Zahn JM, Poosala S, Owen AB et al (2007) AGEMAP: a gene expression database for aging in mice. PLoS Genet 3:e201

Zhang B, Bailey WM, Braun KJ et al (2015) Age decreases macrophage IL-10 expression: implications for functional recovery and tissue repair in spinal cord injury. Exp Neurol 273:83–91

Zhang L, Pitcher LE, Yousefzadeh MJ et al (2022) Cellular senescence: a key therapeutic target in aging and diseases. J Clin Invest 132:e158450

Zou Z, Long X, Zhao Q et al (2021) A single-cell transcriptomic atlas of human skin aging. Dev Cell 56:383–397.e8

Deciphering the Interplay: DNA Methylation Dynamics in Aging and Neurodegenerative Diseases

Young-Gun Lee and Sun-Min Lee

Contents

Abstract Aging is characterized by changes in cellular identity and function over time, with alterations in epigenetic patterns potentially serving as an underlying mechanism that drives human aging. Recent studies have identified sets of individual methylation sites, whose combined DNA methylation status serves as a measure of chronological age, known as the DNA methylation clock. These sites, highly enriched in Polycomb repressive complexes binding locations and associated with chromatin factor localization, are located near genes implicated in mammalian development and longevity. Age-related alterations in the human DNA methylome are believed to contribute to age-related diseases, including neurodegenerative diseases. The two most common neurodegenerative diseases, Alzheimer's disease (AD) and Parkinson's disease (PD), arise from a combination of genetic and envi-

Y.-G. Lee
Department of Neurology, Ilsan Paik Hospital, Inje University College of Medicine, Goyang, South Korea

S.-M. Lee (✉)
Department of Biomedical Science, Hallym University, Chuncheon, South Korea
e-mail: smlee@hallym.ac.kr

J. Barciszewski (ed.), *Decoding Aging and Neurodegeneration*,
RNA Technologies 16, https://doi.org/10.1007/978-3-031-91072-2_2

ronmental factors, with aging being a significant risk factor. DNA methylation changes are among the key hallmarks of neurodegeneration, reflecting both the aging process and environmental influences. In AD, DNA methylation aging accelerates as the disease advances, while in PD, accelerated aging correlates with earlier onset. Widespread epigenetic dysregulation is linked to neurodegenerative changes through accelerated aging and pathogenesis in AD and PD. In this chapter, we explore the intricate relationship between DNA methylation changes, aging, and neurodegenerative diseases.

Keywords DNA methylation · DNA methylation clock · Alzheimer's disease · Parkinson's disease

1 Introduction

Aging is a gradual and complex decline in physiological functions that occurs over time in most living organisms. In humans, advancements in medical care and public health have drastically increased life expectancy in the last few centuries. However, this has also resulted in higher rates of morbidity, as aging is a major risk factor for many diseases, including cancer, metabolic disorders like diabetes, cardiovascular diseases, and neurodegenerative conditions (Kennedy et al. 2014; Niccoli and Partridge 2012). To ensure a healthier and longer life, it is crucial to understand the molecular mechanisms driving aging and identify potential therapeutic targets for age-related diseases.

Several factors have been identified as contributors to aging and longevity, including telomere attrition, nutrient sensing pathways, mitochondrial dysfunction, oxidative stress, impaired DNA repair mechanisms, and the accumulation of DNA damage (Guo et al. 2022). Disruptions in protein homeostasis that lead to misfolded protein buildup also play a significant role. While studies on human longevity suggest that genetic factors explain around 20–30% of the variation in life spans between monozygotic twins, the majority of this variation is believed to result from epigenetic drift occurring throughout life (Herskind et al. 1996; Fraga et al. 2005; Castillo-Fernandez et al. 2014). This epigenetic variability among cells in an aging population leads to transcriptional alterations and genomic instability (Soto-Palma et al. 2022). Over the past two decades, research has demonstrated that DNA methylation changes are a key component of epigenetic regulation in mammals, influencing gene expression and helping to maintain cell lineage differentiation. DNA methyltransferases (DNMTs) induce methylation on cytosine residues, while a key mechanism for removing DNA methylation involves active demethylation driven by ten-eleven translocation (TET)-mediated oxidation of 5-methylcytosine (Lee 2024). Changes in global DNA methylation levels, linked to alterations in DNMT or TET expression, have been reported in association with aging (Wang et al. 2022). DNA methylation is the most extensively studied epigenetic marker in the context of aging (Jones et al. 2015).

Humans, with their highly advanced brains and increased lifespan, are significantly more susceptible to neurodegenerative diseases compared to other species (Chen et al. 2013). Many neurodegenerative diseases involve the misfolding and aggregation of specific proteins. Alzheimer's disease (AD) is the most common neurodegenerative disorder, characterized by neuritic plaques, neurofibrillary tangles, and neuronal loss as its pathological hallmarks (Long and Holtzman 2019). The amyloid cascade hypothesis posits that amyloid-β (Aβ), the primary component of neuritic plaques, plays a central role in propagating tau pathology to other brain regions, which is linked to neurodegeneration (Jagust 2018). According to the revised diagnostic criteria proposed by the Alzheimer's Association workgroup, the accumulation of Aβ pathology is sufficient for the diagnosis of AD, while the extent of tau pathology spread determines the staging of the disease (Jack et al. 2024). Parkinson's disease (PD), the second most common neurodegenerative disorder, is characterized by α-synuclein pathology, a major component of Lewy bodies in the brain (Goedert et al. 2013). While current diagnosis of PD relies on neurological examination and indirect imaging biomarkers, recent studies suggest that the biological diagnosis of PD may be possible through in vivo detection of neuronal α-synuclein (Simuni et al. 2024; Hoglinger et al. 2024).

The shift from a clinicobiological diagnosis of AD and PD to a biological continuum has the potential to extend the diagnosis to asymptomatic individuals. However, concerns have been raised as most individuals with isolated proteinopathy remain asymptomatic throughout their lives, highlighting significant clinical heterogeneity in prognosis (Villain and Planche 2024). Moreover, there is notable variability in clinical symptoms and disease progression even among those in the active stages of neurodegeneration (De Pablo-Fernandez et al. 2019). Integrating biological and clinical staging systems often reveals a mismatch between clinical symptoms and biological markers (Jack et al. 2024). Patients with more severe clinical symptoms than expected based on their biological stage may have copathologies (Godrich et al. 2022), while those with milder symptoms may benefit from cognitive reserve or resilience (Stern et al. 2023). Several factors likely contribute to this clinical heterogeneity in AD and PD, including aging, genetic factors, socioeconomic factors, and environmental influences, such as diet and exposure to neurotoxic agents. As the epigenetic mechanisms, especially DNA methylation, alter gene expression without affecting the DNA sequence may represent the key mediators of the genetic-environmental interactions.

Consistent DNA methylation changes with aging have been observed across various tissues, including the brain, and their correlation with gene expression changes and cellular alterations has been reported. In this chapter, we will summarize these findings and discuss the effect of DNA methylation aging on neurodegenerative diseases such as AD and PD, as well as its physiological significance.

2 DNA Methylation Changes in Human Aging

2.1 *The Genomic Hypomethylation Hypothesis of Aging*

In early studies, fibroblasts from mice, hamsters, and humans were cultured, showing a marked decrease in 5-methylcytosine content in their genomic DNA over time in culture (Wilson and Jones 1983). This decrease appeared to be associated with significant changes in gene expression in aging human fibroblasts, particularly a reduction in DNMT1 mRNA levels (Casillas et al. 2003). In addition to these findings, the DNA methylome of CD4+ T cells was compared between human newborns and centenarians (individuals over 100 years old), revealing that centenarians exhibited lower DNA methylation across all genomic compartments (Heyn et al. 2012). The average methylation level of all analyzed CpG sites was 73% in the centenarian samples, compared to 80% in the newborn samples. The sequences that were most hypomethylated in the centenarian DNA were predominantly found at CpG-poor promoters and in tissue-specific genes. Building on these initial findings, the genomic hypomethylation hypothesis of aging has posited that global DNA methylation decreases overall with age. It has been argued that the loss of DNA methylation in the genome, as individuals age, contributes to the abnormal expression of genes that are typically repressed in cells. This provides a mechanistic explanation for the relaxation of gene expression observed with age, a process referred to as dysdifferentiation.

However, the accumulated results to date indicate that the effects of aging vary by cell type. In particular, recent studies using deep sequencing technology have reported that global hypomethylation associated with aging is not observed in most cases (Unnikrishnan et al. 2018). These findings show that there is no significant decrease in global DNA methylation levels in human cortex, in epidermal cells, or in mouse hippocampus and liver cells.

2.2 *DNA Methylation Clock Across Human Tissues*

In the past decade, researchers have identified consistent patterns of methylation changes at specific CpG sites in the genome that occur throughout the lifespan and are similar across individuals. These discoveries have enabled the creation of "DNA methylation clock," which use methylation levels to accurately estimate an individual's chronological age (Fig. 1) (Horvath 2013). It is also linked to age-related physiological changes (Pal and Tyler 2016). However, the exact degree of aging can vary based on biological age, which is determined by physiological evidence (Oblak et al. 2021). Recent studies have also reported consistent nonlinear patterns in molecular markers of aging (Shen et al. 2024). In spite of these aspects, there is still no definitive measure to accurately assess biological age, so most studies have focused on the correlation between chronological age and DNA methylation

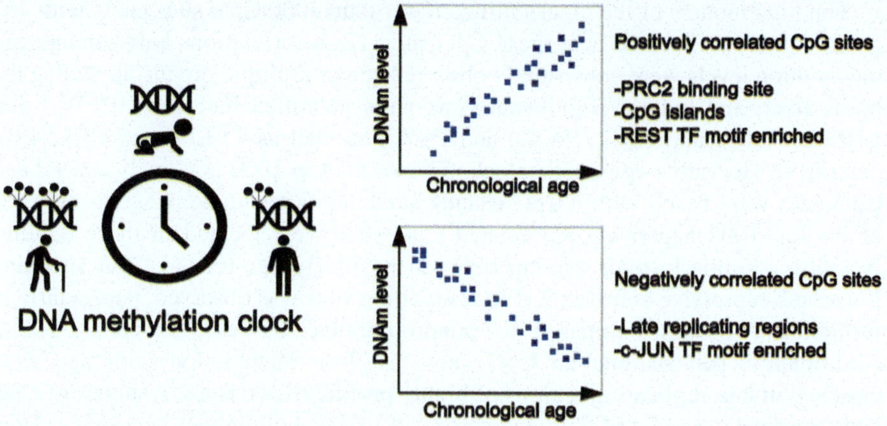

Fig. 1 The DNA methylation clock refers to specific CpG sites that exhibit a consistent tendency to increase or decrease in DNA methylation levels in a linear manner with chronological age. These CpG sites often show localized associations within specific regions of the genome. For instance, in the pan-mammalian clock study, age-related positively and negatively correlated CpG sites demonstrated distinct patterns of clustering within different genomic regions (Lu et al. 2023)

changes. Along with DNA methylation clocks that are applicable across various tissues, tissue-specific clocks, such as those for the brain and blood, have also been identified (Horvath 2013; Pellegrini et al. 2021; Lu et al. 2023).

Horvath's epigenetic clock, which utilizes 353 CpG sites, is a multitissue predictor developed to estimate the age of most tissues and cell types (Horvath 2013). The model is built on data from 51 healthy tissues and cell types. A test set error of 3.6 years indicates that DNA methylation age differs by less than 3.6 years in 50% of individuals. This clock also demonstrated that DNA methylation age remains relatively consistent across different brain regions (temporal cortex, pons, frontal cortex, cerebellum) within the same individuals. Pathway analysis of genes associated with the 353 CpG sites reveals significant enrichment in pathways related to cell death and survival, cellular growth and proliferation, organismal and tissue development, and cancer. Of these CpG sites, 193 are positively correlated with age, showing hypermethylation, while 160 are negatively correlated, showing hypomethylation. Positively correlated CpGs are highly enriched near Polycomb target genes but do not exhibit a strong relationship with CpG island status. In contrast, negatively correlated CpGs are overrepresented in CpG shores and show no significant association with Polycomb target genes.

The Mammalian Methylation Consortium utilized the mammalian methylation array, which they recently developed to assess methylation levels at up to 36,000 CpG sites (Lu et al. 2023). These sites are flanked by DNA sequences highly conserved across the mammalian class. They analyzed methylation levels in 11,754 samples from 59 tissue types, representing 185 mammalian species. As a result, they were able to develop universal pan-mammalian epigenetic clocks, which serve as highly accurate estimators of chronological age ($r \approx 0.96$–0.98).

The final models of the clocks utilize fewer than 1000 CpG sites each, with 401 genes commonly located near these CpG sites. High correlations between age and methylation levels were consistently observed across multiple organs, including the brain, liver, and kidney. Significant gene pairs identified included LHFPL4 and LHFPL3, ZIC1 and ZIC2, PAX2, and PAX5, as well as CELF6 and CELF4. In contrast to Horvath's epigenetic clock, 57% of the top 1000 positively age-related CpG sites were found within CpG islands in the human genome, whereas only 2% of the top 1000 negatively age-related CpG sites were located in these regions. Notably, a similarly high overlap between positively age-related CpG sites and Polycomb repressive complex 2 (PRC2)-binding sites was observed, particularly in proliferative tissues compared to nonproliferative tissues (Lu et al. 2023). Additionally, pan-mammalian CpG sites that lose methylation with age were enriched in late-replicating regions of highly proliferative tissues. Analysis of CpG sites positively correlated with age across all tissues highlighted "nervous system development" as a significantly relevant gene ontology (GO) term. Negatively age-related CpG sites in the brain and muscle were enriched in genes associated with circadian rhythm and in gene sets related to Alzheimer's disease (Lu et al. 2023). Further analysis of the top 1000 positively age-related CpG sites showed that the REST transcription factor motif was significantly enriched. In contrast, fewer significant transcription factor binding enrichments were found for the 1000 negatively age-related CpG sites, with notable enrichments such as c-Jun in blood and brain (Fig. 1) (Lu et al. 2023).

2.3 *Age-Related DNA Methylation Alterations in the Brain*

DNA methylation changes linked to age and sex were observed in four brain regions: the frontal cortex (FC), temporal cortex (TC), entorhinal cortex (ERC), and cerebellum (CRB). The study identified 24,581, 10,077, 404, and 1140 age-dependent differentially methylated positions (aDMPs) in the FC, TC, ERC, and CRB, respectively (Pellegrini et al. 2021). Most aDMPs across all regions displayed hypermethylation with age, representing 76% in the FC, 88% in the TC, 58% in the ERC, and 62% in the CRB.

GO enrichment analysis highlighted several pathways involved in morphogenesis and developmental processes that are common to the FC, TC, and ERC. The study also demonstrated that DNA methylation differences between males and females tend to be consistent across the four brain regions, while aging impacts cortical regions differently compared to the cerebellum (Pellegrini et al. 2021). Interestingly, they found that the proportion of sex-dependent probes also influenced by aging was higher than expected, although methylation differences between males and females were generally preserved.

They also examined DNA methylation differences between AD patients and healthy controls across various brain regions (Pellegrini et al. 2021). The study identified 14 AD-associated differentially methylated positions (AD-DMPs) in the FC, 5405 in the TC, 47 in the ERC, and 1 in the CRB. In all brain regions, the majority of aDMPs were hypermethylated in AD patients compared to controls. The overlap between AD-DMPs and aDMPs revealed 7, 456, 4, and 0 common probes in the FC, TC, ERC, and CRB, respectively. The results also show that AD-related DNA methylation changes are highly concentrated in probes that exhibit age-related methylation variations, with a strong alignment in the direction of changes seen in both aging and AD. This reinforces the concept of accelerated epigenetic aging in AD.

A study comparing DNA methylation in the postmortem orbital frontal cortex from younger (<42 years) and older (>60 years) individuals also found that age-dependent differentially methylated regions (aDMRs) are enriched in genes associated with genetic risk for neurodegenerative diseases, including AD, schizophrenia, and major depressive disorder (MDD) (McKinney et al. 2019). The relationship between age-related DNA methylation changes and neurodegenerative diseases has been consistently observed.

2.4 *Cell-Type-Specific DNA Methylation Alterations in Brain Aging*

Recently, several reports have shown that age-related DNA methylation patterns vary across different cell types in the brain (Chien et al. 2024; Jeong et al. 2024). One such study categorized brain cell types broadly into neurons (NeuN+) and non-neuronal cells, specifically oligodendrocytes (OLIG2+), and examined DNA methylation changes across a wide age range, from 2.4 months to 85 years (Jeong et al. 2024). The findings indicated that age contributed more significantly to the variation in DNA methylation than other factors such as disease status and sex.

The researchers identified 4480 and 2253 CpGs with significant age-related methylation changes in NeuN+ and OLIG2+ cells, respectively. Notably, the aDMRs in neurons and oligodendrocytes were highly distinct, with only 16 overlapping CpGs between the two cell types. In NeuN+ cells, aDMRs were predominantly associated with increased DNA methylation (77%) as age progressed. In contrast, aDMRs in OLIG2+ cells were more evenly distributed between hyper- and hypomethylation changes. Interestingly, the aDMRs showed a strong bias toward CpGs that are specifically methylated in a cell-type-dependent manner. Additionally, age-hypomethylated DMRs were highly enriched in brain enhancers, while age-hypermethylated DMRs were significantly enriched in CpG islands. Genes with increased DNA methylation in their promoters (i.e., containing age-hypermethylated DMRs) exhibited reduced expression with age (Jeong et al. 2024).

Significant variations in cortical cell-type proportions across healthy human aging have been identified using immunohistochemistry and transcriptomic analyses (Soreq et al. 2017; Chen et al. 2023). These studies reported changes in the abundance of both neuronal and glial cells with age. Transcriptome analysis revealed strong negative age correlations with age in the estimated proportions of certain neuronal subtypes, particularly somatostatin (SST) and vasoactive intestinal peptide (VIP)-expressing interneurons, while several glial subtypes, including astrocytes, showed positive correlations (Fig. 2a) (Chen et al. 2023).

Remarkably, single-nucleus multiomic DNA methylation and transcriptome sequencing (snmCT-seq) of the frontal cortex during healthy human aging provided a detailed insights into cell population changes and gene expression alterations associated with DNA methylation change (Chien et al. 2024). Consistent with previous findings, the proportion of inhibitory SST- and VIP-expressing neurons was reduced in the aged frontal cortex, showing a 68% lower proportion in aged samples compared to young samples. Similarly, inhibitory neurons were found to be reduced across multiple brain regions in Alzheimer's disease patients when compared to age-matched controls. In contrast, there was an increase in the proportion of CGE-derived LAMP5 LHX6 neurons and a type of L4–L5 IT excitatory neuron in the aged frontal cortex.

Their findings revealed a greater number of age-related differentially expressed genes (age-DEGs) in excitatory neurons than in inhibitory neurons (Chien et al. 2024). Interestingly, only a small number of age-DEGs were found in SST- and VIP-expressing inhibitory neurons, suggesting that despite the reduced numbers of these cells in aged brains, the remaining neurons retain a gene expression profile similar to that of younger brain. Downregulated genes in aged neurons were

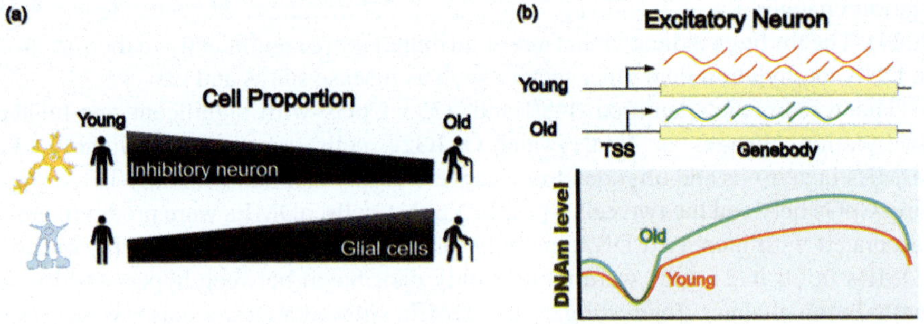

Fig. 2 Age-related cell-type-specific changes in the brain. (**a**) In the human brain, changes in cell proportions during the aging process have been observed across different cell types. Notably, the number of inhibitory neurons tends to decrease with age, while glial cells show an increase in number. (**b**) In excitatory neurons, there is evidence linking age-related changes in DNA methylation to alterations in gene expression. In the aged brain, an increase in DNA methylation was particularly prominent in gene body regions, and this methylation pattern tended to be associated with gene repression

primarily associated with synaptic signaling and cell adhesion pathways. This could be related to reduced synaptic plasticity. In contrast, the upregulated genes may play neuroprotective roles. Interestingly, increased expression of subtelomeric genes was observed in aged brain cells. This change appeared to be linked to telomere damage or shortening, which may disrupt chromatin structure at chromosome ends, rather than changes in DNA methylation. Moreover, shorter telomeres correlated with higher expression of age-related DEGs in subtelomeric regions, supporting the telomere position effect. Additionally, X-linked genes with DNA methylation patterns indicative of escape from X inactivation were identified in aged neurons (Chien et al. 2024).

Similar to transcriptome changes, inhibitory neurons showed highly similar global methylation patterns across all donors. In contrast, the low-dimensional embedding analysis of CG and CH DNA methylation across the genome indicated marked differences between excitatory neurons from young and aged donors. aDMRs were detected in all cell types, with more regions showing increased methylation than decreased methylation. For example, in L4–L5 IT excitatory neurons, 98% of DMRs showed increased methylation in aged samples. They also found that regions with the largest differences in mCG or mCH between aged and young cells within any cell type displayed substantial differences in mean methylation across cell types in young adults. Furthermore, they tested the association between changes in DNA methylation and gene expression. Overall, aDMRs were predominantly located in intronic (69%) and intergenic (25%) regions. Alterations in gene body mCH and mCG were strongly and negatively correlated with changes in mRNA expression across most neuron types (Fig. 2b) (Chien et al. 2024). The relationship between gene body methylation and gene expression varies by cell type. In dividing cells, gene body methylation is positively correlated with gene expression, whereas this trend is not observed in nondividing cells such as adult neurons (Aran et al. 2011). In the mouse frontal cortex, non-CpG site methylation within gene bodies showed a negative correlation with gene expression (Xie et al. 2012). One proposed mechanism involves the methyl-CpG binding protein 2 (MeCP2), which has been shown to repress transcription by binding within transcribed regions of genes, particularly affecting the transcription of long neuronal genes (Kinde et al. 2016).

aDMRs that exhibited increased methylation in aged neurons were enriched in enhancers and regions associated with the PRC2. Motifs of early growth response 1 (EGR1) and WT1 were enriched in excitatory neuron age-DMRs with increased methylation. EGR1 mRNA expression was also significantly reduced with age in both excitatory and inhibitory neurons. EGR1, known to be involved in synaptic structure and function, has been identified as a key mediator and regulator of synaptic plasticity and neuronal activity (Duclot and Kabbaj 2017). Loss of mCG in aged neurons was enriched in transcription factor binding motifs, including those of FOS, c-JUN, and GRE.

3 DNA Methylation Changes in Age-Related Neurodegenerative Diseases

3.1 DNA Methylation Alterations in Alzheimer's Disease

Brain tissue-based epigenome-wide association studies (EWAS) have explored the epigenetic mechanisms underlying the etiology and progression of AD. In the Religious Orders Study/Memory and Aging Project (ROS/MAP) cohort, 71 DMRs, including *ABCA7* and *BIN1*, were associated with the burden of neuritic plaques (De Jager et al. 2014). Enrichment study showed that DNA methylation changes were likely to be in weakly transcribed or inactive regions in the non-AD healthy brain. Similarly, a parallel study from the London Brain Bank for Neurodegenerative Disease demonstrated an association between hypermethylation at the *ANK1* gene and Braak tau staging (Lunnon et al. 2014). Transcriptional repression of hypermethylation of *ANK1* was further evaluated in the independent study (Sanchez-Mut et al. 2016). Another study from the Mount Sinai Alzheimer's Disease and Schizophrenia Brain Bank found a link between hypermethylation of the *HOXA* gene cluster and Braak staging (Smith et al. 2018). Integration analysis of DNA methylation changes and transcriptional changes of nearby genes were conducted by Semick et al. (2019). From the paired DNA methylation and gene expression, data from each brain region showed that AD-associated DNA methylation changes were inversely correlated with gene expression, mostly pronounced in *ANKRD30B*, where hypermethylation in AD brain was associated with repressed transcription.

Interestingly, these studies consistently showed that most DMRs in AD brain tissue exhibit global hypermethylation (Watson et al. 2016; Altuna et al. 2019). This hypermethylation pattern is also a characteristic of aging-related methylation changes in the human brain. A meta-analysis of previous studies revealed 119 DMRs, 73.9% of which were hypermethylated in AD (Zhang et al. 2020). The authors further found that these hypermethylated DMRs were overrepresented in Polycomb-repressed regions and enriched at the binding sites of EZH2 and SUZ12, core subunits of PRC2, suggesting that epigenetic switching to complementary global DNA hypermethylation attributed to PRC2 dysfunction in AD.

The first neuron-type-specific EWAS of AD brain tissue, conducted by Mano et al., revealed that methylation changes in *BRCA1* play a significant role in AD pathogenesis (Mano et al. 2017). Gasparoni et al. fractionated brain tissue into neuronal and non-neuronal cell types and identified DMRs in *LRRC8B* and *SEC14L1* in neurons and in *LRRC8B* and *ANK1* in non-neuronal cells (Gasparoni et al. 2018). A subsequent neuron-specific EWAS study, conducted on 101 brain samples with varying AD pathology, identified 1224 DMRs in enhancer regions, most of which were hypomethylated at non-CpG sites in AD neurons (Li et al. 2019). The authors posited that enhancer hypomethylation occurs naturally in the aging brain but is accelerated in AD neurons. Notably, hypomethylation in enhancers of the *DSCAML1* gene was associated with upregulation of *BACE1* transcripts, which play a crucial role in the amyloidogenic cleavage of APP in AD pathogenesis.

Shireby et al. extended these findings by profiling DNA methylation in different brain cell populations, including neuron-enriched (NeuN+), oligodendrocyte-enriched (SOX10+), and microglia/astrocyte-enriched (NeuN-/SOX10-) subtypes (Shireby et al. 2022). They found that many DMRs identified in previous bulk brain tissue studies were derived from non-neuronal cells. Given the increasing reactivity of microglia and astrocytes as AD progresses, DMRs associated with AD pathology may reflect the crucial role of epigenetic dysregulation in glial populations.

Blood-based EWAS of AD have demonstrated systemic epigenetic dysregulation but distinct from those found in brain tissue. Lunnon et al. identified three differentially methylated positions (DMPs) in the *DAPK1*, *GAS1*, and *NDUFS5* genes, but these blood-based DMPs did not overlap with those identified in brain tissue (Lunnon et al. 2014). Furthermore, methylation changes observed in antemortem blood samples did not reflect the postmortem methylation changes associated with neuritic plaques (Yu et al. 2016). These findings suggest that blood-based DMPs may not be directly linked to the neurodegenerative processes occurring in the brain (Fig. 3).

The utility of blood methylation changes as biomarkers for early diagnosis and progression of AD has been explored in subsequent studies. Kobayashi et al. identified a DMR at the *NCAPH2/LMF2* promoter region, which was associated with disease stage (Kobayashi et al. 2016). Madrid et al. reported that methylation changes in the *B3GALT4* and *ZADH2* genes were correlated with memory scores and cerebrospinal fluid (CSF) biomarkers of AD (Madrid et al. 2018). Additionally, Lardennoije et al. found that DNA methylation at the *OXT* promoter during the pre-dementia stage was associated with future conversion to dementia, suggesting its

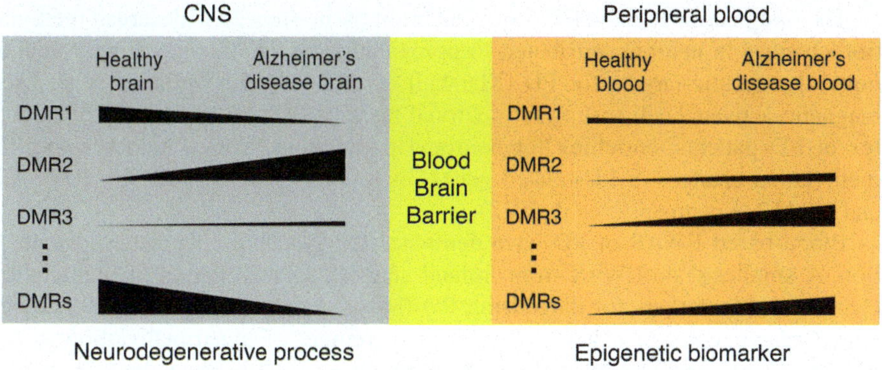

Fig. 3 Distinct role of DMRs from brain tissue and blood in Alzheimer's disease. Due to the tissue-specific nature of DNA methylation changes, methylation alterations in the central nervous system (CNS) reflect neurodegenerative processes, while blood-based methylation changes indicate systemic responses to these neurodegenerative changes, raising the possibility of epigenetic biomarkers. Furthermore, the blood-brain barrier compartmentalizes the CNS and peripheral blood, limiting the ability of blood-based methylation changes to accurately reflect the spatial and temporal alterations occurring within the CNS. Abbreviation: DMR, differentially methylated region

potential as a novel biomarker for disease progression (Lardenoije et al. 2019). Longitudinal studies from cohorts such as NeuroMed (Roubroeks et al. 2020) and Alzheimer's Disease Neuroimaging Initiative (ADNI) (Li et al. 2021) also identified methylation changes in the *HOXB6* and *PM20D1* genes that were associated with clinical diagnosis of AD, further supporting the potential role of methylation as a biomarker for AD.

3.2 DNA Methylation Alterations in Parkinson's Disease

DNA methylation changes in brain tissue related to PD were first evaluated by Masliah et al., who demonstrated concordant methylation changes between blood and brain (Masliah et al. 2013). They reported that approximately 88% of DMRs were hypomethylated. This decreased DNA methylation in PD has been linked to dysfunction of DNMT1, which may contribute to global hypomethylation (Desplats et al. 2011). Of note, hypomethylation of SNCA increased the expression of α-synuclein (Matsumoto et al. 2010). Subsequently, methylation analysis directly from subcortical regions, including the substantia nigra, dorsal motor nucleus of the vagus, and cingulate gyrus, revealed significant DMRs supporting the epigenetic contribution to PD risk, although the cohort size was small (Young et al. 2019). While the previous studies evaluated DMRs related to PD status, one large-scale EWAS study specifically examined DMRs associated with Lewy body pathology, a hallmark neuropathological feature of PD (Pihlstrom et al. 2022). This study identified novel DMRs linked to Braak Lewy body staging, including *TMCC2*, *SFMBT2*, *AKAP6*, and *PHYHIP*.

The first neuron-specific EWAS of PD brain tissue showed widespread methylation changes in neurons, attributed to upregulation of *TET2*, suggesting potential novel therapeutic targets for PD (Marshall et al. 2020). Kochmanski et al. used magnetic-activated cell sorting for neuronal markers (NeuN) from the parietal cortex of PD patients, enriching for neuronal populations. They found sex-specific methylation changes in PD-related genes, including *PARK7*, *SLC17A6*, *PTPRN2*, and *NR4A2* (Kochmanski et al. 2022).

Blood-based EWAS of PD have demonstrated systemic epigenetic dysregulation. A smaller-scale EWAS from Ireland suggested the potential of methylation changes as biomarkers for diagnosing PD (Moore et al. 2014). Follow-up studies with larger cohorts revealed a possible involvement of hypermethylation in the *SLC7A11* gene in PD pathogenesis (Vallerga et al. 2020). From the Parkinson's Disease, Environment, and Genes (PEG) cohort, correlation network analysis demonstrated that methylation changes were significantly enriched in immune system-related genes (Chuang et al. 2017). Longitudinal follow-up studies of methylation changes in PD patients indicated dynamic shifts in DNA methylation as the disease progresses, compared to controls (Henderson-Smith et al. 2019; Gonzalez-Latapi et al. 2024). These findings suggest that the evolution of DNA methylation throughout the disease course might serve as a potential biomarker for PD progression.

3.3 DNA Methylation Clocks in Alzheimer's Disease

The concept of DNA methylation age acceleration (AA) was introduced to evaluate the influence of biological age, which reflects environmental factors, on the development of neurodegenerative disease. This is a comparison of chronological age and age predicted by DNA methylation clock, and if the DNA methylation age is higher than the chronological age of an individual, it is judged that AA has occurred. A recent effort aimed at building a predictive model for AD biomarkers using blood-based methylation clocks was conducted in the ADNI cohort (Sato et al. 2020). The study found a weak positive correlation between AA and CSF Aβ levels but no significant association with CSF tau (Sato et al. 2020). Similarly, another group did not find a significant relationship between AA and the diagnosis of dementia, sex, or CSF tau levels (Inkster et al. 2022). When combining data from the Australian Imaging Biomarkers and Lifestyle study and ADNI databases, encompassing 859 subjects, AA was found to be associated with cross-sectional hippocampal volume, though it was not linked to cognitive status or brain Aβ burden (Milicic et al. 2022). Sugden et al. compared three generations of DNA methylation age algorithms—Horvath and Hannum clocks (first generation), PhenoAge and GrimAge (second generation), and DunedinPACE (third generation)—to predict cognitive scores and clinical stages (Sugden et al. 2022). Their findings indicated that the DunedinPACE algorithm was associated with clinical diagnosis and cognitive scores, whereas earlier algorithms did not show consistent associations (Sugden et al. 2022).

Several studies have examined the effect of epigenetic clock on cognitive decline in population-based cohorts, not specific to AD. A Scottish study revealed that AA was associated with poorer cognitive performance in cross-sectional assessments (Marioni et al. 2015). In a cohort of 1670 African American and White urban middle-aged adults, AA at baseline was linked to greater cognitive decline over a mean follow-up period of 4.7 years (Beydoun et al. 2020). Additionally, another study found a connection between AA and brain MRI connectivity measurements (Chouliaras et al. 2018). However, a study of 486 middle-aged monozygotic twins found no correlation between AA and either cross-sectional or longitudinal cognitive measures (Starnawska et al. 2017). Finally, a study involving older women from the Women's Health Initiative Memory Study demonstrated that AA was associated with faster cognitive decline (Nguyen et al. 2024).

In the ROS/MAP cohort, epigenetic aging assessed from brain tissue samples in 700 dorsolateral prefrontal cortex (DLPFC) specimens was found to be associated with various brain pathologies, including diffuse and neuritic plaques, Aß load, and cognitive scores and clinical staging (Levine et al. 2015). In a similar cohort with 721 samples, Grodstein et al. tested both first- and second-generation DNA methylation clocks, as well as the tissue-specific Cortical Clock (Shireby et al. 2020), to predict neuropathological outcomes (Grodstein et al. 2021). Among the different methods, the Cortical Clock demonstrated the strongest association with amyloid pathology. Furthermore, it was the only clock to show significant associations with tau pathology and neocortical Lewy body pathology. Thrush et al. extended this analysis by applying PCBrainAge, a model trained exclusively on cortical samples, to the same ROS/MAP cohort. When compared to the Cortical Clock, PCBrainAge showed stronger associations with clinical stage and neuropathology (Thrush et al. 2022).

3.4 DNA Methylation Clocks in Parkinson's Disease

Studies investigating the association between blood-derived DNA methylation clocks and PD have demonstrated AA in PD patients compared to controls. The first analysis from the community-based PEG cohort showed a strong association between AA and the risk of PD (Horvath and Ritz 2015). AA measured by epiTOC, which was initially trained for cancer risk assessment, was also increased in the PD group within the PEG cohort (Paul et al. 2021). These studies suggest a potential link between PD risk and AA, leading to the hypothesis that isolated REM sleep behavior disorder (iRBD), a prodromal stage of PD, may also be associated with AA. A study from Italy demonstrated a trend toward AA in individuals with iRBD compared to controls (Baldelli et al. 2023), and a Canadian study found that accelerated aging is associated with earlier onset of iRBD (Senkevich et al. 2023).

However, the association between AA and the age of onset in PD patients remains controversial. A study based on a hospital cohort and the Parkinson's Progression Markers Initiative (PPMI) database demonstrated a relationship between AA and earlier onset of PD (Tang et al. 2022). In contrast, a reanalysis of PPMI data by a Swedish group suggested a potential bias in the estimation of epigenetic aging based on subject age (Gaare et al. 2023). The authors argued that predicting the age of onset of neurodegenerative diseases may require a control population to account for age-related confounding effects, rather than disease-specific effects (Gaare et al. 2023).

Only one study has evaluated the epigenetic clock using brain tissue samples in PD (Kochmanski et al. 2022). The authors employed a cortical epigenetic clock but found no evidence of AA associated with PD status. To explore the possibility that the parietal cortex or neuronal populations may not be suitable for epigenetic clock measurement in PD, further research is necessary to confirm these findings and assess alternative brain regions or cell types.

4 Conclusions

The DNA methylation clock has consistently demonstrated a linear pattern of increased or decreased DNA methylation at several hundred CpG sites with age. In the brain, these clocks predominantly show age-related hypermethylation at CpG sites, many of which are linked to genes involved in cell specificity, synaptic plasticity, and development. Recent findings have also revealed that within the brain, changes in cell number, DNA methylation patterns, and gene expression are cell-type-specific. A strong association with Polycomb target genes suggests that histone modifications, such as H3K27me3 and histone ubiquitination, may play a crucial role in coordinating chromatin structure and regulating gene expression in conjunction with DNA methylation changes during aging.

Epigenetic dysregulation, including alterations in DNA methylation and epigenetic clocks, plays a significant role in the pathogenesis and progression of both AD and PD. There are bound to be limitations to reflect epigenetic changes directly linked to neurodegenerative changes using blood-based epigenetic analysis. Because of the blood-brain barrier, it is difficult to directly reflect the region-specific, cell-type-specific spatial information of the pathology in the brain through blood samples. In addition, there is a time gap due to the difference in the time of research using blood samples and the time of obtaining brain samples through postmortem autopsy, making it difficult to reflect the time of pathology. Recently, various imaging or plasma-based biomarkers that can directly reflect brain pathology have been developed. Further research integrating in-vivo biomarkers and blood-based methylome analysis is needed to unraveling the epigenetic mechanisms of neurodegenerative disease.

Acknowledgments This research was supported by Hallym University Research Fund, 2025(HRFN-202504-003).

References

Altuna M, Urdanoz-Casado A, de Gordoa JS-R et al (2019) DNA methylation signature of human hippocampus in Alzheimer's disease is linked to neurogenesis. Clin Epigenetics 11:91

Aran D, Toperoff G, Rosenberg M et al (2011) Replication timing-related and gene body-specific methylation of active human genes. Hum Mol Genet 20:670–680

Baldelli L, Pirazzini C, Sambati L et al (2023) Epigenetic clocks suggest accelerated aging in patients with isolated REM sleep behavior disorder. NPJ Parkinsons Dis 9:48

Beydoun MA, Shaked D, Tajuddin SM et al (2020) Accelerated epigenetic age and cognitive decline among urban-dwelling adults. Neurology 94:e613–e625

Casillas MA, Lopatina N, Andrews LG et al (2003) Transcriptional control of the DNA methyltransferases is altered in aging and neoplastically-transformed human fibroblasts. Mol Cell Biochem 252:33–43

Castillo-Fernandez JE, Spector TD, Bell JT (2014) Epigenetics of discordant monozygotic twins: implications for disease. Genome Med 6:60

Chen X, Errangi B, Li L et al (2013) Brain aging in humans, chimpanzees (Pan troglodytes), and rhesus macaques (Macaca mulatta): magnetic resonance imaging studies of macro- and microstructural changes. Neurobiol Aging 34:2248–2260

Chen Y, Hunter E, Arbabi K et al (2023) Robust differences in cortical cell type proportions across healthy human aging inferred through cross-dataset transcriptome analyses. Neurobiol Aging 125:49–61

Chien JF, Liu H, Wang BA et al (2024) Cell-type-specific effects of age and sex on human cortical neurons. Neuron 112:2524–2539.e5

Chouliaras L, Pishva E, Haapakoski R et al (2018) Peripheral DNA methylation, cognitive decline and brain aging: pilot findings from the Whitehall II imaging study. Epigenomics 10:585–595

Chuang YH, Paul KC, Bronstein JM et al (2017) Parkinson's disease is associated with DNA methylation levels in human blood and saliva. Genome Med 9:76

De Jager PL, Srivastava G, Lunnon K et al (2014) Alzheimer's disease: early alterations in brain DNA methylation at ANK1, BIN1, RHBDF2 and other loci. Nat Neurosci 17:1156–1163

De Pablo-Fernandez E, Lees AJ, Holton JL et al (2019) Prognosis and neuropathologic correlation of clinical subtypes of Parkinson disease. JAMA Neurol 76:470–479

Desplats P, Spencer B, Coffee E et al (2011) Alpha-synuclein sequesters Dnmt1 from the nucleus: a novel mechanism for epigenetic alterations in Lewy body diseases. J Biol Chem 286:9031–9037

Duclot F, Kabbaj M (2017) The role of early growth response 1 (EGR1) in brain plasticity and neuropsychiatric disorders. Front Behav Neurosci 11:35

Fraga MF, Ballestar E, Paz MF et al (2005) Epigenetic differences arise during the lifetime of monozygotic twins. Proc Natl Acad Sci USA 102:10604–10609

Gaare JJ, Brugger K, Nido GS et al (2023) DNA methylation age acceleration is not associated with age of onset in Parkinson's disease. Mov Disord 38:2064–2071

Gasparoni G, Bultmann S, Lutsik P et al (2018) DNA methylation analysis on purified neurons and glia dissects age and Alzheimer's disease-specific changes in the human cortex. Epigenetics Chromatin 11:41

Godrich D, Martin ER, Schellenberg G et al (2022) Neuropathological lesions and their contribution to dementia and cognitive impairment in a heterogeneous clinical population. Alzheimers Dement 18:2403–2412

Goedert M, Spillantini MG, Del Tredici K et al (2013) 100 years of Lewy pathology. Nat Rev Neurol 9:13–24

Gonzalez-Latapi P, Bustos B, Dong S et al (2024) Alterations in blood methylome as potential epigenetic biomarker in sporadic Parkinson's disease. Ann Neurol 95:1162–1172

Grodstein F, Lemos B, Yu L et al (2021) The association of epigenetic clocks in brain tissue with brain pathologies and common aging phenotypes. Neurobiol Dis 157:105428

Guo J, Huang X, Dou L et al (2022) Aging and aging-related diseases: from molecular mechanisms to interventions and treatments. Signal Transduct Target Ther 7:391

Henderson-Smith A, Fisch KM, Hua J et al (2019) DNA methylation changes associated with Parkinson's disease progression: outcomes from the first longitudinal genome-wide methylation analysis in blood. Epigenetics 14:365–382

Herskind AM, McGue M, Holm NV et al (1996) The heritability of human longevity: a population-based study of 2872 Danish twin pairs born 1870-1900. Hum Genet 97:319–323

Heyn H, Li N, Ferreira HJ et al (2012) Distinct DNA methylomes of newborns and centenarians. Proc Natl Acad Sci USA 109:10522–10527

Hoglinger GU, Adler CH, Berg D et al (2024) A biological classification of Parkinson's disease: the SynNeurGe research diagnostic criteria. Lancet Neurol 23:191–204

Horvath S (2013) DNA methylation age of human tissues and cell types. Genome Biol 14:R115

Horvath S, Ritz BR (2015) Increased epigenetic age and granulocyte counts in the blood of Parkinson's disease patients. Aging (Albany NY) 7:1130–1142

Inkster AM, Duarte-Guterman P, Albert AY et al (2022) Are sex differences in cognitive impairment reflected in epigenetic age acceleration metrics? Neurobiol Aging 109:192–194

Jack CR Jr, Andrews JS, Beach TG et al (2024) Revised criteria for diagnosis and staging of Alzheimer's disease: Alzheimer's Association Workgroup. Alzheimers Dement 20:5143–5169

Jagust W (2018) Imaging the evolution and pathophysiology of Alzheimer disease. Nat Rev Neurosci 19:687–700

Jeong H, Mendizabal I, Yi SV (2024) Human brain aging is associated with dysregulation of cell-type epigenetic identity. bioRxiv. https://doi.org/10.1101/2024.06.01.596981

Jones MJ, Goodman SJ, Kobor MS (2015) DNA methylation and healthy human aging. Aging Cell 14:924–932

Kennedy BK, Berger SL, Brunet A et al (2014) Geroscience: linking aging to chronic disease. Cell 159:709–713

Kinde B, Wu DY, Greenberg ME et al (2016) DNA methylation in the gene body influences MeCP2-mediated gene repression. Proc Natl Acad Sci USA 113:15114–15119

Kobayashi N, Shinagawa S, Nagata T et al (2016) Development of biomarkers based on DNA methylation in the NCAPH2/LMF2 promoter region for diagnosis of Alzheimer's disease and amnesic mild cognitive impairment. PLoS One 11:e0146449

Kochmanski J, Kuhn NC, Bernstein AI (2022) Parkinson's disease-associated, sex-specific changes in DNA methylation at PARK7 (DJ-1), SLC17A6 (VGLUT2), PTPRN2 (IA-2beta), and NR4A2 (NURR1) in cortical neurons. NPJ Parkinsons Dis 8:120

Lardenoije R, Roubroeks JAY, Pishva E et al (2019) Alzheimer's disease-associated (hydroxy) methylomic changes in the brain and blood. Clin Epigenetics 11:164

Lee SM (2024) Detecting DNA hydroxymethylation: exploring its role in genome regulation. BMB Rep 57:135–142

Levine ME, Lu AT, Bennett DA et al (2015) Epigenetic age of the pre-frontal cortex is associated with neuritic plaques, amyloid load, and Alzheimer's disease related cognitive functioning. Aging (Albany NY) 7:1198–1211

Li P, Marshall L, Oh G et al (2019) Epigenetic dysregulation of enhancers in neurons is associated with Alzheimer's disease pathology and cognitive symptoms. Nat Commun 10:2246

Li QS, Vasanthakumar A, Davis JW et al (2021) Association of peripheral blood DNA methylation level with Alzheimer's disease progression. Clin Epigenetics 13:191

Long JM, Holtzman DM (2019) Alzheimer disease: an update on pathobiology and treatment strategies. Cell 179:312–339

Lu AT, Fei Z, Haghani A et al (2023) Universal DNA methylation age across mammalian tissues. Nat Aging 3:1144–1166

Lunnon K, Smith R, Hannon E et al (2014) Methylomic profiling implicates cortical deregulation of ANK1 in Alzheimer's disease. Nat Neurosci 17:1164–1170

Madrid A, Hogan KJ, Papale LA et al (2018) DNA hypomethylation in blood links B3GALT4 and ZADH2 to Alzheimer's disease. J Alzheimers Dis 66:927–934

Mano T, Nagata K, Nonaka T et al (2017) Neuron-specific methylome analysis reveals epigenetic regulation and tau-related dysfunction of BRCA1 in Alzheimer's disease. Proc Natl Acad Sci USA 114:E9645–E9654

Marioni RE, Shah S, McRae AF et al (2015) The epigenetic clock is correlated with physical and cognitive fitness in the Lothian Birth Cohort 1936. Int J Epidemiol 44:1388–1396

Marshall LL, Killinger BA, Ensink E et al (2020) Epigenomic analysis of Parkinson's disease neurons identifies Tet2 loss as neuroprotective. Nat Neurosci 23:1203–1214

Masliah E, Dumaop W, Galasko D et al (2013) Distinctive patterns of DNA methylation associated with Parkinson disease: identification of concordant epigenetic changes in brain and peripheral blood leukocytes. Epigenetics 8:1030–1038

Matsumoto L, Takuma H, Tamaoka A et al (2010) CpG demethylation enhances alpha-synuclein expression and affects the pathogenesis of Parkinson's disease. PLoS One 5:e15522

McKinney BC, Lin CW, Rahman T et al (2019) DNA methylation in the human frontal cortex reveals a putative mechanism for age-by-disease interactions. Transl Psychiatry 9:39

Milicic L, Vacher M, Porte T et al (2022) Comprehensive analysis of epigenetic clocks reveals associations between disproportionate biological ageing and hippocampal volume. Geroscience 44:1807–1823

Moore K, McKnight AJ, Craig D et al (2014) Epigenome-wide association study for Parkinson's disease. Neuromol Med 16:845–855

Nguyen S, McEvoy LK, Espeland MA et al (2024) Associations of epigenetic age estimators with cognitive function trajectories in the Women's Health Initiative Memory Study. Neurology 103:e209534

Niccoli T, Partridge L (2012) Ageing as a risk factor for disease. Curr Biol 22:R741–R752

Oblak L, van der Zaag J, Higgins-Chen AT et al (2021) A systematic review of biological, social and environmental factors associated with epigenetic clock acceleration. Ageing Res Rev 69:101348

Pal S, Tyler JK (2016) Epigenetics and aging. Sci Adv 2:e1600584
Paul KC, Binder AM, Horvath S et al (2021) Accelerated hematopoietic mitotic aging measured by DNA methylation, blood cell lineage, and Parkinson's disease. BMC Genomics 22:696
Pellegrini C, Pirazzini C, Sala C et al (2021) A meta-analysis of brain DNA methylation across sex, age, and Alzheimer's disease points for accelerated epigenetic aging in neurodegeneration. Front Aging Neurosci 13:639428
Pihlstrom L, Shireby G, Geut H et al (2022) Epigenome-wide association study of human frontal cortex identifies differential methylation in Lewy body pathology. Nat Commun 13:4932
Roubroeks JAY, Smith AR, Smith RG et al (2020) An epigenome-wide association study of Alzheimer's disease blood highlights robust DNA hypermethylation in the HOXB6 gene. Neurobiol Aging 95:26–45
Sanchez-Mut JV, Heyn H, Vidal E et al (2016) Human DNA methylomes of neurodegenerative diseases show common epigenomic patterns. Transl Psychiatry 6:e718
Sato K, Mano T, Suzuki K et al (2020) Attempt to predict A/T/N-based Alzheimer's disease cerebrospinal fluid biomarkers using a peripheral blood DNA methylation clock. J Alzheimers Dis Rep 4:287–296
Semick SA, Bharadwaj RA, Collado-Torres L et al (2019) Integrated DNA methylation and gene expression profiling across multiple brain regions implicate novel genes in Alzheimer's disease. Acta Neuropathol 137:557–569
Senkevich K, Pelletier A, Sato C et al (2023) DNA methylation age acceleration as a potential biomarker for early onset of rapid eye movement sleep behavior disorder. Ann Neurol. https://doi.org/10.1002/ana.26811
Shen XT, Wang CC, Zhou X et al (2024) Nonlinear dynamics of multi-omics profiles during human aging. Nat Aging 4:1619
Shireby GL, Davies JP, Francis PT et al (2020) Recalibrating the epigenetic clock: implications for assessing biological age in the human cortex. Brain 143:3763–3775
Shireby G, Dempster EL, Policicchio S et al (2022) DNA methylation signatures of Alzheimer's disease neuropathology in the cortex are primarily driven by variation in non-neuronal cell-types. Nat Commun 13:5620
Simuni T, Chahine LM, Poston K et al (2024) A biological definition of neuronal alpha-synuclein disease: towards an integrated staging system for research. Lancet Neurol 23:178–190
Smith RG, Hannon E, De Jager PL et al (2018) Elevated DNA methylation across a 48-kb region spanning the HOXA gene cluster is associated with Alzheimer's disease neuropathology. Alzheimers Dement 14:1580–1588
Soreq L, UK Brain Expression Consortium, North American Brain Expression Consortium et al (2017) Major shifts in glial regional identity are a transcriptional hallmark of human brain aging. Cell Rep 18:557–570
Soto-Palma C, Niedernhofer LJ, Faulk CD et al (2022) Epigenetics, DNA damage, and aging. J Clin Invest 132:e158446
Starnawska A, Tan Q, Lenart A et al (2017) Blood DNA methylation age is not associated with cognitive functioning in middle-aged monozygotic twins. Neurobiol Aging 50:60–63
Stern Y, Albert M, Barnes CA et al (2023) A framework for concepts of reserve and resilience in aging. Neurobiol Aging 124:100–103
Sugden K, Caspi A, Elliott ML et al (2022) Association of pace of aging measured by blood-based DNA methylation with age-related cognitive impairment and dementia. Neurology 99:e1402–e1413
Tang X, Gonzalez-Latapi P, Marras C et al (2022) Epigenetic clock acceleration is linked to age at onset of Parkinson's disease. Mov Disord 37:1831–1840
Thrush KL, Bennett DA, Gaiteri C et al (2022) Aging the brain: multi-region methylation principal component based clock in the context of Alzheimer's disease. Aging (Albany NY) 14:5641–5668
Unnikrishnan A, Hadad N, Masser DR et al (2018) Revisiting the genomic hypomethylation hypothesis of aging. Ann N Y Acad Sci 1418:69–79

Vallerga CL, Zhang F, Fowdar J et al (2020) Analysis of DNA methylation associates the cystine-glutamate antiporter SLC7A11 with risk of Parkinson's disease. Nat Commun 11:1238

Villain N, Planche V (2024) Disentangling clinical and biological trajectories of neurodegenerative diseases. Nat Rev Neurol 20:693

Wang K, Liu H, Hu Q et al (2022) Epigenetic regulation of aging: implications for interventions of aging and diseases. Signal Transduct Target Ther 7:374

Watson T, Roussos P, Garg P et al (2016) Genome-wide DNA methylation profiling in the superior temporal gyrus reveals epigenetic signatures associated with Alzheimer's disease. Genome Med 8:5

Wilson VL, Jones PA (1983) DNA methylation decreases in aging but not in immortal cells. Science 220:1055–1057

Xie W, Barr CL, Kim A et al (2012) Base-resolution analyses of sequence and parent-of-origin dependent DNA methylation in the mouse genome. Cell 148:816–831

Young JI, Sivasankaran SK, Wang L et al (2019) Genome-wide brain DNA methylation analysis suggests epigenetic reprogramming in Parkinson disease. Neurol Genet 5:e342

Yu L, Chibnik LB, Yang J et al (2016) Methylation profiles in peripheral blood CD4+ lymphocytes versus brain: the relation to Alzheimer's disease pathology. Alzheimers Dement 12:942–951

Zhang L, Silva TC, Young JI et al (2020) Epigenome-wide meta-analysis of DNA methylation differences in prefrontal cortex implicates the immune processes in Alzheimer's disease. Nat Commun 11:6114

The Exposome, DNA Methylation, and Neurodegenerative Disorders

Paola Monti and Luca Ferrari

Contents

Abstract Neurodegenerative disorders (NDs) refer to a diverse group of diseases characterized by the progressive loss of neurons, which leads to severe deficits in memory, cognition, behavior, sensory, and motor functions.

The increasing prevalence of NDs imposes a significant burden on healthcare systems and has driven extensive research into the biological mechanisms underlying these conditions. Recent evidence highlights the critical role of the exposome—the sum of all environmental and lifestyle exposures experienced throughout life—in shaping ND risk. These exposures primarily exert their effects through epigenetic mechanisms rather than direct genetic alterations.

P. Monti
EPIGET LAB, Department of Clinical Sciences and Community Health, Dipartimento di Eccellenza 2023–2027, University of Milan, Milan, Italy

L. Ferrari (✉)
EPIGET LAB, Department of Clinical Sciences and Community Health, Dipartimento di Eccellenza 2023–2027, University of Milan, Milan, Italy

Occupational Health Unit, Fondazione IRCCS Ca' Granda Ospedale Maggiore Policlinico, Milan, Italy
e-mail: luca.ferrari@unimi.it

J. Barciszewski (ed.), *Decoding Aging and Neurodegeneration*,
RNA Technologies 16, https://doi.org/10.1007/978-3-031-91072-2_3

Among epigenetic mechanisms, DNA methylation—the covalent addition of methyl groups to DNA bases—has garnered significant attention. This modification is highly sensitive to external factors, including diet, physical activity, airborne pollutants, pesticides, and viral infections, and is implicated in the onset and progression of NDs.

This chapter provides an overview of the current knowledge linking the exposome, DNA methylation changes, and NDs. First, it outlines epidemiological trends and the main biological alterations, or "hallmarks," of NDs, with a focus on the role of DNA methylation in modulating disease risk. Next, it examines the influence of various environmental and lifestyle factors on ND pathogenesis through DNA methylation changes, integrating findings from population studies and experimental models.

By synthesizing existing research and identifying gaps in the field, this chapter aims to advance understanding of how exposome-driven epigenetic changes contribute to NDs. It also highlights opportunities for leveraging these insights to develop innovative diagnostic tools and therapeutic strategies.

Keywords Exposome · Air pollution · Diet · Exercise · Pesticides · DNA methylation · Neurodegeneration · Alzheimer's disease · Parkinson's disease

1 Introduction

Neurodegenerative disorders (NDs) are a heterogeneous set of diseases that are characterized by the progressive loss of neurons in the central or peripheral nervous system, resulting in the disruption of neural networks and severe impairment in memory, behavior, cognition, sensory, and/or motor functions. Common NDs include Alzheimer's, Parkinson's and Huntington's diseases, and amyotrophic lateral and multiple sclerosis. With the progressive aging of the global population, the burden of NDs is steeply increasing, bringing about inevitable consequences on the economic, social, and healthcare systems. As estimated by the Global Burden of Disease Study, the number of people with dementia will nearly triple between 2019 and 2050, going from 57 million to more than 152 million cases worldwide. Thousands of research papers have contributed to shed light on the molecular alterations occurring in these pathologies, highlighting that many of them share common biochemical alterations, such as the accrual of toxic protein aggregates in the brain, mitochondrial dysfunction, and neuroinflammation (Wilson et al. 2023). Despite the huge efforts, there is currently no therapy to arrest and revert such degenerative changes, with most NDs being considered as incurable. In this framework, there is a growing interest in identifying the etiological agents underlying NDs, with a view to set up preventive strategies and to stratify the population according to ND risk.

Over the past years, numerous studies have identified thousands of genetic loci associated with NDs; nevertheless, the percentage of ND cases that can be attributed to single genetic mutations is very low, as most of them increase disease risk only slightly. Considering the two most prevalent NDs, that is, Alzheimer's disease (AD) and Parkinson's disease (PD), more than 95% of cases are considered as sporadic (i.e., nonmonogenic) (Deng et al. 2018; Bali et al. 2012), whereas high-penetrance pathogenic variants are mostly restricted to rarer early-onset conditions (Pihlstrøm et al. 2018). It is now well established that the genetic contribution to NDs primarily involves numerous low-penetrance variants with medium-to-high allele frequency. These variants cooccur and confer increased susceptibility, playing an integrative role alongside environmental triggers experienced by individuals (Wang et al. 2021; Ogonowski et al. 2024; Migliore and Coppedè 2022). Indeed, it is becoming increasingly evident that the exposome—encompassing the totality of environmental and lifestyle exposures encountered throughout an individual's lifetime and in some cases even prior to conception—plays a pivotal role in shaping ND pathogenesis. The exposome, first conceptualized by Wild in 2005, broadly comprises three groups of exposures, that is, "internal exposures" (factors/processes internal to the body, such as the gut microbiota), "specific external exposures" (individual habits or exposures, such as diet and smoking), and "general external factors" (wider environmental, socioeconomic, and psychological influences, such as the urban environment) (Wild 2012). Unlike genetics, targeting the exposome might be an efficient strategy to reduce ND risk, as the deleterious effect of many exposome-related factors can be potentially dampened by lifestyle interventions or by the implementation of environmental remediation policies. For instance, it has been recently reported that 40% of dementia cases worldwide could be prevented or delayed by tackling 12 modifiable factors all along the lifespan (Livingston et al. 2020).

Of note, several environmental and lifestyle factors have been found to regulate cellular metabolism via epigenetic mechanisms, including DNA methylation, histone modifications, and gene expression regulation by noncoding RNAs. Environmentally induced changes in epigenetic regulation might occur as part of an adaptive response aimed at preserving cell homeostasis in response to the external cue or as a bystander consequence of global dysregulation: in the first case, they might be leveraged as potential biomarkers for risk stratification (with potential protective effects) or as early alterations preceding disease manifestations (if adaptation turns into maladaptation); in the second case, they might serve as diagnostic or prognostic indicators, as well as for treatment monitoring.

Among epigenetic mechanisms, many studies have found DNA methylation to be altered in the brain and peripheral tissues of subjects with NDs. DNA methylation is the covalent attachment of a methyl ($-CH_3$) group to nitrogen bases of the DNA molecule. Most commonly, it involves cytosines (5-methyl cytosines or 5mCs) located at specific DNA sequences, called CpG islands, which are enriched in the promoter region of many protein-coding genes; the methylation of these sites, which is mediated by specific enzymes called DNA methyltransferases (DNMTs), usually results in gene silencing. Of note, DNA methylation can be inherited by mitosis and in some cases also through meiosis; despite this, it is not necessarily a static phenomenon, but it can change through the lifetime in response to physiopathological changes occurring in the organism (such as alterations occurring with aging), as well as to many external stimuli (Wu et al. 2023; Jones et al. 2015). Also, DNA

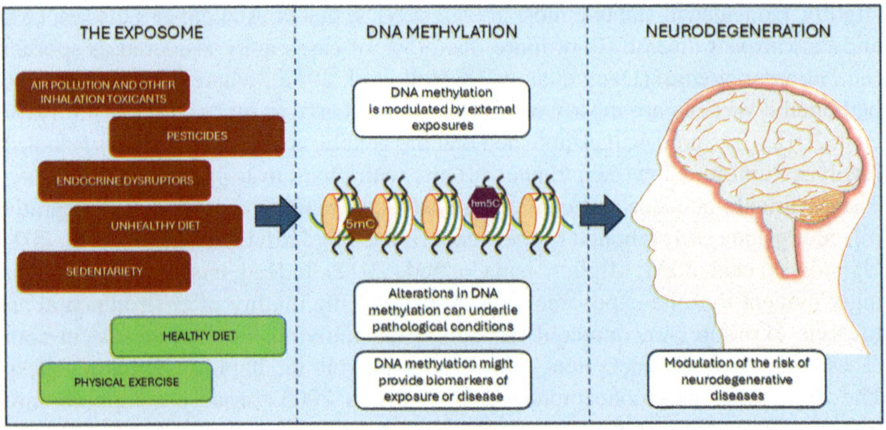

Fig. 1 Role of environmental factors in modulating ND-related DNA methylation changes

methylation is a cell- and tissue-specific process; indeed, it plays a major role during cell differentiation and presents interesting peculiarities in different organs. For instance, postmitotic neurons are tenfold enriched with 5-hydroxymethylcytosine (hm^5C, that is, a cytosine carrying a $-CH_2OH$ group) if compared to peripheral cells; here, the generation of hm^5C by ten-eleven translocase (TET) enzymes is related to a process called "functional demethylation" (Mellén et al. 2017) and has been proposed as a promising environmental biosensor, being modulated by many external cues (Dao et al. 2014).

In this light, it is reasonable to hypothesize that many ND-related changes in DNA methylation might be the results of epigenetic remodeling in response to acute and chronic environmental exposures (Fig. 1).

In the following paragraphs, we provide an overview of the current knowledge concerning the intricate interplay between the exposome, DNA methylation, and NDs. In particular, we focus on two environmental factors that are largely out of individual control, that is, air pollution/occupational inhalable toxicants and pesticides, and two lifestyle factors mostly related to individual decisions, that is, diet and physical exercise. The choice of these factors is due to their relevance for ND risk, mainly for late-onset AD and PD forms, and to the data available in literature.

2 Air Pollution and Occupational Inhalation Toxicants

According to the World Health Organization, air pollution has been defined as the "contamination of the indoor or outdoor environment by any chemical, physical or biological agent that modifies the natural characteristics of the atmosphere." Such a broad definition encompasses thousands of molecules, either derived from anthropic activities or natural sources, which comprise particulate matter (PM), gaseous molecules, volatile organic compounds, and inhalable heavy metals. Notably, chronic exposure to various airborne pollutants has been linked to an increased risk of numerous diseases, including NDs. Even though some gaseous molecules and nanosized PM might directly reach the olfactory bulb or the bloodstream after inhalation (Li et al. 2015; Hopkins et al. 2018), the pathogenic effects of airborne pollutants on the brain are predominantly indirect, mediated by circulating inflammatory mediators released from directly exposed tissues (Kuntić et al. 2024; Costa et al. 2020; Fu et al. 2022). In this regard, many population-based studies have found an association between air pollution exposure, changes in blood DNA methylation, and potential ND-related outcomes. A nonlinear association of chronic exposures to high levels of PM_1, $PM_{2.5}$, PM_{10}, and NO_2 with blood DNA methylation in the promoter region of the *BDNF* gene in 101 subjects with AD has been reported (Huang et al. 2022). Also, long-term coexposure to elevated concentrations of $PM_{2.5}$ and O_3 was found to promote *BDNF* hypermethylation at CpG2 in the same population (Huang et al. 2024). Another study on 64 pregnant African-American women showed that exposure to benzene, toluene, ethylbenzene, and xylenes is associated with differential methylation of 46 DNA regions (containing 201 CpG sites), many of which are located in genes implied in GABA receptor signaling and oxytocin brain signaling (Straughen et al. 2024). A methylome-wide association study conducted on peripheral blood leukocytes from 8397 subjects identified three CpG sites whose methylation level is associated with PM exposure; of note, one of them is localized within the *ARPP21* gene, which encodes a neuronal cAMP-regulated phosphoprotein that is implied in AD-related neurodegeneration (Gondalia et al. 2019).

Although numerous studies have reported an association between exposure to air pollutants and peripheral blood epigenetic alterations, DNA methylation changes occurring in the brain remain largely uninvestigated. To date, the only population-based study available reported that $PM_{2.5}$ exposure is associated with altered DNA methylation profile in the prefrontal cortex of 159 individuals with AD; of note, 24 CpG sites were identified as potential mediators between $PM_{2.5}$ exposure and markers of neuropathology and neuroinflammation (Li et al. 2024). Nevertheless, there are many in vivo and in vitro studies supporting this hypothesis. It has been shown that traffic-related air pollution impairs memory function and promotes microglia activation, neuroinflammation, and oxidative stress in chronically exposed aged mice if compared to nonexposed controls. Such brain alterations are accompanied by increased DNA methylation level at the promoter region of the *Abca7* gene, which is implied in microglial activation in AD; on the contrary, *Pyk*, which is involved in synaptic plasticity, was hypomethylated (Xu et al. 2022). Besides, a study on human neuronal cells showed that treatment with $PM_{2.5}$ alters redox balance, reduces the levels of methyl donor S-adenosylmethionine (SAM), and results in global DNA hypomethylation; nevertheless, *NRXN1* and *NLGN3*, two genes related to synapse formation, were found to be hypermethylated and downregulated in terms of mRNA and protein levels (Wei et al. 2016). Interestingly, $PM_{2.5}$-induced neurotoxicity might be also mediated by alterations in DNA hydroxymethylation (hm^5C); indeed, after fine PM treatment, hm^5C levels were increased both globally and in the promoter region of *MeCP2*, *GRIN1*, *GABRB3*, *NRXN1*, and *NLGN3*, which are implied in neuronal growth, synapse plasticity, and memory functions (Wei et al. 2017). Such methylation changes might be mediated by $PM_{2.5}$-induced oxidative stress, which also results in the upregulation of Tet1 expression; indeed, treatment with antioxidants was found to counter Tet1 mRNA increases, as well as alterations in the profile of DNA methylation and hydroxymethylation (Wei et al. 2016, 2017). Prenatal exposure to air pollution can also shape DNA methylation during CNS development (Wei et al. 2016; Feil et al. 2023; Broséus et al. 2024); nevertheless, it is not clear if it might also exert a long-lasting effect on the brain, consequently modulating the risk of age-related NDs.

Besides air pollution, also occupational inhalation exposures to heavy metals can also foster epigenetic alterations that damage the nervous system. In particular, many studies have focused on the effect of welding fumes (WFs) rich in manganese (Mn) on DNA methylation, whose deregulation might promote NDs. A study conducted on welders showed that recent exposure to WFs was associated with reduced blood DNA methylation levels of the *NOS2* gene, which codes for the inducible nitric oxide synthase, and that *NOS2* was hypomethylated in PD cases ($n = 49$) if compared to controls ($n = 103$) (Searles et al. 2015). Another study reported a reduction in global DNA methylation in the brain of rats exposed to WFs if compared to controls exposed to filtered air (Shoeb et al. 2020). ND-related DNA methylation alterations caused by excessive exposure to Mn were also supported by an in vitro study by Liu an colleagues, showing that Mn enhances *Tp53* methylation in microglial cells, resulting in increased levels of COX-2, an enzyme that plays a key role in neuroinflammation (Liu et al. 2022).

The cited studies are summarized in Table 1.

3 Pesticides

Pesticides include a wide range of chemicals commonly employed to protect crops from various "pests," including insects, infesting plants, fungi, and pathogenic microbes. Despite their undoubtedly valuable impact on agricultural yield and food production/conservation, long-term exposure to these compounds has been associated with detrimental effects on the CNS. In particular, many pesticides have been linked to increased PD risk, mainly by interfering with the ubiquitin-proteasome system and with mitochondrial functions in dopaminergic neurons of the substantia nigra (Islam et al. 2021). As numerous population-based studies have found an association between pesticides and DNA methylation, it is possible that this epigenetic mechanism might contribute to pesticide-induced neurotoxicity. A study conducted on 342 subjects with PD and 238 elderly controls showed that chronic low-level exposure to organophosphates (OPs) is associated with global DNA hypomethylation only in PD cases; also, the authors identified blood 70 CpG sites that are differentially methylated in response to OP exposure in the whole population, of which 7 were specifically altered in subjects with PD. Notably, the top genes regulated by

Table 1 Studies concerning the association between air pollution, DNA methylation, and ND risk. Exposure or treatment type, experimental approach, organism and biological matrix where DNA methylation/hydroxymethylation was measured, and potential impact on ND risk are reported

Exposure	Experimental approach	Organism	Biological matrix	Effect on ND risk	References
PM_1, $PM_{2.5}$, PM_{10}, and NO_2	Population-based study	Human	Peripheral blood	Potentially increased risk	Huang et al. (2022)
$PM_{2.5}$ and O_3	Population-based study	Human	Peripheral blood	Potentially increased risk	Huang et al. (2024)
Benzene, toluene, ethylbenzene, and xylenes	Population-based study	Human	Peripheral blood	Potentially increased risk	Straughen et al. (2024)
$PM_{2.5}$ and PM_{10}	Population-based study	Human	Peripheral blood	Potentially increased risk	Gondalia et al. (2019)
$PM_{2.5}$	Population-based study	Human	Brain	Potentially increased risk	Li et al. (2024)
TRAP	In vivo	Mouse	Brain	Increased risk	Xu et al. (2022)
$PM_{2.5}$	In vitro	Human	Cultured neurons	Potentially increased risk	Wei et al. (2016)
$PM_{2.5}$	In vitro	Human	Cultured neurons	Potentially increased risk	Wei et al. (2017)
Prenatal PM_{10}	Population-based study	Human	Cord blood	Unknown	Feil et al. (2023)
Prenatal $PM_{2.5}$, PM_{10}, and NO_2	Population-based study	Human	Placenta	Unknown	Broséus et al. (2024)
Welding fumes	Population-based study	Human	Peripheral blood	Potentially increased risk	Searles et al. (2015)
Welding fumes	In vivo	Rat	Brain	Potentially increased risk	Shoeb et al. (2020)
Manganese	In vitro	Mouse	Cultured microglia	Potentially increased risk	Liu et al. (2022)

OP-induced methylation changes were implied in the muscarinic and nicotinic acetylcholine receptor signaling (Paul et al. 2018). Also, chronic exposure to pyrethroid insecticides was associated with differential blood methylation at 415 CpG sites, which were enriched in genes implied in calcium ion binding, cancer, and neurological diseases, in a population of 237 PD-free elderly subjects (Furlong et al. 2020).

Notably, the DNA methylation-mediated neurotoxic effect of pesticides has been observed not only in humans but also in involuntarily targeted animals living in contaminated areas, such as pigeons (Cao et al. 2014). Besides, many studies conducted on animal models support the role of pesticide-induced DNA methylation changes in PD. Tail injection with the insecticide fenpropathrin (Fen) resulted in degeneration of dopaminergic neurons and impairment of motor functions in treated mice. Such effects were accompanied by increased mRNA and protein levels of Dnmt3b, whereas Dnmt1 and Dnmt3a showed an opposite trend, suggesting that

DNA methylation might be a mediator of Fen-induced neurotoxicity. Indeed, Fen-injected mice had higher DNA methylation/decreased transcript levels of *Ambra1*, a gene known to be implied in mitophagy and PD risk, while treatment with 5-Aza-dC rescued *Ambra1* downregulation and countered Fen-induced neuronal damage (He et al. 2024). Of note, the effect of some pesticides on DNA methylation might be sex specific. In this regard, repeated dietary administration of dieldrin, a highly toxic organochlorine pesticide, to expecting female mice significantly affected DNA methylation in the ventral mesencephalon of the offspring, with female pups having 478 differentially methylated sites (mostly hypomethylated) and males having only 115 (mostly hypermethylated). In particular, differential methylation was found at two genes (viz., *Nr4a2* and *Lmx1b*) related to dopaminergic neuron development and potentially PD (Kochmanski et al. 2019). Also, prenatal exposure to pesticides has been associated with increased risk of NDs later in life. A study showed that dopaminergic pathway impairment in rats treated with permethrin during neonatal brain development was paralleled by increased Tet and Mecp2 protein levels, as well as higher global DNA methylation and hydroxymethylation in the striatum if compared to controls (Bordoni et al. 2019). Besides permethrin-induced changes in striatal DNA methylation can be also transmitted over generations, with a study showing that females treated in early life with this pesticide and their offspring had global DNA hypomethylation if compared to untreated controls (Bordoni et al. 2015).

Finally, the noxious effect of pesticides on PD-related processes is supported by extensive literature on cell cultures, to the point that many pesticides or derivatives are commonly employed as neurotoxic agents for in vitro studies. In this regard, it has been shown that rotenone, a neurotoxic pesticide and insecticide, can affect DNMT1-dependent allele-specific DNA methylation and RNA levels of 14 genes, which are mainly expressed in the brain and implied in intercellular signaling; of these, two genes previously associated to neuronal plasticity and PD risk (*HCN2* and *NEFM*) were found to be hypomethylated (Freeman et al. 2020). Instead, the toxic effect of MPP+ (whose chloride salt, known as "cyperquat," has been widely employed as an herbicide in the 1970s) on dopaminergic neurons was exacerbated by 5-Aza-dC treatment; indeed, both agents exert a demethylating effect in the promoter region of *SNCA*, which codes for α-synuclein (Wang et al. 2013). In a subsequent study, they also demonstrated that MPP+ enhanced *SNCA* transcription and altered DNA methyltransferase levels, with DNMT3a and DNMT3b being downregulated and DNMT1 upregulated, possibly promoting α-synuclein accumulation in PD (Yang et al. 2017). Another study found that MPP+ enhances Tet2 expression and alters the DNA hydroxymethylation in the promoter of *Cdkn2A* in treated dopaminergic neurons; instead, downregulation of Tet2 can attenuate MPP+-induced cell damage and potentially PD symptoms (Wu et al. 2020). Also, treatment of cultured neurons with the herbicide paraquat (a structural analog of cyperquat) results in global hypermethylation, which in turn downregulates the proapoptotic miR-17-5p and causes cell damage (Zhan et al. 2020).

The cited studies are summarized in Table 2.

Table 2 Studies concerning the association between pesticides, DNA methylation, and ND risk. Exposure or treatment type, experimental approach, organism and biological matrix where DNA methylation/hydroxymethylation was measured, and potential impact on ND risk are reported

Exposure	Experimental approach	Organism	Biological matrix	Effect on ND risk	References
Organophosphates	Population-based study	Human	Peripheral blood	Potentially increased risk	Paul et al. (2018)
Pyrethroids	Population-based study	Human	Peripheral blood	Potentially increased risk	Furlong et al. (2020)
Avermectin	In vivo	Pigeons	Brain	Potentially increased risk	Cao et al. (2014)
Fenpropathrin	In vivo	Mouse	Brain	Increased risk	He et al. (2024)
Dieldrin	In vivo	Mouse	Brain	Potentially increased risk	Kochmanski et al. (2019)
Prenatal exposure to permethrin	In vivo	Rat	Brain	Potentially increased risk	Bordoni et al. (2019)
Prenatal exposure to permethrin	In vivo	Rat	Brain	Potentially increased risk	Bordoni et al. (2015)
Rotenone	In vitro	Human	Cultured kidney cell	Potentially increased risk	Freeman et al. (2020)
MPP+	In vitro	Human, rat, and mouse	Cultured neurons	Potentially increased risk	Wang et al. (2013)
MPP+	In vitro	Human	Cultured neurons	Potentially increased risk	Yang et al. (2017)
MPTP and MPP+	In vivo/in vitro	Mouse and human	Brain, cultured neurons	Increased risk	Wu et al. (2020)
Paraquat	In vitro	Mouse	Cultured neurons	Potentially increased risk	Zhan et al. (2020)

4 Diet

There is accumulating evidence that dietary habits can modulate the risk of several NDs; in particular, low-calorie dietary regimens, as well as diets rich in antioxidants and antiinflammatory molecules, have been found to delay age-related cognitive decline via different mechanisms, including DNA methylation. In 1644 elderly participants from the Framingham Heart Study, long-term adherence to a healthy diet resulted in lower pace of biological aging, as measured by a blood DNA methylation score, and reduced risk of dementia (Thomas et al. 2024). In a mouse model of vascular dementia, intermittent fasting was found to ameliorate cognitive and learning deficits and hallmarks of neuropathology induced by chronic cerebral hypoperfusion by modulating DNA methylation in the cerebral cortex (Selvaraji et al. 2022). Similarly, caloric restriction can exert a neuroprotective effect by countering age-related changes in DNA methylation in murine hippocampus (Hadad et al. 2018).

Instead, Western diet and other unhealthy dietary patterns rich in highly processed foods and saturated fats have been associated with DNA methylation alterations that might foster NDs. High-fat diet (HFD) interacts with aging to regulate the hippocampal methylome of mice, mainly affecting the expression of genes related to metabolism and neurodegeneration (Vander Velden and Osborne 2022). HFD-induced obesity was also reported to be associated with impaired synaptic plasticity and spatial memory, as well as promoter hypermethylation and reduced expression of *Ppp1cb*, *Reln*, and *Sirt1* in the mouse hippocampus. In particular, Sirt1 might play an important role as the promoter hm^5C levels and catalytic activity of this deacetylase were found to be reduced; however, mice fed with HFD supplemented with resveratrol had restored Sirt1 activity and hippocampus-dependent memory (Heyward et al. 2016). Alterations in DNA methylation and age-related cognitive deficits are also found in the offspring of female mice fed with HFD; nevertheless, as in the previous case, these noxious effects can be countered by maternal dietary supplementation with resveratrol (Izquierdo et al. 2021). Indeed, many studies reported a DNA-methylation and/or hm^5C-based mechanism of neuroprotection due to plant-derived antioxidative compounds, such as curcumin and cordycepin (Izquierdo et al. 2019; Cao et al. 2020; Deng et al. 2014).

Among dietary antioxidants, multiple studies have focused on polyunsaturated fatty acids (PUFAs), as they are implied in many signaling pathways related to neurotransmission, synaptic function, and neuroinflammation (Bazinet and Layé 2014). Such a regulatory role might be at least partially mediated by changes in DNA methylation, with potential consequences on ND-related processes. Here, a study conducted on 63 subjects with AD showed that participants receiving dietary supplementations with a marine oil rich in ω-3 PUFAs had lower global DNA methylation in circulating leukocytes if compared to placebo controls (Karimi et al. 2017). Also, diet supplementation with PUFAs leads to enhanced cognitive performances in mice, which might be partially mediated by modulation of the DNA methylation landscape in the brain (Chakraborty et al. 2017). Similarly, eicosapentaenoic acid (a ω-3 PUFA) was found to counter PD-related cellular/mitochondrial damage and apoptosis in cultured neurons, by upregulating *BDNF* and *GDNF* via changes in DNA methylation/hydroxymethylation (Ceccarini et al. 2022). On the contrary, ω-3 PUFA deficiency experienced in utero and during lactation leads to persistent changes in DNA methylation and reduced expression of *Bdnf* in the brain cortex of the offspring, which cannot be reversed by ω-3 PUFA administration later in life (Fan et al. 2016).

Unsurprisingly, unbalanced diets deprived or exceeding in compounds implied in the one-carbon cycle (1CC) also interfere with DNA methylation processes in the brain, resulting in increased risk of ND and related cognitive deficits (Pogribny et al. 2008; Grossi et al. 2016). Besides the benefits of SAM-induced DNA methylation changes (Jalgaonkar et al. 2023; Fuso et al. 2011), there is corroborated evidence supporting the neuroprotective role of folic acid (FA, vitamin B9) and other B vitamins implied in 1CC; instead, excessive intake of methionine-rich food (such as red meat) can result in increased levels of homocysteine, a neurotoxic molecule which has been associated with AD pathogenesis (Ungvari et al. 2023).

In a mouse model of AD, folate-deficient diet (FDD) resulted in brain DNA hypomethylation if compared to control diet or FA gavage; instead, folate supplementation was associated with downregulation of multiple genes implied in the JAK-STAT and LTD signaling pathways, which may be relevant for AD pathogenesis (Li et al. 2016). In the same transgenic model, FDD also decreased promoter DNA methylation and increased APP, PS1, and amyloid β (Aβ) protein levels in the hippocampus, while folate administration had the opposite effect (Li et al. 2015b; Liu et al. 2016). The DNA methylation-mediated beneficial effect of FA is also consolidated by numerous in vitro studies (Li et al. 2015a, 2016; Liu et al. 2016; Yu et al. 2014). It has been shown that FA treatment can counter the toxic effect of Aβ oligomers on hippocampal neurons by stimulating DNMT activity, modulating PS1 and APP promoter methylation and RNA levels, and increasing cell viability (Liu et al. 2016). Of note, DNMT inhibition strongly reduces the beneficial effect of FA on Aβ production (Li et al. 2015a). Besides, mice models of AD fed with a diet deficient in B vitamins (FA, vitamins B6 and B12) had worsened cognitive performances and exacerbated AD-related molecular alterations, which were mediated by the hypomethylation in the promoter region of *Alox5*. This gene codes for an enzyme that participates in Aβ formation and tau phosphorylation and that is known to be upregulated in AD brains; indeed, treatment with an Alox5 inhibitor prevented such diet-induced neurodegenerative effects (Li et al. 2017). Similarly, B vitamin deficiency lowers DNA methylation and increases expression of *PSEN1* (Fuso et al. 2011). Besides mice subjected to a diet enriched in methionine and with low folate and vitamin B content had impaired memory performances if compared to controls. Such cognitive were possibly mediated by brain DNA hypermethylation, resulting in the downregulation of the netrin-1 gene, which encodes an antiinflammatory and neuroprotective glycoprotein; of note, administration of netrin was found to ameliorate diet-induced memory deficits (Kalani et al. 2019). The DNA methylation-mediated detrimental effect of methyl-donor-rich diets was also found to be transgenerational (Ryan et al. 2018).

In addition to folate and B-group vitamers, other vitamins have been found to modulate DNA methylation in NDs. Their important role might be attributed to their antioxidant properties or being cofactors of DNMTs or other enzymes related to one-carbon metabolism (e.g., vitamin C is an activator of TETs). In this context, two vitamin K derivatives, namely, menaquinone-4 and reduced menaquinone-7, were found to increase DNA methylation and downregulate the expression of genes implied in inflammation and amyloid neuropathology in treated neuroblastoma cells (Orticello et al. 2023). Also, treatment with vitamin E and circulating vitamin E levels (measured in 860 healthy subjects) were positively associated with DNA methylation level of *MAPT* in cultured neuroblasts and in blood leukocytes, respectively; as *MAPT* is a susceptibility gene for idiopathic PD, vitamin E might exert a neuroprotective effect (Coupland et al. 2014). Notably, vitamins could also play an important role for ND therapy; indeed, a study showed that vitamin C can modify the DNA methylation profile of treated neural stem cells, prompting their differentiation into authentic midbrain-type dopamine neurons with enhanced therapeutic potential after transplantation in PD model rats (Wulansari et al. 2017).

The cited studies are summarized in Table 3.

5 Physical Exercise

Physical exercise (PE) can counter many age-related degenerative processes occurring in the brain, significantly reducing the risk of dementia and AD. The neuroprotective role of regular PE can be attributed to modulation of several ND-related pathways and biological processes that are progressively altered in the aging brain, including oxidative stress, autophagy, neuronal plasticity, synthesis of neurotransmitters, and neuroendocrine factors (Mahalakshmi et al. 2020). Importantly, such PE-induced modifications in brain physiology might be mediated by epigenetic changes, including DNA methylation; in particular, numerous studies have reported that PE can slow down biological aging (estimated by a DNA methylation-based score) and delay age-related neurodegeneration (Xu et al. 2021). However, it is important to note that different types of PE can target different disease-related genes and exert a different and sometimes opposite effect on DNA methylation (Vasileva et al. 2024). Also, the beneficial effect of PE-induced DNA methylation changes on age-related diseases might be influenced by the intensity and frequency of physical activity. Here, a study conducted on power athletes ($n = 94$), endurance athletes ($n = 82$), and controls ($n = 128$) showed that very intense PE promotes hypermethylation of *TRIM59* and *KLF14*, two genes implied in cellular senescence and inflammation, and that very intense sport might promote accelerated aging (Spólnicka et al. 2018).

Instead, long-term moderate PE has been generally associated with DNA methylation changes that are beneficial for age-related ND. A study conducted on 40 elderly women free of AD showed that a 14-week combined exercise training resulted in improved blood pressure and motor skills, as well as altered methylation in 12 AD-related genes in peripheral blood. Although the majority of the CpG sites analyzed were hypomethylated, *ABCA7*, *BDNF*, and *WWOX*, which are implied in the development of AD, were hypermethylated (da Silva Rodrigues et al. 2024). Another study evaluated the effect of a 6-month aerobic exercise on genome-wide DNA methylation in elderly African American subjects with mild cognitive

Table 3 Studies concerning the association between diet, DNA methylation, and ND risk. Exposure or treatment type, experimental approach, organism and biological matrix where DNA methylation/hydroxymethylation was measured, and potential impact on ND risk are reported

Exposure	Experimental approach	Organism	Biological matrix	Effect on ND risk	References
Health diet	Population-based study	Human	Peripheral blood	Potentially decreased risk	Thomas et al. (2024)
Intermittent fasting	In vivo	Mouse	Brain	Potentially decreased risk	Selvaraji et al. (2022)
Caloric restriction	In vivo	Mouse	Brain	Potentially decreased risk	Hadad et al. (2018)
High-fat diet	In vivo	Mouse	Brain	Potentially increased risk	Vander Velden and Osborne (2022)
High-fat diet (+resveratrol[a])	In vivo	Mouse	Brain	Potentially increased risk (countered by [a])	Heyward et al. (2016)
High-fat diet (+resveratrol[a])	In vivo	Mouse	Brain	Potentially increased risk (countered by [a])	Izquierdo et al. (2021)
Resveratrol	In vivo	Mouse	Brain	Decreased risk	Izquierdo et al. (2019)
Cordycepin, cordycepic acid, and resveratrol	In vitro, in vivo	Human and mouse	Cultured neurons	Decreased risk	Cao et al. (2020)
Curcumin	In vitro	Mouse	Cultured neurons	Potentially decreased risk	Deng et al. (2014)
ω-3 PUFAs	Population-based study	Human	Peripheral blood	Potentially decreased risk	Karimi et al. (2017)
PUFAs	In vivo	Mouse	Brain	Potentially decreased risk	Chakraborty et al. (2017)
Eicosapentaenoic acid	In vitro	Human	Cultured neurons	Potentially decreased risk	Ceccarini et al. (2022)
ω-3 PUFA deficiency	In vivo	Mouse	Brain	Potentially increased risk	Fan et al. (2016)
Folate/methyl-deficient diet	In vivo	Rat	Brain	Potentially increased risk	Pogribny et al. (2008)

(continued)

Table 3 (continued)

Exposure	Experimental approach	Organism	Biological matrix	Effect on ND risk	References
Low folate and homocysteine	Population-based study	Human	Peripheral blood	Potentially increased risk	Grossi et al. (2016)
SAM	In vivo	Rat	Brain	Decreased risk	Jalgaonkar et al. (2023)
B vitamin deficiency (+SAM[b])	In vitro and in vivo	Human and mouse	Cultured neurons and brain	Potentially increased risk (countered by [b])	Fuso et al. (2011)
Folate-deficient diet (+folate supplementation[c])	In vivo and in vitro	Mouse	Brain and cultured neurons	Potentially increased risk (countered by [c])	Li et al. (2016)
Folate-deficient diet (+folate supplementation[c])	In vivo	Mouse	Brain	Potentially increased risk (countered by [c])	Li et al. (2015b)
Folate-deficient diet (+folate supplementation[c])	In vivo and in vitro	Mouse and rat	Brain and cultured neurons	Potentially increased risk (countered by [c])	Liu et al. (2016)
Folate	In vitro	Rat	Cultured neural stem cells	Potentially decreased risk	Yu et al. (2014)
Folate	In vitro	Mouse	Cultured neurons	Potentially decreased risk	Li et al. (2015a)
B vitamin deficiency	In vivo	Mouse	Brain and cultured neurons	Potentially increased risk	Li et al. (2017)
Methionine-rich, folate-deficient diet	In vivo	Mouse	Brain	Potentially increased risk	Kalani et al. (2019)
Methyl donor-rich diet	In vivo	Mouse	Brain	Potentially increased risk	Ryan et al. (2018)
Vitamin K2	In vitro	Human	Cultured neurons	Potentially decreased risk	Orticello et al. (2023)
Vitamin E	In vitro	Human	Peripheral blood and cultured neurons	Potentially decreased risk	Coupland et al. (2014)
Vitamin C	In vivo and in vitro	Human and rat	Cultured neural stem cells	Potentially decreased risk	Wulansari et al. (2017)

impairment. If compared to controls ($n = 8$), subjects who underwent training ($n = 11$) showed hypomethylation and upregulation of the *VPS52* gene, which is implied in the trafficking of amyloid precursor protein (Ngwa et al. 2022). In addition, regular exercise was found to decrease whole blood DNA methylation at a single nucleotide polymorphism (SNP) within *GPNMB*, which codes for a glycoprotein implicated in PD pathogenesis, only in males (Chen et al. 2023).

The effect of PE on DNA methylation in the context of NDs was also reported by numerous in vivo studies. It has been shown that rat models of cognitive impairment subjected to moderate-intensity intermittent training had improved memory and learning ability, which were paralleled by hypermethylation and downregulation in the hippocampus of the *PDE4D* gene, which is considered a potential target for the treatment of NDs (Zhang et al. 2023). In the same animal model, similar beneficial effects on cognition were reported for swimming exercise, which was found to reduce hippocampal methylation and increase mRNA/protein levels of CREB, a transcription factor with a crucial role in modulating synaptic plasticity (Li et al. 2019). Instead, it has been found that voluntary wheel running (VWR) stimulates DNA demethylation at *Bdnf* exon IV promoter and increases the levels of Bdnf mRNA and protein in the rat hippocampus (Gomez-Pinilla et al. 2011). The same trend in hippocampal DNA methylation and expression of *Bdnf* following to VWR was observed in mice, which also had higher DNA methylation and lower transcript levels of Nos1, a gene implied in neurotransmission and behavioral processes (Tomiga et al. 2021). Another study reported hippocampal demethylation in the promoter of *VegfA*, which encodes a proangiogenic and neuroprotective growth factors, in exercised rats if compared to sedentary ones; also, the former had altered expression levels of enzymes modulating DNA methylation, with Tet1 mRNA being increased and Dnmt3b decreased (Sølvsten et al. 2018).

Indeed, PE can also modulate brain levels and posttranslational modifications of DNA methylation-related enzymes, such as the phosphorylation of MeCP2 (which results in its dissociation from *Bdnf* promoter) (Gomez-Pinilla et al. 2011); as shown by Cechinel and colleagues, DNMT3b levels are increased in the prefrontal cortex of aged rats submitted to acute treadmill exercise (Cechinel et al. 2016). De Meireles and coworkers showed that different training modalities (repeated aerobic, acrobatic, resistance, or combined PE sessions) increase hippocampal H3K4 trimethylation at the *Bdnf* and *Dnmt3a* promoters, which might be related to increased Dnmt3a expression and consequently global hypermethylation, to ameliorate aversive memory performance and survival rate in aged male rats (de Meireles et al. 2019). Another study reported that voluntary running wheel exercise can counter the age-dependent decline in *Tet1* and *Tet2* expression, which might be related to improved memory functions in mice and to increased hippocampal hm^5C in the promoter region of miR-137, a miRNA involved in adult neurogenesis (Jessop and Toledo-Rodriguez 2018). Interestingly, not only PE regulates DNA-methylation related enzymes but also vice versa. Indeed, overexpression of *Dnmt3a2/3L* in dopaminergic neurons of transgenic mice leads to higher firing frequency and excitability, resulting in increased spontaneous PE and improved motor performances (Cui et al. 2020). Instead, *Dnmt3a* knockout in hypothalamic neurons leads to DNA methylation reshaping and enhanced sedentary behavior in transgenic mice (MacKay et al. 2019).

Besides the beneficial effect of PE on the CNS can also be transmitted across generations, and this might me mediated—at least partially—by DNA methylation changes occurring in the offspring in response to parental training. Of note, such DNA methylation changes do not only modulate neurodevelopmental processes but can also persist throughout the lifespan of the offspring, posing as potential modulatory factors on ND risk in late life. Two studies showed that pregestational aerobic or resistance exercise was associated with significant reshaping of hippocampal global DNA methylation and improved spatial cognitive ability in the rat male offspring (Segabinazi et al. 2019). Similar beneficial effects on spatial learning were found to be elicited by paternal preconception training, which resulted in decrease global DNA methylation in the hippocampus of the male offspring (Spindler et al. 2019).

The cited studies are summarized in Table 4.

6 Future Directions and Conclusions

Although the concept of exposome has been proposed for the first time only 20 years ago, the need for an integrated and unifying view that goes beyond the health effect of single exposures is becoming more and more pressing. This is of particular importance to gain a comprehensive knowledge of the etiological and pathogenic mechanisms underlying multifactorial, age-related noncommunicable diseases (NCDs), which nowadays represent the greatest burden for the healthcare system in middle- and high-income countries.

A handful of in vivo studies have explored the impact of combined lifestyle interventions on DNA methylation in the context of NDs, demonstrating the potential of specific interventions to mitigate the harmful effects of environmental exposures (Kretzschmar et al. 2022; Benassi et al. 2019; Rodrigues et al. 2015; Zhang et al. 2020). For example, swimming exercise was found to modulate global DNA methylation changes in response to acute stress in rat brain (Rodrigues et al. 2015), while forced running dampened the effect of radiation-induced cognitive dysfunction by increasing hydroxymethylation and *Bdnf* expression levels in murine hippocampus (Zhang et al. 2020).

However, the complexity of studying the combined effects of multiple exposures on health, particularly in human populations, presents substantial methodological challenges. One of the primary obstacles lies in the difficulty of isolating and quantifying the individual contributions of diverse environmental, lifestyle, and genetic factors in real-world scenarios. Humans are constantly exposed to a range of environmental influences, including pollution, diet, and physical activity, which interact in ways that are difficult to replicate or predict under controlled experimental settings. Moreover, the impact of these exposures may vary based on genetic predispositions, making it challenging to generalize findings across different population

Table 4 Studies concerning the association between physical exercise, DNA methylation, and ND risk. Exposure or treatment type, experimental approach, organism and biological matrix where DNA methylation/hydroxymethylation and related processes was measured, and potential impact on ND risk are reported

Exposure	Experimental approach	Organism	Biological matrix	Effect on ND risk	References
Intense PE	Population-based study	Human	Peripheral blood	Potentially increased risk	Spólnicka et al. (2018)
Long-term, moderate PE	Population-based study	Human	Peripheral blood	Potentially decreased risk	da Silva Rodrigues et al. (2024)
Long-term, aerobic PE	Population-based study	Human	Peripheral blood	Potentially decreased risk	Ngwa et al. (2022)
Long-term PE	Population-based study	Human	Peripheral blood	Potentially decreased risk	Chen et al. (2023)
Moderate PE	In vivo	Rat	Brain	Potentially decreased risk	Zhang et al. (2023)
Swimming	In vivo	Rat	Brain	Potentially decreased risk	Li et al. (2019)
Voluntary wheel running	In vivo	Rat	Brain	Potentially decreased risk	Gomez-Pinilla et al. (2011)
Voluntary wheel running	In vivo	Mouse	Brain	Potentially decreased risk	Tomiga et al. (2021)
PE (vs. sedentary lifestyle)	In vivo	Rat	Brain	Potentially decreased risk	Sølvsten et al. (2018)
Acute treadmill exercise	In vivo	Rat	Brain	Unknown	Cechinel et al. (2016)
Different PE types	In vivo	Rat	Brain	Potentially decreased risk	de Meireles et al. (2019)
Voluntary wheel running	In vivo	Mouse	Brain	Potentially decreased risk	Jessop and Toledo-Rodriguez (2018)
Prenatal aerobic or resistance PE	In vivo	Rat	Brain	Potentially decreased risk	Segabinazi et al. (2019)
Paternal treadmill running	In vivo	Rat	Brain	Potentially decreased risk	Spindler et al. (2019)

groups. While in vivo studies, particularly in animal models, have begun to shed light on the potential mechanisms underlying the effects of combined exposures, the results cannot always be directly translated to humans due to differences in biological processes, lifespan, and environmental conditions.

In addition, human studies are often confounded by the difficulty of accounting for the cumulative and long-term nature of exposures, especially as they relate to age-related NCDs, including the most common NDs. These diseases typically develop over decades, with multiple genetic and environmental factors contributing at various stages of life. As a result, research must consider not only the immediate impact of exposures but also their cumulative effects over time. This requires sophisticated longitudinal studies, ideally involving large, diverse cohorts, that can track changes in DNA methylation, gene expression, and other biomarkers across the lifespan.

Furthermore, there is a need for integrative approaches that combine genetic, epigenetic, and exposomic data to better understand the complex interactions between environmental, lifestyle, and genetic factors. Advances in high-throughput technologies, such as genome-wide association studies, DNA methylation profiling, and metabolomics, have made it possible to generate vast amounts of data; however, interpreting such information requires new bioinformatic tools and statistical models that consider the intricate relationships between these variables. Only by overcoming these methodological hurdles can we hope to gain a comprehensive understanding of the underlying mechanisms driving NCDs and identify the most effective prevention and intervention strategies.

Ultimately, bridging these gaps in our understanding is crucial for developing targeted, personalized approaches to prevent and mitigate the effects of NCDs. This could involve the implementation of specific lifestyle or environmental interventions aimed at counteracting the detrimental effects of genetic susceptibility or exposure to harmful environmental factors. Additionally, such research has the potential to uncover new therapeutic targets or biomarkers for early detection and intervention, offering hope for more effective treatments and better health outcomes, particularly for aging populations. The need for further research in this field is clear, as it holds promise for providing valuable insights into the interplay between genetics, environment, and lifestyle, which could significantly improve public health strategies for managing NDs.

References

Bali J, Gheinani AH, Zurbriggen S et al (2012) Role of genes linked to sporadic Alzheimer's disease risk in the production of β-amyloid peptides. Proc Natl Acad Sci USA 109:15307–15311

Bazinet RP, Layé S (2014) Polyunsaturated fatty acids and their metabolites in brain function and disease. Nat Rev Neurosci 15:771–785

Benassi B, Santangeli S, Merla C et al (2019) 50-Hz MF does not affect global DNA methylation of SH-SY5Y cells treated with the neurotoxin MPP. Bioelectromagnetics 40:33–41

Bordoni L, Nasuti C, Mirto M et al (2015) Intergenerational effect of early life exposure to permethrin: changes in global DNA methylation and in Nurr1 gene expression. Toxics 3:451–461

Bordoni L, Nasuti C, Fedeli D et al (2019) Early impairment of epigenetic pattern in neurodegeneration: additional mechanisms behind pyrethroid toxicity. Exp Gerontol 124:110629

Broséus L, Guilbert A, Hough I et al (2024) Placental DNA methylation signatures of prenatal air pollution exposure and potential effects on birth outcomes: an analysis of three prospective cohorts. Lancet Planet Health 8:e297–e308

Cao Y, Chen L-J, Zhang Z et al (2014) Global DNA hypomethylation: a potential mechanism in king pigeon nerve tissue damage induced by avermectin. Chem Biol Interact 219:113–122

Cao D, Jiang D, Zhou D et al (2020) A comparative study on 5hmC targeting regulation of neurons in AD mice by several natural compounds. Biomed Res Int 2020:5016706

Ceccarini MR, Ceccarelli V, Codini M et al (2022) The polyunsaturated fatty acid EPA, but not DHA, enhances neurotrophic factor expression through epigenetic mechanisms and protects against parkinsonian neuronal cell death. Int J Mol Sci 23:16176

Cechinel LR, Basso CG, Bertoldi K et al (2016) Treadmill exercise induces age and protocol-dependent epigenetic changes in prefrontal cortex of Wistar rats. Behav Brain Res 313:82–87

Chakraborty N, Muhie S, Kumar R et al (2017) Contributions of polyunsaturated fatty acids (PUFA) on cerebral neurobiology: an integrated omics approach with epigenomic focus. J Nutr Biochem 42:84–94

Chen YC, Liaw YC, Nfor ON et al (2023) Epigenetic regulation of Parkinson's disease risk variant GPNMB Cg17274742 methylation by sex and exercise from Taiwan biobank. Front Aging Neurosci 15:1235840

Costa LG, Cole TB, Dao K et al (2020) Effects of air pollution on the nervous system and its possible role in neurodevelopmental and neurodegenerative disorders. Pharmacol Ther 210:107523

Coupland KG, Mellick GD, Silburn PA et al (2014) DNA methylation of the MAPT gene in Parkinson's disease cohorts and modulation by vitamin E in vitro. Mov Disord 29:1606–1614

Cui D, Mesaros A, Burdeos G et al (2020) Dnmt3a2/Dnmt3L overexpression in the dopaminergic system of mice increases exercise behavior through signaling changes in the hypothalamus. Int J Mol Sci 21:6297

da Silva Rodrigues G, Noronha NY, Yonehara Noma IH et al (2024) 14-week exercise training modifies the DNA methylation levels at gene sites in non-Alzheimer's disease women aged 50 to 70 years. Exp Gerontol 186:112362

Dao T, Cheng RYS, Revelo MP et al (2014) Hydroxymethylation as a novel environmental biosensor. Curr Environ Health Rep 1:1–10

de Meireles LCF, Galvão F, Walker DM et al (2019) Exercise modalities improve aversive memory and survival rate in aged rats: role of hippocampal epigenetic modifications. Mol Neurobiol 56:8408–8419

Deng Y, Lu X, Wang L et al (2014) Curcumin inhibits the AKT/NF-κB signaling via CpG demethylation of the promoter and restoration of NEP in the N2a cell line. AAPS J 16:649–657

Deng H, Wang P, Jankovic J (2018) The genetics of Parkinson disease. Ageing Res Rev 42:72–85

Fan C, Fu H, Dong H et al (2016) Maternal N-3 polyunsaturated fatty acid deprivation during pregnancy and lactation affects neurogenesis and apoptosis in adult offspring: associated with DNA methylation of brain-derived neurotrophic factor transcripts. Nutr Res 36:1013–1021

Feil D, Abrishamcar S, Christensen GM et al (2023) DNA methylation as a potential mediator of the association between indoor air pollution and neurodevelopmental delay in a south African birth cohort. Clin Epigenetics 15:31

Freeman DM, Lou D, Li Y et al (2020) The conserved DNMT1-dependent methylation regions in human cells are vulnerable to neurotoxicant rotenone exposure. Epigenetics Chromatin 13:17

Fu C, Kuang D, Zhang H et al (2022) Different components of air pollutants and neurological disorders. Front Public Health 10:959921

Furlong MA, Paul KC, Yan Q et al (2020) An epigenome-wide association study of ambient pyrethroid pesticide exposures in California's Central Valley. Int J Hyg Environ Health 229:113569

Fuso A, Nicolia V, Pasqualato A et al (2011) Changes in Presenilin 1 gene methylation pattern in diet-induced B vitamin deficiency. Neurobiol Aging 32:187–199

Gomez-Pinilla F, Zhuang Y, Feng J et al (2011) Exercise impacts brain-derived neurotrophic factor plasticity by engaging mechanisms of epigenetic regulation. Eur J Neurosci 33:383–390

Gondalia R, Baldassari A, Holliday KM et al (2019) Methylome-wide association study provides evidence of particulate matter air pollution-associated DNA methylation. Environ Int 132:104723

Grossi E, Stoccoro A, Tannorella P et al (2016) Artificial neural networks link one-carbon metabolism to gene-promoter methylation in Alzheimer's disease. J Alzheimers Dis 53:1517–1522

Hadad N, Unnikrishnan A, Jackson JA et al (2018) Caloric restriction mitigates age-associated hippocampal differential CG and non-CG methylation. Neurobiol Aging 67:53–66

He S, Qu Q, Chen X et al (2024) Downregulation of Ambra1 by altered DNA methylation exacerbates dopaminergic neuron damage in a fenpropathrin-induced Parkinson-like mouse model. Ecotoxicol Environ Saf 271:115995

Heyward FD, Gilliam D, Coleman MA et al (2016) Obesity weighs down memory through a mechanism involving the neuroepigenetic dysregulation of Sirt1. J Neurosci 36:1324–1335

Hopkins LE, Laing EA, Peake JL et al (2018) Repeated iron-soot exposure and nose-to-brain transport of inhaled ultrafine particles. Toxicol Pathol 46:75–84

Huang H, Yang L, Liu Y et al (2022) Long-term ambient air pollution exposure and DNA methylation of peripheral brain-derived neurotrophic factor promoter. Ecotoxicol Environ Saf 244:114061

Huang H, Xie B, Liu Y et al (2024) Long-term exposure to PM2.5 compositions and O3 and their interactive effects on DNA methylation of peripheral brain-derived neurotrophic factor promoter. Int J Environ Health Res 34:2957–2968

Islam MS, Azim F, Saju H et al (2021) Pesticides and Parkinson's disease: current and future perspective. J Chem Neuroanat 115:101966

Izquierdo V, Palomera-Ávalos V, López-Ruiz S et al (2019) Maternal resveratrol supplementation prevents cognitive decline in senescent mice offspring. Int J Mol Sci 20:1134

Izquierdo V, Palomera-Ávalos V, Pallàs M et al (2021) Resveratrol supplementation attenuates cognitive and molecular alterations under maternal high-fat diet intake: epigenetic inheritance over generations. Int J Mol Sci 22:1453

Jalgaonkar S, Gajbhiye S, Sayyed M et al (2023) S-adenosyl methionine improves motor coordination with reduced oxidative stress, dopaminergic neuronal loss, and DNA methylation in the brain striatum of 6-hydroxydopamine-induced neurodegeneration in rats. Anat Rec (Hoboken) 306:820–830

Jessop P, Toledo-Rodriguez M (2018) Hippocampal TET1 and TET2 expression and DNA hydroxymethylation are affected by physical exercise in aged mice. Front Cell Dev Biol 6:45

Jones MJ, Goodman SJ, Kobor MS (2015) DNA methylation and healthy human aging. Aging Cell 14:924–932

Kalani A, Chaturvedi P, Kalani K et al (2019) A high methionine, low folate and vitamin B6/B12 containing diet can be associated with memory loss by epigenetic silencing of netrin-1. Neural Regen Res 14:1247–5124

Karimi M, Vedin I, Levi YF et al (2017) DHA-rich n-3 fatty acid supplementation decreases DNA methylation in blood leukocytes: the OmegAD study. Am J Clin Nutr 106:1157–1165

Kochmanski J, VanOeveren SE, Patterson JR et al (2019) Developmental dieldrin exposure alters DNA methylation at genes related to dopaminergic neuron development and Parkinson's disease in mouse midbrain. Toxicol Sci 169:593–607

Kretzschmar GC, Targa ADS, Soares-Lima SC et al (2022) Folic acid and vitamin B12 prevent deleterious effects of rotenone on object novelty recognition memory and Kynu expression in an animal model of Parkinson's disease. Genes 13:2397

Kuntić M, Hahad O, Münzel T et al (2024) Crosstalk between oxidative stress and inflammation caused by noise and air pollution-implications for neurodegenerative diseases. Antioxidants (Basel) 13:266

Li G, Chen Y, Wang J et al (2015) Direct imaging of biological sulfur dioxide derivatives in vivo using a two-photon phosphorescent probe. Biomaterials 63:128–136

Li W, Jiang M, Zhao S et al (2015a) Folic acid inhibits amyloid β-peptide production through modulating DNA methyltransferase activity in N2a-APP cells. Int J Mol Sci 16:25002–25013

Li W, Liu H, Yu M et al (2015b) Folic acid administration inhibits amyloid β-peptide accumulation in APP/PS1 transgenic mice. J Nutr Biochem 26:883–891

Li W, Liu H, Yu M et al (2016) Folic acid alters methylation profile of JAK-STAT and long-term depression signaling pathways in Alzheimer's disease models. Mol Neurobiol 53:6548–6556

Li J-G, Barrero C, Merali S et al (2017) Five lipoxygenase hypomethylation mediates the homocysteine effect on Alzheimer's phenotype. Sci Rep 7:46002
Li X, Wang L, Zhang S et al (2019) Timing-dependent protection of swimming exercise against d-galactose-induced aging-like impairments in spatial learning/memory in rats. Brain Sci 9:236
Li Z, Liang D, Ebelt S et al (2024) Differential DNA methylation in the brain as potential mediator of the association between traffic-related PM2.5 and neuropathology markers of Alzheimer's disease. Alzheimers Dement 20:2538–2551
Liu H, Li W, Zhao S et al (2016) Folic acid attenuates the effects of amyloid β oligomers on DNA methylation in neuronal cells. Eur J Nutr 55:1849–1862
Liu X, Yao C, Tang Y et al (2022) Role of P53 methylation in manganese-induced cyclooxygenase-2 expression in BV2 microglial cells. Ecotoxicol Environ Saf 241:113824
Livingston G, Huntley J, Sommerlad A et al (2020) Dementia prevention, intervention, and care: 2020 report of the lancet commission. Lancet 396:413–446
MacKay H, Scott CA, Duryea JD et al (2019) DNA methylation in AgRP neurons regulates voluntary exercise behavior in mice. Nat Commun 10:5364
Mahalakshmi B, Maurya N, Lee S-D et al (2020) Possible neuroprotective mechanisms of physical exercise in neurodegeneration. Int J Mol Sci 21:5895
Mellén M, Ayata P, Heintz N (2017) 5-Hydroxymethylcytosine accumulation in postmitotic neurons results in functional demethylation of expressed genes. Proc Natl Acad Sci USA 114:E7812–E7821
Migliore L, Coppedè F (2022) Gene–environment interactions in Alzheimer disease: the emerging role of epigenetics. Nat Rev Neurol 18:643–660
Ngwa JS, Nwulia E, Ntekim O et al (2022) Aerobic exercise training-induced changes on DNA methylation in mild cognitively impaired elderly African Americans: gene, exercise, and memory study—GEMS-I. Front Mol Neurosci 14:752403
Ogonowski NS, García-Marín LM, Fernando AS et al (2024) Impact of genetic predisposition to late-onset neurodegenerative diseases on early life outcomes and brain structure. Transl Psychiatry 14:185
Orticello M, Cavallaro RA, Antinori D (2023) Amyloidogenic and neuroinflammatory molecular pathways are contrasted using Menaquinone 4 (MK4) and reduced Menaquinone 7 (MK7R) in association with increased DNA methylation in SK-N-BE neuroblastoma cell line. Cells 13:58
Paul KC, Chuang Y-H, Cockburn M et al (2018) Organophosphate pesticide exposure and differential genome-wide DNA methylation. Sci Total Environ 645:1135–1143
Pihlstrøm L, Wiethoff S, Houlden H (2018) Chapter 22: Genetics of neurodegenerative diseases: an overview. Handb Clin Neurol 145:309–323
Pogribny IP, Karpf AR, James SR et al (2008) Epigenetic alterations in the brains of fisher 344 rats induced by long-term Administration of Folate/methyl-deficient diet. Brain Res 1237:25–34
Rodrigues GM, Toffoli LV, Manfredo MH et al (2015) Acute stress affects the global DNA methylation profile in rat brain: modulation by physical exercise. Behav Brain Res 279:123–128
Ryan DP, Henzel KS, Pearson BL et al (2018) A paternal methyl donor-rich diet altered cognitive and neural functions in offspring mice. Mol Psychiatry 23:1345–1355
Searles N, Checkoway SH, Criswell SR et al (2015) Inducible nitric oxide synthase gene methylation and parkinsonism in manganese-exposed welders. Parkinsonism Relat Disord 21:355–360
Segabinazi E, Spindler C, Ferreira de Meireles AL et al (2019) Effects of maternal physical exercise on global DNA methylation and hippocampal plasticity of rat male offspring. Neuroscience 418:218–230
Selvaraji S, Efthymios M, Sik Yin Foo R et al (2022) Time-restricted feeding modulates the DNA methylation landscape, attenuates hallmark neuropathology and cognitive impairment in a mouse model of vascular dementia. Theranostics 12:3007–3023
Shoeb M, Mustafa GM, Kodali VK (2020) A possible relationship between telomere length and markers of neurodegeneration in rat brain after welding fume inhalation exposure. Environ Res 180:108900

Sølvsten CAE, de Paoli F, Christensen JH et al (2018) Voluntary physical exercise induces expression and epigenetic remodeling of VegfA in the rat hippocampus. Mol Neurobiol 55:567–582
Spindler C, Segabinazi E, de Meireles ALF et al (2019) Paternal physical exercise modulates global DNA methylation status in the hippocampus of male rat offspring. Neural Regen Res 14:491–500
Spólnicka M, Pośpiech E, Adamczyk JG et al (2018) Modified aging of elite athletes revealed by analysis of epigenetic age markers. Aging 10:241–252
Straughen JK, Loveless I, Chen Y et al (2024) The impact of environmental benzene, toluene, ethylbenzene, and xylene exposure on blood-based DNA methylation profiles in pregnant African American women from Detroit. Int J Environ Res Public Health 21:256
Thomas A, Ryan CP, Caspi A et al (2024) Diet, pace of biological aging, and risk of dementia in the Framingham heart study. Ann Neurol 95:1069–7109
Tomiga Y, Sakai K, Ra S-G et al (2021) Short-term running exercise alters DNA methylation patterns in neuronal nitric oxide synthase and brain-derived neurotrophic factor genes in the mouse hippocampus and reduces anxiety-like behaviors. FASEB J 35:e21767
Ungvari A, Gulej R, Csik B et al (2023) The role of methionine-rich diet in unhealthy cerebrovascular and brain aging: mechanisms and implications for cognitive impairment. Nutrients 15:4662
Vander Velden JW, Osborne DM (2022) Prolonged diet-induced obesity modifies DNA methylation and gene expression in the hippocampus. Neurosci Lett 780:136656
Vasileva F, Hristovski R, Font-Lladó R et al (2024) Physical exercise–induced DNA methylation in disease-related genes in healthy adults—a systematic review with bioinformatic analysis. J Strength Cond Res 38:384
Wang Y, Wang X, Li R et al (2013) A DNA methyltransferase inhibitor, 5-Aza-2′-deoxycytidine, exacerbates neurotoxicity and upregulates Parkinson's disease-related genes in dopaminergic neurons. CNS Neurosci Ther 19:183–190
Wang H, Yang F, Zhang S et al (2021) Genetic and environmental factors in Alzheimer's and Parkinson's diseases and promising therapeutic intervention via fecal microbiota transplantation. NPJ Parkinsons Dis 7:70
Wei H, Liang F, Meng G et al (2016) Redox/methylation mediated abnormal DNA methylation as regulators of ambient fine particulate matter-induced neurodevelopment related impairment in human neuronal cells. Sci Rep 6:33402
Wei H, Feng Y, Liang F et al (2017) Role of oxidative stress and DNA hydroxymethylation in the neurotoxicity of fine particulate matter. Toxicology 380:94–103
Wild CP (2012) The exposome: from concept to utility. Int J Epidemiol 41:24–32
Wilson DM, Cookson MR, Van Den Bosch L et al (2023) Hallmarks of neurodegenerative diseases. Cell 186:693–714
Wu T-T, Liu T, Li X et al (2020) TET2-mediated Cdkn2A DNA hydroxymethylation in midbrain dopaminergic neuron injury of Parkinson's disease. Hum Mol Genet 29:1239–1252
Wu H, Eckhardt CM, Baccarelli AA (2023) Molecular mechanisms of environmental exposures and human disease. Nat Rev Genet 24:332–344
Wulansari N, Kim E-H, Sulistio YA et al (2017) Vitamin C-induced epigenetic modifications in donor NSCs establish midbrain marker expressions critical for cell-based therapy in Parkinson's disease. Stem Cell Rep 9:1192–1206
Xu M, Zhu JY, Liu X-D et al (2021) Roles of physical exercise in neurodegeneration: reversal of epigenetic clock. Transl Neurodegener 10:30
Xu C, Zhang J, Zhou Q et al (2022) Exposure to a real traffic environment impairs brain cognition in aged mice. Environ Res 215:114181
Yang J, Yang Z, Wang X et al (2017) CpG demethylation in the neurotoxicity of 1-methyl-4-phenylpyridinium might mediate transcriptional up-regulation of α-synuclein in SH-SY5Y cells. Neurosci Lett 659:124–132
Yu M, Li W, Luo S et al (2014) Folic acid stimulation of neural stem cell proliferation is associated with altered methylation profile of PI3K/Akt/CREB. J Nutr Biochem 25:496–502

Zhan Y, Guo Z, Zheng F et al (2020) Reactive oxygen species regulate miR-17-5p expression via DNA methylation in paraquat-induced nerve cell damage. Environ Toxicol 35:1364–1373

Zhang J, Li J, Zhu Y et al (2020) Forced running exercise mitigates radiation-induced cognitive deficits via regulated DNA hydroxymethylation. Epigenomics 12:385–396

Zhang J, Gao Q, Gao J et al (2023) Moderate-intensity intermittent training alters the DNA methylation pattern of PDE4D gene in hippocampus to improve the ability of spatial learning and memory in aging rats reduced by D-galactose. Brain Sci 13:422

Regulation of RNA Transcriptomics During Ageing- and Neurodegeneration-Associated Cognitive Changes

Faiza Gilani and Akash Gautam

Contents

Abstract Cognitive decline caused by ageing and neurodegeneration is a considerable healthcare challenge in our rapidly ageing population. This chapter delves into the critical function of RNA transcriptomics in understanding and treating age- and neurodegeneration-related cognitive alterations. We provide a detailed overview of numerous RNA modifications and their regulatory mechanisms, including changes to messenger RNA (mRNA), transfer RNA (tRNA), and ribosomal RNA (rRNA). Specific tRNA changes, including 5-methylcytosine (m^5C), queuosine (Q), and pseudouridine (Ψ), have been identified as key participants in neurodegenerative

F. Gilani · A. Gautam (✉)
Centre for Neural and Cognitive Sciences, School of Medical Sciences, University of Hyderabad, Hyderabad, India
e-mail: akash@uohyd.ac.in

J. Barciszewski (ed.), *Decoding Aging and Neurodegeneration*,
RNA Technologies 16, https://doi.org/10.1007/978-3-031-91072-2_4

processes. The chapter also looks at the regulatory roles of noncoding RNAs, including microRNAs and long noncoding RNAs, in cognitive function and dysfunction. We address how RNA alterations affect synaptic plasticity and memory formation and their changed patterns in diverse neurodegenerative diseases. We also look at the therapeutic potential of targeting RNA changes and the promising function of epigenetic medicines in cognitive health.

Keywords RNA modifications · Cognitive decline · Ageing · Neurodegeneration · Epitranscriptomics · tRNA modifications · Noncoding RNA · Therapeutic targets

1 Introduction

As our global population lives longer than ever, we face an unprecedented challenge: preserving the cognitive vitality that allows us to engage fully with the world, even in advanced age. The heartbreaking sight of an elderly grandmother struggling to recall the names of her beloved grandchildren or a once-brilliant professor gradually losing the ability to deliver the eloquent lectures that had once come so naturally represents the all-too-human face of age-related cognitive decline and neurodegeneration. These conditions, which affect thousands worldwide, pose one of the finest healthcare-demanding situations of the twenty-first century. Grappling with this challenge is not only a matter of improving quality of life but also of maintaining the intellectual and creative capital that powers our societies. Uncovering the biological mechanisms underlying cognitive changes in ageing and developing effective interventions has never been more crucial.

Senescence is an innate process in the life cycle of all living organisms, characterized by gradual physiological deterioration and high mortality (Shabir and Singh 2022). Crucially, it is regarded as a significant risk factor for most neurodegenerative diseases (NDs), including frontotemporal dementia, Parkinson's disease (PD), Alzheimer's disease (AD), and Huntington's disease (HD). The chronicity of these ailments in older populations constitutes a considerable danger, as the WHO forecasts that the number of individuals nearing 60 and over will be twofold by 2050 (United Nations 2020), increasing the incidence of dementia and neurodegenerative diseases. There is a strong relationship between ageing and neurodegeneration. In older adults, NDs have become prevalent, and brains free of disease are rare. This implies that the ageing of the brain may be regarded as an indicator of the development of neurodegeneration. ND onset, as well as progression, may be predicted by human genetic and environmental factors (Wyss-Coray 2016).

Cognitive ageing is a complicated combination of healthy physiological modifications and neurological disorders. While some degree of cognitive change is regarded as a standard element of ageing, the distinction between natural ageing and pathological deterioration remains a topic of active scientific research

(Buckley and Pascual-Leone 2020). Cognitive decline is a prevalent side effect of both typical ageing and other neurodegenerative diseases. While neuronal loss leads to cognitive impairment, changes in synaptic functioning and signaling pathways are expected to play a more important impact. Age-related cognitive alterations often manifest in several important domains. Novotný et al. (2022) found that the ability to digest information and respond quickly reduces over time. Working memory, or the ability to temporarily store and manipulate data, may also decline. Furthermore, the effectiveness of executive functions, which include planning, decision-making, and mental flexibility, can decline. One of the most noticeable changes in this case is the inability to remember specific events or personal experiences (episodic memory).

Genetic variables significantly influence people's cognitive trajectories as they age. A study analyzing DNA patterns from disease-free octogenarians and older people discovered that certain single nucleotide polymorphisms (SNPs) are linked to cognitive function regulation. The findings imply a connection between brain health and cognitive performance, meaning genetic factors may influence whether a person ages healthily or develops neurodegenerative diseases. Normal ageing and neurodegeneration include complex chemical changes affecting neuronal activity and viability. These include alterations in neural plasticity, mitochondrial failure, oxidative stress, and inflammation (Cenini et al. 2019). Recent research has demonstrated that quite many of these alterations are governed at the RNA level, emphasizing the importance of transcriptional and posttranscriptional pathways in cognitive health and disease (Butler et al. 2016).

Advanced sequencing technology and powerful computational approaches have transformed our understanding of RNA regulation in the ageing brain (Ham and Lee 2020). During ageing and neurodegeneration, the transcriptome, the complete set of RNA molecules in a cell or tissue, undergo major modifications. These modifications influence coding as well as noncoding RNAs, including miRNA, circRNA, and lncRNAs (Calzoni et al. 2021). Understanding these RNA-level changes is important for a variety of reasons. First, they offer mechanistic insights into how ageing and disease processes influence cognitive function at the cellular level. Second, they may be helpful indicators for early cognitive impairment detection and monitoring. Third, possibly most crucially, they represent viable therapeutic targets for therapies to prevent or halt cognitive decline.

Current research suggests that RNA changes such as methylation, editing, and alternative splicing are critical for synaptic plasticity, memory formation, and neuronal survival. The multifariousness of these changes can reach a variety of processes, including protein synthesis rate and the creation of stress granules, ultimately reducing cognitive capacities (Qiu et al. 2023). Furthermore, anomalies in these RNA alterations have been corelated to an assortment of neurodegenerative disorders, indicating their potential as therapeutic targets.

2 Importance of RNA Transcriptome for Human Cognition

Transcriptome is defined as the complete complement of mRNA molecules produced by a cell or population of cells, serving as a critical mediator of human cognitive function via diverse molecular mechanisms. This dynamic system entails a complex interplay of multiple RNAs, encompassing messenger RNA, long noncoding RNA, and countless tiny RNAs, each contributing uniquely to cognitive processes and brain function. The large amounts of noncoding RNAs and transposons, once considered garbage, now offer the ability to regulate the plasticity needed to program human ontogeny and cognition (Assis et al. 2022).

The human brain has extremely complex transcriptional patterns, including region- and cell-type-specific expression profiles that support cognitive function (Yang et al. 2024). The temporal and spatial control of the transcriptome is essential for brain growth and differentiation, synaptic plasticity maintenance, memory acquisition and consolidation, responsiveness to environmental stimuli, and cognitive flexibility. This precise regulation ensures adequate neuronal function across diverse brain regions and at distinct developmental stages (Weyn-Vanhentenryck et al. 2018).

The messenger RNAs are the principal protein transcripts throughout the neural transcriptome, driving the production of proteins required for brain function. Neurotransmitter receptors, ion channels, synaptic proteins, transcription factors, and structural proteins all play a role in maintaining brain architecture and function. These mRNAs' appropriate expression and regulation are critical for regular cognitive processes and neural circuit operation (Bhat et al. 2022).

Long noncoding RNAs, or transcripts that exceed 200 nucleotides, have emerged as essential regulators of brain activity. These compounds play important roles in gene expression control, chromatin remodeling, and brain development. A recent study has revealed their significant participation in synaptic plasticity and cognitive processing, indicating a more complex role in neuronal function than previously thought. LncRNAs can operate as scaffolding for protein complexes, guides for chromatin-modifying enzymes, and gene expression regulators, all of which contribute to the complicated control of neuronal function and cognitive processes (Gionco and Bernstein 2024).

MicroRNAs and other short RNAs play important cognitive functions by regulating posttranscriptional processes and controlling protein synthesis. These chemicals are critical for fine-tuning synapse function and memory formation processes. Their ability to swiftly and precisely control gene expression makes them important regulators of activity-dependent changes in brain circuits (Grinkevich 2020). The brain's transcriptome is remarkably adaptable to varied inputs and situations. Environmental influences, including learning experiences, stress reactions, circadian rhythms, and social interactions, can cause considerable alterations in the transcriptional landscape (Park et al. 2020). The table mentions key RNA modifications and their role in cognition (Table 1).

Table 1 Key RNA modifications and their role in cognition

Modification type	Chemical structure	RNA type	Role in memory and cognition	References
m^6A (*N^6*-methyladenosine)	Methylation at *N^6* position of adenosine	mRNA, lncRNA, and rRNA	Modulation of learning and memory formation required for axon regeneration	Li et al. (2022) and Chatterjee et al. (2021)
m^1A (*N^1*-methyladenosine)	Methylation at *N^1* position of adenosine	tRNA and mt-tRNA	Supports tRNA stability and impacts translation of mitochondria	Chatterjee et al. (2021)
Ψ (Pseudouridine)	Isomerization of uridine	All RNA types	Promotes RNA stability, which is essential for ribosome activity, and modulates stress response	Jung and Goldman (2018)
m^5C (5-methylcytosine)	Methylation at C^5 position of cytosine	tRNA, rRNA, and mRNA	Manages neural stem cells, controls memory development, and impacts anxiety behaviors	Jung and Goldman (2018)
Nm (2′-*O*-methylation)	Methylation at 2′-OH of any nucleoside	All RNA types	Safeguards short RNAs and regulates miRNA function, which is critical for plasticity	Jung and Goldman (2018)

2.1 *RNA in Synaptic Plasticity and Memory Formation*

Synaptic plasticity, or synapses' ability to strengthen or weaken over time, provides the biological foundation for learning and memory formation. Complex circuits connected by synapses are required for higher-level cognitive processes in the brain, including memory and learning. The molecular mechanisms that underpin these processes entail complicated interactions among numerous cellular components, with RNA molecules playing critical regulatory functions. Recent breakthroughs in molecular neuroscience have revealed that various RNA species contribute significantly to activity-dependent synapse changes and memory consolidation (Kennedy 2016).

The formation of different types of memory involves distinct RNA-based mechanisms. Short-term memory formation typically involves rapid transcriptional responses through immediate early genes, whereas long-term memory necessitates sustained gene expression variability patterns (Li et al. 2017). The CREB-dependent transcription pathway is pivotal in this mechanism, initiating waves of gene expression that support long-lasting synaptic modifications. The synaptic tag and capture hypothesis provides a framework for understanding how specific synapses maintain their strength over time. This process depends heavily on RNA-dependent mechanisms, including local protein synthesis and the trafficking of specific mRNAs to activated synapses. These mechanisms ensure that synaptic modifications remain specific to the activated circuits while allowing for the integration of information across time (Hildebrandt 2022).

The regulation of synaptic plasticity through RNA-based mechanisms encompasses multiple layers of complexity, beginning with messenger RNAs (mRNAs) and extending to various noncoding RNA species (Lakhotia et al. 2020). Local protein synthesis at synapses, mediated by mRNAs, is essential for long-lasting forms of synaptic plasticity. This process requires sophisticated mechanisms for mRNA transport and localization within dendrites. De novo protein synthesis utilizing localized mRNAs as templates and translation-dependent neuronal plasticity constitute two significant biochemical processes that control synaptic structure and function. RNA-binding proteins, especially HuD that have been related to learning, synaptic plasticity, neural differentiation, and regrowth not only bind to mRNAs but also network and associate with other RNA types in the cell, such as circular RNAs, a new group of noncoding RNAs generated through back-splicing pre-mRNAs, as well as certain long noncoding RNAs (Rajgor et al. 2021).

Emerging evidence suggests a pivotal role of various noncoding RNA species, like microRNA, in maintaining synaptic function and memory formation. MicroRNAs are a class of small, noncoding RNAs that adhere to the 3′ untranslated regions of target mRNAs that induce translational repression or degradation, thereby modifying gene expression at the posttranscriptional level (Baptista et al. 2021). These regulatory RNAs are considered essential facilitators in synaptic development and plasticity, along with cellular mechanisms involved in memory formation (Soutschek and Schratt 2023).

Numerous research has been conducted on the involvement of microRNAs in neuronal differentiation, long-term memory formation, its role in the pathogenesis of diseases on neurodegenerative ailments, and dementia. Extended dendritic tree restructuring involves miR-135, miR-137, and miR-191 (Grinkevich 2020; Siedlecki-Wullich et al. 2021). miR-26a and miR-384-5p are involved in LTP formation, and their expression is reduced during tetanization in an RSK3-dependent manner. Synapses of structural as well as signalling proteins are targets of these microRNAs. Research has determined that receptor formation, proteins involved in synaptic vesicle movement, along with proteins essential for spine growth, can occur in neurites, as neurites possess all the necessary components for mRNA translation, consisting of ribosomes, as well as pre-microRNA, microRNA, and the Dicer enzyme, enabling microRNA for regulating local biosynthesis (Grinkevich 2020; Hu and Li 2017).

Synaptic plasticity proteins were found to be elevated in forebrain neurons with a Dicer knockout, indicating its role in learning and memory; however, its direct correlation to individual microRNA or its effect on the global RNA level is still unclear. DGCR8 variation, which impacts drosha-dependent miRNAs, worsens performance in the Morris water maze while also causing distinct effects on the excitability of neurons and neurogenesis (Tassinari et al. 2023).

Long noncoding RNAs (lncRNAs) contribute to synaptic function through various mechanisms, including transcriptional regulation and scaffold functions. These molecules can interact with proteins and other RNA species to influence synaptic plasticity. Notable examples include NEAT1, which plays a role in synaptic maintenance, and MALAT1, which influences synapse formation (Srinivas et al. 2023). BDNF-AS (brain-derived neurotrophic factor antisense transcript) represents another important example of lncRNA function in synaptic regulation (Bhattacharyya et al. 2021). Among the identified functionally characterized lncRNAs, BC1 (known as BC200 in humans) is localized to both axons and dendrites, interacting with several protein partners, encompassing FMRP, PABP1, eIF4A, and eIF4B (Srinivas et al. 2023). The lack of BC1 RNA in mice leads to elevated spine density, enlarged PSD (postsynaptic density) size in cortical neurons, and elevated spontaneous activity (Briz et al. 2017).

These mRNAs express proteins required for the plasticity of synapses and memory, including ARC, Argonaute2, and BDNF. According to Abrishamdar et al. (2022), the Bdnf antisense transcript Bdnf OS-Bdnf mRNA is a remarkable pair demonstrating a noncorrelated expression variation with neural activity. It increases BDNF levels at the transcript and protein levels when knocked out. This increase eventually leads to neuronal outgrowth, typically related to synaptic plasticity and memory. In the hippocampus, overexpression of LoNA led to decreased spine density and long-term potentiation (LTP), two memory-related processes. These results indicate that loss of LoNA enhances hippocampus-dependent spatial memory. However, the injection of LoNA has a reverse impact (Li et al. 2018). The BC1-KO mice exhibited spatial learning and memory deficits, as demonstrated by active spot aversion tests involving the hippocampus. This observation aligns with the enhanced mGluR1-mediated long-term depression. The efficacy of the prefrontal cortex in discrimination and conflict learning has been demonstrated to be diminished in mice deficient in BC1. This finding indicates that the BC1-KO mouse exhibits no deficits in memory recall but cannot distinguish between recently acquired memories. These mice cannot implement an altered response strategy when presented with a situational context that contradicts the prior situational context while knowing the components of the context (Iacoangeli et al. 2017).

The numerous lncRNAs expressed in the brain indicate that these limited instances of lncRNA-mediated regulation of synaptic plasticity and memory represent a fraction of the overall phenomenon. Utilizing advanced sequencing techniques to detect low copy numbers of lncRNAs from synaptic divisions, effective manipulation of lncRNA function via CRISPR-Cas13, and animal models to assess cell-type specific translation, we anticipate a swift enhancement of our knowledge of memory mechanisms influenced by this novel regulator.

2.2 *Epitranscriptomic Signatures in Cognitive Disorders*

The advent of epitranscriptomics, a field dedicated to studying the chemical and structural features of RNA alterations, has changed the landscape of molecular neuroscience (Cerneckis et al. 2024). This fast-evolving profession has grown more intertwined with other areas of life science, ranging from cancer biology and stem cell research to vaccine development and neurology. Studying these RNA changes has specifically revolutionized our understanding of cognitive diseases, revealing critical insights into neuronal function and malfunction at the molecular level (Mathoux et al. 2021).

Epitranscriptomic regulation is a sophisticated layer of posttranscriptional control in which chemical alterations to RNA molecules dynamically affect their metabolism, stability, and function. These changes eventually influence cognitive functions via intricate biological mechanisms. Recent scientific breakthroughs have revealed that dysregulation of the epitranscriptome has significant links with a range of illnesses, including neuropsychiatric, neurodevelopmental, and neurodegenerative problems (Peer et al. 2017).

The complexity of RNA modifications in neuronal function is astounding, with over a hundred unique chemical changes discovered across multiple RNA types. RNA methylation is particularly significant, accounting for over 60% of all RNA modifications. *N*6-methyladenosine (m^6A) has become eukaryotes' most significant posttranscriptional RNA modification. The enrichment of m^6A in adult brain tissue is particularly notable, indicating its crucial function in neurogenesis, neurodevelopment, and NDs (Wang et al. 2022). The intricate machinery that governs m^6A modifications includes writers, erasers, and readers, all of which collaborate to ensure proper brain function. Alongside 5-methylcytosine (m^5C), pseudouridine (Ψ), m^6A, and *N*1-methyladenosine (m^1A) perform essential functions in regulating RNA metabolism in the brain (Zhao 2017).

In the case of Alzheimer's disease (AD), distinct epitranscriptomic patterns emerge that are markedly distinct from a healthy brain. The m^6A landscape significantly alters transcripts encoding proteins crucial for synaptic function and amyloid processing (Calzoni et al. 2021). Recent studies have revealed decreased expression of m^6A methyltransferase METTL13 in AD, while the m^6A reader YTHDC1 has demonstrated mitigating effects in ischemic stroke. These results underline the beneficial impact of m^6A on brain health (Huang et al. 2020).

Understanding epitranscriptomic markers in cognitive diseases holds significant therapeutic potential. The capacity to change RNA opens up new treatment options, particularly for illnesses where effective medicines are unavailable. Identifying specific epitranscriptomic patterns associated with distinct cognitive illnesses could help develop therapeutic targets or biomarkers for the early detection of diseases.

3 mRNA Modifications During Ageing- and Neurodegeneration-Associated Cognitive Changes

The intricate relationship between ageing, neurodegeneration, and cognitive decline is increasingly understood through the lens of mRNA modifications (Jiapaer et al. 2022). These modifications represent critical regulatory mechanisms that undergo significant alterations during ageing and neurodegenerative processes, contributing to cognitive changes and neuronal dysfunction. Understanding these modifications provides crucial insights into both normal ageing and pathological conditions affecting cognitive function. From early embryonic development to ageing, RNA modifications are crucial for almost every biological process (Tassinari et al. 2023). RNA modifications significantly influence transcript stability, translation initiation, splicing, nuclear export, localization, and folding (Wilkinson et al. 2022) (Fig. 1).

RNA methylation, particularly m^6A modification, represents a dynamic regulatory mechanism that shows significant alterations during ageing and neurodegeneration. The distribution and abundance of m^6A marks change with age, affecting the stability and translation of key neuronal transcripts. Alteration m^6A was initially recognized in the 1970s. Recent advancements in identification as well as quantification of m^6A in transcriptome at single-base resolution established m^6A as the most well-characterized as well as prevalent internal RNA modification, with approximately 0.2–0.6% of adenosines in mammalian mRNAs exhibiting m^6A (Jiapaer et al. 2022). Functionally, m^6A engages in practically every step of the mRNA life cycle, encompassing splicing and processing in the nucleus, as well as translation and degradation in the cytoplasm. At the cellular level, m^6A is essential for the transition of cellular identity in different states throughout differentiation or stress response by impacting transcriptome output (Wang et al. 2022).

Dendritic structure, brain development, neurogenesis, axon regeneration, learning memory, and spinogenesis are all influenced by m^6A modifications (Yen and Chen 2021). ND phenotypes emerge from dysregulated m^6A pathways, resulting from disease-specific mutations or alterations in levels of various m^6A components. Despite the incomplete understanding of underlying mechanisms, RNA modifications at m^6A positions are associated with intellectual disability, schizophrenia, and epilepsy (Livneh et al. 2020). By altering the protein levels of transcripts leading to disease phenotypes, m^6A modification also plays an essential role in AD. In the development of age-related disorders, consisting of AD, PD, FTLD, etc., the m^6A

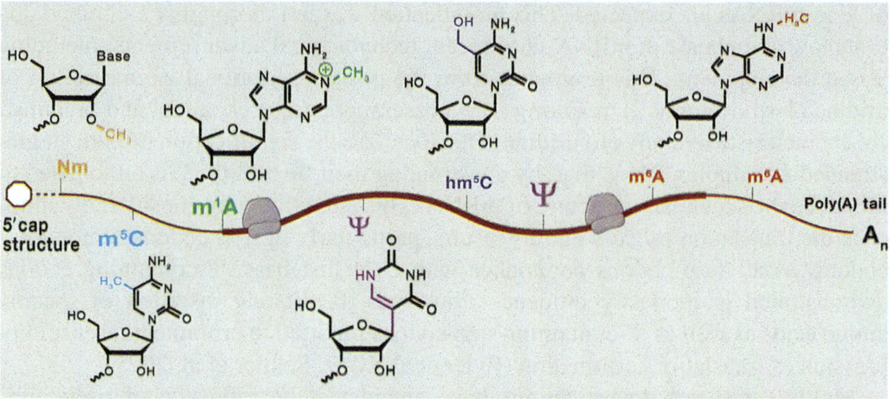

Fig. 1 Site-specific chemical modifications of mRNA. Key regulatory modifications found in eukaryotic messenger RNA. These modifications form part of the epitranscriptome that modulates mRNA function and fate. (Reprinted with permission from Roundtree et al. 2017)

modifications at the 3′-UTR of mRNAs appear to have essential mechanistic roles, mainly through modulation of the translation of these transcripts. Dysregulated m^6A modification has also been associated with NDs in several RNA metabolic processes, especially splicing (Mendel et al. 2021).

The most prevalent mRNA modification, m^5C, was initially identified in 1975 and is caused by methylation at the fifth position of cytosine. With ribose exception, this modification is comparable to DNA methylation m^5C. This mRNA alteration is related to NSUN2 (NOP2/Sun RNA methyltransferase family member 2). Additionally, the mRNA export adaptor protein ALYREF (ALY/REF export factor) RNA could detect m^5C modification (Sokpor et al. 2021). Because m^5C methyltransferases are essential for RNA metabolism, several diseases may be caused by decreased expression of these genes. For instance, mutations in genes encoding NSUN proteins have been associated with a number of neurodevelopmental disorders (Mendel et al. 2021). A homozygous mutation of the NSUN2 gene in humans has been associated with the development of autosomal-recessive intellectual disability. This mutation causes arginine to be substituted for glycine at position 679, which is nonsynonymous and causes the protein to accumulate mislocalized in the nucleolus. A recent investigation that examined the conditional disruption of NSUN2 in mice's prefrontal cortex demonstrated that depression-related behaviors altered in either direction. In addition to decreased translation efficiency linked to a glycine-codon defect, NSUN2 deficiency produced alterations in 1488 proteins in the prefrontal cortex. As a result, the mice exhibited impaired contextual fear memory and impaired pyramidal neuron synaptic signaling in the prefrontal cortex (Blaze et al. 2021).

Initially identified in 1957, Ψ is an isomerization of uridine, also referred to as 5-ribosyl uracil Ψ. Because of its high abundance in cellular RNA, it initially emerged as a fifth base in RNA. Several tRNA and rRNA PUS (Pseudouridine synthase) conserved throughout eukaryotes are partly responsible for this mRNA modification (Eyler et al. 2019). Additional biochemical investigation and comprehension of Ψ in mRNAs are expected. This modification was not thoroughly examined due to its low abundance in mRNA until recent technological advancements, including Ψ-seq development. This originates from the posttranscriptional isomerization of uridine (1-ribosyl uracil), resulting in Ψ possessing unique chemical and biophysical characteristics relative to uridine. This localizes the rigidification of both single-stranded and duplex RNA, thereby constraining their flexibility. Ψ is anticipated to influence the secondary structure of mRNA. This alteration may considerably influence the translation process and its results, particularly in stop codons or nonsense codons, as all stop codons commence with U at first base. Ψ-containing codons demonstrated to modestly influence ribosomes, facilitating insertion of specific amino acids as well as Ψ-containing stop codons reported to promote nonsense suppression of translation termination (Eyler et al. 2019; Sokpor et al. 2021).

Multiple research investigations have elucidated the influence of pseudouridylation on neuronal functions. Nociceptor neuron-specific loss of RluA-1 and RluA-2, which encode Drosophila PUS enzymes, results in hypersensitive nociception phenotypes, including thermal hyperalgesia (Song et al. 2020). Elevated amounts are recognized in AD patients' urine, though this feature has not been symptomatically linked to AD. Consequently, pseudouridylation could be a therapeutic target for NDs, including DM2. Mutations in PSU genes are recognized to induce neuronal dysfunction. Two-point mutations in PUS1 were identified in a patient exhibiting mild cognitive impairment (Cao et al. 2016).

Alternative splicing undergoes profound changes during ageing and neurodegeneration, representing a key mechanism in cognitive decline. Notably, synaptic function and plasticity genes often show altered splicing patterns in ageing brain tissue. In neurodegenerative conditions, these splicing defects become more pronounced, affecting crucial proteins involved in neuronal survival and function. The accumulation of aberrant splice variants can contribute to cellular stress and dysfunction, accelerating cognitive decline. The primary constituent of Lewy bodies as well as PD defining feature, α-synuclein, has multiple isoforms, and alternative splicing controls their expression (Dick et al. 2020). Similarly, AD, another well-known ND, is influenced by the ratio of alternatively spliced products to tau gene product MAPT, specifically 3R tau (exon-10 formation) and 4R tau (formed upon exon-10 inclusion). According to some recent research, splicing defects and alternative splicing contribute to various NDs. One of the primary causes of HD is identified as splicing defects. The expansion of CAG repeats in SCA (spinocerebellar ataxia)-type 6-linked genes induces modifications to mRNA splicing patterns, leading to the development of polyglutamine-containing proteins resulting in disease (Perrone et al. 2020).

Polyadenylation represents another crucial modification that undergoes significant changes during ageing and neurodegeneration. The length and regulation of poly(A) tails directly influence mRNA stability and translation efficiency, which are increasingly dysregulated with age. Age-related changes in polyadenylation patterns impact gene expression related to synaptic plasticity and cognitive function. The complex machinery governing polyadenylation, including poly(A) polymerases and regulatory factors, shows altered activity profiles in ageing neurons (Skariah and Todd 2021). These changes can lead to inappropriate mRNA turnover rates and protein expression patterns, potentially contributing to cognitive decline. Disrupted polyadenylation may accelerate the loss of important neuronal proteins in neurodegenerative conditions, exacerbating cellular dysfunction. APA of mRNAs involves utilizing multiple polyadenylation sites within primary transcripts, often in conjunction with alternative splicing. APA enhances cellular transcriptomic diversity by producing unique mRNA isoforms (Tian and Manley 2017) (Table 2).

Recent technological advances have enhanced our ability to study these modifications in unprecedented detail. High-throughput sequencing approaches combined with sophisticated computational analysis allow comprehensive mapping of modification patterns across the transcriptome. Single-cell analyses reveal cell-type-specific changes in RNA modifications during ageing and neurodegeneration, providing new insights into cellular vulnerability and disease progression (Jackson et al. 2024).

Table 2 Dysregulated RNA modifications in major neurodegenerative diseases

Disease	RNA type	Modification	Regulation	References
Alzheimer's disease (AD)	mRNA	m^6A	Upregulated	Wang et al. (2022) and Huber et al. (2023)
	tRNA	m^5C	Downregulated	
	rRNA	Ψ (pseudouridine)	Downregulated	
Parkinson's disease (PD)	mRNA	m^6A	Unregulated	Gionco and Bernstein (2024)
	tRNA	mcm^5s^2U	Downregulated	
	lncRNA	m^5C	Upregulated	
Huntington's disease (HD)	mRNA	m^6A	Upregulated	Wild and Tabrizi (2017)
	tRNA	Q (queuosine)	Downregulated	
Frontotemporal dementia (FTLD)	mRNA	m^6A	Upregulated	Calzoni et al. (2021)

4 tRNA Modifications During Ageing- and Neurodegeneration-Associated Cognitive Changes

The progressive decline in neuronal function and cognitive abilities is a hallmark of NDs, which include PD and AD, along with progressive supranuclear palsy. Recent research has focused on the complex function of tRNA modifications, even though oxidative stress, mitochondrial dysfunction, and protein aggregation have all been recognized as essential contributors to their development. Transfer-RNA (tRNA) is an adaptor molecule that associates a specific codon in mRNA with its associated amino acid during protein synthesis. tRNAs are subjected to enzymatic modifications posttranscriptionally, guaranteeing the precision and efficacy of protein translation and contributing substantially to gene regulation expression and preserving cellular homeostasis. Recent evidence suggests that irregularities in tRNA modifications perform a vital function in the pathophysiology of various human diseases, highlighting their broader importance in cellular dysfunction (Lv et al. 2024).

tRNA is the most heavily modified RNA molecule inside the cell, with residues differentially distributed among different tRNAs and across the various species. tRNA modifications are highly linked to a plethora of physiological and pathological neurological phenotypes (Chujo and Tomizawa 2021). A vast array of modifications on tRNAs have been identified, including N^4-acetylcysteine (ac^4C), galactosyl-Queosine (galQ), N^2-dimethyl guanosine ($m^{2,2}G$), N^2-methylguanosine (m^2G), 1-methylguanosine (m^1G), m^1A, dihydrouridine, 3-(3-amino-3-carboxypropyl)uridine (acp^3U), methoxycarbonyl methyl uridine (mcm^5U), 3-methylcytidine (m^3C), 2-methylthio-N^6-threonylcarbamoyladenosine (ms^2t^6A), mannosyl-queuosine (manQ), 5-methoxy carbonylmethyl-2-thiouridine (mcm^5s^2U), inosine, 1-methylinosine, queuosine (Q), 5-formyl-2′-*O*-methylcytidine (f^5Cm), N^6-methyl-N^6-threonylcarbamoyladenosine (m^6t^6A), N^6-threonylcarbamoyladenosine (t^6A), N^6-isopentenyl adenosine (i^6A), cytosine (yW), 1-methyl pseudouridine (m^1J), peroxywybutosine (o^2yW), m^7G, m^5C, 5-methyluridine (m^5U), 2-methyladenosine (m^2A), and 2′-*O*-methyl pseudouridine (Ψm) (Suzuki 2021).

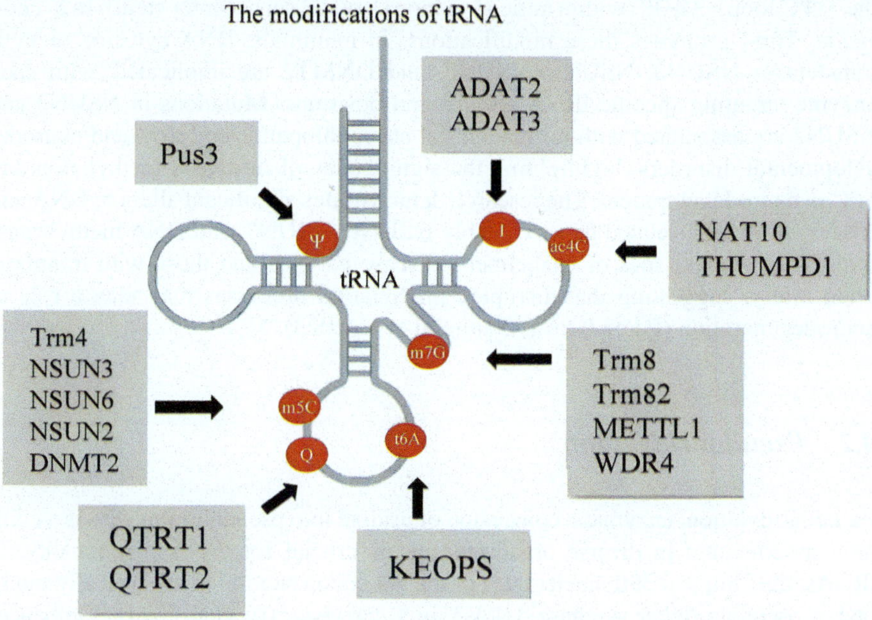

Fig. 2 Enzymatic modification of tRNA and their associated enzymes. The diagram illustrates various modification sites on tRNA molecules and the corresponding enzyme responsible. Key modifications include pseudouridine (Ψ), inosine (I), m^7G, m^5C, queuosine (Q), and t^6A. Each modification is catalyzed by specific enzymes such as Pus3, ADAT2/ADAT3, Trm4, and NSUN family protein, which are shown in gray boxes pointing to their respective modification sites on the tRNA structure. (Reprinted from Lv X et al. (2024), originally published under a Creative Commons Attribution (CC BY) license)

The particular purpose of each modification is contingent upon its location within the tRNA molecule, influencing tRNA stability, aminoacylation, and decoding. While modifications of the tRNA body augment its structural stability, the modified residues in the anticodon loop mainly interact with the codons on the mRNA and with the ribosomes in the decoding center, resulting, therefore, crucial for decoding during translation (Lv et al. 2024) (Fig. 2). Alterations in the anticodon region predominantly happen at positions 34 and 37, which are crucial for the precision as well as the effectiveness of mRNA decoding. Recent research results indicate a higher frequency of mutations in the genes encoding human tRNA-modifying enzymes. Such mutations frequently result in diseases primarily impacting CNS (Lei and Burton 2022).

4.1 *m⁵C in tRNA*

Adding a methyl group (CH3) to tRNA nucleotides, referred to as methylation, is essential for structural stability and molecular interactions. In eukaryotic tRNAs, m^5C typically located at specific positions: 48, 49, 50 (from the variable loop and the T-ΨC loop), 34, 38 (within anticodon loop), and 72 (in acceptor stem). In *S. cerevisiae*, Trm4 promotes these modifications. In mammals, RNA cytosine methyltransferases NSUN3, NSUN6, NSUN2, and DNMT2 are implicated, with each enzyme targeting specific tRNA species and positions. Mutations in NSUN3 and NSUN2 are associated with mitochondrial encephalopathy, epilepsy, and neurodevelopmental disorders, highlighting the significance of mt-tRNA-methylation and m^5C in neurodevelopment. The research demonstrates significant alters in DNA and RNA methylation-related genes (notably NSUN6, NSUN7, and DNA methyltransferases) in brain tissues of Alzheimer's disease patients and those with traumatic brain injury, suggesting that disrupted methylation processes may play a role in neurodegeneration (Perez Grovas-Saltijeral et al. 2023).

4.2 Pseudouridylation

Pseudouridylation, enzymatic conversion of uridine into pseudouridine (Ψ), is a common modification in diverse organisms and is crucial for tRNA functionality. In tRNAs, the 38th and 39th nucleotides of the anticodon stem-loop are typically modified by pseudouridylate synthase-3 (Pus3) in *S. cerevisiae* (Lv et al. 2024). Ψ enhances the thermal stability of anticodon stem-loop by amplifying base-stacking interactions, a characteristic observed in yeast tRNA-Phe and human tRNA-Lys. This modification emphasizes the importance of Ψ in maintaining tRNA structural integrity. The advancement of the reverse transcriptase-active DNA polymerase variant RT-KTq I614Y signifies a significant breakthrough in RNA modification research. Utilizing this technology alongside next-generation sequencing may yield novel insights into the impact of pseudouridylation on cellular mechanisms, as it facilitates direct identification of pseudouridine and queuosine modifications in RNA without necessitating prior treatment (Huber et al. 2023). Although elevated levels of Ψ have been reported in AD patients' urine, this characteristic is not symptomatically associated with AD. Neuronal dysfunction is related to abnormalities in PSU genes. For instance, a patient with mild cognitive impairment, sideroblastic anemia from childhood, adult-onset hepatopathy, cardiomyopathy, and insulin-dependent diabetes—all prevalent clinical symptoms of myopathy-lactic acidosis-sideroblastic anemia (MLASA) syndrome—was discovered to have two-point mutations in PUS1. Additionally, the research that correlates PUS3 truncation to intellectual disability in humans validates the concept that PUS3 mRNA expression in the nervous system of mouse embryos might serve an essential part in neuronal development (Chujo and Tomizawa 2021).

4.3 Queuosine (Q)

At the tRNAs wobble anticodon position reading codons of amino acids Tyr, His, Asn, and Asp, there is a conserved tRNA modification termed queuosine (Q). The control of translation fidelity and efficiency is significantly influenced by

Q-modification in tRNA (Katanski et al. 2022). Although queuosine aids in maintaining the integrity of the tRNA machinery, its physiological significance and function in eukaryotes have not been thoroughly determined. It has also been demonstrated that human HepG2 cells lacking queuosine and mice lacking Q-tRNA, which disrupts tRNA TGTase, have a reduced capacity to convert phenylalanine into tyrosine. This is due to a reduction in cofactor BH4 supply caused by increased tetrahydrobiopterin oxidation, which elevates BH2 (dihydrobiopterin) through an as-yet-unidentified mechanism (Lv et al. 2024). Given that dopamine (DA) biosynthesis requires a steady supply of BH4, this has important ramifications. Therefore, the possibility that such a Q-tRNA deficiency could resemble PD is interesting.

4.4 *t⁶A*

Another recognized alteration is *N*6-threonyl-carbamoyl-adenosine (t^6A) modification. A complex adenosine alteration appears at position 37 (t^6A37), adjacent to the anticodon stem loop of several tRNAs that decode ANN codons. Recent research reported a neurodegenerative phenotype developed in two patients as a result of a biosynthesis defect in the t^6A molecule resulting from a mutation in the kinase-associated endopeptidase gene of highly conserved kinase, endopeptidase as well as other proteins of small size (KEOPS) complex (Han et al. 2020a). This contributes to the expanding list of neurological disorders related to abnormalities in cytoplasmic tRNA modification enzymes.

While the current literature on tRNA modifications in neurodegenerative disorders remains relatively nascent, emerging evidence suggests their profound implications in neuronal function and dysfunction. The paucity of direct mechanistic studies linking specific tRNA modifications to neurodegenerative pathologies represents a significant gap in our understanding of disease progression. Recent advancements in epitranscriptomics, encompassing high-throughput sequencing techniques and mass spectrometry-based methodologies, are being utilized to clarify the complex landscape of tRNA modifications within neuronal systems. However, the field awaits more comprehensive investigations to elucidate the precise molecular mechanisms through which altered tRNA modification patterns contribute to neurodegeneration. The intricate interplay between various tRNA modifications, their writers, readers, and erasers in the context of neuronal homeostasis presents an unexplored frontier with substantial therapeutic potential. Understanding these relationships offers novel insights into disease pathogenesis and potentially reveals new therapeutic targets for neurodegenerative disorders. This emerging field warrants extensive investigation using advanced molecular tools and systematic approaches to bridge the current knowledge gap between tRNA epitranscriptomics and neurodegenerative disease mechanisms.

5 Regulation by Noncoding RNAs During Ageing- and Neurodegeneration-Associated Cognitive Changes

Despite the initial dismissal of numerous noncoding RNA (ncRNA) species as mere byproducts of random transcription, a diverse range of ncRNA regulatory mechanisms has become apparent. The expression of ncRNA in the brain is dynamically regulated in an activity-dependent and spatiotemporally controlled manner, suggesting that ncRNAs perform specific regulatory functions in brain development and operation (Valizadeh et al. 2024).

miRNAs demonstrate significant expression patterns in NDs, contributing substantially to neurodegeneration progression. Overexpression of miR-9-5p activates AD's Nrf2/Keap1 signalling pathway (Liu et al. 2020b). The downregulation of miR-15/107 elevates CDK5R1/p35 and CDK5 activities in AD (Hu et al. 2024). miR-21 has its protecting contribution to AD via targeting PDCD4/PI3K/AKT/GSK-3β in AD (Bai and Bian 2022). Increased miR-126 could play a functional role in the pathogenesis of PD via IGF-1/PI3K/AKT signalling downregulation (Xue et al. 2022).

miR-20a-5p is associated with apoptosis in ND. miR-20a-5p contributes to MPP+-induced apoptosis by reducing elevating cleaved-caspase 3, Bcl-2 levels and Bax, AIF, and cytochrome-C levels. Furthermore, miR-20b-5p inhibition reduced Aβ25-35-induced apoptosis by targeting RhoC. miR-34a declined in the cortex of a transgenic mouse model of AD. Overexpression of miR-34a precipitated swift cognitive decline. miR-34a influences bcl2 through a single miRNA recognition element (MRE) located in 3′UTR of bcl2 in an animal model of AD (Gentile et al. 2022).

Research has identified miR-124 as a negative regulator of BACE1 in AD, potentially contributing to its pathogenesis. miR-124 may modulate cell death in AD by targeting BACE1 and is proposed as a novel therapeutic target in treating AD. Researchers additionally validated the involvement of miR-124 in neurogenesis, neurotransmission, synaptic morphology, mitochondrial dysfunction, cellular survival, autophagy, neuroinflammation, and oxidative damage in PD. Furthermore, plasma miR-124 levels could be utilized as a diagnostic biomarker for PD (Han et al. 2020b). In the SH-SY5Y cells' in vitro model of PD induced by MPP, miR-124-3p expression decreased. By targeting STAT3, overexpression of miR-124-3p reduced MPP-induced neuronal damage, cell death, and Caspase-3 activity. In MPTP-induced PD models, miR-124 inhibited the apoptosis of dopamine cells. According to some researchers, AD is associated with downregulated expression of miR-124 (Liu et al. 2020a). miR-124 downregulation decreased SH-SY5Y cell apoptosis. By blocking Bim along with preventing apoptosis, miR-124 reduces the translocation of Bax to mitochondria as well as lysosomes, thereby providing neuroprotective effects. In the SH-PC12 and SY5Y cells used in the in vitro model of PD induced by 6-OHDA, miR-124-3p expression dropped. The caspase-3 activities in PC12 and 6-OHDA-treated SH-SY5Y cells decreased by overexpressing miR-124-3p. Cells transfected with miR-124-NPs exhibited reduced apoptosis (Esteves et al. 2022).

NDs may have miR-107 as one of their biomarkers. The target gene for lnc-MALAT1 and lnc-BDNF-AS in PD was miR-125b-5p. Astrogliosis and cell cycle dysfunction brought on by miRNA-125b overexpression led to neurodegeneration. In primary neurons, miR-125b overexpression activated the cdk5, p35, and p44/42-MAPK pathway and produced tau hyperphosphorylation. AD patients also had higher levels of miR-125b expression. Through CDK5 and p35/25 activation, miR-125b significantly increased neuronal apoptosis and Tau phosphorylation (Fan et al. 2020). According to Fan et al. (2020), Bax's relative mRNA level also increased.

Recent studies have revealed that approximately 40% of long noncoding RNA (lncRNA) genes show specific expression patterns in brain tissue. While the functionality of all lncRNAs remains a subject of debate, with some potentially representing transcriptional noise, mounting evidence suggests that many lncRNAs serve crucial biological functions, particularly in modifying chromatin structure and regulating protein levels.

One of the most well-characterized examples is BACE1-AS in AD. BACE1-AS upregulates BACE1 mRNA through duplex formation, subsequently increasing BACE1 protein levels. Notably, in vivo knockdown of BACE1-AS via siRNA infusion directly in the brain demonstrates reduced β-amyloid synthesis and aggregation, suggesting therapeutic potential. Brain cytoplasmic RNA (BCYRN1) shows differential expression in AD-affected regions, particularly Brodmann's area 9. While some studies report elevated BCYRN1 levels correlating with AD severity, others note significant reductions, possibly due to variations in sampling locations or disease stages (Garbo et al. 2021). LncRPPH1, a catalytic component of RNase P, has emerged as a significant regulator in Alzheimer's disease (AD). Studies demonstrate elevated LncRPPH1 expression in high-activity neocortical areas and AD brains, exhibiting protective functions against Aβ25-35-induced apoptosis. This protection operates through multiple mechanisms, including modulation of the Wnt/β-catenin pathway, downregulation of caspase-12, and regulation of PKM2 expression, ultimately reducing endoplasmic reticulum stress (Valizadeh et al. 2024) (Fig. 3).

Primate genomes are the only ones containing GDNF-AS transcribed from the opposite strand of the GDNF gene. The downregulation of mature GDNF peptide in AD patients suggests potential involvement in disease progression. Additionally, 17A, a novel ncRNA within the GPR51 genes, influences GABAB signaling and amyloid-β secretion during inflammatory responses in AD brains (Valizadeh et al. 2024).

In Huntington's disease (HD), BDNF-AS has emerged as a significant player through its posttranscriptional inhibition of BDNF expression. This interaction presents a potential therapeutic target, given that HD patients exhibit reduced brain BDNF levels. Recent studies of Abhd11os (ABHD11-AS1 in humans) demonstrate its neuroprotective properties against mutant Huntington protein. In Parkinson's disease (PD), NaPINK1, a human-specific ncRNA transcribed from the PINK1 locus, stabilizes PINK1 expression. This interaction is crucial given PINK1's role in

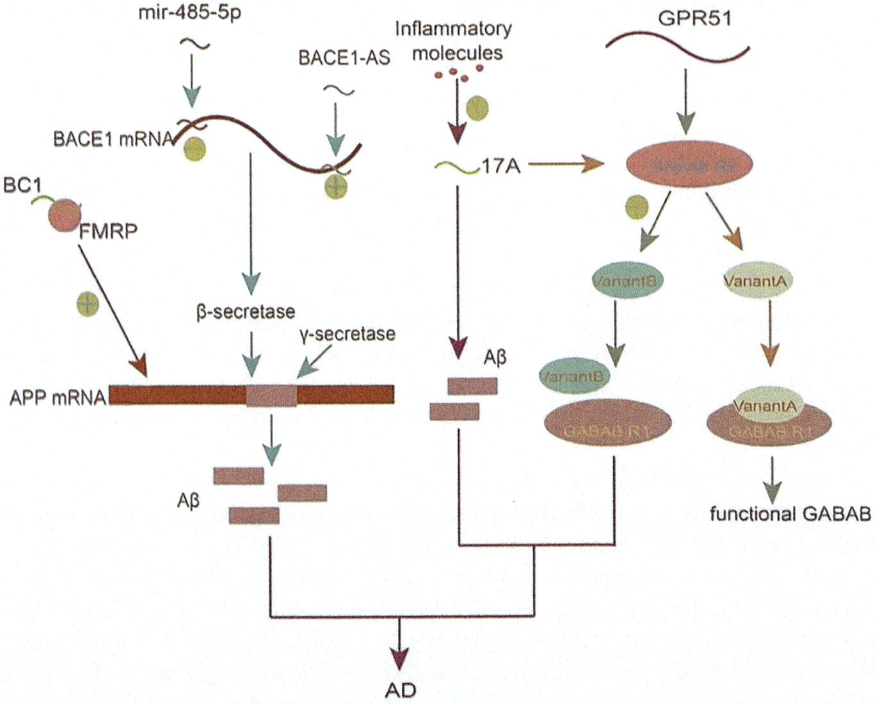

Fig. 3 Regulatory role of lncRNAs in AD. The lncRNA BACE1-AS stabilizes BACE1 mRNA, promoting amyloid-beta production, while mir-385-5p inhibits BACE1 expression. The lncRNA BC1 enhance amyloid precursor protein translation. Under inflammation, lncRNA-17A induces GABAB receptor splicing, altering inhibitory neurotransmission and increasing amyloid-beta levels. These complex lncRNA-mediated mechanisms underscore their crucial roles in Alzheimer's disease pathogenesis. (Reprinted from Wang et al. (2018), originally published under a Creative Commons Attribution (CC BY) license)

dopamine release and motor function, with NaPINK1 silencing leading to decreased neuronal PINK1 expression (Lv et al. 2024).

MEG3, a 1.6 kb nucleotide transcript, demonstrates complex involvement in both AD and Parkinson's disease (PD). In AD, MEG3 shows elevated neuronal expression, inducing necroptosis, while in PD, decreased plasma levels correlate with disease progression and cognitive dysfunction. Research has revealed its functionality through regulation of LRRK2 expression and p53 protein degradation, NEAT1, located on chromosome 11q13.1.

It functions as a molecular hub in NDs through its role in forming nuclear paraspeckles. These structures are crucial for cellular pathway regulation, with NEAT1 modulating apoptosis through miR-124/BACE1 interaction in AD and influencing Tau protein levels and amyloid protein production. NEAT1 demonstrates significant regulatory control over cell survival in PD through miR-212-5p targeting (Valizadeh et al. 2024).

Frontotemporal lobar degeneration (FTLD) research has revealed interesting interactions between lncRNAs and FTLD/ALS-related proteins TDP-43 and FUS/TLS. These interactions appear bidirectional, with lncRNAs both regulating and being regulated by these proteins, suggesting complex regulatory networks in FTLD pathogenesis. The 8779 bp transcript MALAT1 exhibits varied neuroprotective effects across neurodegenerative conditions. In AD, it suppresses neuronal apoptosis through miR-30b/CNR1 targeting, while in PD, it influences disease progression through the miR-135b-5p/GPNMB axis and regulates dopaminergic neuron survival via miR-124 interaction. This diverse functionality underscores its potential therapeutic significance (Li et al. 2020; Lv et al. 2021; Lu et al. 2020).

The complex involvement of these lncRNAs in neurodegenerative processes highlights their potential as therapeutic targets and biomarkers. Their tissue-specific expression patterns and regulatory roles in critical pathways make them particularly promising for targeted therapeutic development. Understanding these intricate RNA-based regulatory networks is crucial to developing more effective treatments for neurodegenerative diseases. Future research should focus on translating these molecular insights into practical therapeutic strategies while considering the complex nature of lncRNA-mediated regulation in the central nervous system. The continued investigation of lncRNA functions may provide crucial breakthroughs in treating these devastating neurological conditions (Table 3).

6 Potential Therapeutic Targets Based on RNA Transcriptomics Study

The advent of high-throughput RNA sequencing technologies has revolutionized our understanding of transcriptional regulation in ageing and neurodegeneration. These technological advances have unveiled a complex landscape of RNA-based therapeutic targets that hold promise for addressing cognitive decline associated with ageing and neurodegenerative disorders (Malla et al. 2021). These discoveries have highlighted several promising therapeutic targets that could potentially ameliorate cognitive decline. Identifying these targets represents a significant step forward in developing targeted interventions for age-related cognitive impairment and neurodegenerative conditions, offering new perspectives for therapeutic development (Shademan et al. 2021).

Alternative splicing regulators have emerged as critical therapeutic targets, given their fundamental role in generating protein diversity and maintaining cellular homeostasis. Of particular interest is NOVA1 (neurooncological ventral antigen 1), a neural-specific splicing factor that shows significant downregulation during ageing and neurodegeneration (Malla et al. 2021). Therapeutic strategies targeting NOVA1 could involve developing small molecules to enhance its activity or expression, potentially restoring proper splicing patterns of synaptic proteins. Similarly, the SRSF (serine/arginine-rich splicing factors) family, essential for regulated

Table 3 Key long noncoding RNAs in neurodegenerative diseases

lncRNA name	Disease association	Expression	Target/mechanism	References
BACE1-AS	AD	Upregulated	Makes an RNA duplex with BACE1 mRNA, enhancing BACE1 stability and Aβ synthesis	Olufunmilayo and Holsinger (2023) and Yang et al. (2021)
NEAT1	AD, HD, and ALS	Upregulated	Modulates RNA splicing, forms nuclear paraspeckles, and modulates neuroinflammation	Idda et al. (2018), Riva et al. (2016) and Yang et al. (2021)
MEG3	AD and PD	Downregulated	Regulates the p53 pathway, influences mitochondrial activity, and affects apoptosis	Olufunmilayo and Holsinger (2023) and Valizadeh et al. (2024)
MALAT1	AD and HD	Upregulated	Contributes to synapse disruption and cognitive decline	Riva et al. (2016) and Yang et al. (2021)
BDNF-AS	AD and HD	Upregulated	Inhibits BDNF expression and affects neuronal survival	Yang et al. (2021) and Valizadeh et al. (2024)
HAR1	HD	Downregulated	Relevant to cortical development and affected by REST dysfunction	Yang et al. (2021) and Valizadeh et al. (2024)
UCHL1-AS	PD	Downregulated	Controls UCHL1 protein production and influences protein degradation	Yang et al. (2021) and Valizadeh et al. (2024)

splicing events in neurons, presents valuable therapeutic opportunities by modulating their phosphorylation status (Sandoval-Castellanos et al. 2023).

The field of noncoding RNAs has opened novel therapeutic avenues, particularly through lncRNAs and miRNAs. Notable examples include NEAT1 and MALAT1, which demonstrate altered expression patterns in the ageing brain. Age-associated miRNAs, including miR-34a, miR-146a, and miR-181a, present additional opportunities for intervention through miRNA mimics, antagomiRs, and targeted delivery systems designed to restore normal miRNA function in affected brain regions (Ilieva 2024).

RNA modification enzymes constitute another crucial category of therapeutic targets. The m^{6}A writers and erasers, including the METTL3/METTL14 complex and FTO, demonstrate altered activity with age, suggesting potential interventions through small molecule inhibitors, targeted protein degradation approaches, or RNA-targeted CRISPR systems. Pseudouridine synthases, which influence RNA modification patterns that change with age, represent additional targets amenable to chemical modulation or substrate-specific targeting strategies (Wang 2021).

Several technological challenges must be addressed to realize the therapeutic potential of these targets. Delivery systems require optimization for efficient blood-brain barrier penetration and targeted tissue distribution. The successful development of RNA-based therapeutics requires careful consideration of multiple factors, including delivery methods, target specificity, individual patient variations, and the potential for combined therapeutic approaches (Liang et al. 2024). As our understanding of RNA regulation in ageing and neurodegeneration continues to expand, these therapeutic targets offer increasingly promising avenues for intervention in cognitive decline. The key challenge lies in developing safe and effective methods to modulate these targets in a tissue-specific and temporally controlled manner while maintaining cellular RNA homeostasis (Winkle et al. 2021). The future of RNA-based therapeutics in treating age-related cognitive decline and neurodegenerative disorders appears promising. Continued advances in delivery technologies, combined with an improved understanding of RNA regulatory networks, may lead to breakthrough treatments.

7 Conclusion and Future Perspective

The correlation between aberrant RNA modifications and their dysregulation in neurological disorders facilitates a comprehensive understanding of the pathophysiology of these diseases, enabling the identification of specific molecular markers that may be utilized for precise targeting in potential treatments. RNA modifications provide an additional regulatory layer in preserving the mammalian brain. Additional research is required to elucidate mechanisms through RNA modifications that influence brain development and disease. Mapping RNA modifications, particularly at single-nucleotide resolution, significantly enhances comprehension of the functions of such modifications. What cellular or environmental cues determine whether a transcript is modified or unmodified, and how do various modifications synergistically interact?

Single-cell sequencing investigations commenced to gain insight into the transcriptome of distinct cell types in the brain, while this methodology is being broadened to explore the epitranscriptome. The advent of single-molecule direct RNA sequencing could explain the diverse RNA modifications present on a particular transcript. This allows us to evaluate the impact of modifications on gene expression thoroughly. Technologies produced by nanopore sequencing will be essential for detecting, with minimal RNA input, the diverse modifications on individual transcripts within a specific cell type at a given time.

References

Abrishamdar M, Jalali MS, Rashno M (2022) MALAT1 lncRNA and Parkinson's disease: the role in the pathophysiology and significance for diagnostic and therapeutic approaches. Mol Neurobiol 59:5253–5262

Assis AF, Oliveira EH, Donate PB et al (2022) What is the transcriptome and how it is evaluated. In: Transcriptomics in health and disease. Springer International Publishing, Cham, pp 3–50

Bai X, Bian Z (2022) MicroRNA-21 is a versatile regulator and potential treatment target in central nervous system disorders. Front Mol Neurosci 15:842288

Baptista B, Riscado M, Queiroz JA et al (2021) Noncoding RNAs: emerging from the discovery to therapeutic applications. Biochem Pharmacol 189:114469

Bhat VD, Jayaraj J, Babu K (2022) RNA and neuronal function: the importance of post-transcriptional regulation. Oxf Open Neurosci 1:kvac011

Bhattacharyya N, Pandey V, Bhattacharyya M et al (2021) Regulatory role of long non coding RNAs (lncRNAs) in neurological disorders: from novel biomarkers to promising therapeutic strategies. Asian J Pharm Sci 16:533–550

Blaze J, Navickas A, Phillips HL et al (2021) Neuronal Nsun2 deficiency produces tRNA epitranscriptomic alterations and proteomic shifts impacting synaptic signaling and behavior. Nat Commun 12:4913

Briz V, Restivo L, Pasciuto E et al (2017) The noncoding RNA BC1 regulates experience-dependent structural plasticity and learning. Nat Commun 8:293

Buckley R, Pascual-Leone A (2020) Age-related cognitive decline is indicative of neuropathology. Ann Neurol 87:813–815

Butler AA, Webb WM, Lubin FD (2016) Regulatory RNAs and control of epigenetic mechanisms: expectations for cognition and cognitive dysfunction. Epigenomics 8:135–151

Calzoni E, Argentati C, Cesaretti A et al (2021) RNA modifications in neurodegenerations. In: Epitranscriptomics. Springer, pp 23–77

Cao M, Donà M, Valentino L et al (2016) Clinical and molecular study in a long-surviving patient with MLASA syndrome due to novel PUS1 mutations. Neurogenetics 17:65–70

Cenini G, Lloret A, Cascella R (2019) Oxidative stress in neurodegenerative diseases: from a mitochondrial point of view. Oxidative Med Cell Longev 2019:2105607

Cerneckis J, Ming GL, Song H et al (2024) The rise of epitranscriptomics: recent developments and future directions. Trends Pharmacol Sci 45:24–38

Chatterjee B, Shen CKJ, Majumder P (2021) RNA modifications and RNA metabolism in neurological disease pathogenesis. Int J Mol Sci 22:11870

Chujo T, Tomizawa K (2021) Human transfer RNA modopathies: diseases caused by aberrations in transfer RNA modifications. FEBS J 288:7096–7122

Dick F, Nido GS, Alves GW et al (2020) Differential transcript usage in the Parkinson's disease brain. PLoS Genet 16:e1009182

Esteves M, Abreu R, Fernandes H et al (2022) MicroRNA-124-3p-enriched small extracellular vesicles as a therapeutic approach for Parkinson's disease. Mol Ther 30:3176–3192

Eyler DE, Franco MK, Batool Z et al (2019) Pseudouridinylation of mRNA coding sequences alters translation. Proc Natl Acad Sci USA 116:23068–23074

Fan Y, Zhao X, Lu K et al (2020) LncRNA BDNF-AS promotes autophagy and apoptosis in MPTP-induced Parkinson's disease via ablating microRNA-125b-5p. Brain Res Bull 157:119–127

Garbo S, Zwergel C, Battistelli C (2021) m6A RNA methylation and beyond–the epigenetic machinery and potential treatment options. Drug Discov Today 26:2559–2574

Gentile G, Morello G, La Cognata V et al (2022) Dysregulated miRNAs as biomarkers and therapeutical targets in neurodegenerative diseases. J Pers Med 12:770

Gionco JT, Bernstein AI (2024) Emerging role of environmental epitranscriptomics and RNA modifications in Parkinson's disease. J Parkinsons Dis 14(4):643–656

Grinkevich LN (2020) The role of microRNAs in learning and long-term memory. Vavilovskii Zhurnal Genet Selektsii 24:885–896

Ham S, Lee SJV (2020) Advances in transcriptome analysis of human brain aging. Exp Mol Med 52:1787–1797

Han D, Dong X, Zheng D et al (2020a) MiR-124 and the underlying therapeutic promise of neurodegenerative disorders. Front Pharmacol 10:1555

Han M, Liu Z, Xu Y et al (2020b) Abnormality of m6A mRNA methylation is involved in Alzheimer's disease. Front Neurosci 14:98

Hildebrandt RP (2022) Elucidating mechanisms of RNA localization mediated by modular domains of RNA binding proteins. Doctoral dissertation. University of Florida

Hu GJ, Jiang XY, Du SY, Zhang K, Chen Z (2024) miR-107-5p ameliorates neurological damage, oxidative stress, and immune responses in mice with Alzheimer's disease by suppressing the Toll-like receptor 4 (TLR4)/nuclear factor-kappaB(NF-κB) pathway. Kaohsiung J Med Sci. 40(2):119–130. https://doi.org/10.1002/kjm2.12797

Hu Z, Li Z (2017) miRNAs in synapse development and synaptic plasticity. Curr Opin Neurobiol 45:24–31

Huang H, Camats-Perna J, Medeiros R et al (2020) Altered expression of the m6A methyltransferase METTL3 in Alzheimer's disease. eNeuro 7. https://doi.org/10.1523/ENEURO.0125-20.2020

Huber LB, Kaur N, Henkel M et al (2023) A dual-purpose polymerase engineered for direct sequencing of pseudouridine and queuosine. Nucleic Acids Res 51:3971–3987

Iacoangeli A, Dosunmu A, Eom T et al (2017) Regulatory BC1 RNA in cognitive control. Learn Mem 24:267–277

Idda, M. L., Munk, R., Abdelmohsen, K., & Gorospe, M. (2018). Noncoding RNAs in Alzheimer's disease. Wiley Interdisciplinary Reviews: RNA, 9(2), e1463

Ilieva MS (2024) Noncoding RNAs in neurological and neuropsychiatric disorders: unraveling the hidden players in disease pathogenesis. Cells 13:1063

Jackson WS, Bauer S, Kaczmarczyk L et al (2024) Selective vulnerability to neurodegenerative disease: insights from cell type-specific translatome studies. Biology 13:67

Jiapaer Z, Su D, Hua L et al (2022) Regulation and roles of RNA modifications in aging-related diseases. Aging Cell 21:e13657

Jung Y, Goldman D (2018) Role of RNA modifications in brain and behavior. Genes Brain Behav 17:e12444

Katanski CD, Watkins CP, Zhang W et al (2022) Analysis of queuosine and 2-thio tRNA modifications by high throughput sequencing. Nucleic Acids Res 50:e99–e99

Kennedy MB (2016) Synaptic signaling in learning and memory. Cold Spring Harb Perspect Biol 8:a016824

Lakhotia SC, Mallick B, Roy J (2020) Noncoding RNAs: ever-expanding diversity of types and functions. In: RNA-based regulation in human health and disease, vol 19. Academic Press, pp 5–57

Lei L, Burton ZF (2022) "Superwobbling" and tRNA-34 wobble and tRNA-37 anticodon loop modifications in evolution and devolution of the genetic code. Life 12:252

Li B, Zhang XQ, Liu SR et al (2017) Discovering the interactions between circular RNAs and RNA-binding proteins from CLIP-seq data using circScan. BioRxiv:115980. https://doi.org/10.1101/115980

Li D, Zhang J, Wang M et al (2018) Activity dependent LoNA regulates translation by coordinating rRNA transcription and methylation. Nat Commun 9:1726

Li L, Xu Y, Zhao M et al (2020) Neuro-protective roles of long noncoding RNA MALAT1 in Alzheimer's disease with the involvement of the microRNA-30b/CNR1 network and the following PI3K/AKT activation. Exp Mol Pathol 117:104545

Li P, Wang Y, Sun Y et al (2022) N6-methyladenosine RNA methylation: from regulatory mechanisms to potential clinical applications. Front Cell Dev Biol 10:1055808

Liang W, Luo Y, Xu A et al (2024) Advances in carrier-delivered small interfering RNA based therapeutics for treatment of neurodegenerative diseases. Biomater Sci 12:4927–4945

Liu J, Liu D, Zhao B et al (2020a) Long noncoding RNA NEAT1 mediates MPTP/MPP+-induced apoptosis via regulating the miR-124/KLF4 axis in Parkinson's disease. Open Life Sci 15:665–676

Liu J, Zuo X, Han J et al (2020b) MiR-9-5p inhibits mitochondrial damage and oxidative stress in AD cell models by targeting GSK-3β. Biosci Biotechnol Biochem 84:2273–2280

Livneh I, Moshitch-Moshkovitz S, Amariglio N et al (2020) The m6A epitranscriptome: transcriptome plasticity in brain development and function. Nat Rev Neurosci 21:36–51

Lu Y, Gong Z, Jin X et al (2020) LncRNA MALAT1 targeting miR-124-3p regulates DAPK1 expression contributes to cell apoptosis in Parkinson's disease. J Cell Biochem 121:4838–4848
Lv K, Liu Y, Zheng Y et al (2021) Long noncoding RNA MALAT1 regulates cell proliferation and apoptosis via miR-135b-5p/GPNMB axis in Parkinson's disease cell model. Biol Res 54:10
Lv X, Zhang R, Li S et al (2024) tRNA modifications and dysregulation: implications for brain diseases. Brain Sci 14:633
Malla B, Guo X, Senger G et al (2021) A systematic review of transcriptional dysregulation in Huntington's disease studied by RNA sequencing. Front Genet 12:751033
Mathoux J, Henshall DC, Brennan GP (2021) Regulatory mechanisms of the RNA modification m6A and significance in brain function in health and disease. Front Cell Neurosci 15:671932
Mendel M, Delaney K, Pandey RR et al (2021) Splice site m6A methylation prevents binding of U2AF35 to inhibit RNA splicing. Cell 184:3125–3142
Novotný JS, Gonzalez-Rivas JP, Vassilaki M et al (2022) Natural pattern of cognitive aging. J Alzheimers Dis 88:1147–1155
Olufunmilayo EO, Holsinger RD (2023) Roles of noncoding RNA in Alzheimer's disease pathophysiology. Int J Mol Sci 24:12498
Park CW, Lee SM, Yoon KJ (2020) Epitranscriptomic regulation of transcriptome plasticity in development and diseases of the brain. BMB Rep 53:551
Peer E, Rechavi G, Dominissini D (2017) Epitranscriptomics: regulation of mRNA metabolism through modifications. Curr Opin Chem Biol 41:93–98
Perez Grovas-Saltijeral A, Rajkumar AP, Knight HM (2023) Differential expression of m5C RNA methyltransferase genes NSUN6 and NSUN7 in Alzheimer's disease and traumatic brain injury. Mol Neurobiol 60:2223–2235
Perrone B, La Cognat V, Sprovieri T et al (2020) Alternative splicing of ALS genes: misregulation and potential therapies. Cell Mol Neurobiol 40:1–14
Qiu L, Jing Q, Li Y et al (2023) RNA modification: mechanisms and therapeutic targets. Mol Biomed 4:25
Rajgor D, Welle TM, Smith KR (2021) The coordination of local translation, membranous organelle trafficking, and synaptic plasticity in neurons. Front Cell Dev Biol 9:711446
Riva P, Ratti A, Venturin M (2016) The long noncoding RNAs in neurodegenerative diseases: novel mechanisms of pathogenesis. Curr Alzheimer Res 13:1219–1231
Roundtree IA, Evans ME, Pan T et al (2017) Dynamic RNA modifications in gene expression regulation. Cell 169:1187–1200
Sandoval-Castellanos AM, Bhargava A, Zhao M et al (2023) Serine and arginine rich splicing factor 1: a potential target for neuroprotection and other diseases. Neural Regen Res 18:1411–1416
Shabir S, Singh MP (2022) The aging: introduction, theories, principles, and future prospective. In: Anti-aging drug discovery on the basis of hallmarks of aging. Academic Press, pp 1–17
Shademan B, Biray Avci C, Nikanfar M et al (2021) Application of next-generation sequencing in neurodegenerative diseases: opportunities and challenges. Neuromol Med 23:225–235
Siedlecki-Wullich D, Miñano-Molina AJ, Rodríguez-Álvarez J (2021) microRNAs as early biomarkers of Alzheimer's disease: a synaptic perspective. Cells 10:113
Skariah G, Todd PK (2021) Translational control in aging and neurodegeneration. Wiley Interdiscip Rev RNA 12:e1628
Sokpor G, Xie Y, Nguyen HP et al (2021) Emerging role of m6A methylome in brain development: implications for neurological disorders and potential treatment. Front Cell Dev Biol 9:656849
Song W, Ressl S, Tracey WD (2020) Loss of pseudouridine synthases in the RluA family causes hypersensitive nociception in Drosophila. G3 (Bethesda) 10:4425–4438
Soutschek M, Schratt G (2023) Noncoding RNA in the wiring and remodeling of neural circuits. Neuron 111:2140–2154
Srinivas T, Mathias C, Oliveira-Mateos C et al (2023) Roles of lncRNAs in brain development and pathogenesis: emerging therapeutic opportunities. Mol Ther 31:1550–1561
Suzuki T (2021) The expanding world of tRNA modifications and their disease relevance. Nat Rev Mol Cell Biol 22:375–392

Tassinari V, La Rosa P, Guida E et al (2023) Contribution of A-to-I RNA editing, M6A RNA methylation, and alternative splicing to physiological brain aging and neurodegenerative diseases. Mech Ageing Dev 212:111807

Tian B, Manley JL (2017) Alternative polyadenylation of mRNA precursors. Nat Rev Mol Cell Biol 18:18–30

United Nations (2020) Department of Economic and Social Affairs PD

Valizadeh M, Derafsh E, Abdi Abyaneh F et al (2024) Noncoding RNAs and neurodegenerative diseases: information of their roles in apoptosis. Mol Neurobiol 61:4508–4537

Wang DO (2021) RNA modifications in the central nervous system. In: The Oxford handbook of neuronal protein synthesis, vol 153. Oxford University Press

Wang DQ, Fu P, Yao C et al (2018) Long non-coding RNAs, novel culprits, or bodyguards in neurodegenerative diseases. Mol Ther Nucleic Acids 10:269–276

Wang S, Lv W, Li T et al (2022) Dynamic regulation and functions of mRNA m6A modification. Cancer Cell Int 22:48

Weyn-Vanhentenryck SM, Feng H, Ustianenko D et al (2018) Precise temporal regulation of alternative splicing during neural development. Nat Commun 9:2189

Wild EJ, Tabrizi SJ (2017) Therapies targeting DNA and RNA in Huntington's disease. Lancet Neurol 16:837–847

Wilkinson E, Cui YH, He YY (2022) Roles of RNA modifications in diverse cellular functions. Front Cell Dev Biol 10:828683

Winkle M, El-Daly SM, Fabbri M et al (2021) Noncoding RNA therapeutics—challenges and potential solutions. Nat Rev Drug Discov 20:629–651

Wyss-Coray T (2016) Ageing, neurodegeneration and brain rejuvenation. Nature 539:180–186

Xue B, Qu Y, Zhang X et al (2022) miRNA-126a-3p participates in hippocampal memory via alzheimer's disease-related proteins. Cereb Cortex 32:4763–4781

Yang S, Yang H, Luo Y et al (2021) Long noncoding RNAs in neurodegenerative diseases. Neurochem Int 148:105096

Yang A, Zhao X, Yang YT et al (2024) Deciphering causal relationships between cell type-specific genetic factors and brain imaging-derived phenotypes and disorders. medRxiv. https://doi.org/10.1101/2024.08.30.24312836

Yen YP, Chen JA (2021) The m6A epitranscriptome on neural development and degeneration. J Biomed Sci 28:40

Zhao B (2017) The exuberant vine of epitranscriptome: RNA methylation as a new epigenetic regulatory mechanism. The University of Chicago

DNA Methylation Control over Aging: A Focus on Repetitive Elements

Viviana Barra and Pietro Salvatore Carollo

Contents

Abstract DNA methylation is an epigenetic modification that has been gaining particular attention among researchers in the field of aging and aging-related diseases. Specific DNA methylation changes, indeed, characterize aged cells and potentially predict age and lifespan, suggesting that they may play a key role during the aging process. In particular, DNA demethylation of repetitive elements (RE) is a common observation in aging cells and correlates with the increase of transcription of repeat-derived noncoding RNAs. Demethylation of satellites and loss of suppression of retrotransposons generate genome instability, can increase cell stress and affect cell viability, and also can impact the regulation of genes associated with cell-lineage specificity, thus contributing to neurodegeneration and susceptibility to cancer. In this chapter, we discuss recent work on the DNA methylation dynamics

V. Barra (✉) · P. S. Carollo
Department of Biological, Chemical and Pharmaceutical Sciences and Technologies (STEBICEF), University of Palermo, Palermo, Italy
e-mail: viviana.barra@unipa.it

J. Barciszewski (ed.), *Decoding Aging and Neurodegeneration*,
RNA Technologies 16, https://doi.org/10.1007/978-3-031-91072-2_5

at RE during the aging process outlining the potential roles played during aging. A better understanding of these roles could help in the advancement of approaches to enhance healthy aging and of therapies for aging-related diseases.

Keywords DNMTs · Satellite DNA · LINEs · Transposons · Progeria · Amyotrophic lateral sclerosis · Dementia

1 Introduction

Aging can be defined as the progressive functional decline of tissues and organs in organisms with the passing of time. This decline results more or less slowly in the reduction of the ability to cope with any kind of external event including environmental insults, stress, infections, diseases (i.e., cancer, cardiovascular diseases, and neurodegenerative disorders), and inflammation, which lead to a constantly increasing risk of death. At the cellular level, aging correlates with molecular and biochemical changes that result in the hallmarks of aging which include telomere shortening and genomic instability, epigenetic alterations, mitochondrial dysfunction, chronic inflammation, and cellular senescence.

Determining the true cause of aging remains one of biology's most intriguing challenges. Various environmental, genetic, and chemical factors have been identified in the last 30 years that affect organisms' lifespan. In general, stressful conditions tend to accelerate aging and shorten lifespan by causing DNA mutations and damaging essential cellular components, such as proteins and cellular organelles. This damage disrupts normal cellular function, leading to the concept that aging and age-related diseases result from the gradual accumulation of unrepaired cellular damage during lifetime. However, merely enhancing repair mechanisms or minimizing damage-causing factors does not effectively slow aging, implying that the primary causes of aging are driven by forces beyond what cells are able to repair.

Epigenetic changes, particularly in CpG dinucleotides and histones, have been linked to various cellular states, including senescence and aging, along with aging-related diseases. Within the nucleus, chromatin structure can vary from a tight and inaccessible conformation known as heterochromatin to a relaxed and open organization called euchromatin. The open chromatin allows transcription, while the compacted chromatin represses transcription. Epigenetic modifications allow this switch by enhancing or inhibiting the interactions between adjacent nucleosomes. Also, epigenetic modifications can favor the binding of specific DNA binding proteins to DNA which can increase the compaction level of the chromatin and constitute a molecular signal within the cell as well. For these reasons, the term epigenetics includes every reversible chemical modification in histones, DNA, and RNA that alters gene expression without changing the DNA sequence. Thus, among the epigenetic modifications, we can distinguish the posttranslational modifications on the histones (i.e., acetylation, methylation, phosphorylation, ubiquitylation), DNA chemical modifications (mainly 5-methylcytosine, 5mC), and RNA modifications (i.e., N^6-methyladenosine). Some epigenetic modifications including H3K27ac and H3K4m1/2/3 are active markers of transcription as they induce the opening of the chromatin, while others, such as H3K27m3, H3K9m3, and 5mC, are repressive markers of transcription as they keep chromatin in a condensed state.

It has been shown that fibroblasts from older donors (including septuagenarians, octogenarians, and nonagenarians) exhibited a significant reduction in the heterochromatic histone marker H3K9m3 when compared to fibroblasts from very young donors (under 10 years old) (Scaffidi and Misteli 2006). Additionally, fibroblasts from older individuals showed higher levels of DNA damage relative to their younger counterparts. Remarkably, similar reductions in H3K9m3, as well as increased DNA damage, were observed in Hutchinson-Gilford progeria syndrome (HGPS) patients that are characterized by accelerated aging (discussed below), suggesting shared defects between HGPS patients and normal aging individuals. During the same period, another group investigated epigenetic changes associated with HGPS and observed a reduction in histone marks for heterochromatin (H3K27m3, H3K9m3, and H4K20m3) in fibroblasts from a female HGPS patient compared to age-matched control cells from her sibling (Shumaker et al. 2006). Furthermore, the researchers detected transcripts from chromosome 9 sat III in HGPS fibroblasts, which were absent in normal fibroblasts, indicating that the silencing of pericentric heterochromatin in DNA repeats is altered in progeria.

Altogether, these findings reveal that aging cells undergo heterochromatin changes. Similarly, cellular replicative senescence, which plays a fundamental role in the aging phenotype (discussed below), is characterized by alterations of the heterochromatin markers. In this context, the effects of senescence in IMR90 fibroblasts induced by the expression of the RAS oncogene have been investigated. It has been discovered that compared to proliferating cells, senescent fibroblasts formed senescence-associated heterochromatic Foci (SAHFs), which were notably enriched for heterochromatin markers such as H3K9m3 (Narita et al. 2006). Also, during replicative senescence, there is a reduction in histone biosynthesis in late-passage cultured cells (O'Sullivan et al. 2010). Specifically, it was observed that IMR90 and WI38 fibroblasts at passage 75 have decreased levels of histone proteins H3 and H4 compared to younger cells at passage 30. This reduction in histone levels was accompanied by changes in histone methylation patterns, with increased monomethylation and decreased di- and trimethylation of H3K9, decreased monomethylation and increased demethylation of H4K20, and loss of H4K20 trimethylation. Furthermore, a reduction in H3 and H4 synthesis was also observed in cells from older individuals (92 years old) compared to those from younger individuals (9 years old), reinforcing the connection between aging and epigenetic alterations (O'Sullivan et al. 2010). Epigenetic alterations, in turn, have been closely related to nuclear lamina components which also vary during senescence, natural aging, and accelerated aging (Shah et al. 2013).

Due to the influence of epigenetics in the aging process and considering the reversibility of epigenetic modifications, it could be inferred that aging can be reversed. In this context, there is emerging evidence suggesting that it may be possible to counteract aging-related effects. For instance, it has been demonstrated that the natural polyamine spermidine can extend the lifespan of cells across various species, including human peripheral blood mononuclear cells (PBMCs) (Eisenberg et al. 2009). Treatment with spermidine led to reduced acetylation levels of lysines 16 and 18 on histone H3 in PBMCs, which was associated with increased cell lifespan and reduced senescence. Further supporting this notion, another study highlighted that directly inhibiting histone acetyltransferases (HATs) can lead to similar outcomes (Huang et al. 2020), reinforcing the link between histone acetylation, cellular aging, and senescence. A very recent publication also incredibly showed how it is possible to revert murine aging features by reprogramming the epigenome by exogenously expressing the "Yamanaka factors" Oct4, Sox2, Klf4, and Myc (OSKM) in the mice (Yang et al. 2023). These findings suggest that interventions targeting epigenetic modifications, such as those induced by spermidine, could have potential therapeutic applications in counteracting the effects of aging.

A key epigenetic factor in the aging and senescence process seems to be DNA methylation, whose specific changes can predict age. In this chapter, we will indeed focus on the DNA methylation reduction that specifically targets most of the genomes at the level of the repetitive elements (REs) during the aging process. We will limit our review to those findings regarding humans, and we will start by summarizing what DNA methylation is and its role in the genome, describing the different types of repetitive elements as well. This chapter will finally address the consequences of methylation loss at REs both at the cellular and the organism levels that can lead to age-related diseases.

2 DNA Methylation and DNA Methyltransferases

DNA methylation was the first epigenetic modification discovered in humans and is involved in many cellular processes including cell differentiation and chromosome X inactivation in females. DNA methylation, together with the trimethylation of lysine 9 of histone H3 (H3K9m3), is an epigenetic marker of heterochromatin. It consists in the addition of a methyl group to the C-5 position of the cytidine in the context of a CpG dinucleotide. DNA methylation leads to chromatin compaction and consequent repression directly by facilitating nucleosome assembly through DNA rigidification and in an indirect way by providing the binding site for proteins with the methyl binding domain (MBD) that subsequently trigger the formation of the heterochromatin.

The chemical reaction is catalyzed by DNA methyltransferases (DNMTs), which are distinguished in *de novo* and *maintenance* DNMTs depending if they methylate a fully unmethylated DNA or a DNA methylated only in one strand (hemimethylated), respectively. *De novo* DNMTs, DNMT3A, and DNMT3B in humans, methylate a DNA that is not methylated, thus establishing cell-specific DNA methylation patterns. This event is usually restricted to embryo development and cell differentiation when the establishment of the methylation pattern determines the genes that are transcribed in that cell lineage. On the other hand, *maintenance* DNMTs, DNMT1 in humans, methylate the hemimethylated DNA during DNA replication, by adding the methyl group in the newly synthetized strand, which is not methylated, following the methylation pattern still present in the parental DNA strand. This process guarantees that the methylation pattern of the cell is perpetuated at each cell division, which is fundamental to preserving cell healthy conditions (Pappalardo and Barra 2021).

CpG islands are small regions where there is an extremely high content of GC frequently associated with gene promoters, but often they are not methylated. Their methylation, though, leads to stable gene repression. During gametogenesis, embryo development, and cell differentiation, CpG islands are differentially methylated in order to determine both the imprinted genes and the cell-lineage-specific gene transcription pattern. DNA methylation at gene bodies can regulate gene transcription as well, however mainly with an active effect (Wang et al. 2022).

DNA methylation can occur not only in gene promoters or gene bodies but also at repetitive elements, specific DNA sequences that are recurrent along the whole genome. REs constitute about two-thirds of the human genome and are distinguished into two different families that will be discussed in the next chapter.

3 Repetitive Elements in the Human Genome

3.1 Tandem Repeats (Satellite DNA)

The term satellite DNA covers long arrays—from a few hundred base pairs to several megabases—of nearly identical sequences that are highly repeated in a head-to-tail fashion, called tandem repeats. Tandem repeats also include simple smaller repeats 1–6 base pairs and <20 base pairs long which are called microsatellites and minisatellites, respectively. All together tandem repeats constitute about 10% of the human genome (Altemose 2022). Satellites are mainly defined by their position along the chromosome, such as centromeric and telomeric satellites.

Human centromeres are composed of different types of satellites, both short and long, of whom the most abundant are the α-satellites. The α-satellite DNA consists of monomers of 171 base pairs, 60% divergent in the sequence, that are arranged into higher-order repeats (HORs). The core centromere, the functional region that connects the chromosomes to the microtubules arising from the spindle poles, is mainly made up of highly homogenous HORs. Here, chromatin is characterized by markers of active transcription (H3K4m1, H3K4m2 H4K20m1, H3K36m2, H3K36m3, histone acetylation, low DNA methylation) and is indeed transcribed at specific timing during the cell cycle to ensure the correct loading of CENP-A, the histone H3 variant that identifies the centromere (Smurova and De Wulf 2018). However, H3K9m3, a marker of transcriptional repression, is also found at the core centromere, suggesting that a balance between activation and repression of transcription is usually needed in this region. The regions flanking the core centromere, known as pericentromeres, have lesser and lesser homogenous HORs and predominantly monomers of the α-satellites which are mixed with interspersed repeats, and different types of micro- and mini-satellites, such as satellites 1, 2, and 3 and γ-satellites that characterize specific chromosomes (Thakur et al. 2021). Pericentromeres are the chromosomal regions bound by cohesin which has a fundamental role in preventing the premature separation of sister chromatids during chromosome segregation in mitosis. These regions are normally maintained repressed with all the repressive markers of constitutive heterochromatin, mainly H3K9m3 and DNA methylation (Smurova and De Wulf 2018). The different epigenetic states of the core centromere and pericentromere reflect an intricate mechanism of transcriptional regulation whose fine orchestration ensures chromosome stability in cycling cells (Barra and Fachinetti 2018).

Telomeres, the terminal regions of each chromosome, are constituted by thousands of repeats of the hexamer TTAGGG (G-strand) which is bound by a six-protein complex named Shelterin. The G-strand is longer than the complementary C-strand, and thus, a single-stranded overhang protrudes and invades the double-stranded region. This specialized architecture at the end of the linear chromosome creates a cap capable of protecting the DNA from degradation and recombination. Due to the fact that DNA polymerase replicates DNA from 5′-to-3′, the end of each linear chromosome cannot be fully replicated, causing telomere shortening at each cell cycle. This event triggers the replicative senescence, a condition where cells no longer divide leading to tissue aging. The presence of the reverse transcriptase telomerase in germline and stem cells, however, prevents this phenomenon by adding de novo telomeric repeats to the G-strand overhang, thus ensuring the preservation of the genome and the cell cycling. Subtelomeric domains immediately adjacent to telomeres also contain tandemly repeated satellites; however, they are of variable copy number and degree of divergence both among individuals and between chromosomes (Ambrosini et al. 2007).

3.2 *Interspersed Repeats (Transposable Elements)*

Transposable elements (TEs) are abundant repetitive sequences of the human genome which are characterized by the ability to move and change position within the genome. Due to the fact that copies of these sequences can be transposed, their repeats are scattered throughout the genome, unlike tandem repeats that are clustered together (see *Tandem Repeats* (*Satellite DNA*) paragraph) and, thus, are called interspersed repeats. TEs can be distinguished into two classes, DNA and RNA transposons. DNA transposons can transpose through a "cut and paste" mechanism based on the enzymatic activity of a transposase. RNA transposons (also known as retrotransposons or endogenous retroelements), instead, can be multiplied and moved through a "copy-and-paste" mechanism mediated by a reverse transcriptase. The latter class can be further divided into long terminal repeats (LTRs) and non-LTR retrotransposons. LTR retrotransposons, mainly HERVs (human endogenous retroviruses), are reminiscent of ancient retroviral infections and represent about 8% of the human genome. Among the non-LTR group, the long interspersed nuclear elements (LINEs) and the short interspersed nuclear elements (SINEs) are the most recurrent in the human genome, being 17% and 11% of the genome respectively. LINEs are autonomously able to mobilize because they contain two open reading frames (ORFs) encoding a protein with endonuclease and reverse transcriptase activity (ORF2) and a protein with nucleic acid chaperone activity (ORF1), both required for retrotransposition. Instead, SINEs relate to the proteins encoded by the LINEs to jump to another part of the genome.

For a thorough guide about TEs, we suggest a review paper (Wells and Feschotte 2020). The LINE-1 family is the most represented type of LINEs in humans, while Alu elements are the most common SINEs in the human genome.

Mobilization of TEs can endanger genome stability as it can induce chromosomal rearrangements and alterations of gene regulation if TEs are inserted in the coding region or in the regulatory portion of the gene (Gorbunova et al. 2021). To avoid this possibility, cells maintain TEs in a repressed state by using several mechanisms. The most common mechanism to silence TEs in humans is dependent on small noncoding RNAs, such as PIWI-interacting RNAs (piRNAs) and endogenous-small interfering RNAs (endo-siRNAs), which have sequence homology with the target TEs and induce their suppression both at the posttranscriptional and transcriptional level (Gorbunova et al. 2021). Importantly, silencing of TEs can also occur epigenetically, mainly through DNA methylation (Pappalardo and Barra 2021). Any modification of the elements that ensure TE repression has a profound impact on genome stability.

4 DNA Methylation Changes During Aging

Early studies conducted in the 1980s examined DNA methylation in cells from mice, hamsters, and humans. Their findings indicated that a loss of 5mC serves as a hallmark of aging; specifically, the percentage of DNA methylation decreases with successive population doublings (PDs) across all species studied and in the liver, brain, and small intestine mucosa of mice over time (Wilson and Jones 1983; Wilson et al. 1987; Hoal-van Helden and van Helden 1989). Notably, this loss of DNA methylation is not observed in immortal cell lines (Wilson and Jones 1983) and exhibits an inverse relationship with maximum lifespan potential (Wilson et al. 1987). An interesting study specifically showed that the levels of total 5mC in mice decreased from day 1 to day 8 postbirth in the liver, heart, skeletal muscle, and brain, with the liver showing the most significant loss. After day 8, the liver continued to lose methylation until day 180, while the heart, skeletal muscle, and brain reached a plateau in 5mC levels. Despite the low proliferation rates of adult liver cells, they maintained regenerative capacity throughout life, unlike cardiac cells, skeletal muscle cells, and neurons. This led the authors to conclude that methylation decreases as cells transition from actively dividing embryonic cells to young adult cells (Hoal-van Helden and van Helden 1989). An increase in 5mC loss was similarly noted in human bronchial epithelial cells with the increasing age of the donors (Wilson et al. 1987), suggesting that progressive reduction of DNA methylation is a feature of humans as well.

Aberrant levels of 5mC are thus recognized as a key feature of aging, which is closely linked to cancer, often considered an age-related pathology. In accordance with this, changes in DNA methylation have been implicated in tumorigenesis. For example, pyrosequencing of specimens from normal prostate tissue revealed that 5mC levels in CpG islands increased with donor age. These CpG islands are located within genes such as RARb2, RASSF1A, GSTP1, and ESR1, which have been shown to undergo increased methylation in prostate cancer (RARb2, RASSF1A, GSTP1) or to be affected by aging (ESR1). The authors concluded that alterations in CpG island methylation are due to the aging process in normal prostate tissue and become substantially more pronounced in cancer (Kwabi-Addo et al. 2007).

A further study provided additional evidence linking DNA methylation alterations to aging (Bjornsson et al. 2008). The researchers conducted a genome-wide DNA methylation analysis using the LUminometric Methylation Assay (LUMA) on peripheral blood DNA samples from individuals in Iceland and Utah, collected during two visits that were 11 and 16 years apart, respectively. Their findings revealed that the methylation status of CpG dinucleotides within the same individual exhibited significant changes over time. These changes in 5mC levels indicated both increase and decrease in methylation, highlighting the dynamic nature of epigenetic modifications as individuals age (Bjornsson et al. 2008). Instead, by employing Illumina's Golden Gate methylation platform to analyze 217 human samples from 10 distinct anatomical sites, it has been found that the levels of 5mC increased with age in loci located within CpG islands, while 5mC levels decreased in loci situated outside of CpG islands. Furthermore, the study revealed that CpG methylation patterns are tissue-dependent. Thus, the authors concluded that the methylation patterns change within an aging tissue and this is CpG island context-dependent (Christensen et al. 2009).

Epigenetic changes associated with aging also exhibit notable differences in monozygotic twins. CpG methylation as well as H3 and H4 acetylation variations have been investigated in monozygotic twins using high-performance liquid chromatography (HPLC) and high-performance capillary electrophoresis on DNA and RNA samples extracted from peripheral lymphocytes (Fraga et al. 2005). These analyses revealed that the level of CpG methylation decreased (in microsatellites) with age, particularly among older twins compared to younger counterparts. The most pronounced differences were observed in twins who had distinct medical histories and had lived together for only a short period of their lifetime. To further explore these changes, the authors employed AIMS, a global DNA methylation fingerprint technique where anonymous bands, representing DNA sequences flanked by two methylated sites, can be separated through gel electrophoresis. This analysis demonstrated that older twins (50 years old) exhibited significantly greater changes in methylation patterns compared to a younger pair (3 years old) who had not lived together for most of their lives nor shared the same medical history. Additionally, a bisulfite assay revealed variability in the methylation status of CpG sites within Alu sequences between the two older twins, with one twin showing hypermethylation while the other displayed hypomethylation. These methylation changes were further reflected in differing patterns of DNA methylation signals, particularly at telomeres, in the metaphase chromosomes of older twins compared to younger twins, whose patterns remained similar due to their comparable lifestyles. Notably, the methylation status of CpG islands also varied significantly; in 50-year-old twins, these regions exhibited 2.5 times more variability compared to 3-year-old twins (Fraga et al. 2005). Lastly, this study identified a distinct gene expression profile between young and old twins, as well as variations within the older twins. Notably, the twin exhibiting the highest levels of DNA hypomethylation and hyperacetylation of histones H3 and H4 showed a corresponding increase in the number of overexpressed genes. This suggests a strong correlation between epigenetic modifications and gene expression dynamics associated with aging. This study primarily focused on young and old pairs, overlooking the epigenetic modifications occurring at the extremes of lifespan, such as in newborns and centenarians. This gap was addressed by conducting a comprehensive analysis of the methylome in both newborns and nonagenarians/centenarians using whole genome bisulfite sequencing (WGBS) and a 450,000 CpG DNA methylation microarray analysis (Heyn et al. 2012). These findings revealed a significant decrease in methylation at CpG sites located outside of islands, while an increase in methylation was observed in CpG islands among the elderly compared to newborns. Specifically, the overall methylation percentage of all covered CpG sites in DNA extracted from a centenarian sample was lower (73.0%) than that from a newborn's (80.5%). This reduced methylation of CpG sites in the centenarian sample was consistent across various genomic compartments, including promoter sequences, exons, introns, and intergenic regions. When analyzing the promoters based on their CpG content, it was noted that CpG-poor promoters in the centenarian sample were unmethylated, whereas methylation was retained in CpG island promoters. Furthermore, the majority of specific differentially methylated regions (DMRs) lost methylation (87%), with only a small fraction (13%) retaining it. Additionally, transcript levels of Alu elements and mammalian interspersed repeats (MIRs) were found to be elevated in the centenarian sample compared to the newborn's one. Unmethylated CpG sites were also identified in nonagenarian samples. In conclusion, the study highlights distinct differences in the methylomes of individuals at the extremes of the human lifespan, demonstrating that the DNA of the elderly is generally less methylated compared to that of newborns (Heyn et al. 2012).

The studies mentioned above indicate that specific chromatin regions undergo methylation changes as organisms age. These regions have been proposed as "aging-associated differentially methylated regions" (aDMRs), which exhibit hypermethylation as age increases. Notably, this hypermethylation occurs early in precursor and stem cells at bivalent chromatin domain promoters. These same promoters are also hypermethylated in various cancers, establishing a link between age-related hypermethylation and cancer development (Rakyan et al. 2010).

In 2013, the concept of the "epigenetic clock" was introduced as a method designed to correlate DNA methylation age with chronological age (Horvath 2013). By analyzing over 8000 healthy samples from 82 Illumina DNA methylation array datasets and considering more than 20,000 CpGs, 353 specific CpGs were identified, in the epigenetic clock, as aging predictors across multiple tissue types. Using this approach, it has been demonstrated that DNA methylation age correlates with chronological age in certain cell types, such as brain cells, and emphasized that this correlation reflects changes in the methylome rather than alterations in cell composition. Notably, the 353 CpGs in the epigenetic clock can be categorized based on their methylation changes with age: 193 CpGs exhibit hypermethylation, while 160 CpGs show hypomethylation as individuals age (Horvath 2013). This methodology thus offers a valuable tool for understanding the complex relationship between DNA methylation and aging.

In addition to the concept of the "epigenetic clock," the "epigenetic drift" has gained attention in recent years. Unlike the epigenetic clock, which focuses on specific methylation changes that correlate with chronological age across individuals, epigenetic drift refers to stochastic methylation changes that occur within a single individual over time. While the epigenetic clock identifies consistent patterns at particular sites that can predict age across different individuals, epigenetic drift highlights the variability in methylation patterns that can arise due to individual experiences and environmental influences. Importantly, the epigenetic clock can exhibit tissue specificity, meaning that certain tissues may better reflect chronological age than others. Despite the clear distinction between these concepts, "epigenetic drift" and "epigenetic clock" are often used interchangeably in literature. The phenomenon of epigenetic drift was initially observed in cell cultures and later in studies involving monozygotic twins, where discrepancies in methylation patterns were noted upon aging (Jones et al. 2015). This underscores the complexity of how aging influences the epigenome.

The specific alterations of DNA methylation at the different types of repetitive elements and their consequences will be discussed below.

5 Transcription of REs During Aging

A recent analysis of published RNA-seq datasets from dermal fibroblasts of healthy subjects 1–94 years old revealed a strong correlation between the transcription of most RE families and aging. This study showed that RNAs arising from DNA transposons, LTRs, satellites, and LINEs positively correlated with age. The correlation was so strong for specific REs that it was possible to predict age with better accuracy than with the panel of the most expressed genes in aged people (LaRocca et al. 2020). Similarly, another study that collected the RNA-seq of peripheral blood mononuclear cells (PBMCs) from young and old subjects disclosed the increase of RE expression with age. Interestingly, by comparing sedentary and exercising adults, it was also shown how physical exercise correlates with a reduction of RE transcripts. These results suggested both the importance of healthy aging interventions and that RE dysregulation might be a key event in the aging process which can be reversed (Wahl et al. 2020). In addition, the reversibility of the RE dysregulation associated with particular behaviors evokes the involvement of epigenetic modifications in this process. Details of the engagement of epigenetic modifications, specifically DNA methylation, in the expression of different types of REs during aging will be discussed in the next paragraphs.

5.1 *Satellite DNA Distension During Replicative Senescence and Aging*

Increased transcription of satellite DNA was pointed out in physiologically aged human fibroblasts (LaRocca et al. 2020). Conversely, though, it has been recently showed that in BJ skin fibroblasts, where senescence was induced by low doses of doxorubicin, there is a decreased expression of α-satellites and only a slight, not significant, increase of RNAs arising from satellite 3. This difference in centromere/pericentromere transcription is considered by the authors as the cause of the observed defective CENP-A deposition at aging centromeres, probably the origin of the genome instability that characterizes aged cells (Sikder et al. 2023). These results also suggest that the processes of replicative senescence and of induced senescence are substantially different.

Interestingly, the increased transcription of satellites was matched with the distention of the satellite DNA sequences in senescent cells (Enukashvily et al. 2007; Swanson et al. 2013), meaning that the chromatin loci are relaxed, decompacted, and lacking DNA methylation. Satellites of centromeres (α-satellites) and pericentromeres (satellites 2 and 3) are concerned with active transcription in both natural aging and Hutchinson-Gilford progeria syndrome (Enukashvily et al. 2007). Also, (peri-)centromere satellites exhibit senescence-associated distention which has been demonstrated to be an early event in the cell senescence process (Swanson et al. 2013). Indeed, satellites' relaxation preceded in cell culture the loss of the common histone markers of heterochromatin such as H3K9m3 and H3K27m3, which are, instead, lost at late passages in culture. This study did not monitor the DNA methylation changes at centromeric/pericentromeric satellites during cell senescence progression, and we cannot claim, in this case, whether DNA methylation loss is an early or a late event during cell senescence. However, the authors showed that long treatments with 5-azacytidine, a hypomethylation inducer, lead to satellites' distention in the U2OS cancer cell line, suggesting that DNA methylation loss could be the trigger. Nevertheless, these and other authors indicated that nuclear lamin proteins and, in particular, Lamin B1, which is reduced in senescent cells exhibiting satellites' distention, may contribute to satellites' dysregulated architecture through its role in anchoring heterochromatin to the nuclear lamina (Swanson et al. 2013). This hypothesis has been again recently proposed given that Lamin B1 can indirectly interact with precisely methylated DNA (Carollo and Barra 2023).

5.2 *Telomere Dysfunctions Drive Cellular Senescence and Aging*

Telomere erosion activates DNA damage response (DDR) that progressively leads to cell cycle arrest and senescence. However, another layer of regulation of senescence and aging comes from telomeres. Long noncoding RNAs called TERRA (telomeric repeat-containing RNAs) can arise from telomeric repeats which have a wide range of sizes, usually spanning from 5 to 15 kb in humans. Most of these RNAs have the 5′ capping and 7% of them also the polyadenylation at 3′, all characteristics of stable RNAs, meaning that they have a role in the cell. They indeed contribute to telomere maintenance and genome stability. TERRA transcription is regulated by TRF1 (member of the shelterin complex) from one side that binds and recruits RNA polymerase II and on the other side by heterochromatin that represses transcription (H3K9m and Heterochromatin Protein 1 are present). DNA methylation usually characterizes subtelomeric repeats where TERRA promoters are located to suppress TERRA transcription. However, upon DDR response at telomeres and apparently, during aging, TERRA is actively transcribed. The first study describing TERRA involvement in human aging has been done in immunodeficiency, centromeric region instability, and facial anomalies (ICF) syndrome patients bearing the mutation in DNMT3B gene (ICF type 1). These patients showed accelerated telomere shortening and significant telomere erosion leading to a premature replicative senescence. Interestingly, they also showed elevated levels of TERRA (Yehezkel et al. 2013), suggesting a strong correlation between TERRA expression and the aging process. The DNA methylation loss at the telomeric and subtelomeric region is most probably the cause of TERRA expression (Yehezkel et al. 2008) in ICF1. Similarly to ICF1 patients, fibroblasts derived from Hutchinson-Gilford progeria syndrome patients also exhibit overexpression of TERRA (Aguado et al. 2019). HGPS is a rare genetic condition that causes accelerated aging beginning in childhood (see Hypomethylation of REs in Age-Related Diseases paragraph), reinforcing the idea that TERRA expression strongly contributes to human aging. However, the methylation status of telomeres and subtelomeric DNA has not been evaluated yet. Despite the evidence of TERRA expression in dysfunctional conditions that trigger aging, its involvement in physiological aging has to be clarified. Moreover, the methylation loss at subtelomeric repeats during natural aging has also been experimentally debated (Maeda et al. 2009; Bacalini et al. 2021). The discrepancy between the published results is, though, probably dependent on the different approaches used to quantify the methylation of these complicated DNA regions and especially the target of these analyses, if few chromosomes (Bacalini et al. 2021) are considered or all of them (Maeda et al. 2009).

Apart from TERRA, there are also other RNAs arising from telomeres whose transcription is induced by DNA double-strand breaks at telomeres (damage-induced long noncoding RNAs, tdilncRNAs). These RNAs are enriched by both G- and C- and colocalize with the site of damage. Strikingly, both fibroblasts expressing the mutant form of Lamin A typical of HGPS, Progerin, and HGPS-derived dermal fibroblasts, together with HGPS model mice, exhibit increased transcription of tdilncRNAs (Aguado et al. 2019). The depletion of these RNAs by using antisense oligonucleotides targeting telomeric repeats (tASOs) was able to partially rescue cell senescence in both human cell and mouse models of HGPS, suggesting that tdilncRNAs expression plays a major role in the senescent phenotype of HGPS. However, despite tdilncRNA inhibition, tASOs could not rescue the loss of heterochromatin markers, the increase of TERRA, and the nuclear lamina alterations typical of HGPS. This consideration, together with the observation that tdilncRNA expression begins only at late passages during the culture of HGPS-derived dermal fibroblasts (Aguado et al. 2019), could suggest that tdilncRNAs are involved only in the late events of cell senescence. It would be interesting to understand if their transcriptional regulation depends on DNA methylation similarly to TERRA activation.

5.3 *Activation of TEs During Cell Senescence and Aging*

Derepression of the heterochromatin compartments in senescent cells, aged tissues, and cells isolated from patients exhibiting accelerated aging (with HGPS or Werner syndrome) has been linked to the increase of transcription from TEs and, consequently, of transposition (De Cecco et al. 2019; LaRocca et al. 2020; Della Valle et al. 2022; Ramini et al. 2022; Senapati et al. 2023). Multiple classes of TEs are expressed in human fibroblasts with age. For example, LINE-1 RNA is upregulated in fibroblasts derived from patients with HGPS. This deregulation preceded both the loss of the H3K9m3 heterochromatin marker and the senescent phenotype that appeared at later passages in culture (LaRocca et al. 2020). Strikingly, transcripts derived from LINE-1s have been directly correlated with the activation of the type 1 interferon (IFN-1) response that, in turn, promotes the typical aging-associated inflammation (De Cecco et al. 2019; Gorbunova et al. 2021). The authors of this study also demonstrated that LINE-1 activation and IFN-1 response are part of a late phase of the cell senescence process (De Cecco et al. 2019). Interestingly, ASOs complementary to LINE-1 transcripts rescued heterochromatin and reduced the senescent phenotype in mesenchymal cells derived from patients with HGPS and Werner syndrome (Della Valle et al. 2022), similar to tASOs. In this work, DNA methylation loss was not detected during cell passages in culture, making it difficult to understand if it is the cause of LINE-1 expression. However, the treatment with ASOs targeting LINE-1s induced a decreased DNA methylation age according to both the Horvath clock and the Pheno-Age clock, meaning that there is an alteration of DNA methylation. LINE-1 and Alu RNAs were also found to be upregulated in replicative senescence human umbilical vein endothelial cells (HUVECs) and human dermal fibroblasts (NHDFs), and this was strictly associated with DNA methylation loss (Ramini et al. 2022). Intriguingly, a recent work by Nam and colleagues clearly showed a positive correlation, especially in the colorectal epithelium, between LINE-1 retrotransposition and aging in a collection of different cell types of 28 individuals. By combining whole genome methylation with RNA expression profiles, this study also demonstrated a negative correlation between LINE-1 promoter methylation and transcription, suggesting a main role for DNA methylation as a regulator of LINE-1 mobilization with age (Nam et al. 2023).

Similarly to LINE-1s, transcription of the HERV family of TEs has also been associated with aging. Different subtypes of HERVs (HERV-K, HERV-W, and HERV-H) were recently found upregulated in both tissues and sera of aged individuals (Balestrieri et al. 2015; Nevalainen et al. 2018) and also in hematopoietic stem cells and mesenchymal stem cells undergoing senescence (Capone et al. 2018; Liu et al. 2023). Importantly, it was demonstrated that the inhibition of HERV-Ks by CRISPR-dCas9-KRAB transcriptional inactivation or by interfering short hairpin RNAs (shRNAs) alleviated the senescent phenotype of human mesenchymal cells derived from both HGPS and Werner syndrome. Conversely, the treatment with 5-azacytidine by reducing the methylation of HERV-Ks caused accelerated cellular senescence in primary mesenchymal cells derived from a young individual (Liu et al. 2023). This study also demonstrated that HERV-K expression correlated with the formation of HERV-K retrovirus-like particles (RVLPs) with the ability to be released extracellularly and infect young cells inducing their senescence. These results suggested how the expression of HERV-Ks can be a driving event not only of cell senescence but also of tissue aging.

Interestingly, transcription of TEs, specifically of Alu elements, was also associated with impairing the self-renewal capacity of human adult stem cells that undergo senescence upon *ex vivo* expansion (Wang et al. 2011). Adult adipose-derived mesenchymal stem cells (hADSCs) can keep their stem capacity until several population doublings after isolation when they enter cell senescence. This state is accompanied by the expression of TEs, mainly Alu elements arising from the pericentromeric regions which are the loci of persistent DNA damage (γH2AX-positive foci) in hADSCs undergoing senescence. The study showed that the transduction of lentivirus carrying shRNAs suppressing TEs' expression is sufficient to revert the senescent phenotype, reestablishing cell proliferation and self-renewal in hADSCs (Wang et al. 2011). These results indicate a cellular toxicity of Alu transcripts which can be involved in the process of aging and age-associated diseases.

Several studies analyzing the DNA methylation pattern in the peripheral blood of healthy volunteers of different ages showed a decrease of the methylation in TEs with age (Bollati et al. 2009; Cho et al. 2015). All these studies agree in the anticorrelation between aging and the DNA methylation of Alu elements in particular; however, one work also showed that LINE-1 methylation decreases with age in a cohort of healthy male individuals ranging from 21 to 57 years old (Cho et al. 2015). Loss of methylation of HERV-Ks was also observed in parallel with the increased transcription of these elements in senescent human mesenchymal cells (Liu et al. 2023).

The variation of the DNA methylation control over the expression of TEs during aging can have multiple explanations. One above all is probably the exposure to environmental insults whose effects inevitably add up over time. In addition, lifestyle and diet have been demonstrated to impact the methylation of TEs, such as the case of LINE1 methylation which is increased in the blood cells of individuals with a healthy lifestyle (i.e., no smokers, people practicing sports) and a fat-poor diet (Marques-Rocha et al. 2016).

6 Hypomethylation of REs Contributes to Genome Instability

Genome instability refers to the tendency of the genome of a cell to undergo mutations, those permanent, inheritable changes to the DNA sequence (i.e., base substitutions, insertions, deletions, chromosomal aberrations) that can have harmful effects on cell functions contributing to cancer and genetic diseases and that are the basis of evolution as well. The causes of mutations are multiple, and above all, DNA damage is considered among the main factors that drive the degeneration process typical of aging.

There is a close relationship between hypomethylation of virtually all types of REs and genomic instability. An emblematic example of this connection is seen in the cells derived from patients with ICF syndrome. The loss of DNA methylation especially at the pericentromeric and centromeric satellites of specific chromosomes (mainly chr 1 and chr 16) is associated with chromatin decondensation and a high rate of chromosomal rearrangements of these regions. Centromere breakage, chromosome arm missegregation, and chromosomal translocations with breakpoints at the centromere were observed in these cells which lead to the formation of multiradial chromosomes, a typical feature that allows the diagnosis of the ICF syndrome. Centromere instability associated with DNA methylation loss is also a hallmark of cancer. The passive DNA demethylation occurring at centromeric regions has been also correlated with whole chromosome missegregation and consequent aneuploidy (Costa et al. 2016). These pieces of evidence and the observed involvement of the pericentromeric satellite repeats of the pericentromere in this passive demethylation (Scelfo et al. 2024) suggest that the DNA methylation of the satellites, at least, contributes to the assembly of a functional kinetochore or to the binding of the microtubules of the mitotic spindle to the kinetochore, or both, thus ensuring chromosome stability. Another possible mechanism of genome instability induced by DNA hypomethylation of REs of the (peri-)centromere region could be, indeed, by increased transcription of these elements. Transcription of the pericentromeric region has been correlated with replication stress during DNA synthesis due to the collision between DNA and RNA polymerases resulting in replication fork stalling (Zaratiegui et al. 2011). Replication stress, in turn, is known to generate DNA breaks and impair chromosome segregation leading to genome instability.

Due to their intrinsic ability to transpose, TEs are also a cause of genome instability. It has been demonstrated that the transposition, and thus the expression, of LINE-1s induced double-stranded breaks with the formation of γ-H2AX foci, fragmented DNA, and interstitial deletions within the nucleus of human cells (Han et al. 2005; Gasior et al. 2006). A similar behavior has been documented for Alu repeats as well (Han et al. 2005). These effects have been identified as the cause of several genetic diseases in humans, including neurofibromatosis type I (Vogt et al. 2014).

Interestingly, a recent study demonstrated that, during aging, demethylation affected TEs involved in the regulation of genes deputed to the luminal epithelial cell lineage specificity. This event not only undermined the maintenance of lineage specificity (see Regulation of REs and Cell-Lineage Specification paragraph) but can also increase the susceptibility to breast cancer. Indeed, the authors found a strong correlation between age-dependent demethylation of some TEs with overexpression of the same TEs in breast cancer cells (Senapati et al. 2023). Another work similarly showed that epithelial ovarian cancers are characterized by passive DNA demethylation occurring at LINE-1s which leads to LINE-1s overexpression and copy number alterations accompanied by increased chromosomal instability (Zhang et al. 2020).

For an in-depth excursus on human genetic diseases where the expression of TEs is involved, without a specific consideration of their methylation status, we suggest reading the following review (Payer and Burns 2019).

7 Regulation of REs and Cell-Lineage Specification

In humans, the process of cell differentiation that drives the totipotent stem cells to gradually acquire cell lineage specificities following fertilization is finely orchestrated by epigenetic modifications. DNA methylation is a leading player in this phenomenon known as "epigenetic reprogramming," where genes are activated or repressed depending on the cell type. However, it is becoming clear that also REs are programmed as well as the genes during cell differentiation. A very interesting study showed that the different classes of REs undergo differential DNA methylation changes during the transition from human embryonic stem cells (hESCs) to fetal fibroblasts (IMR90) (Su et al. 2012). Globally, REs were demethylated in the stem cells compared to the differentiated cells. Specifically, the major changes were found in LINEs, LTRs, and satellite repeats. Almost all differentially methylated REs located near genes were found in introns. Strikingly, downregulated genes in IMR90 cells compared to hESCs were associated with hypomethylated REs in their gene body. This suggested both that hypomethylation within the gene body leads to gene silencing and that methylation of REs contributes to the regulation of cell lineage specificity.

By matching the binding sites of transcription factors (NANOG, OCT4, and CTCF) (TF-binding sites) in hESCs with the mapping of REs, it was noted that 99.8% of TF-binding sites were embedded within REs (mainly LINEs and LTRs) and located in genomic regions that are hypermethylated in hESCs and hypomethylated in differentiated IMR90 cells (Glinsky 2015). All this suggested a regulatory role for REs and their methylation during differentiation. A recent fine research also illustrated that satellite DNA methylation is dynamically regulated during human embryo development. For example, centromeric and pericentromeric satellites peaked their expression at the four-cell stage in agreement with their hypomethylated state. In contrast, telomeric satellites were upregulated at the eight-cell stage. This study also analyzed TEs. While different LINE repeats exhibited different expression patterns between four cells and implantation stages, the expression of SINEs was decreased at both stages (Yandım and Karakülah 2019). Promoters of the genes specific for each developmental stage resulted in being enriched with the differentially expressed REs, suggesting, once again, a role for REs and their methylation in gene regulation during development and cell differentiation.

In agreement with these observations, it was described that there is a cell type-specific hypomethylation involved in the regulation of the genes establishing cell identity. This specific hypomethylation mainly occurs at the "superenhancers," a large cluster of enhancers regulating tissue-specific genes. TEs have been frequently found within gene enhancers, and in particular, the expression of LTRs and Alu repeats has been associated with the expression of cell-type specific genes (Zhang et al. 2019; Karttunen et al. 2023). Consequentially, another study demonstrated that Alu elements that were not expressed in a specific tissue sample displayed highly methylated levels, conversely, those Alu elements that were expressed were not methylated. Also, the expression of Alu repeats correlated with that of neighboring genes specific to the corresponding tissue sample, reinforcing the idea of a function as cell-type-specific enhancers (Zhang et al. 2019).

8 Hypomethylation of REs in Age-Related Diseases

The activity of TEs has been often tied to age-related diseases, especially neurodegenerative disorders. It is the case of amyotrophic lateral sclerosis (ALS) and frontotemporal dementia (FTD), both neurodegenerative disorders that share the nuclear loss and consequent cytoplasmic aggregation of TAR DNA-binding protein-43 (TDP-43), an RNA-binding protein that regulates transcription and alternative splicing. Aging is considered the major risk factor for these diseases. It has been observed that loss of nuclear TDP-43 is associated with DNA demethylation and consequent decondensation and expression/retrotransposition of LINEs (Savage et al. 2020). Considering that TDP-43 had been shown to bind transcripts derived from TEs, including LINEs, it has been proposed that it usually protects from TE expression which can be detrimental to genome stability, as we discussed above. As such, TE expression could contribute to ALS and FTD. We could also hypothesize that when TDP-43 activity is hampered, it is not able to counteract the increased expression of TEs that naturally occurs due to aging-dependent DNA demethylation. Other common degenerative diseases including Alzheimer's and Parkinson's diseases are marked with the overexpression of TEs and often of satellites as well. DNA methylation plays a fundamental role in neurodevelopment (Martino et al. 2023), and taking into account its deregulation during aging, it could be involved in virtually all neurodegenerative disorders, especially if they are correlated with age. However, no definitive evaluation of the loss of DNA methylation of the REs has clarified yet whether this epigenetic change is the cause of TE overexpression in other degenerative diseases. Hence, we preferred not to include these diseases in this chapter. However, we wish to discuss the progeroid syndromes and in particular the Hutchinson-Gilford progeria syndrome, where opposing results have been obtained regarding the DNA methylation changes. HGPS is a genetic disease caused by the mutation of the *LMNA* gene encoding a component of the nuclear lamina (Lamin-A) that leads to very fast degeneration of organs and bones, resembling an accelerated form of aging. Despite several groups describing the DNA methylation evaluation of the whole genome, there is no ad hoc analysis of the specific conditions of the REs. However, overall changes in DNA methylation have been documented, in terms of both hypo- and hypermethylation. In addition, by comparing the methylation pattern of cells derived from aged healthy donors and HGPS cells, some authors concluded that the DNA methylation pattern of HGPS is distinct from that established during natural aging (Köhler et al. 2020). Intriguingly, though, overexpression of TEs has been clearly assessed. Both LINE-1s and HERV-Ks are interested in this overexpression, although the responsible mechanism has not been clarified yet (Della Valle et al. 2022; Liu et al. 2023). Considering the close relationship between nuclear lamina and heterochromatin, as previously proposed (Carollo and Barra 2023), it is reasonable to think that dysfunctions of the nuclear lamina, such as in the case of HGPS, can destabilize heterochromatin markers, including DNA methylation, leading to the derepression of the TEs, similarly to what occurs during natural aging.

9 Conclusions

Aging is a multifaceted biological process influenced by both genetic and epigenetic factors. The progressive loss of DNA methylation that occurs physiologically during aging and senescence, particularly at repetitive elements, like tandem repeats and transposable elements, plays a critical role in genome instability, a hallmark of cellular aging. Telomere erosion is considered a major driver of cellular senescence and aging. However, the hypomethylation of the repetitive elements, including those at telomeres, can further exacerbate genome instability, thus representing a key element in the senescence and aging process. Age-associated hypomethylation has been correlated with the transcription activation of both types of REs, transposable elements, and satellites, endangering genome stability and therefore cell viability and function (Fig. 1). REs' hypomethylation and consequent unscheduled transcripts arising from REs can be threatening for the cells by inducing, in the first place, DNA breaks, chromosome rearrangements, and replication stress. Also, these transcripts have been correlated with the cell-lineage determination and reprogramming, potentially impacting the aging of organs and tissues, such as the heart and muscles, leading to their degeneration. All this can contribute not only to the aging process but also to age-related diseases (Fig. 1).

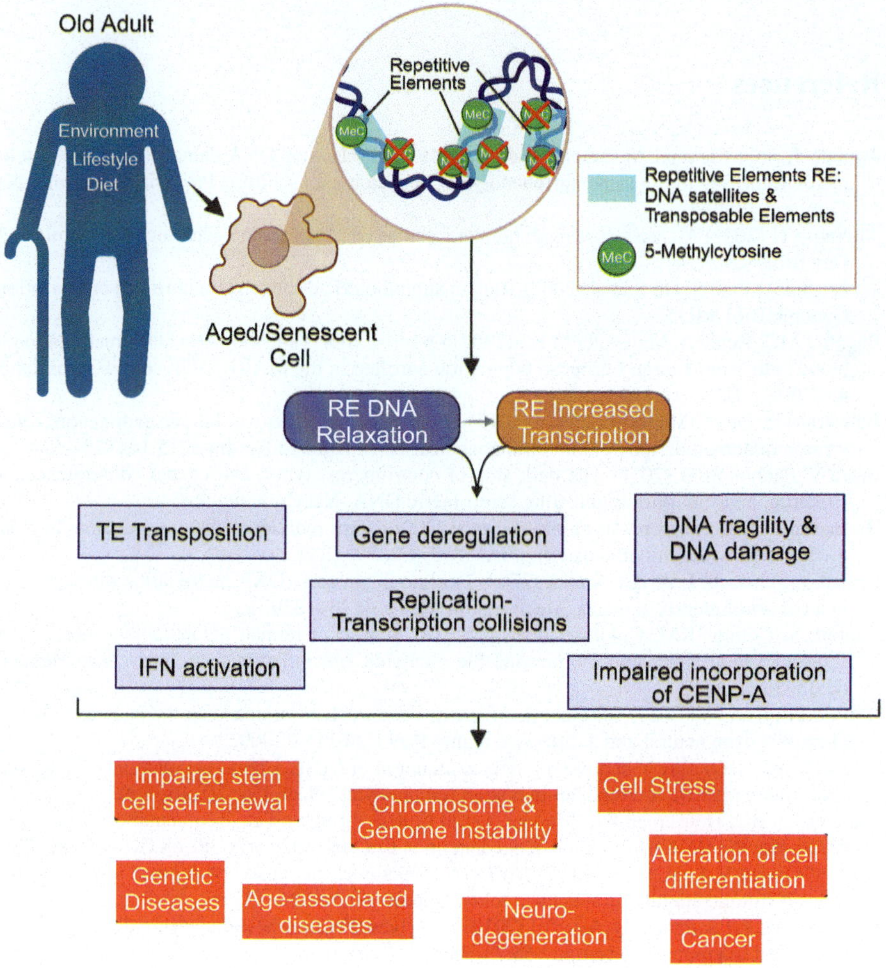

Fig. 1 DNA methylation of repetitive elements contributes to human aging and aged-related diseases. Human aging is associated with the reduction of DNA methylation at repetitive elements (REs), both satellites and transposable elements (TEs). REs constitute a major part of the human genome and are normally hypermethylated. Natural aging process, together with external factors, including lifelong lifestyle, diet, and environment, impact on RE methylation levels. The hypomethylation is the cause of the relaxation and consequent transcription of REs. This event can trigger many alterations such as gene expression deregulation, TE transposition, collision between replication and transcription machineries, DNA damage, and IFN-1 response. All these concur to the aged phenotype and to age-related pathological conditions including neurodegeneration and cancer. (*Created in BioRender. Barra, V. (2024)* https://BioRender.com/p49c825)

These considerations highlight the importance of maintaining genome stability during aging through the regulation of repetitive elements by DNA methylation. Future research into the precise molecular mechanisms controlling these processes may offer insights into therapeutic strategies that could mitigate age-associated cellular damage and extend healthy lifespan.

References

Aguado J, Sola-Carvajal A, Cancila V et al (2019) Inhibition of DNA damage response at telomeres improves the detrimental phenotypes of Hutchinson-Gilford Progeria Syndrome. Nat Commun 10:4990

Altemose N (2022) A classical revival: human satellite DNAs enter the genomics era. Semin Cell Dev Biol 128:2–14

Ambrosini A, Paul S, Hu S et al (2007) Human subtelomeric duplicon structure and organization. Genome Biol 8:R151

Bacalini MG, Reale A, Malavolta M et al (2021) Ageing affects subtelomeric DNA methylation in blood cells from a large European population enrolled in the MARK-AGE study. Geroscience 43:1283–1302

Balestrieri E, Pica F, Matteucci C et al (2015) Transcriptional activity of human endogenous retroviruses in human peripheral blood mononuclear cells. Biomed Res Int 2015:164529

Barra V, Fachinetti D (2018) The dark side of centromeres: types, causes and consequences of structural abnormalities implicating centromeric DNA. Nat Commun 9:4340

Bjornsson HT, Sigurdsson MI, Fallin MD et al (2008) Intra-individual change over time in DNA methylation with familial clustering. JAMA 299:2877–2883

Bollati V, Schwartz J, Wright R et al (2009) Decline in genomic DNA methylation through aging in a cohort of elderly subjects. Mech Ageing Dev 130:234–239

Capone S, Connor KM, Colombo A et al (2018) Senescent human hematopoietic progenitors show elevated expression of transposable elements and inflammatory genes. Exp Hematol 62:33–38.e6

Carollo PS, Barra V (2023) Chromatin epigenetics and nuclear lamina keep the nucleus in shape: examples from natural and accelerated aging. Biol Cell 115:2200023

Cho YH, Woo HD, Jang Y et al (2015) The association of LINE-1 hypomethylation with age and centromere positive micronuclei in human lymphocytes. PLoS One 10:e0133909

Christensen BC, Houseman EA, Marsit CJ et al (2009) Aging and environmental exposures alter tissue-specific DNA methylation dependent upon CpG Island context. PLoS Genet 5:e1000602

Costa G, Barra V, Lentini L et al (2016) DNA demethylation caused by 5-aza-2′-deoxycytidine induces mitotic alterations and aneuploidy. Oncotarget 7:3726–3739

De Cecco M, Ito T, Petrashen AP et al (2019) L1 drives IFN in senescent cells and promotes age-associated inflammation. Nature 566:73–78

Della Valle F, Reddy P, Yamamoto M et al (2022) LINE-1 RNA causes heterochromatin erosion and is a target for amelioration of senescent phenotypes in progeroid syndromes. Sci Transl Med 14:eabl6057

Eisenberg T, Knauer H, Schauer A et al (2009) Induction of autophagy by spermidine promotes longevity. Nat Cell Biol 11:1305–1314

Enukashvily NI, Donev R, Waisertreiger IS-R et al (2007) Human chromosome 1 satellite 3 DNA is decondensed, demethylated and transcribed in senescent cells and in A431 epithelial carcinoma cells. Cytogenet Genome Res 118:42–54

Fraga MF, Ballestar E, Paz MF et al (2005) Epigenetic differences arise during the lifetime of monozygotic twins. Proc Natl Acad Sci 102:10604–10609

Gasior SL, Wakeman TP, Xu B et al (2006) The human LINE-1 retrotransposon creates DNA double-strand breaks. J Mol Biol 357:1383–1393

Glinsky GV (2015) Transposable elements and DNA methylation create in embryonic stem cells human-specific regulatory sequences associated with distal enhancers and noncoding RNAs. Genome Biol Evol 7:1432–1454

Gorbunova V, Seluanov A, Mita P et al (2021) The role of retrotransposable elements in ageing and age-associated diseases. Nature 596:43–53

Han K, Sen SK, Wang J et al (2005) Genomic rearrangements by LINE-1 insertion-mediated deletion in the human and chimpanzee lineages. Nucleic Acids Res 33:4040–4052

Heyn H, Li N, Ferreira HJ et al (2012) Distinct DNA methylomes of newborns and centenarians. Proc Natl Acad Sci 109:10522–10527

Hoal-van Helden EG, van Helden PD (1989) Age-related methylation changes in DNA may reflect the proliferative potential of organs. Mutat Res 219:263–266. https://doi.org/10.1016/0921-8734(89)90027-1

Horvath S (2013) DNA methylation age of human tissues and cell types. Genome Biol 14:3156

Huang B, Zhong D, Zhu J et al (2020) Inhibition of histone acetyltransferase GCN5 extends lifespan in both yeast and human cell lines. Aging Cell 19:e13129

Jones MJ, Goodman SJ, Kobor MS (2015) DNA methylation and healthy human aging. Aging Cell 14:924–932

Karttunen K, Patel D, Xia J et al (2023) Transposable elements as tissue-specific enhancers in cancers of endodermal lineage. Nat Commun 14:5313

Köhler F, Bormann F, Raddatz G et al (2020) Epigenetic deregulation of lamina-associated domains in Hutchinson-Gilford progeria syndrome. Genome Med 12:46

Kwabi-Addo B, Chung W, Shen L et al (2007) Age-related DNA methylation changes in normal human prostate tissues. Clin Cancer Res 13:3796–3802

LaRocca TJ, Cavalier AN, Wahl D (2020) Repetitive elements as a transcriptomic marker of aging: evidence in multiple datasets and models. Aging Cell 19:e13167

Liu X, Liu Z, Wu Z et al (2023) Resurrection of endogenous retroviruses during aging reinforces senescence. Cell 186:287–304.e26

Maeda T, Guan JZ, Higuchi Y et al (2009) Aging-related alterations of subtelomeric methylation in sarcoidosis patients. J Gerontol A 64A:752–760

Marques-Rocha JL, Milagro FI, Mansego ML et al (2016) LINE-1 methylation is positively associated with healthier lifestyle but inversely related to body fat mass in healthy young individuals. Epigenetics 11:49–60

Martino S, Carollo PS, Barra V (2023) A glimpse into chromatin organization and nuclear lamina contribution in neuronal differentiation. Genes 14:1046

Nam CH, Youk J, Kim JY et al (2023) Widespread somatic L1 retrotransposition in normal colorectal epithelium. Nature 617:540–547

Narita M, Narita M, Krizhanovsky V et al (2006) A novel role for high-mobility group a proteins in cellular senescence and heterochromatin formation. Cell 126:503–514

Nevalainen T, Autio A, Mishra BH et al (2018) Aging-associated patterns in the expression of human endogenous retroviruses. PLoS One 13:e0207407

O'Sullivan RJ, Kubicek S, Schreiber SL et al (2010) Reduced histone biosynthesis and chromatin changes arising from a damage signal at telomeres. Nat Struct Mol Biol 17:1218–1225

Pappalardo XG, Barra V (2021) Losing DNA methylation at repetitive elements and breaking bad. Epigenetics Chromatin 14:25

Payer LM, Burns KH (2019) Transposable elements in human genetic disease. Nat Rev Genet 20:760–772

Rakyan VK, Down TA, Maslau S et al (2010) Human aging-associated DNA hypermethylation occurs preferentially at bivalent chromatin domains. Genome Res 20:434–439

Ramini D, Latini S, Giuliani A et al (2022) Replicative senescence-associated LINE1 methylation and LINE1-Alu expression levels in human endothelial cells. Cells 11:3799
Savage AL, Lopez AI, Iacoangeli A et al (2020) Frequency and methylation status of selected retrotransposition competent L1 loci in amyotrophic lateral sclerosis. Mol Brain 13:154
Scaffidi P, Misteli T (2006) Lamin A-dependent nuclear defects in human aging. Science 312:1059–1063
Scelfo A, Barra V, Abdennur N et al (2024) Tunable DNMT1 degradation reveals DNMT1/DNMT3B synergy in DNA methylation and genome organization. J Cell Biol 223:e202307026
Senapati P, Miyano M, Sayaman RW et al (2023) Loss of epigenetic suppression of retrotransposons with oncogenic potential in aging mammary luminal epithelial cells. Genome Res 33:1229–1241
Shah PP, Donahue G, Otte GL et al (2013) Lamin B1 depletion in senescent cells triggers large-scale changes in gene expression and the chromatin landscape. Genes Dev 27:1787–1799
Shumaker DK, Dechat T, Kohlmaier A et al (2006) Mutant nuclear lamin A leads to progressive alterations of epigenetic control in premature aging. Proc Natl Acad Sci USA 103:8703–8708
Sikder S, Baek S, McNeil T, Dalal Y (2023) Unraveling the mechanism of centromere inactivation in human aging. Preprint at bioRxiv. https://doi.org/10.1101/2023.12.30.573721
Smurova K, De Wulf P (2018) Centromere and pericentromere transcription: roles and regulation … in sickness and in health. Front Genet 9:674
Su J, Shao X, Liu H et al (2012) Genome-wide dynamic changes of DNA methylation of repetitive elements in human embryonic stem cells and fetal fibroblasts. Genomics 99:10–17
Swanson EC, Manning B, Zhang H, Lawrence JB (2013) Higher-order unfolding of satellite heterochromatin is a consistent and early event in cell senescence. J Cell Biol 203:929–942
Thakur J, Packiaraj J, Henikoff S (2021) Sequence, chromatin and evolution of satellite DNA. Int J Mol Sci 22:4309
Vogt J, Bengesser K, Claes KB et al (2014) SVA retrotransposon insertion-associated deletion represents a novel mutational mechanism underlying large genomic copy number changes with non-recurrent breakpoints. Genome Biol 15:R80
Wahl D, Cavalier AN, Smith M et al (2020) Healthy aging interventions reduce repetitive element transcripts. J Gerontol A Biol Sci Med Sci 76:805–810
Wang J, Geesman GJ, Hostikka SL et al (2011) Inhibition of activated pericentromeric SINE/Alu repeat transcription in senescent human adult stem cells reinstates self-renewal. Cell Cycle Georget Tex 10:3016–3030
Wang Q, Xiong F, Wu G et al (2022) Gene body methylation in cancer: molecular mechanisms and clinical applications. Clin Epigenetics 14:154
Wells JN, Feschotte C (2020) A field guide to eukaryotic transposable elements. Annu Rev Genet 54:539–561
Wilson VL, Jones PA (1983) DNA methylation decreases in aging but not in immortal cells. Science 220:1055–1057. https://doi.org/10.1126/science.6844925
Wilson VL, Smith RA, Ma S, Cutler RG (1987) Genomic 5-methyldeoxycytidine decreases with age. J Biol Chem 262:9948–9951. https://doi.org/10.1016/S0021-9258(18)61057-9
Yandım C, Karakülah G (2019) Expression dynamics of repetitive DNA in early human embryonic development. BMC Genomics 20:439
Yang J-H, Hayano M, Griffin PT et al (2023) Loss of epigenetic information as a cause of mammalian aging. Cell 186:305–326.e27
Yehezkel S, Segev Y, Viegas-Péquignot E et al (2008) Hypomethylation of subtelomeric regions in ICF syndrome is associated with abnormally short telomeres and enhanced transcription from telomeric regions. Hum Mol Genet 17:2776–2789
Yehezkel S, Shaked R, Sagie S et al (2013) Characterization and rescue of telomeric abnormalities in ICF syndrome type I fibroblasts. Front Oncol 3:35

Zaratiegui M, Castel SE, Irvine DV et al (2011) RNAi promotes heterochromatic silencing through replication-coupled release of RNA Pol II. Nature 479:135–138
Zhang X-O, Gingeras TR, Weng Z (2019) Genome-wide analysis of polymerase III–transcribed Alu elements suggests cell-type–specific enhancer function. Genome Res 29:1402
Zhang W, Klinkebiel D, Barger CJ et al (2020) Global DNA hypomethylation in epithelial ovarian cancer: passive demethylation and association with genomic instability. Cancers 12:764

DNA Methylation in a Systemic Framework for Healthy Aging and Longevity

Paolo Abondio

Contents

Abstract Extreme longevity, which manifests as an exceptionally prolonged physical and cognitive performance into very old age, can be considered a model of successful aging. Studies of this phenomenon have delineated a complex interaction of genetic, lifestyle, and environmental factors contributing to increased longevity and healthspan. Centenarians and supercentenarians (those living to 100 and 110 years of age, respectively) display distinct genetic profiles when compared to younger cohorts, and such profiles include favorable mutations that are protective of cardiovascular health, support DNA damage repair, and promote inflammatory regulation. Lifestyle-related nongenetic factors (diet, physical activity, socialization) also play a role in healthy aging, often through the filter of epigenetic mechanisms, to which DNA methylation belongs. It is clear that complex interactions between genetic background, lifestyle, and environmental exposures during the lifespan impact the aging process by altering methylation patterns that activate or silence genes relevant to crucial metabolic functions, sometimes not even directly related to aging. In this framework, an omnigenic and multiomic view of aging and longevity encompassing functional genomics, epigenetic markers, and environmental/life-

P. Abondio (✉)
Department of Biology, University of Rome Tor Vergata, Rome, Italy
e-mail: paolo.abondio@uniroma2.it

J. Barciszewski (ed.), *Decoding Aging and Neurodegeneration*,
RNA Technologies 16, https://doi.org/10.1007/978-3-031-91072-2_6

style influences could allow for a holistic interpretation of this complex phenotype and instigate targeted, personalized solutions to maximize healthy aging and healthspan.

Keywords DNA · Methylation · Healthspan · Longevity · Centenarian

1 Introduction

Extreme longevity is an area of notable scientific interest, viewed as a paradigm of successful aging where individuals not only reach a very advanced age but also retain unusually high levels of physical and cognitive function (Garagnani et al. 2013; Giuliani et al. 2017). This phenomenon can be construed as a desirable aging phenotype that contrasts sharply with the typical age-related decline experienced by most of the human population. Research into extreme longevity is focused on understanding why certain individuals appear resilient to ailments commonly associated with aging, such as cardiovascular disease, cancer, diabetes, and neurodegenerative disorders (Borras et al. 2020), while others are not. This highly complex phenotype emerges from an intersection of genetic, lifestyle, and environmental factors that promote both an extended lifespan and unusually long healthspan (Puca et al. 2018).

Centenarians and supercentenarians—individuals who reach 100 and 110 years of age respectively—are particularly unique subsets of the general population that exhibit extreme longevity, offering valuable insights into what successful aging actually looks like. The study of centenarian cohorts is increasingly significant as the global elderly population continues to rise, stimulating research into the attainment of a high quality of life in advanced age. Indeed, there are indications that the life expectancy and health outcomes of centenarians differ remarkably from those of the general elderly population, particularly in their tendency to delay illness until the final years or even months of life: the so-called compression of morbidity (an effective, although sometimes criticized concept) is particularly prominent in this group of individuals, as they often experience a relatively short period of sickness and disability immediately before death (Lantz 2020; Pimlott 2023). While most individuals face an extended phase of declining health due to chronic diseases, centenarians appear to maintain a high quality of life for much longer (Fig. 1). This phenomenon is essential to the phenotype under investigation, because it underscores the idea that it is possible to reach advanced ages without experiencing a prolonged period of impairment. Therefore, this delay in serious health decline is considered a hallmark of successful aging and highlights the potential for extending not just life expectancy but also healthspan, the part of one's life in which individuals remain healthy and self-sufficient, as many centenarians are able to live independently and remain cognitively intact until the very end (Islam et al. 2024; Martin et al. 2024). This contrasts sharply with the general elderly population, where rates of dementia, cardiovascular diseases, and other chronic conditions tend to increase

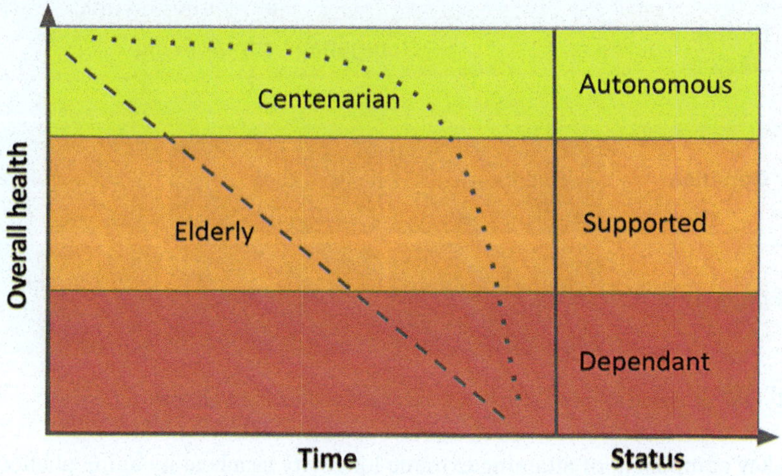

Fig. 1 Health trajectories for the general elderly population and centenarian subjects

sharply and significantly with age (Zhao et al. 2024; Livingston et al. 2024). For these reasons, the ability of centenarians to compress morbidity has become a focal point for aging research, as it suggests pathways by which medical interventions, lifestyle changes, and possibly pharmacological developments could help in extending the healthspan of the broader population.

2 Nongenetic Determinants of Healthy Aging and Longevity

Genetic factors play a crucial role in extreme longevity, as studies of centenarians and supercentenarians have identified several genetic variants that may confer protective effects against age-related diseases (Caruso et al. 2022; Smulders and Deelen 2024). For example, mutations in genes associated with cellular repair mechanisms and the maintenance of DNA structure integrity have been linked to this phenotype, as well as genes related to inflammation and immune response, which appear to be significant as lower levels of chronic inflammation are associated with healthier aging (Anaya et al. 2024). Several beneficial mutations have also been consistently associated with the attainment of longer lifespans and are significantly more prevalent among centenarians, as they reduce the risk of common age-related ailments such as cardiovascular disease, Alzheimer's disease, and cancer (Garagnani et al. 2021; Tesi et al. 2024; Ying et al. 2024). However, genetics alone does not fully explain the phenomenon: studies have shown that, conservatively, around 20–30% of longevity is heritable (Ruby et al. 2018; Van Den Berg et al. 2019), meaning that contextual nongenetic factors play a substantial role in determining who reaches such an advanced age (Fig. 2).

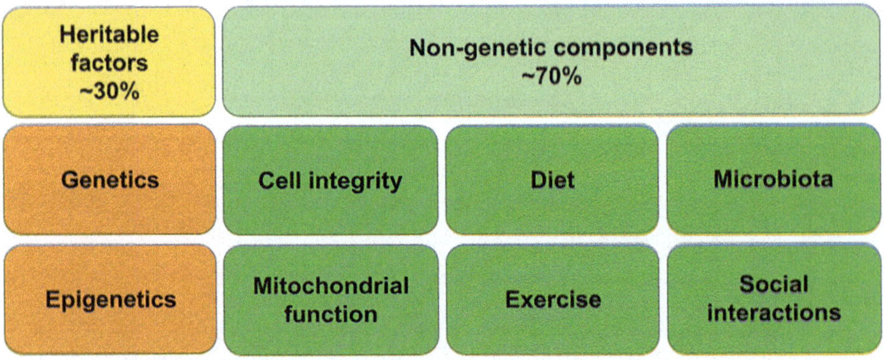

Fig. 2 Genetic and nongenetic factors influencing healthspan and lifespan

A key component of attaining extreme longevity as a healthy aging phenotype is the body's ability to maintain functional capacity at cellular and tissue levels. Cellular senescence, a process by which cells cease to divide and accumulate with age, plays a critical role in physiological age-related decline (Borghesan et al. 2020). Senescent cells also release inflammatory molecules that can damage the surrounding tissue, a phenomenon known as "senescence-associated secretory phenotype" (SASP) (Wang et al. 2024), which is also one of the hallmark components of chronic low-grade inflammation associated with aging, or "inflammaging": individuals with extreme longevity often show lower levels of cellular senescence, possibly due to superior cellular repair mechanisms and efficient cell clearance (Franceschi et al. 2000; Fulop et al. 2023; Ajoolabady et al. 2024). Pharmacological interventions with senolytic drugs designed to selectively eliminate senescent cells are being explored as potential therapies to improve healthspan by delaying cell accumulation and mimicking the characteristics observed in long-lived individuals (Rad and Grillari 2024; Calabrò et al. 2024).

Another interesting factor contributing to the extreme longevity phenotype is mitochondrial function: while generating the energy necessary for all cellular processes, mitochondria release reactive oxygen species (ROS) as a byproduct. High levels of ROS can damage cellular components including nucleic acids, leading to oxidative stress, structural DNA changes, and point mutations, therefore contributing to age-related diseases (Giorgi et al. 2018). However, it has been noticed that an increased concentration of some ROS in a spatiotemporal manner may work as a signaling device in aging and may even favor lifespan extension (Karagianni and Bazopoulou 2024). Concurrently, therapies aimed at enhancing mitochondrial function, such as mitochondria-targeted antioxidants, are currently under investigation for their potential to reduce neuronal damage, extend cellular longevity, and, therefore, slow the aging process (Jiang et al. 2020; Fields et al. 2023).

Beyond genetics and cellular components, lifestyle and environmental factors are crucial for healthy aging: diet, exercise, and social engagement are all elements that have been linked to longevity. In particular, engaging with diets rich in vegetables, lean proteins, and healthy fats, with a lower intake of processed foods and sugars, constitutes a nutritional pattern that, combined with regular physical activity and movement throughout the day, does contribute to cardiovascular health and reduces the risk of metabolic disorders regardless of age (Dai et al. 2024). Social relations, such as strong community connections, a sense of purpose, and mutual help, have been also proposed as relevant psychosocial variables that may increase longevity by reducing stress and improving general mental health (Friedman et al. 2024). This is notable, as the impact of lifestyle factors on longevity further suggests that, while genetics provides a baseline, behavior and environment can profoundly and unpredictably modify health trajectories (Fig. 2).

The study of gut microbiota has also emerged as a promising field in understanding extreme longevity, as a consistent body of research has shown that the composition of gut bacteria influences immune function, metabolism, and inflammation, all of which are critical for healthy aging (Pang et al. 2023). In centenarians, the gut microbiota tends to display youth-associated characteristics and shows a relationship with eating habits throughout the lifespan (Pang et al. 2023). This diverse microbiota may contribute to the resilience seen in long-lived individuals by enhancing their ability to withstand infections and preserve gut integrity, which often deteriorates with age. Emerging therapies targeting the gut microbiota, such as the use of probiotics and prebiotics, can therefore be investigated as means to promote healthier aging (Badal et al. 2020).

It is now clear that extreme longevity as a healthy phenotype in aging emerges from the complex intersection of genetic predispositions, lifestyle choices, and environmental variables. While genetic factors do provide a foundation for an unusually long life, behavioral characteristics, including diet, physical activity, and social engagement, are crucial in promoting healthspan and reducing the risk of age-related diseases. Advances in biomedical research hinted at in this section (senolytics, mitochondrial-targeted therapies, and microbiome modulation) offer promising avenues to mimic the health benefits observed in long-lived individuals, potentially extending healthy lifespan for the general population. Furthermore, a holistic approach that includes psychological and social well-being alongside physical health will likely be essential in cultivating the characteristics of healthy longevity.

3 Genetic Biomarkers and the Heritability of Extreme Longevity

Heritability studies on centenarians and their offspring provide crucial insights into the genetic and nongenetic factors that contribute to extreme longevity. Centenarians represent a unique population segment that has successfully avoided or delayed common age-related diseases, making them ideal candidates for studying factors that may protect against these conditions and promote an exceptionally long and healthy lifespan. Heritability in this context refers to the proportion of observed variation in lifespan attributable to genetic factors as opposed to environmental or lifestyle influences. Research indicates in general that the heritability of extreme longevity is moderate, with reasonable estimates reaching around 30% heritability (Giuliani et al. 2018). The offspring of centenarians are particularly valuable in longevity studies because they can directly demonstrate a strong genetic predisposition to towards a longer lifespan. Studies have consistently shown that centenarians' offspring not only have lower rates of mortality compared to the general population, but they also experience delayed onset of age-related diseases, which suggests that longevity-related genetic characteristics may be passed from one generation to the next: just like their parents, offspring of centenarians are less likely to develop cardiovascular disease, diabetes, and other metabolic disorders, and they often maintain higher cognitive and physical functioning levels well into advanced age (Thalén and Ledberg 2024).

Regarding examples of beneficial gene variants, studies in centenarians have led to the identification of specific genes that appear to directly provide a strong contribution to longevity (Table 1). For instance, polymorphisms in genes associated with cholesterol metabolism, such as *APOE* and *CETP*, are commonly found in centenarians. The APOE gene, which encodes the apolipoprotein E protein involved in lipid transport, produces three main protein isoforms, designated E2, E3, and E4. While the E4 isoform is associated with an increased risk of Alzheimer's disease and cardiovascular issues, the E2 isoform is relatively more common in centenarians and is thought to have a protective effect (Abondio et al. 2019; Tesi et al. 2024). Another gene, *CETP*, plays a role in lipid transfer and high-density lipoprotein (HDL) metabolism. Specific variants of this gene, such as the polymorphism converting the amino acid isoleucine to valine in position 405 of the resulting protein product (I405V), are linked to increased HDL levels and reduced risk of cardiovascular disease, which may partially explain the extended health span observed in centenarians (Vergani et al. 2006). Several other gene variants, such as those involved in insulin signaling (*IGF1*) (Vitale et al. 2019), DNA repair (*WRN*) (Sebastiani et al. 2013), and inflammatory pathways (*IL-6*) (Franceschi et al. 2020), have also been implicated in longevity. The FOXO3A gene, in particular, has garnered significant attention in longevity research. Studies have shown that peculiar mutations of *FOXO3A* are overrepresented in centenarians, and this gene plays a crucial role in cellular stress resistance, metabolism, and aging processes (Bae et al.

Table 1 Genes presenting polymorphisms strongly associated with extreme longevity

Gene name	Function	Reference
APOE	Lipid transport and metabolism	Tesi et al. (2024)
CETP	Lipoprotein metabolism	Vergani et al. (2006)
IGF1	Insulin signaling pathway	Vitale et al. (2019)
WRN	DNA repair	Sebastiani et al. (2013)
FOXO3A	Insulin signaling pathway	Bae et al. (2018)
IL-6	Immune response	Franceschi et al. (2020)

2018). *FOXO3A* belongs to the Forkhead box O family of transcription factors, which regulate the expression of genes involved in oxidative stress response, apoptosis, and DNA repair: the presence of longevity-associated *FOXO3A* alleles may contribute to enhanced cellular resilience in centenarians, allowing them to better withstand age-related physiological stressors (Zhao and Liu 2021). These findings also underscore the importance of stress response pathways in longevity and suggest that interventions targeting *FOXO3A* activity could hold promise for promoting health span in the general population.

Another area of heritability studies that offers insights into longevity involves the function and inheritance of mitochondrial DNA (mtDNA). Mitochondria, the cellular organelles responsible for energy production, have their own DNA, which is maternally inherited and carries 37 genes involved in energy production and protein biosynthesis (Lee et al. 2023a). Variations in mtDNA and nuclear-encoded mitochondrial genes have been linked to differences in lifespan, and studies have shown that certain variations are more prevalent in centenarians (Gonzalez et al. 2023). Mitochondrial efficiency and resistance to oxidative stress are important factors in cellular aging, and centenarians as well as their offspring often display superior mitochondrial function and similar structural characteristics in familial longevity (Tamvaka et al. 2024). This suggests that mitochondrial genetics may contribute to maternally inherited lifespan and that maintaining organelle health could be a critical aspect of delaying age-related decline in a cross-generational way.

MicroRNAs (miRNAs) have also become promising biomarkers for aging and longevity research (Koch 2023). These small, noncoding RNAs regulate gene expression by targeting messenger RNAs (mRNAs) for degradation or translational repression, influencing various cellular processes. Specific miRNAs, such as miR-21 and miR-146a, have been found to modulate pathways related to inflammation, stress response, and cellular senescence, all of which are relevant to healthy aging (Olivieri et al. 2021). Another miRNA, miR-34a, is known to regulate genes associated with cell cycle arrest and apoptosis, which are critical for managing cellular senescence—another hallmark of aging (Fulzele et al. 2019). Levels of these miRNAs fluctuate with age and are often altered in age-associated diseases, including cardiovascular diseases and neurodegenerative disorders. Interestingly, centenarians show distinct miRNA expression profiles, suggesting that these molecules may contribute to longevity by preserving cellular homeostasis and reducing inflammation (Borrás et al. 2017). The modulation of miRNA expression represents a potential avenue for therapeutic intervention, as synthetic miRNA mimics or inhibitors could theoretically restore youthful gene expression patterns.

4 The Role of DNA Methylation in Aging and Longevity

4.1 Nuclear DNA

Methylation, a biochemical process involving the addition of a methyl group (-CH_3) to DNA, proteins, or other molecules, plays a crucial role in the regulation of gene expression and is increasingly recognized as a significant factor in general health, healthy aging, and extreme longevity (Salameh et al. 2020). In humans, there are specific clusters of two-base sequences called "CpG islands" (sequences of cytosine immediately followed by a guanine) where epigenetic methylation of the cytosine residue tends to occur more often and, as a consequence, added methyl groups cluster together in these regions (Fig. 3). Since CpG islands are located more frequently at the transcription start site of housekeeping genes (which are constitutively active and support the basic functions of the organism), this epigenetic modification can heavily influence cellular function and aging by affecting fundamental gene activity without altering the underlying DNA sequence. In the context of aging, one of the most studied aspects of methylation is its impact on gene expression patterns associated with cellular senescence, inflammation, and stress response: in healthy centenarians, for example, optimal methylation patterns are often characterized by a

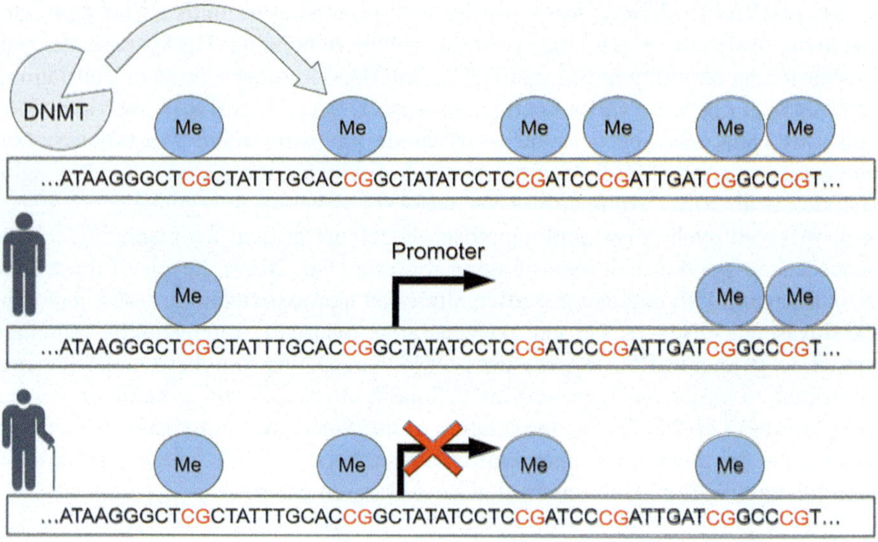

Fig. 3 Methylation profiles and their change during aging. Top: the enzyme DNA methyltransferase (DNMT) adds and removes methyl groups (Me) from cytosine residues in CpG couples (highlighted in red). Center: a healthy, biologically young methylome allows for the activation of a promoter for the transcription of a downstream gene. Bottom: an aged methylome has a different distribution of methylated sites, and some may compromise the transcription of genes relevant to the attainment of healthy longevity

balanced expression of genes involved in cell proliferation, DNA repair, and apoptosis (Winnefeld and Lyko 2012; Gentilini et al. 2013). It is also paramount to highlight that dysregulation of DNA methylation is in itself a biomarker of disease, indicative of immune exhaustion and metabolic imbalance (Caldwell et al. 2024): epigenetic modifications may introduce pathological alterations that may themselves alter epigenetic homeostasis, in an ideally catastrophic cycle of functional disruption over time (Fig. 3).

Indeed, research has demonstrated that the overall methylation landscape changes with age, leading to a phenomenon often referred to as "epigenetic drift." This drift involves both hypermethylation and hypomethylation of specific genes and regions of the genome, ultimately contributing to a decline in cellular function and loss of homeostasis (Li and Tollefsbol 2016). For instance, hypermethylation of tumor suppressor genes can promote cancer progression, while hypomethylation of oncogenes can enhance cancer risk (Yu et al. 2021; Allegra et al. 2023). The interplay between methylation and aging is complex, with studies indicating that interventions aimed at modulating methylation patterns could have profound effects on healthspan and lifespan (Fig. 3). In fact, lifestyle factors such as diet, exercise, and exposure to toxins can influence DNA methylation as a mediator of the impact of environmental fluctuations on DNA (Rajado et al. 2023). As an example, a diet rich in folate and other methyl donors has been associated with healthier methylation patterns and sustained recovery during illness, while deficiencies in these nutrients can contribute to adverse methylation changes (Anderson et al. 2012; De Conti and Pogribny 2017).

The role of methylation in aging has also provided the shape for a novel interpretation of time, where chronological age and biological age at the level of single tissues do not match anymore. Indeed, the concept of "epigenetic clocks" has emerged as a powerful tool for measuring biological age, providing insights into the aging process and the effectiveness of antiaging interventions (Kabacik et al. 2022; Duan et al. 2022). These clocks are essentially biochemical tests that entail the systematic analysis of DNA methylation patterns at specific loci in the genome, revealing that certain methylation signatures are well correlated with chronological age (Levine 2020).

The development of epigenetic clocks, such as the Horvath clock and the Hannum clock, has revolutionized the study of aging and longevity. The Horvath clock (Horvath and Raj 2018), for instance, was constructed using a large dataset encompassing diverse tissues and species, identifying 353 CpG sites which methylation levels serve as reliable predictors of biological age. This clock demonstrates remarkable predictive accuracy and can be applied to blood, skin, brain, and other tissue samples, enabling researchers to estimate an individual's biological age relative to their chronological age. The Hannum clock, on the other hand, focuses specifically on blood-based methylation data, providing insights into aging as it pertains to hematopoietic tissues (Hannum et al. 2013). Both clocks have proven effective in elucidating the relationships between biological age and health outcomes, revealing that deviations from expected biological age—termed "epigenetic age acceleration"—are associated with a higher risk of age-related diseases (Faul et al. 2023; Zavala et al. 2024).

Centenarian-based epigenetic clocks are being continually developed (Dec et al. 2023) and utilized in clinical practice: research has begun to incorporate epigenetic age measurements to evaluate the impact of diseased states, pharmacological agents, supplements, and lifestyle modifications on biological aging (Nwanaji-Enwerem and Mair 2022; Lo and Lin 2022). For instance, the use of senolytic drugs, which selectively target and eliminate senescent cells, has been shown to induce significant changes in epigenetic age, suggesting that rejuvenation at the cellular level may translate into a slower biological aging process (Lee et al. 2024; Martínez-Magaña et al. 2024). Similarly, studies investigating the effects of compounds like resveratrol and metformin have reported changes in DNA methylation patterns that correspond to decreased epigenetic age in specific tissues, illustrating the potential of these compounds as longevity-promoting agents (Marra et al. 2023; Zhang and Kiarasi 2024).

In addition to therapeutic implications, the concept of epigenetic aging poses intriguing questions about the interplay between genetics and environment in determining lifespan and healthspan. The dynamic nature of the epigenome suggests that aging is not a predetermined outcome but rather a process influenced by a myriad of factors, including lifestyle choices, stress exposure, and socioeconomic conditions. The idea of epigenetic plasticity allows for the possibility that interventions during critical periods of development or throughout life can lead to lasting changes in epigenetic marks, potentially altering the trajectory of aging. This concept aligns with the emerging field of epigenetic reprogramming, where researchers are exploring methods to reverse age-related epigenetic changes, offering promising avenues for future therapies aimed at maintaining and extending a healthy lifespan (Zhang et al. 2023; Pereira et al. 2024).

4.2 Mitochondrial Methylation and Longevity

The mitochondrion plays a crucial role in cellular energy metabolism and has emerged as a significant area of study in the context of healthy aging. Unlike nuclear DNA, however, mtDNA is inherited maternally and is located in this cellular organelle. Research has shown that the accumulation of mtDNA mutations correlates with aging and age-related diseases and that the processes of aging and mutation accumulation have mutual effects (Chocron et al. 2019; Kong et al. 2022). Age-related increases in mtDNA mutations contribute to mitochondrial dysfunction, leading to decreased ATP production and increased oxidative stress (Nissanka and Moraes 2018); these changes can further impair cellular function and promote both cellular senescence and the manifestation of various age-related pathologies (Miwa et al. 2022; Lee et al. 2023b). The concept of mitochondrial dysfunction in the context of the lifespan has been framed through the mitochondrial theory of aging, which posits that mitochondrial malfunction is the primary contributor to the aging process and age-related diseases, independently of ROS generation during mitochondrial respiration (Sun et al. 2016; Amorim et al. 2022). Indeed, while excessive ROS can cause damage to mtDNA, proteins, and lipids, leading to cellular senescence and death, it has also been noted that the presence of ROS does stimulate the repair mechanisms of the cell, therefore inducing regular cell clearance and delaying senescence (Maldonado et al. 2023; Karagianni and Bazopoulou 2024). This provides a double interpretation for the role of ROS and mitochondrial respiration in determining the aging phenotype.

Mitochondrial methylation in itself is a rare and still debated process in humans, as the addition of methyl groups to mtDNA appears to be almost impossible (Mechta et al. 2017). However, mtDNA methylation patterns seem to influence gene expression and mitochondrial function, as recent studies suggest that minute differences between individuals and alterations in their mtDNA methylation status may contribute to the aging process and the onset of diseases (Bacalini et al. 2017; Stoccoro and Coppedè 2021; Mongelli et al. 2023). The decline in mitochondrial function can be further exacerbated by increased methylation of mtDNA in older individuals, indicating a potential mechanism through which methylation may influence aging and mitochondrial dysfunction (Stoccoro and Coppedè 2021). So far, however, few studies have attempted to define biomarkers of aging on mtDNA, highlighting a lack of specific literature on centenarians that could help in determining overall methylation patterns suggestive of a functional trend in these cohorts, with respect to younger individuals.

5 An Omnigenic, Multiomic, and Systemic Understanding of Aging and Longevity

The study of aging and longevity is at the forefront of genetics and genomic research, with polygenic and omnigenic theories offering foundational frameworks for understanding the intricate networks of genes and environmental interactions affecting these complex traits (Lappalainen et al. 2024). Aging and longevity are indeed influenced by numerous biological pathways, genetic predispositions, and environmental factors (Castruita et al. 2022), and the current polygenic model suggests that these traits result from the immediate cumulative effects of many small genetic variants across the genome, with each variant contributing modestly to the overall phenotype (Crouch and Bodmer 2020). This model aligns well with observations in aged cohorts, where numerous genes have been linked to pathways such as DNA repair, cellular senescence, and metabolic processes that cumulatively influence lifespan in a direct manner (Jabalameli et al. 2024). Meanwhile, the more recent omnigenic theory expands on this view and posits that in highly complex traits like aging and longevity, nearly all genes in the genome may have some level of influence (Boyle et al. 2017; Mathieson 2021). This theory widens its scope beyond the traditionally implicated genes, suggesting that peripheral genes, which are indirectly involved through their interactions with core aging-related genes, also play a significant role in modulating complex phenotypes. More specifically, the polygenic model of aging is supported by studies indicating that specific genetic variants, often with small effect sizes, can cumulatively affect lifespan and aging-related phenotypes, as genome-wide association studies (GWAS) have identified numerous common loci associated with aging-related traits, including genes involved in lipid metabolism, immune response, and inflammatory pathways (Deelen et al. 2019; Bae et al. 2022). These findings emphasize the polymorphic and polygenic nature of aging, as the cumulative effect of variants in these genes contributes to an individual's aging trajectory and susceptibility to age-associated diseases: although each variant may only marginally impact longevity, their combined effects produce observable differences in aging phenotypes. The omnigenic model pushes the limit even further and suggests that, beyond genes directly affecting longevity with variable intensity, an extensive network of peripheral genes may also contribute to aging-related phenotypes by influencing core genes through regulatory and protein-protein interactions. This model proposes that genetic variants with no obvious connection to aging pathways can still influence longevity by modulating the cellular environment in which core aging genes operate. For instance, genes involved in basic cellular functions like transcriptional regulation, RNA processing, and protein folding may indirectly influence the stability and efficiency of proteins central to aging processes. Through complex regulatory networks, these peripheral genes exert downstream effects on metabolic stability, DNA repair fidelity, and inflammation, which are pivotal in aging biology. This is particularly relevant given the central role of cellular homeostasis in aging, where even subtle disruptions in unrelated pathways can lead to age-related cellular dysfunction (Hartl 2016). Therefore, considering the vast network of indirect genetic effects, the omnigenic model provides a broader view that encapsulates the systemic nature of aging as an emergent property of an interconnected gene network, where nearly every gene in the genome has the potential to influence the aging process through a cascade of molecular interactions.

In applying the polygenic and omnigenic theories to aging research, one critical area of focus is the identification and mapping of genetic variants associated with the longevity phenotype across diverse populations. Polygenic risk scores (PRS), derived from GWAS data, can help quantify an individual's genetic predisposition to aging-related diseases and longevity (Revelas et al. 2023; Don et al. 2024). These scores aggregate the effects of numerous variants, generating a cumulative measure of risk or resilience that can be applied across populations. Although PRS has been informative, especially in understanding the genetic architecture of aging-related diseases, its utility in predicting longevity outcomes is limited. This limitation is due, in part, to the modest effect sizes of most aging-related variants and the context-dependent nature of these effects, which are often influenced by environmental factors such as diet, exercise, and stress. Incorporating a broader view into PRS could improve predictive accuracy by accounting for indirect genetic effects and gene-gene interactions, and this could be achieved by integrating functional genomic data, such as gene expression and regulatory networks, into PRS calculations. Such an approach would be able to include not only the contribution of core aging genes but also the broader network of peripheral genes that interact with aging pathways, enhancing the ability of PRS to capture complex aging-related phenotypes.

Another critical application of these theories is in understanding the role of epigenetics in aging. Aging is characterized by dynamic changes in the epigenome, including DNA methylation, histone modifications, and chromatin remodeling, which regulate gene expression patterns across the lifespan. Epigenetic changes, while not altering the DNA sequence, do modulate the expression of both core and peripheral genes, making epigenetic regulation a key interface between genes and the longevity phenotype. Furthermore, DNA methylation age is a widely studied biomarker that reflects biological age more accurately than chronological age, correlating with health outcomes and mortality risk. By integrating epigenomic data with PRS in an omnigenic and multiomic framework, the broader context of the variability of individual complex aging phenotypes could be better elucidated.

The application of an omnigenic setup also extends to the investigation of environmental influences on aging, filtered through the lens of methylation. The gene-environment interactions that shape aging outcomes can be studied by examining how lifestyle factors, such as diet, physical activity, and exposure to toxins, influence the expression and function of aging-related genes through their methylation patterns. In a polygenic/omnigenic context, specific genetic variants coupled with peculiar methylation profiles may predispose individuals to benefit more from healthy behaviors or be more vulnerable to environmental stressors, as environmental factors may influence peripheral genes, which, in turn, affect core aging pathways, and this interaction may occur indirectly, through alterations in liminal gene expression networks, protein-protein contacts, and metabolic pathways unrelated to aging itself.

In practice, the insights from an omnigenic approach are critical for developing personalized approaches to aging and longevity: the ability to predict an individual's aging trajectory and susceptibility to age-related diseases could become a routine part of precision medicine, as integrated omnigenic models may facilitate early identification of high-risk individuals, guiding interventions to mitigate age-related outcomes. In conclusion, what emerges is the importance of a systems-level approach to aging, suggesting that rapid interventions targeting broad cellular processes through large-scale modulation of methylation, rather than individual genes, may be more effective in promoting healthy aging as a desirable complex phenotype.

References

Abondio P, Sazzini M, Garagnani P et al (2019) The genetic variability of APOE in different human populations and its implications for longevity. Genes (Basel) 10:E222

Ajoolabady A, Pratico D, Tang D et al (2024) Immunosenescence and inflammaging: mechanisms and role in diseases. Ageing Res Rev 101:102540

Allegra A, Caserta S, Mirabile G et al (2023) Aging and age-related epigenetic drift in the pathogenesis of leukemia and lymphomas: new therapeutic targets. Cells 12:2392

Amorim JA, Coppotelli G, Rolo AP et al (2022) Mitochondrial and metabolic dysfunction in ageing and age-related diseases. Nat Rev Endocrinol 18:243–258
Anaya J-M, Lozada-Martinez ID, Torres I et al (2024) Autoimmunity in centenarians. A paradox. J Transl Autoimmun 8:100237
Anderson OS, Sant KE, Dolinoy DC (2012) Nutrition and epigenetics: an interplay of dietary methyl donors, one-carbon metabolism and DNA methylation. J Nutr Biochem 23:853–859
Bacalini MG, D'Aquila P, Marasco E et al (2017) The methylation of nuclear and mitochondrial DNA in ageing phenotypes and longevity. Mech Ageing Dev 165:156–161
Badal VD, Vaccariello ED, Murray ER et al (2020) The gut microbiome, aging, and longevity: a systematic review. Nutrients 12:3759
Bae H, Gurinovich A, Malovini A et al (2018) Effects of FOXO3 polymorphisms on survival to extreme longevity in four centenarian studies. J Gerontol A Biol Sci Med Sci 73:1439–1447
Bae H, Gurinovich A, Karagiannis TT et al (2022) A genome-wide association study of 2304 extreme longevity cases identifies novel longevity variants. Int J Mol Sci 24:116
Borghesan M, Hoogaars WMH, Varela-Eirin M et al (2020) A senescence-centric view of aging: implications for longevity and disease. Trends Cell Biol 30:777–791
Borrás C, Serna E, Gambini J et al (2017) Centenarians maintain miRNA biogenesis pathway while it is impaired in octogenarians. Mech Ageing Dev 168:54–57
Borras C, Ingles M, Mas-Bargues C et al (2020) Centenarians: an excellent example of resilience for successful ageing. Mech Ageing Dev 186:111199
Boyle EA, Li YI, Pritchard JK (2017) An expanded view of complex traits: from polygenic to omnigenic. Cell 169:1177–1186
Calabrò A, Accardi G, Aiello A et al (2024) Senotherapeutics to counteract senescent cells are prominent topics in the context of anti-ageing strategies. Int J Mol Sci 25:1792
Caldwell BA, Wu Y, Wang J et al (2024) Altered DNA methylation underlies monocyte dysregulation and immune exhaustion memory in sepsis. Cell Rep 43:113894
Caruso C, Ligotti ME, Accardi G et al (2022) How important are genes to achieve longevity? Int J Mol Sci 23:5635
Castruita PA, Piña-Escudero SD, Rentería ME et al (2022) Genetic, social, and lifestyle drivers of healthy aging and longevity. Curr Genet Med Rep 10:25–34
Chocron ES, Munkácsy E, Pickering AM (2019) Cause or casualty: the role of mitochondrial DNA in aging and age-associated disease. Biochim Biophys Acta Mol Basis Dis 1865:285–297
Crouch DJM, Bodmer WF (2020) Polygenic inheritance, GWAS, polygenic risk scores, and the search for functional variants. Proc Natl Acad Sci USA 117:18924–18933
Dai Z, Lee SY, Sharma S et al (2024) A systematic review of diet and medication use among centenarians and near-centenarians worldwide. Geroscience 46:6625–6639
De Conti A, Pogribny IP (2017) Epigenetics of dietary methyl-group donor deficiency and liver cancer. In: Patel V, Preedy V (eds) Handbook of nutrition, diet, and epigenetics. Springer International Publishing, Cham, pp 1–16
Dec E, Clement J, Cheng K et al (2023) Centenarian clocks: epigenetic clocks for validating claims of exceptional longevity. Geroscience 45:1817–1835
Deelen J, Evans DS, Arking DE et al (2019) A meta-analysis of genome-wide association studies identifies multiple longevity genes. Nat Commun 10:3669
Don J, Schork AJ, Glusman G et al (2024) The relationship between 11 different polygenic longevity scores, parental lifespan, and disease diagnosis in the UK Biobank. Geroscience 46:3911–3927
Duan R, Fu Q, Sun Y et al (2022) Epigenetic clock: a promising biomarker and practical tool in aging. Ageing Res Rev 81:101743
Faul JD, Kim JK, Levine ME et al (2023) Epigenetic-based age acceleration in a representative sample of older Americans: associations with aging-related morbidity and mortality. Proc Natl Acad Sci USA 120:e2215840120

Fields M, Marcuzzi A, Gonelli A et al (2023) Mitochondria-targeted antioxidants, an innovative class of antioxidant compounds for neurodegenerative diseases: perspectives and limitations. Int J Mol Sci 24:3739

Franceschi C, Bonafè M, Valensin S et al (2000) Inflamm-aging. An evolutionary perspective on immunosenescence. Ann N Y Acad Sci 908:244–254

Franceschi C, Garagnani P, Olivieri F et al (2020) The contextualized genetics of human longevity. J Am Coll Cardiol 75:968–979

Friedman E, Franks M, Teas E et al (2024) Social connectedness, functional capacity, and longevity: a focus on positive relations with others. Soc Sci Med 340:116419

Fulop T, Larbi A, Pawelec G et al (2023) Immunology of aging: the birth of Inflammaging. Clin Rev Allergy Immunol 64:109–122

Fulzele S, Mendhe B, Khayrullin A et al (2019) Muscle-derived miR-34a increases with age in circulating extracellular vesicles and induces senescence of bone marrow stem cells. Aging 11:1791–1803

Garagnani P, Giuliani C, Pirazzini C et al (2013) Centenarians as super-controls to assess the biological relevance of genetic risk factors for common age-related diseases: a proof of principle on type 2 diabetes. Aging (Albany NY) 5:373–385

Garagnani P, Marquis J, Delledonne M et al (2021) Whole-genome sequencing analysis of semi-supercentenarians. eLife 10:e57849

Gentilini D, Mari D, Castaldi D et al (2013) Role of epigenetics in human aging and longevity: genome-wide DNA methylation profile in centenarians and centenarians' offspring. Age 35:1961–1973

Giorgi C, Marchi S, Simoes ICM et al (2018) Mitochondria and reactive oxygen species in aging and age-related diseases. Int Rev Cell Mol Biol 340:209–344

Giuliani C, Pirazzini C, Delledonne M et al (2017) Centenarians as extreme phenotypes: an ecological perspective to get insight into the relationship between the genetics of longevity and age-associated diseases. Mech Ageing Dev 165:195–201

Giuliani C, Garagnani P, Franceschi C (2018) Genetics of human longevity within an eco-evolutionary nature-nurture framework. Circ Res 123:745–772

Gonzalez B, Tare A, Ryu S et al (2023) High-throughput sequencing analysis of nuclear-encoded mitochondrial genes reveals a genetic signature of human longevity. Geroscience 45:311–330

Hannum G, Guinney J, Zhao L et al (2013) Genome-wide methylation profiles reveal quantitative views of human aging rates. Mol Cell 49:359–367

Hartl FU (2016) Cellular homeostasis and aging. Annu Rev Biochem 85:1–4

Horvath S, Raj K (2018) DNA methylation-based biomarkers and the epigenetic clock theory of ageing. Nat Rev Genet 19:371–384

Islam MA, Sehar U, Sultana OF et al (2024) SuperAgers and centenarians, dynamics of healthy ageing with cognitive resilience. Mech Ageing Dev 219:111936

Jabalameli MR, Lin J-R, Zhang Q et al (2024) Polygenic prediction of human longevity on the supposition of pervasive pleiotropy. Sci Rep 14:19981

Jiang Q, Yin J, Chen J et al (2020) Mitochondria-targeted antioxidants: a step towards disease treatment. Oxidative Med Cell Longev 2020:1–18

Kabacik S, Lowe D, Fransen L et al (2022) The relationship between epigenetic age and the hallmarks of aging in human cells. Nat Aging 2:484–493

Karagianni C, Bazopoulou D (2024) Redox regulation in lifespan determination. J Biol Chem 300:105761

Koch L (2023) microRNAs as systemic regulators of ageing. Nat Rev Genet 24:415–415

Kong M, Guo L, Xu W et al (2022) Aging-associated accumulation of mitochondrial DNA mutations in tumor origin. Life Med 1:149–167

Lantz PM (2020) Revisiting compression of morbidity and health disparities in the 21st century. Milbank Q 98:664–667

Lappalainen T, Li YI, Ramachandran S et al (2024) Genetic and molecular architecture of complex traits. Cell 187:1059–1075

Lee W, Zamudio-Ochoa A, Buchel G et al (2023a) Molecular basis for maternal inheritance of human mitochondrial DNA. Nat Genet 55:1632–1639

Lee YH, Kuk MU, So MK et al (2023b) Targeting mitochondrial oxidative stress as a strategy to treat aging and age-related diseases. Antioxidants 12:934

Lee E, Carreras-Gallo N, Lopez L et al (2024) Exploring the effects of Dasatinib, Quercetin, and Fisetin on DNA methylation clocks: a longitudinal study on senolytic interventions. Aging (Albany NY) 16:3088–3106

Levine ME (2020) Assessment of epigenetic clocks as biomarkers of aging in basic and population research. J Gerontol A Biol Sci Med Sci 75:463–465

Li Y, Tollefsbol TO (2016) Age-related epigenetic drift and phenotypic plasticity loss: implications in prevention of age-related human diseases. Epigenomics 8:1637–1651

Livingston G, Huntley J, Liu KY et al (2024) Dementia prevention, intervention, and care: 2024 report of the Lancet standing Commission. Lancet 404:572–628

Lo Y-H, Lin W-Y (2022) Cardiovascular health and four epigenetic clocks. Clin Epigenetics 14:73

Maldonado E, Morales-Pison S, Urbina F et al (2023) Aging hallmarks and the role of oxidative stress. Antioxidants 12:651

Marra PS, Yamanashi T, Crutchley KJ et al (2023) Metformin use history and genome-wide DNA methylation profile: potential molecular mechanism for aging and longevity. Aging 15:601–616

Martin P, Poon LW, Georgia Centenarian Study (2024) Healthy aging among centenarians and near-centenarians: findings from the Georgia Centenarian Study. Maturitas 185:108001

Martínez-Magaña JJ, Hurtado-Soriano J, Rivero-Segura NA et al (2024) Towards a novel frontier in the use of epigenetic clocks in epidemiology. Arch Med Res 55:103033

Mathieson I (2021) The omnigenic model and polygenic prediction of complex traits. Am J Hum Genet 108:1558–1563

Mechta M, Ingerslev LR, Fabre O et al (2017) Evidence suggesting absence of mitochondrial DNA methylation. Front Genet 8:166

Miwa S, Kashyap S, Chini E et al (2022) Mitochondrial dysfunction in cell senescence and aging. J Clin Invest 132:e158447

Mongelli A, Mengozzi A, Geiger M et al (2023) Mitochondrial epigenetics in aging and cardiovascular diseases. Front Cardiovasc Med 10:1204483

Nissanka N, Moraes CT (2018) Mitochondrial DNA damage and reactive oxygen species in neurodegenerative disease. FEBS Lett 592:728–742

Nwanaji-Enwerem JC, Mair WB (2022) Redefining age-based screening and diagnostic guidelines: an opportunity for biological aging clocks in clinical medicine? Lancet Health Longev 3:e376–e377

Olivieri F, Prattichizzo F, Giuliani A et al (2021) miR-21 and miR-146a: the microRNAs of inflammaging and age-related diseases. Ageing Res Rev 70:101374

Pang S, Chen X, Lu Z et al (2023) Longevity of centenarians is reflected by the gut microbiome with youth-associated signatures. Nat Aging 3:436–449

Pereira B, Correia FP, Alves IA et al (2024) Epigenetic reprogramming as a key to reverse ageing and increase longevity. Ageing Res Rev 95:102204

Pimlott N (2023) Compression of morbidity is dead; long live compression of morbidity. Can Fam Physician 69:78

Puca AA, Spinelli C, Accardi G et al (2018) Centenarians as a model to discover genetic and epigenetic signatures of healthy ageing. Mech Ageing Dev 174:95–102

Rad AN, Grillari J (2024) Current senolytics: mode of action, efficacy and limitations, and their future. Mech Ageing Dev 217:111888

Rajado AT, Silva N, Esteves F et al (2023) How can we modulate aging through nutrition and physical exercise? An epigenetic approach. Aging (Albany NY) 15:3191–3217

Revelas M, Thalamuthu A, Zettergren A et al (2023) High polygenic risk score for exceptional longevity is associated with a healthy metabolic profile. Geroscience 45:399–413

Ruby JG, Wright KM, Rand KA et al (2018) Estimates of the heritability of human longevity are substantially inflate due to assortative mating. Genetics 210:1109–1124

Salameh Y, Bejaoui Y, El Hajj N (2020) DNA methylation biomarkers in aging and age-related diseases. Front Genet 11:171

Sebastiani P, Bae H, Sun FX et al (2013) Meta-analysis of genetic variants associated with human exceptional longevity. Aging (Albany NY) 5:653–661

Smulders L, Deelen J (2024) Genetics of human longevity: from variants to genes to pathways. J Intern Med 295:416–435

Stoccoro A, Coppedè F (2021) Mitochondrial DNA methylation and human diseases. IJMS 22:4594

Sun N, Youle RJ, Finkel T (2016) The mitochondrial basis of aging. Mol Cell 61:654–666

Tamvaka N, Heckman MG, Johnson PW et al (2024) Associations of mitochondrial genomic variation with successful neurological aging. Mitochondrion 78:101948

Tesi N, van der Lee S, Hulsman M et al (2024) Cognitively healthy centenarians are genetically protected against Alzheimer's disease. Alzheimers Dement 20:3864–3875

Thalén A, Ledberg A (2024) Consequences of heterogeneity in aging: parental age at death predicts midlife all-cause mortality and hospitalization in a Swedish national birth cohort. BMC Geriatr 24:207

Van Den Berg N, Rodríguez-Girondo M, Van Dijk IK et al (2019) Longevity defined as top 10% survivors and beyond is transmitted as a quantitative genetic trait. Nat Commun 10:35

Vergani C, Lucchi T, Caloni M et al (2006) I405V polymorphism of the cholesteryl ester transfer protein (CETP) gene in young and very old people. Arch Gerontol Geriatr 43:213–221

Vitale G, Pellegrino G, Vollery M et al (2019) ROLE of IGF-1 system in the modulation of longevity: controversies and new insights from a centenarians' perspective. Front Endocrinol 10:27

Wang B, Han J, Elisseeff JH et al (2024) The senescence-associated secretory phenotype and its physiological and pathological implications. Nat Rev Mol Cell Biol 25:958–978

Winnefeld M, Lyko F (2012) The aging epigenome: DNA methylation from the cradle to the grave. Genome Biol 13:165

Ying K, Castro JP, Shindyapina AV et al (2024) Depletion of loss-of-function germline mutations in centenarians reveals longevity genes. Nat Commun 15:9030

Yu C, Wong EM, Joo JE et al (2021) Epigenetic drift association with cancer risk and survival, and modification by sex. Cancers 13:1881

Zavala DV, Dzikowski N, Gopalan S et al (2024) Epigenetic age acceleration and chronological age: associations with cognitive performance in daily life. J Gerontol A Biol Sci Med Sci 79:glad242

Zhang S, Kiarasi F (2024) Therapeutic effects of resveratrol on epigenetic mechanisms in age-related diseases: a comprehensive review. Phytother Res 38:2347–2360

Zhang J, Wang S, Liu B (2023) New insights into the genetics and epigenetics of aging plasticity. Genes (Basel) 14:329

Zhao Y, Liu Y-S (2021) Longevity factor FOXO3: a key regulator in aging-related vascular diseases. Front Cardiovasc Med 8:778674

Zhao D, Wang Y, Wong ND et al (2024) Impact of aging on cardiovascular diseases: from chronological observation to biological insights: JACC family series. JACC Asia 4:345–358

Epigenetic Parameters as Potential Prognostic and Therapeutic Factors in Neurodegenerative Diseases

Bartosz Słowikowski, Wojciech Owecki, Joanna Poszwa, Michał Jezierski, Michał Draguła, Ulyana Goutor, Paweł Jagodziński, Wojciech Kozubski, and Jolanta Dorszewska

Contents

Abstract Genetic and epigenetic factors are crucial for the proper functioning of the nervous system. Neurodegeneration is a consequence of dysregulation of both the genome and epigenome. Epigenetic processes, such as DNA methylation, histone modifications, or RNA interference, can lead to changes in protein expression, which results in the induction of pathological processes. Studies confirm the participation of these processes in the pathogenesis of diseases such as Parkinson's disease, Alzheimer's disease, stroke, epilepsy, and other neurological diseases. Moreover, epigenetic factors, due to their ability to regulate protein expression, rep-

B. Słowikowski · P. Jagodziński
Department of Biochemistry and Molecular Biology, Poznan University of Medical Sciences, Poznan, Poland

W. Owecki · J. Poszwa · M. Jezierski · M. Draguła · U. Goutor · J. Dorszewska (✉)
Laboratory of Neurobiology, Department of Neurology, Poznan University of Medical Sciences, Poznan, Poland
e-mail: jolanta.dorszewska@ump.edu.pl

W. Kozubski
Department of Neurology, Poznan University of Medical Sciences, Poznan, Poland

J. Barciszewski (ed.), *Decoding Aging and Neurodegeneration*,
RNA Technologies 16, https://doi.org/10.1007/978-3-031-91072-2_7

resent a promising therapeutic target for neurological diseases. In this chapter, we will discuss epigenetic factors with particular emphasis on their role in modifying the expression of genes involved in the pathogenesis of the mentioned diseases.

Keywords Epigenetics · DNA methylation · Histone modification · RNA activity · Neurodegeneration · Neurological diseases

1 Introduction

Neurodegenerative diseases are progressive disorders characterized mostly by the loss of neuronal function and structure, leading to several disabilities, including cognitive, motor, and sensory impairments. As the global population ages, the prevalence of these disorders has steadily increased, creating an urgent need for more effective diagnostic and therapeutic strategies (Van Schependom and D'haeseleer 2023). Despite significant advances in understanding the molecular mechanisms underlying neurodegeneration, current treatment strategies, although developing rapidly, mainly address symptoms without inhibiting or reversing disease progression. Moreover, early diagnostic biomarkers remain elusive, limiting the possibilities of accurate and timely actions. In this context, epigenetic factors have emerged as key players in the pathogenesis of neurodegenerative disorders and represent a promising target for both as prognostic and therapeutic approach.

Epigenetics refers to heritable changes in gene expression that do not involve alterations in the DNA sequence itself. These changes are mediated by a variety of mechanisms, including DNA methylation, histone modifications, and activity of noncoding RNAs such as microRNA (miRNA), which collectively regulate gene activity in response to environmental and cellular signals (Fig. 1). Unlike genetic mutations, epigenetic modifications are reversible, making them attractive targets for therapeutic intervention.

In neurodegenerative diseases, growing body of evidence suggests that aberrant epigenetic regulation plays a critical role in neuronal dysfunction and degeneration. For example, abnormal patterns of DNA methylation have been observed in key genes involved in neuroinflammation, synaptic plasticity, and mitochondrial function, all of which may be disrupted in diseases like Alzheimer's disease (AD) and Parkinson's disease (PD) (Komada and Nishimura 2022). Similarly, dysregulation of histone modifications and noncoding RNA networks has been implicated in the progression of these diseases (Jużwik et al. 2019). The dynamic and reversible nature of epigenetic modifications holds great promise for their application in both the prognosis and therapy of neurodegenerative diseases.

Using epigenetic markers as prognostic biomarkers seems to be an inventible tool; thus DNA methylation, histone modifications, and noncoding RNAs can provide valuable insights into disease progression and responses to environmental factors. For instance, it has been demonstrated that global levels of both DNA methylation and hydroxymethylation are decreased in patients with

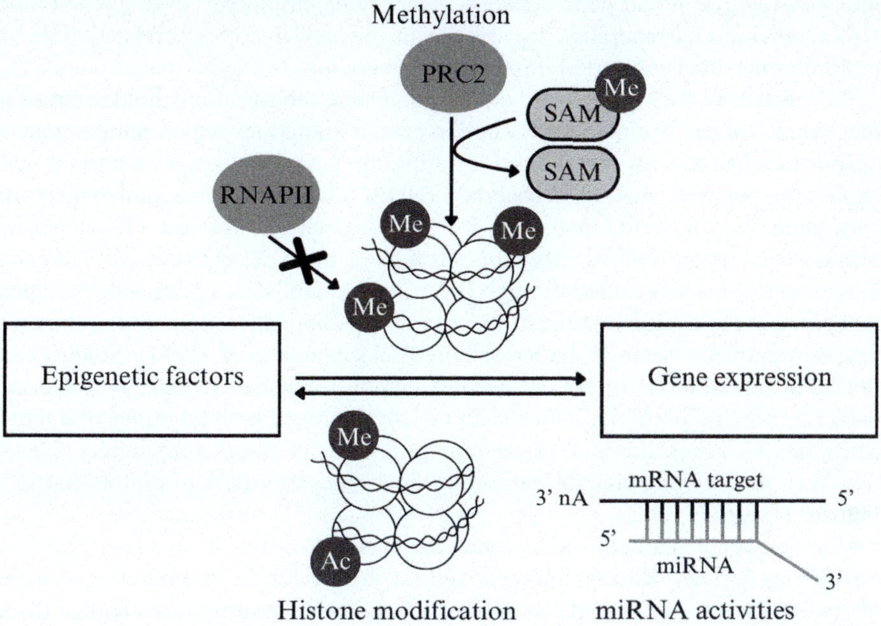

Fig. 1 The role of epigenetic factors in changing gene expression. *RNAPII* RNA polymerase II, *PRC2* polycomb repressive complex 2, *SAM S*-adenosylmethionine, *Me* methylation, *Ac* acetylation. (Created with Canva Pty Ltd.)

neurodegenerative disorders and age-related cerebrovascular disease (Martínez-Iglesias et al. 2020). Additionally, a significant correlation between age and DNA methylation in PD subjects has been identified.

Other studies investigating the status of DNA methyltransferase 1 (DNMT1)—the enzyme responsible for maintaining methylation patterns during DNA replication—have revealed alterations in AD. In postmortem brain tissue, DNMT1 expression was reduced in neurons within the entorhinal cortex layer II and hippocampus. These changes are accompanied by an overall reduction in two key epigenetic DNA modifications: 5-methylcytosine (5mC) and 5-hydroxymethylcytosine (5hmC), both of which are crucial for gene expression regulation (Chouliaras et al. 2013).

Another promising group of biomarkers are miRNAs, small noncoding RNAs that regulate gene expression posttranscriptionally. Alterations in miRNA expression profiles have been identified in neurodegenerative diseases. The most frequently dysregulated miRNAs in neurological diseases and their models include miR-9-5p, miR-21-5p, the miR-29 family, miR-132-3p, miR-124-3p, miR-146a-5p, miR-155-5p, and miR-223-3p (Juźwik et al. 2019). Analysis of the pathways targeted by aforementioned miRNAs reveals significant functional overlap across different neurodegenerative conditions. Additionally, each miRNA appears to play a pivotal role in both the neural and immune components of these diseases. Since miRNAs regulate broad gene networks, identifying commonly dysregulated miRNAs across neurodegenerative diseases can inspire new hypotheses related to shared molecular mechanisms underlying neurodegeneration.

In addition to their prognostic utility, epigenetic modifications hold promise as therapeutic targets. As mentioned, unlike genetic mutations which are permanent, epigenetic changes can be reversed, offering the potential to restore normal gene expression patterns. Histone deacetylase (HDAC) inhibitors have garnered significant attention as potential treatments for neurodegenerative diseases. By increasing histone acetylation, HDAC inhibitors promote a more open chromatin structure. Targeting histone acetylation through HDACi has emerged as a promising therapeutic approach for cancer treatment; therefore their application in neurodegenerative disease treatments seemed matter of time (Eckschlager et al. 2017). Studies conducted in animal models have consistently demonstrated that HDACi treatment can enhance synaptic plasticity, along with improvements in both cognitive and motor functions. However, the lack of selectivity in many of these compounds, coupled with their neurotoxic potential, raises concerns about the safety of prolonged use in humans (Coppede 2022).

Another epigenetically associated therapeutic approach involves targeting miRNA molecules. Recent findings suggest that restoring or inhibiting disease-altered miRNAs may provide promising strategies for treating neurological disorders. Techniques such as viral delivery and the administration of modified oligonucleotides designed to mimic or inhibit specific miRNAs have demonstrated potential in model systems. For example, miRNAs have been effectively utilized to target and inhibit specific transcripts, such as the huntingtin gene (*HTT*), which encodes a protein associated with Huntington's disease (HD) (Sogorb-Gonzalez et al. 2024). Additionally, employing "miRNA sponges"—structures that bind to overabundant miRNAs in certain conditions—represents another viable therapeutic strategy (Fu et al. 2019).

2 Epigenetic Factors in Parkinson's Disease

PD is one of the most common neurodegenerative disorders with a globally rising prevalence. The epidemiology of PD presents marked variations in age, geography, ethnicity, sex, or time. The global prevalence of PD has risen from 2.5 million in the 1990s to 6.1 million, and the prognosis indicates that this number will further increase. Recent estimates show that approximately 2–3% of the population above 65 years old is affected by PD (Kumar et al. 2022). Despite a 200-year-long research on this disorder, the current status of knowledge about PD pathogenesis and clinical phenotype remains limited. Evidence indicates that PD pathogenesis is associated with pathological aggregation of α-synuclein combined with the spreading of pathology into the gut, brainstem, or higher regions of the brain. On the other hand, focusing on the cellular level, abnormal lysosomal, endosomal, or mitochondrial functions were demonstrated in the PD (Morris et al. 2024). Furthermore, oxidative stress, decreased performance of the ubiquitin-proteasome system, neurotoxic substances, or energy disturbances in the cells have been linked with neurodegeneration in PD (Dorszewska et al. 2021). Another important factor of this disease is genetics. Usually, only 3–5% of PD cases can be associated with known PD genes (monogenic PD), while 90 genetic risk variants jointly elucidate up to 36% of the heritable risk of nonmonogenic PD (Bloem et al. 2021). Recently, an increasing number of studies show that epigenetic factors also participate in the pathogenesis of PD (Słowikowski et al. 2024). Unfortunately, although multiple mechanisms of PD pathogenesis are uncovered, currently, there is a lack of cure for this disorder and disease-modifying agents. Nevertheless, available treatment options include pharmacological therapy (mainly levodopa or a dopamine agonist), strenuous aerobic exercise, rehabilitation, complementary or alternative medicine substances, and surgery. Here, we will discuss epigenetic features as potential prognostic and therapeutic factors in PD.

The three main epigenetic alterations include DNA methylation, histone modification, and abnormal RNA (Słowikowski et al. 2024) (Fig. 2). A recent study focusing on DNA methylation analyzed patients over a period of time. At the initial phase, 5178 cytosine-guanine dinucleotides (CpGs) were differentially methylated in comparison to sporadic PD patients and healthy controls. Over 3 years, 579 CpGs of these previously mentioned underwent significant methylation changes. Multiple alterations in methylation were present especially near *CYP2E1* and *NDRG4* genes. The findings indicate that changes in DNA methylation reflect processes of ongoing neurodegeneration in PD and may serve as potential epigenetic biomarkers in this disease (Gonzalez-Latapi et al. 2024). Similarly, a differential methylation in seven

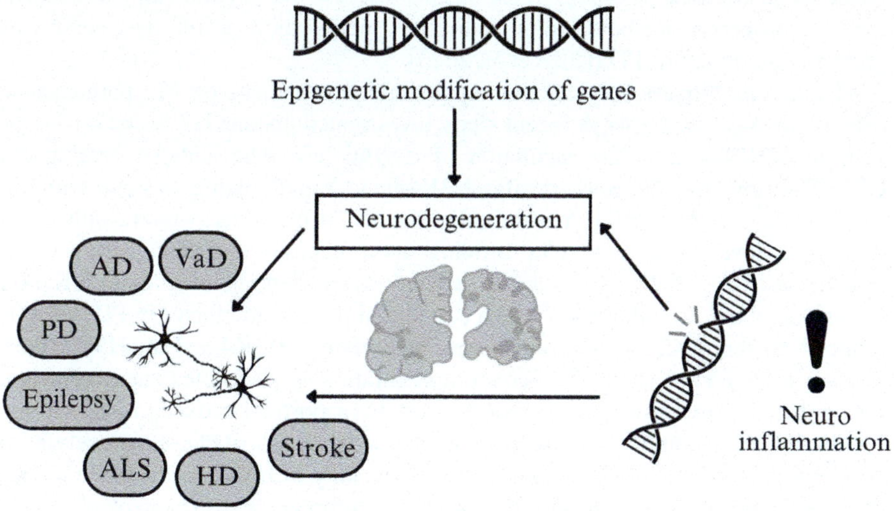

Fig. 2 The role of epigenetic factors in neurodegeneration. *AD* Alzheimer's disease, *PD* Parkinson's disease, *VaD* vascular dementia, *HD* Huntington's disease, *ALS* amyotrophic lateral sclerosis. (Created with Canva Pty Ltd.)

CpGs associated with faster decline and eight CpGs linked to faster motor progression have been identified, indicating that epigenetic alterations may be associated with PD progression and prognosis (Chuang et al. 2019). Another research utilized a fenpropathrin-exposed PD mouse model. Fenpropathrin is a pyrethroid insecticide used in agriculture and is known as a risk factor for PD. In this mouse model, researchers observed that hypermethylation-low expression of the Ambra1 gene intensified the neurodegeneration process via the Ambra1/Parkin/LC3B-mediated mitophagy pathway. Administration of 5-aza-2′-deoxycytidine (5-Aza-dC) increased expression of Ambra1, reducing mitophagy and exhibiting neuroprotective effects. Moreover, it proved that regulation of DNA methylation may contribute to the development of a novel treatment strategy for PD (He et al. 2024). Other approaches concentrated on Synuclein Alpha (*SNCA*) overexpression, specifically on DNA methylation at *SNCA* intron 1, known for its implication in PD pathogenesis. A system consisting of an all-in-one lentiviral vector based on CRISPR-deactivated Cas9 (dCas9) joined with the catalytic domain of DNA-methyltransferase 3A (DNMT3A) has been developed (Kantor et al. 2018). Five years later, a similar system comprising the same vector with CRISPR-dSaCas9 and guide RNA (gRNA) targeted at *SNCA* intron 1 combined with a synthetic repressor molecule of Krüppel-associated box (KRAB)/methyl CpG binding protein 2 (MeCp2) transcription repression domain (TRD) has been utilized (Sun et al. 2024). In both cases, *SNCA* expression decreased leading to the alleviation of disease-related cellular phenotypes (Kantor et al. 2018; Sun et al. 2024). Interesting findings also describe exercise and PD in the context of epigenetic alterations. Exercise significantly impacted methylation levels of the glycoprotein nonmetastatic melanoma protein B (GPNMB) cg17274742 in PD males but not in PD females. Furthermore, hypomethylation was observed in exercise PD men. These results suggest that a personalized diet with exercise programs attributed to an individual methylation profile may serve as a precision treatment of PD (Chen et al. 2023).

Histone modification is another epigenetic mechanism of PD pathogenesis (Słowikowski et al. 2024). A recent study investigated histone H3 lysine 4 trimethylation (H3K4me3) in the regulation of α-synuclein. The authors developed a CRISPR/dCas9-based locus-specific H3K4me3 demethylating system with the catalytic domain of JARID1A recruited to the *SNCA* promoter, which resulted in a decreased level of α-synuclein proving its potential therapeutic use in PD (Guhathakurta et al. 2021). On the other hand, it was demonstrated that GSK-J4, a histone demethylase inhibitor, may be a potential therapeutic factor in PD. GSK-J4 reduced the abnormal alterations of histone methylation (H3K4me3 and H3K27me3) in PD rats (Mu et al. 2020). Moreover, upregulation of H3K4me3 caused elevated ferroportin-1 expression and neuroprotection. Ferroportin-1 reduced increased ferrous iron (Fe^{2+}) accumulation, which is a recognized hallmark of PD. Hyperacetylation at several histone sites in PD brain tissues, especially at H3K27, which is a marker of active enhancers and promoters, has been demonstrated (Toker et al. 2021). Concomitantly, the authors observed a strong predilection of H3K27 hyperacetylation for genes related to PD (*SNCA*, Parkin RBR E3 Ubiquitin Protein Ligase, *PRKN*; Parkinsonism-associated deglycase, *DJ-1*; and microtubule-associated protein Tau, *MAPT*). These findings implicate an association between H3K27 hyperacetylation and PD, opening up possibilities for potential pharmacological treatment with histone deacetylase inhibitors (HDACi). Indeed, recently HDACi have been of great interest in terms of their potential therapeutic effects in PD (Zhang et al. 2024). Pan-HDACi and cinnamyl sulphonamide hydroxamate derivatives (NMJ-2 and NMJ-3) alleviated neurodegenerative effects after oral administration in PD rats. This result was achieved through the reduction of inflammation as well as oxidative stress and restoring the dopamine levels (Meka et al. 2023). Another study utilized a combined treatment comprising the p300/H3K27ac inhibitor GNE-049, resulting in reduced p300 expression and H3K27ac deposition, decreased inflammation-associated transcription factor Nos2, and elevated IRF4 and ARG1 levels. Furthermore, GNE-049 exhibited protective effects on mitochondria, strengthening its potential as a novel therapeutic strategy in PD (Huang et al. 2023). Neuroprotection was also observed in the treatment of PD cell models with LMK235, an HDAC4/5 inhibitor, suggesting its potential as a pharmacological agent (Mazzocchi et al. 2021). Yet another HDACi studied in PD mice models was valproic acid (VPA). VPA administration increased histone acetylation level, reduced activated microglia, and improved PD symptoms in mice, suggesting its possible therapeutic role in PD (Kim et al. 2019). Similarly, sodium butyrate, another HDACi, has been utilized. In a dopaminergic neuronal cell model, sodium butyrate mitigated DNA damage induced by wild-type α-synuclein, probably through upregulation of genes participating in DNA repair. On the other hand, another research investigated the relationship between histone acetylation and DNA methylation in PD. The authors observed that a histone deacetylase, silence information regulator 1 (SIRT1), deacetylates ten-eleven translocation 2 (TET2), a DNA hydroxylase upregulated in PD. This interaction leads to TET2 degradation and neuroprotection. Moreover, SIRT1 activation promotes the cyclin-dependent kinase inhibitor 2A (CDKN2A) DNA hypermethylation, enabling cell growth and proliferation. The complex SIRT1-TET2-CDKN2A axis provides a novel potential target for therapeutic intervention in PD (Li et al. 2021).

RNA is considered an important epigenetic factor in PD with potential therapeutic effects (Słowikowski et al. 2024). Long noncoding RNA (lncRNA) plays a crucial role in microglial activation in the pathogenesis of PD. LncRNA SOX21-AS1 intensified LPS/ATP-induced activation of BV2 cells, causing consequent nerve damage. On the other hand, SOCS3 was epigenetically inhibited by SOX21-AS1 via recruiting EZH2 to SOCS3's promoter. Inhibition of SOX21-AS1 alleviated PD features, indicating a potential target for therapy (Feng et al. 2024). Another research suggests lncRNA, HOXA-AS2, as a potential therapeutic option in PD. HOXA-AS2 induces neuroinflammation through the regulation of microglial polarization. The latter process occurs via HOXA-AS2 interaction with polycomb repressive complex 2 (PRC2) and histone methylation modulation of the promoter of peroxisome proliferator-activated receptor gamma coactivator-1a (PGC-1α) (Yang et al. 2021). Micro-RNA (miRs) also participate in the epigenetics of PD (Słowikowski et al. 2024). MiR-124 is considered a potential therapeutic target due to its action through calpain 1/p25/cyclin-dependent kinases 5 (CDK5), signal transducer and activator of transcription 3 (STAT3), nuclear factor-kappa B (NF-κB), 5′ adenosine monophosphate-activated protein kinase (AMPK), Bcl-2-interacting mediator of cell death (Bim), and extracellular signal-regulated kinase (ERK)-mediated pathways. These interactions enable the regulation of cell survival and apoptosis, oxidative damage, mitochondrial dysfunction, and neuroinflammation in PD (Angelopoulou et al. 2019). On the other hand, miR-17 mediates the downregulation of DNMT1 leading to abnormal DNA methylation in PD. This implicates a potential therapeutic role of miR-17 in modulating DNA methylation (Zhang et al. 2021). An interesting interaction between miR and lncRNA has been described. LncRNA-T199678 ameliorated the neuron damage generated by α-synuclein through regulating the cell cycle, oxidative stress, and apoptosis. Moreover, lncRNA-T199678 mediated α-synuclein-induced neuronal injury through upregulating miR-101-3p, which promotes the improvement of PD pathology. Other miRs related to potential treatment of PD are miR-182-5p and miR-183-5p. Both miRs contribute to the neuroprotection of dopaminergic neurons and increase neurite outgrowth combined with decreased expression of transcription factors (FOXO1 and FOXO3) and increased PI3K-Akt signaling (Roser et al. 2018). Both lncRNA MAPT-AS1 and DNMTs were analyzed in the context of the pathogenic microtubule-associated protein tau (*MAPT*) gene in PD. The results indicate that both lncRNA MAPT-AS1 and DNMT1 exhibit epigenetic regulatory effects on *MAPT* expression across four brain regions (Coupland et al. 2016). Interestingly, this lncRNA may serve as a disease stage biomarker or as a potential therapeutic target inhibiting *MAPT* expression, although *MAPT* expression does not reflect neuropathology severity.

Several different epigenetic therapeutic or prognostic options were also investigated. An interesting pathway was described by Srivastava et al. (2021). In a PD model, hyperphosphorylated epigenetic polycomb repressor complex-1 subunit BMI-1 loses function of the canonical polycomb E3 ligase, leading to the reduction of monoubiquitylation of histone 2A at lysine 119 (H2AK119ub). This corroborates the cellular accumulation of α-synuclein phosphorylated at serine 129. Furthermore, BMI-1 is crucial for melatonin-induced degradation of α-synuclein since accumulated phosphorylated α-synuclein is priorly complexed with BMI-1, which suggests potential therapeutic options of this interaction. Another process associated with neurodegeneration is SUMOylation, a posttranslational modification involving the binding of the small ubiquitin-like modifier (SUMO) to target proteins. Since SUMOylation plays a crucial role in PD pathogenesis, impacting degradation, subcellular targeting, and protein-protein interaction of α-synuclein, regulating SUMOylation may provide therapeutic benefits in PD (Mishra et al. 2024). Yet another possibility for potential PD treatment constitutes phytochemicals (Açar et al. 2023). Evidence indicates that phytochemicals can be considered epigenetic regulators in synucleinopathies, including PD (Surguchov et al. 2021). Apart from the previously mentioned VPA and sodium butyrate, multiple other natural-based epigenetic modulators are known, such as curcumin, epigallocatechin gallate (EGCG), resveratrol, or chlorogenic acid (Rathore et al. 2021). Resveratrol exhibits numerous epigenetic effects, including inhibition of DNMTs, HDACs, or histone acetyltransferases (HATs), regulation of miR expression, mainly miR-132, miR-134, miR-15a, miR-124, miR-146, and miR-let7A, which influences inflammatory factors. Curcumin presents similar functions, regulating HDACs, HATs, miRs, and inhibiting DNMTs (Surguchov et al. 2021). On the other hand, chlorogenic acid modulates the expression of autophagy- and PD-related genes (α-synuclein, p62, atg5, lc3b, ulk1b, and atg7), proving its potential to induce the autophagy interrupted in the PD pathology (Gao et al. 2023). Finally, EGCG has been shown as an inhibitor of DNMTs and HATs (Rathore et al. 2021).

Over the years, multiple studies have described epigenetic pathways involved in PD pathogenesis (Song et al. 2023). Although there is still no breakthrough in PD treatment methods, novel findings regarding epigenetic mechanisms open up new opportunities for PD therapy and prognosis (Zhang et al. 2023).

3 Epigenetics of Dementia

Alterations in epigenetic mechanisms are related to Alzheimer's disease and other dementias that lack a known cause. The relationship established between genes and the environment in the context of disease pathogenesis has driven research toward epigenetic-based treatment strategies aimed at effectively addressing these ailments. Although research on uncovering epigenetic factors in various dementias is still in its infancy, significant progress has been made in understanding the epigenetic aspects of AD (Fig. 2).

AD represents a neurodegenerative condition that advances progressively and stands as the most prevalent type of dementia worldwide, predominantly impacting older adults. The hallmark of AD involves a relentless deterioration of brain structures, with a significant impact on the temporal lobe, frontal lobes, frontal cortex, and hippocampus. The deterioration of neural structures in these specific areas results in the clinical presentation of AD, characterized by dementia, deficits in cognitive function, and a progressive decline in mental capabilities.

Modifications at the molecular level in DNA that affect gene regulation are distinct from the underlying DNA sequence, as characterized in the study of epigenetics (Li 2021). Factors in the environment induce epigenetic changes, which differentiates them from genetic changes that are primarily resistant to outside influences. DNA methylation, histone modification, and noncoding RNA molecules are three significant epigenetic processes that control the relationship between genes and their environmental contexts.

DNMTs initiate the DNA methylation process, leading to the creation of 5-methylcytosine (5mC), which attaches to the 5′ carbon position of the CpG dinucleotide within the genome. Among various epigenetic alterations, 5mC has garnered significant focus due to its critical role in the development of the brain. Investigations have shown that the levels of 5mC in the middle temporal gyrus and middle frontal gyrus are considerably heightened in patients with AD, and this increase correlates positively with Aβ, neurofibrillary tangles, and ubiquitin levels

(Coppieters et al. 2014). Through the enzymatic action of TET family members, 5mC undergoes oxidation to form 5-hydroxymethylcytosine (5hmC). The presence of 5-hmC is notably significant in the central nervous system and correlates with various neurodegenerative diseases and developmental disorders of the nervous system. Evidence suggests that neurodegeneration in mice models of AD is directly linked to the decline in 5hmC levels, attributed to the suppressed activity of TET (Zhang et al. 2020). The significance of mitochondrial DNA methylation in the development of AD has been established, confirming that epigenetic modifications can affect mitochondrial DNA, and these changes are crucial for the disease's advancement (Mposhi et al. 2017). In summary, the interplay between methylation and the development of AD cannot be overlooked, as substantial evidence reinforces the significant epigenetic connection to AD. This calls for in-depth studies on epigenetic interventions and their potential as treatments for AD.

Histone proteins are closely linked to DNA, encircling the double-stranded structure, thereby reinforcing the DNA's stability and serving as a platform for the transcription of genes. An octamer is formed by the assembly of histones, around which DNA coils to create the nucleosome structure. Covalent modifications of histones and DNA can alter chromosomal architecture and influence gene expression patterns. Alterations in chromatin configuration arise from histone modifications after translation, primarily through methylation, acetylation, phosphorylation, ubiquitination, glycosylation, and ADP ribosylation, leading to variations in gene expression (Ramazi et al. 2020).

Histone methylation occurs through the action of histone methyltransferases (HMTs) and is regulated by histone demethylases (HDMs). Researchers extensively study lysine methylation in its various forms—mono-, di-, and trimethylation—focusing particularly on the fourth lysine position of histone H3 (H3K4) because of its significant role in activating gene expression. The correlation between H3K4 methylation and synaptic transmission, axonal firing, and neuronal development highlights their significance in the onset of AD. Evidence suggests also that AD cell pathology manifests through heightened concentrations of H3K4me3 in the cytoplasmic area (Mastroeni et al. 2015).

Compared to methylation, histone acetylation is significantly more effective in promoting the expression of genes. Research findings highlight that the onset of amyloid protein aggregation in the brain is characterized by a rise in acetylation of histone H4 at lysine 12 (H4K12ac). The presence of acetylated histone H4 at lysine 12 (H4K12ac) was identified in individuals with amnestic mild cognitive impairment, suggesting its potential role as an early biomarker for AD. A comprehensive analysis of H3K27ac across the epigenome in the entorhinal cortex of individuals with AD and age-matched controls exhibiting low pathology demonstrated a significant accumulation of H3K27ac in genes associated with amyloid-beta and tau-related pathologies, as well as in regions containing variants linked to late-onset AD (Marzi et al. 2018). Moreover, studies have shown that AD patients exhibit reduced H4K16ac levels across their genomes relative to healthy individuals. The findings suggest that the pathology of AD is more complex, showing unique mechanisms that differ from the typical aging process (Nativio et al. 2018). Investigations have demonstrated that the concentration of phosphorylated H3 rises in the hippocampal neurons of those suffering from AD, with the energized form of H3 largely restricted to the cytoplasm. This restriction results in disruptions in mitosis, neuronal degeneration, and the manifestation of apraxia.

The modification of histones through phosphorylation is considered to be a significant factor in the pathological processes associated with AD. The histone variant H2AX has been found to be phosphorylated in brain samples taken from those diagnosed with AD. The occurrence of DNA damage, particularly in the form of a double-strand break (DSB), activates the phosphorylation of a certain histone on Ser-139, which in turn leads to the production of γH2AX. The presence of elevated γH2AX levels in astrocytes from the hippocampus and cortex of AD patients, absent in age-equivalent controls, highlights the crucial role of astrocytes and their DNA damage response in AD, while also contributing to our knowledge regarding the significance of histone phosphorylation in the context of the disease (Myung et al. 2008).

Research on histone modifications such as ubiquitylation, SUMOylation, and various other histone posttranslational modifications (hPTMs) remains significantly limited in the context of AD. Evidence from specific studies suggests that H2B K120 ubiquitination in the brains of individuals diagnosed with AD is 91% higher than in the control population (Anderson and Turko 2015). The role of SUMOylation in AD is derived from its impact on histones, leading to changes in the function of HDAC proteins. Research has shown that the SUMOylation process occurs on Lys-444 and Lys-476 of HDAC1, seemingly regulating its functional activity.

In conjunction with histone changes and DNA methylation, studies indicate that long and short nonprotein coding RNAs (ncRNAs) might significantly impact the epigenetic processes associated with AD. Research indicates that microRNAs play a significant role in the regulation of BACE1, the enzyme that catalyzes the production of the Aβ peptide in the brain, with findings showing that these RNAs exhibit distinct alterations in the blood and brain of individuals diagnosed with AD. Studies also reveal that the presence of miR-29c positively influences memory performance and learning skills in hippocampal neuronal cells (Yang et al. 2015). Moreover, miR-33, known for its involvement in lipid metabolism, together with miR-34a and miR-155, which are pertinent to neuroinflammation, has been shown to correlate with the epigenetic advancement of AD (Guedes et al. 2014). Observational data suggests that noncoding RNA may serve as a viable option for the early detection and management of AD; however, additional studies are required to confirm this potential.

Vascular dementia (VaD) stands as the second leading category of dementia, succeeding AD, and comprises around 20% of the overall dementia population. Extensive investigations have not yielded clarity on the origins of vascular dementia, leaving its causes still unidentified. Previous research in genetics related to VaD has pinpointed single nucleotide polymorphisms (SNPs) as crucial factors influencing its development (Sun et al. 2015). In contrast to AD, factors related to lifestyle, including age, diabetes, hypertension, and metabolic syndrome, exert a more significant influence on the onset of VaD. To date, research on epigenetics related to vascular dementia has been quite sparse, underscoring the importance of advancing this field of study. The goal of the research conducted by Park JM was to discover genes that are modified in terms of DNA methylation in the hippocampus due to the effects of VaD. Scientists executed a wide-ranging assessment of DNA methylation across the genome, applying sequencing strategies that leverage methylated DNA-binding domains. The analysis of rats in the VaD model revealed a significant decrease in promoter region methylation and an upsurge in mRNA expression of hippocampal genes, particularly vascular endothelial growth factor (*VEGFA*) and kinase insert domain receptor. Notably, the only statistically significant variations were observed in the expression of VEGFA mRNA (Park et al. 2018). Findings from another study demonstrate that subjecting mice to 1 h of environmental hypoxia each day, 3 days a week for a span of 2 months, can prevent the loss of recognition memory and restore the integrity of nerve communication within the hippocampus. Researchers observed that the adult progeny of a different group of mice, whose parents experienced repeated episodes of hypoxia before reproduction, exhibited similar resistance to memory deficits, suggesting an epigenetic response that spans generations. This study's discoveries suggest that epigenetic strategies may play a crucial role in halting the progression of memory decline and other dementia indicators, with the remarkable possibility that the protective effects could be passed on to subsequent generations (Belmonte et al. 2022).

4 Epigenetics of Stroke

Stroke still to be one of the major neurological diseases, which leads to serious disability that worsen the quality of patient life. There are two types of stroke; an ischemic stroke, when the artery supplying the brain with oxygen is blocked, and the hemorrhagic stroke, which occurs when there is a rupture in the continuity of the brain vessels (Stanzione et al. 2020). The activation of anaerobic processes due to tissue ischemia results in an increase in pH and proinflammatory cytokine activity, while restoring perfusion to damaged areas leads to decreased cellular ATP levels, increased calcium levels, impaired proton pump function, and elevated free oxygen radicals, collectively known as reperfusion injury, ultimately causing irreversible neuronal damage. However, it is the inflammatory processes that play the most crucial role in determining the condition of stroke patients (Tang and Zhuang 2019; Peng et al. 2022). Understanding the epigenetic changes caused by strokes in damaged tissues is crucial for developing effective drugs and new therapy methods to help stroke patients (Tang and Zhuang 2019). The link between strokes and epigenetic mechanisms has been confirmed through DNA methylation, histone modifications (methylation and acetylation), and the impact of microRNA (Fig. 2).

DNA methylation plays a key role in how transcription factors interact with their binding sites. Presently, it is the most heavily researched topic in the area of epigenetics associated with strokes (Peng et al. 2022). DNA methylation typically leads to a reduction in gene expression; however, depending on the tissue conditions and the genomic location of the gene, it may also result in an increase in expression. In animal studies involving middle cerebral artery occlusion (MCAO), a higher level of DNA methyltransferase (DNMT) was observed. Additionally, the pharmacological inhibition of DNMT—such as through 5-aza-2′-deoxycytidine or zebularine—resulted in diminished DNA methylation, leading to a smaller region of damage caused by brain ischemia (Stanzione et al. 2020; Peng et al. 2022). Lowering DNMT levels influences chromatin structure, which enables ischemia-related factors to attach to neuroprotective genes, consequently enhancing their expression. As a result, DNA hypomethylation could play a role in creating treatments for ischemic strokes (Peng et al. 2022). Ischemic stroke may elevate the methylation levels of the *CDKN2B* gene, which is linked to atherosclerosis and arterial calcification (Tang and Zhuang 2019). Disorders in the metabolism of homocysteine, influenced by crucial factors like methylenetetrahydrofolate reductase (MTHFR), cystathionine-beta-synthase (CBS), and methionine synthase, are linked to a heightened risk of stroke; notably, hypermethylation of the CBS promoter results in elevated plasma homocysteine levels, thereby further increasing the likelihood of stroke (Stanzione et al. 2020). In ischemic stroke patients, elevated plasma homocysteine levels contribute to DNA hypermethylation of the thrombomodulin gene, affecting blood coagulation. In patients undergoing antiplatelet therapy with clopidogrel or acetylsalicylic acid, evidence of hypomethylation in tumor necrosis factor receptor-associated factor 3 (TRAF3) and protein phosphatase 1A (PPM1A) key regulators of inflammation was observed, indicating that dysfunction in this regulatory system may heighten inflammation, consequently elevating the risk of ischemic stroke. Epigenetic modifications influence blood coagulability, as evidenced by the hypermethylation of the thrombospondin 1 (THBS1) sequence—an essential factor in platelet aggregation, neoangiogenesis, and neuroprotection, observed in an in vitro model of ischemia, which subsequently impacts reperfusion injury (Stanzione et al. 2020; Ciarambino et al. 2022). Research indicates that apolipoprotein E (ApoE), which plays a significant role in cholesterol metabolism, is implicated in stroke pathogenesis; specifically, hypermethylation of the *APOE* gene promoter results in diminished expression and elevated LDL cholesterol levels, potentially serving as a contributing factor in atherosclerosis-related ischemic stroke (Stanzione et al. 2020). A reduction in the methylation of repeatable LINE-1 sequences found in the blood not only heightens the risk of stroke but is also correlated with increased mortality rates and a markedly poorer prognosis for patients, further complicated by elevated levels of VCAM-1 that may intensify inflammation following a stroke and expand the extent of tissue damage (Ng et al. 2018). Matrix metalloproteinase-2 (MMP-2) serves as a significant enzyme marker during both the acute and chronic phases of stroke symptoms, and demethylation of the MMP-2 promoter has been observed in males with ischemic stroke due to small vessel occlusion (Lin et al. 2017). Matrix metalloproteinase-2 (MMP-2) plays a crucial role in neuronal development, as evidenced by a study conducted on a mouse model that, when supplemented with ascorbate (vitamin C), exhibited enhanced expression of neuroprotective genes and notable recovery of damaged tissue functions following a stroke (Peng et al. 2022).

Histone modifications are also involved in determining chromatin structure, which subsequently influences DNA accessibility for transcription factors (Ciarambino et al. 2022). The acetylation and deacetylation processes are mediated by two distinct enzyme groups: histone acetylases (HACs) and HDACs. Research has demonstrated that ischemic stroke leads to histone deacetylation, which promotes the proliferation of inflammatory factors and compromises the blood-brain barrier; however, the application of HDAC inhibitors like VPA can mitigate these effects. Inhibition of HDAC through sodium butyrate enhances the expression of antiinflammatory mediators such as IL-10 and STAT3 while concurrently reducing the levels of proinflammatory mediators including TNF-α, NOS2, IL-1, and IL-18; furthermore, it promotes the expression of insulin-like growth factor (IGF-1), which has demonstrated neuroprotective properties (Ng et al. 2018). In investigating the MCAO model, the findings reveal that the administration of HDAC inhibitors leads to a significant decrease in infarct volume. This promising potential for enhancing therapeutic outcomes highlights the efficacy of HDAC inhibition in stroke management. However, despite these advantages, HDAC inhibitors are not commonly integrated into routine stroke prevention protocols. This reluctance stems from several critical challenges, including their low bioavailability, difficulty in effectively crossing the blood-brain barrier, and a range of associated adverse effects. These limitations underscore the need for ongoing research to develop more effective and safer alternatives for stroke prevention (Stanzione et al. 2020; Ciarambino et al. 2022). Histone methylation represents a significant modification mechanism, facilitated by histone methyltransferases (HMTs). This process, akin to DNA methylation, possesses the capacity to both enhance and repress gene expression. In vitro investigations utilizing animal models have elucidated the impact of histone methylation on the progression of stroke, as well as its correlation with the expression of genes associated with proinflammatory cytokines.

RNA modifications are another regulators of gene expression. The concentration of miR-124 serves as a valuable biomarker for assessing neuronal damage post-stroke. In the middle cerebral artery occlusion (MCAO) model, a notable elevation of miRNA-124 levels was observed in both cerebral tissue and plasma, with peak concentrations occurring 24 h following the ischemic event (Stanzione et al. 2020). MiRNAs play a significant role in the processes of blood coagulation, which may be intricately linked to the initial inflammatory response that occurs following a stroke. Research indicates that following a stroke, there is a notable reduction in the levels of miR-19. This particular miRNA is associated with key regulators such as the tissue factor pathway inhibitor (TFPI) and plasminogen activator inhibitor 1 (SERPIN1), as well as tissue factor III, all of which are critical in the formation of thrombi. In addition, there is a corresponding decrease in other miRNAs, including miR-let-7i, miR-122, and miR-148, which also function as regulators of coagulation. Implementing therapeutic strategies designed to elevate the concentrations of these specific miRNAs could potentially mitigate the risk of vessel occlusion during subsequent strokes (Ng et al. 2018). In a study utilizing the SHRSP rat model, which serves as an analog for human stroke conditions, it was observed that levels of miR-122 were significantly diminished. This reduction correlates with compromised endothelial integrity and a heightened inflammatory response. Additionally, this model demonstrated an elevated expression of miR-503, a microRNA implicated in the regulation of the mitochondrial protein UCP2. UCP2 is known for its antioxidant properties across various tissues. Notably, a decrease in UCP2 expression has been associated with an increased prevalence of stroke occurrences (Stanzione et al. 2020). Long noncoding RNAs (LncRNAs) appear to play a significant role in the pathophysiology of ischemic stroke. Specifically, several LncRNAs, including MALAT1, MEG3, TUG1, and ANRIL, which are implicated in the regulation of cellular processes such as proliferation, apoptosis, and inflammation, have shown markedly elevated expression levels in ischemic stroke models (Ciarambino et al. 2022).

All those mechanism suggest that these molecular entities may be crucial in understanding the underlying mechanisms of stroke development and could potentially serve as targets for future therapeutic interventions. Despite that the above information regards ischemic strokes, the epigenetics of hemorrhagic strokes is not sufficiently developed. Few studies were conducted on DNA methylation, involvement of miRNA, and histone modifications, although the results were similar to that presented in ischemic strokes, even in the process of histone modifications, providing the HDAC inhibitors did not produce the intended effects, in stark contrast to the favorable prospects observed for these inhibitors in the context of ischemic strokes (Giralt-Steinhauer et al. 2022).

5 Epigenetic Factors in Epilepsy

Epilepsy affects about 50 million people worldwide. It is a chronic brain disease involving repeated epileptic seizures. The pathogenesis underlying epilepsy remains still elusive. However, studies confirm that epigenetic mechanisms, such as DNA methylation, histone modifications, and chromatin remodeling are involved in the epileptogenesis or progression of epilepsy (Fig. 2) (Van Loo et al. 2022).

DNA methylation was observed in temporal lobe epilepsy (TLE). Research shows that CpGs methylation biomarkers may be used for both diagnosis and pharmacoresistance prediction in TLE since drug response-related CpGs are strongly associated with DNA methylation (Xiao et al. 2020). Drug-resistant temporal lobe epilepsy (DR-TLE) affects about 30% of all patients suffering from epilepsy. Researchers have identified as many as 3210 differentially methylated probes (DMPs) in DR-TLE patient cortex: 1827 hypomethylated and 1383 hypermethylated. Furthermore, they identified 59 DMPs in amygdala: 27 hypomethylated and 32 hypermethylated and 32 DMPs in hippocampus including 14 hypomethylated and 18 hypermethylated (Sánchez-Jiménez et al. 2023). There is some evidence that DNA methylation regulated by adenosine may have therapeutic applications for epilepsy patients that are refractory to conventional treatment. Adenosine is a precursor for *S*-adenosylmethionine (SAM) and thereby provides the methyl groups that are transferred to DNA by DNMTs in the transmethylation pathway (3B,C). A crucial role in adenosine metabolism is played by adenosine kinase (ADK), which level is significantly upregulated in patients with epilepsy (Fig. 3a–c). In consequence, level of adenosine is decreased, which leads to increased methylation of DNA and occurring seizures. Research shows that exogenous application of adenosine through ketogenic diet augments adenosine signaling, ultimately decreasing DNA methylation (Murugan and Boison 2020).

Transcriptional effects of posttranscriptional histone modifications are still poorly understood. In animal model of tuberous sclerosis (TSC), characterized by occurring seizures, histone H3 acetylation was reduced in hippocampus. Inhibition of HDAC activity led to increased acetylation in hippocampal histone H3, normalizing synaptic plasticity and decreasing seizure threshold (Basu et al. 2019). Research suggest association between dimethylation of histone 3 lysine 9 (H3K9m2) and its enzyme euchromatic histone-lysine *N*-methyltransferase 2 (G9a) and transcriptional regulation of the potassium channel 10 encoded by *KCNJ10* gene. It shows that while G9a and H3K9m2 were downregulated, *KCNJ10* was upregulated. Same study confirmed that G9a and H3K9m2 levels in hippocampus and insular cortex were downregulated 5 h after seizure. The initial expression level was restored at 24 h (Zhang et al. 2018) (Fig. 4a, b). This indicates that they are sensitive to epileptic activity and can impact transcriptional regulation of this potassium channel.

Studies shows that circulating miRNA could be diagnostic biomarker in TLE due to three miRNAs (miR-142, miR-146a, miR-223) being significantly upregulated in TLE patients (De Benedittis et al. 2021). They can also be used in order to distinguish drug-sensitive from drug-resistant patients. Analyses demonstrate that both

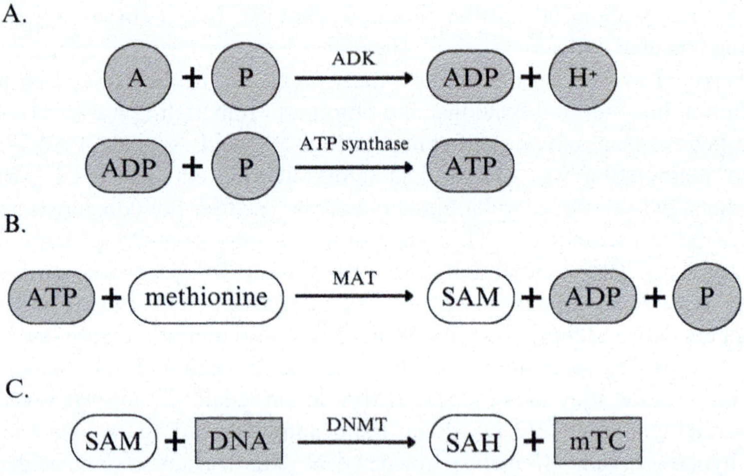

Fig. 3 Adenosine metabolism and transmethylation pathway. *A* adenosine, *P* phosphate residue, *ADP* adenosine diphosphate, *ATP* adenosine triphosphate, *SAM S*-adenosylmethionine, *SAH S*-adenosylhomocysteine, *MAT* methionine adenosyltransferase, *DNMT* DNA methyltransferase, *mTC* methylated thymine-cytosine. (Created with Canva Pty Ltd.)

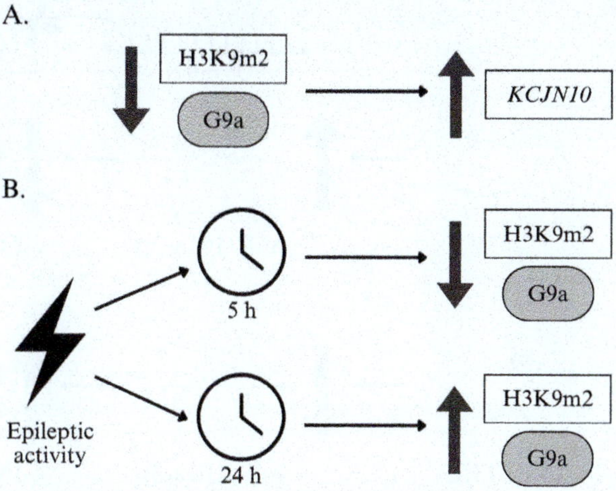

Fig. 4 Transcriptional regulation of the *KCNJ10* gene. *H3K9m2* histone 3 lysine 9, *G9a* histone-lysine *N*-methyltransferase 2, *KCNJ10* potassium inwardly rectifying channel subfamily J member 10. (Created with Canva Pty Ltd.)

miR-142 and miR-223 were increased in the sera of drug-resistant TLE patients. Other studies show that miR-22 may have a protective role and antiepileptogenic function due to its involvement in early inflammatory responses to status epilepticus (SE). miR-22-deficient animal models show suppression of inflammatory response during epileptogenesis leading to more frequent and longer seizures (Almeida Silva et al. 2020). There is also evidence that miR-101a-3p may suppress pilocarpine-induced epilepsy. In animal pilocarpine-induced epilepsy models, miR-101a-3p expression was decreased, while c-FOS protein level was elevated. Research confirmed that miR-101a-3p negatively regulates c-FOS expression, resulting in decreased epileptic activity (Geng et al. 2021) (Fig. 5). Therefore, miRNAs may be the potential biomarkers for the diagnosis of epilepsy.

Studies show that hippocampal and serum ILF3-AS1 levels were higher in TLE patients than in controls. Moreover, ILF3-AS1 is a suppressing expression of miRNA-212 leading to decreased miRNA-212 levels in hippocampus and serum of TLE patients than control. Overexpression of ILF3-AS1 promotes expression on matrix metalloproteinases (MMPs), which was previously confirmed to be increased in TLE patients' hippocampus (Cai et al. 2020). In another study, UCA1 expression was found to be higher in both refractory and nonrefractory epileptic patients than in controls with highest expression in nonrefractory patients (Irwin et al. 2021). Study highlights that these results point on lncRNA as a potential peripheral biomarker for epilepsy.

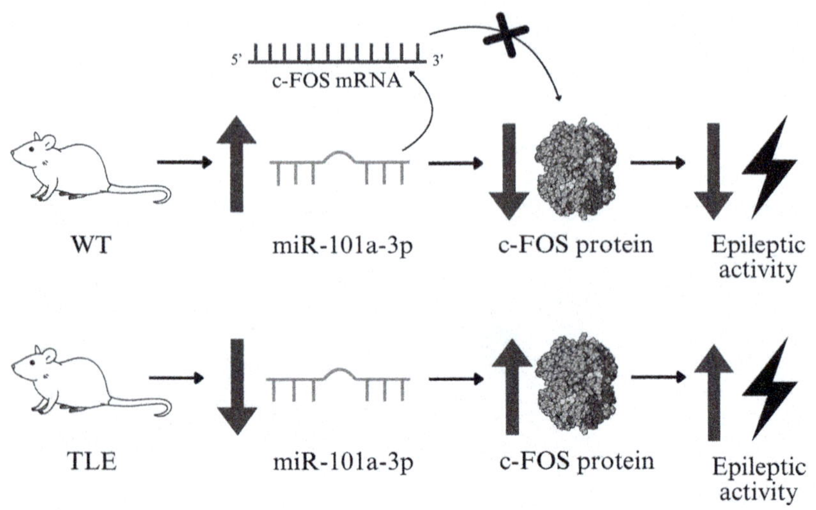

Fig. 5 Regulatory relationship between miR-101a-3p and c-FOS protein. *WT* wild type, *TLE* temporal lobe epilepsy. (Created with Canva Pty Ltd.)

6 Epigenetic of Other Neurological Diseases

Epigenetic mechanisms play a vital role in the pathology of both amyotrophic lateral sclerosis (ALS), characterized by the progressive loss of motor neurons, and HD, a neurodegenerative disorder that typically presents in midlife with symptoms including chorea, dementia, and personality changes. These mechanisms, which include DNA methylation, histone modifications, and noncoding RNA regulation, collectively influence gene expression without altering the underlying DNA sequence (Fig. 2).

DNA methylation, a well-studied epigenetic modification primarily occurring in CpG, plays a key role in various brain processes such as neural stem cell development, synaptic function, and memory. Abnormal methylation patterns have been identified in genes associated with ALS, including C9orf72, where reduced gene expression correlates with hypermethylation (Belzil et al. 2013). It has been demonstrated through both cell culture and animal models that increased DNA methyltransferase (Dnmt) activity can promote neuronal apoptosis. These findings indicated that overexpression of Dnmt3a, but not Dnmt1, triggered degeneration in spinal cord neurons, suggesting that abnormal DNA methylation could contribute to ALS pathology.

Additionally, 38 regions of hyper- or hypomethylation were identified in brain samples from sporadic ALS patients, although these methylation patterns varied across individual samples (Morahan et al. 2009). This genome-wide methylation study revealed new candidate genes linked to calcium regulation, excitotoxicity, and oxidative stress, all of which are processes associated with sporadic

ALS. Furthermore, differential methylation profiles have been observed in monozygotic twins and triplets discordant for ALS, indicating that epigenetic factors may be pivotal in disease onset and progression (Tarr et al. 2019). In addition to the DNA methylation changes observed in ALS, significant alterations in 5-methylcytosine DNA methylation have also been reported in fibroblasts from HD patients. Inhibiting HDAC has shown promise in partially reversing these methylation effects. Furthermore, the DNA modification 7-methylguanine (7-mG), which is essential for transcription regulation, has been found to be significantly altered in both human brain samples and animal models of HD, suggesting a broader dysfunction in the DNA methylation machinery (Thomas et al. 2023).

The adenosine A2A receptor (A2AR), encoded by the *ADORA2A* gene, is predominantly expressed in the striatum, particularly in vulnerable GABAergic medium-sized spiny neurons that are significantly impacted in HD. The epigenetic regulation of the A2AR gene was investigated by analyzing levels of 5-mC and 5-hydroxymethylcytosine (5-hmC) in the gene's 5′-UTR region, focusing on samples from the putamen of HD patients and the striatum of mice models that express the toxic expanded exon-1 of the huntingtin protein. These findings revealed that reduced A2AR levels are associated with increased 5-mC in the 5′-UTR region of the *ADORA2A* gene in HD patients. Additionally, lower levels of 5-hmC correlated with decreased A2AR expression in the striatum of HD transgenic mice. Thus, altered DNA methylation patterns in the *ADORA2A* gene appear to contribute to the pathologically reduced expression of A2AR observed in HD (Bassi et al. 2017).

Histones undergo various posttranslational modifications, such as acetylation, methylation, and phosphorylation, which impact chromatin structure and gene accessibility. In ALS, for instance, trimethylation at specific lysine residues on histones H3 (H3K9, H3K27, H3K79) and H4 (H4K20) has been associated with decreased C9orf72 expression in cases with expanded hexanucleotide repeats due to increased DNA-histone binding (Belzil et al. 2013). Studies have also linked the histone acetyltransferase (HAT) protein ELP3 to motor neuron degeneration in ALS through microsatellite-based genetic associations. Additionally, postmortem analyses of ALS brains show elevated histone deacetylase 2 (HDAC2) mRNA and lower histone deacetylase 11 (HDAC11) mRNA levels, suggesting that the increased HDAC2 limits transcriptional accessibility, while the decreased HDAC11 may impair immune function (Bosch 2021).

In HD, studies highlight similar dysregulation in histone modifications, affecting neurodegeneration. Huntingtin was found interacting with protein arginine methyltransferase 5 (PRMT5), which catalyzes symmetrical dimethylation of arginine on key cellular proteins, including histones and Sm proteins (Ratovitski et al. 2015). While normal huntingtin enhanced PRMT5 activity, the mutant form was deficient in facilitating this dimethylation on native substrates. Expanding on chromatin-modulating enzymes in HD, it has been examined how reducing the expression of JARID1C, an H3K4me3 demethylase, affects HD models in cells and drosophila. Lowering JARID1C expression restored the downregulated neuronal genes, like *BDNF*, providing neuroprotection and highlighting chromatin-modulating enzymes like JARID1C as promising therapeutic targets.

Noncoding RNAs (ncRNAs) play a significant role in the epigenetics of ALS and HD by modulating gene expression posttranscriptionally. In sporadic ALS patients, an analysis of miRNA expression in leukocytes revealed substantial dysregulation of several miRNAs linked to neurodegeneration. Notably, hsa-miR-451 and hsa-miR-3935 were downregulated in both ALS and PD, while hsa-miR-183 showed downregulation in ALS but upregulation in PD (Noches et al. 2024). Furthermore, widespread downregulation of miRNAs was detected in the cerebrospinal fluid of ALS patients, except for hsa-miR-181a-5p, which exhibited increased levels. Among the 8 significantly altered miRNAs, 6 were listed among the 50 most prevalent miRNAs in the cerebrospinal fluid (Benigni et al. 2016). Although many miRNAs have been identified as dysregulated in HD and its animal models, only a few have been shown to influence chromatin complexes, either directly or indirectly. For example, it was found that miR-9 and miR-9* were significantly downregulated in the HD cortex, and this reduction correlated positively with the severity of the disease. Similarly, miR-124 was downregulated in both the caudate and motor cortex of HD patients (Han et al. 2020).

Additionally, aberrant regulation of the polycomb repressive complex 2 (PRC2) is associated with increased levels of five specific mRNAs (miR-10b-5p, miR-196a-5p, miR-196b-5p, and miR-615-3p) in the prefrontal cortex of HD brains (Hoss et al. 2014). These miRNAs, which are involved in cell death and nervous system development, are located in intergenic regions of Hox gene clusters. In a feedback loop, 11 Hox genes are targets of these miRNAs, and they also exhibit altered expression in HD Hox genes, crucial for body development and regulation, that are typically suppressed by PRC2, suggesting that mutant huntingtin may disrupt PRC2's function. This disruption could impact miRNA regulation and, subsequently, neuronal development and brain function (Hoss et al. 2014).

6.1 Epigenetic-Based Therapeutic Strategies

As research progresses, the potential for therapeutic targeting of epigenetic components in ALS becomes increasingly evident. Epigenetic therapies, such as histone deacetylase inhibitors (HDACi), are being explored for their ability to restore the expression of neuroprotective genes and reduce neuronal loss. For instance, treatment with the HDACi sodium phenylbutyrate led to improvements in the dysregulated levels of histone acetylation. These therapies aim to counteract the negative effects of epigenetic dysregulation and enhance neuronal resilience, ultimately improving outcomes for ALS patients (Klingl et al. 2021).

Given the crucial role of acetylation in gene expression changes observed in HD, several clinical trials have explored HDACi as a potential treatment. Early studies demonstrated that broad-spectrum HDACi, such as sodium butyrate (NaB), suberoylanilide hydroxamic acid (SAHA), and trichostatin A (TSA), provided neuroprotective effects in HD model systems, which paved the way for phase I clinical trials. Oral administration of sodium phenylbutyrate led to moderate yet significant improvements in mRNA levels across various blood biomarkers, offering a promising outlook for HDACi in HD therapy (Shukla and Tekwani 2020).

7 Conclusion

In conclusion, the growing body of evidence linking epigenetic modifications to neurodegenerative diseases highlights their potential as both prognostic biomarkers and therapeutic targets. By unraveling the complex interplay between genetic and epigenetic factors in these disorders, researchers are opening new possibilities for early diagnosis and innovative treatments. Future research aimed at deciphering the specific epigenetic changes driving neurodegeneration will be crucial in advancing our understanding of these diseases and developing more effective interventions.

References

Açar Y, Ağagündüz D, De Cicco P et al (2023) Flavonoids: their putative neurologic roles, epigenetic changes, and gut microbiota alterations in Parkinson's disease. Biomed Pharmacother 168:115788

Almeida Silva LF, Reschke CR, Nguyen NT et al (2020) Genetic deletion of microRNA-22 blunts the inflammatory transcriptional response to status epilepticus and exacerbates epilepsy in mice. Mol Brain 13:114

Anderson KW, Turko IV (2015) Histone post-translational modifications in frontal cortex from human donors with Alzheimer's disease. Clin Proteomics 12:26

Angelopoulou E, Paudel YN, Piperi C (2019) miR-124 and Parkinson's disease: a biomarker with therapeutic potential. Pharmacol Res 150:104515

Bassi S, Tripathi T, Monziani A et al (2017) Epigenetics of Huntington's disease. Adv Exp Med Biol 978:277–299

Basu T, O'Riordan KJ, Schoenike BA et al (2019) Histone deacetylase inhibitors restore normal hippocampal synaptic plasticity and seizure threshold in a mouse model of Tuberous Sclerosis Complex. Sci Rep 9:5266

Belmonte KCD, Holmgren EB, Wills TA et al (2022) Epigenetic conditioning induces intergenerational resilience to dementia in a mouse model of vascular cognitive impairment. Alzheimers Dement 18:1711–1720

Belzil VV, Bauer PO, Prudencio M et al (2013) Reduced C9orf72 gene expression in c9FTD/ALS is caused by histone trimethylation, an epigenetic event detectable in blood. Acta Neuropathol 126:895–905

Benigni M, Ricci C, Jones AR et al (2016) Identification of miRNAs as potential biomarkers in cerebrospinal fluid from amyotrophic lateral sclerosis patients. Neuromolecular Med 18:551–556

Bloem BR, Okun MS, Klein C (2021) Parkinson's disease. Lancet 397:2284–2303

Bosch L (2021) Opportunities for histone deacetylase inhibition in amyotrophic lateral sclerosis. Br J Pharmacol 178:1353–1372

Cai X, Long L, Zeng C et al (2020) LncRNA ILF3-AS1 mediated the occurrence of epilepsy through suppressing hippocampal miR-212 expression. Aging (Albany NY) 12:8413–8422

Chen Y-C, Liaw Y-C, Nfor ON et al (2023) Epigenetic regulation of Parkinson's disease risk variant GPNMB cg17274742 methylation by sex and exercise from Taiwan Biobank. Front Aging Neurosci 15:1235840

Chouliaras L, Mastroeni D, Delvaux E et al (2013) Consistent decrease in global DNA methylation and hydroxymethylation in the hippocampus of Alzheimer's disease patients. Neurobiol Aging 34:2091–2099

Chuang Y-H, Lu AT, Paul KC et al (2019) Longitudinal epigenome-wide methylation study of cognitive decline and motor progression in Parkinson's disease. J Parkinsons Dis 9:389–400

Ciarambino T, Crispino P, Mastrolorenzo E et al (2022) Stroke and etiopathogenesis: what is known? Genes 13:978

Coppede F (2022) Targeting the epigenome to treat neurodegenerative diseases or delay their onset: a perspective. Neural Regen Res 17:1745–1747

Coppieters N, Dieriks BV, Lill C et al (2014) Global changes in DNA methylation and hydroxymethylation in Alzheimer's disease human brain. Neurobiol Aging 35:1334–1344

Coupland KG, Kim WS, Halliday GM et al (2016) Role of the long non-coding RNA MAPT-AS1 in regulation of microtubule associated protein tau (MAPT) expression in Parkinson's disease. PLoS One 11:e0157924

De Benedittis S, Fortunato F, Cava C et al (2021) Circulating microRNA: the potential novel diagnostic biomarkers to predict drug resistance in temporal lobe epilepsy, a pilot study. Int J Mol Sci 22:702

Dorszewska J, Kowalska M, Prendecki M et al (2021) Oxidative stress factors in Parkinson's disease. Neural Regen Res 16:1383–1391

Eckschlager T, Plch J, Stiborova M et al (2017) Histone deacetylase inhibitors as anticancer drugs. Int J Mol Sci 18:1414

Feng D, Liu Y, Zuo F et al (2024) LncRNA SOX21-AS1 promotes activation of BV2 cells via epigenetical silencing of SOCS3 and aggravates Parkinson's disease. Gerontology 70:1–11

Fu Y, Chen J, Huang Z (2019) Recent progress in microRNA-based delivery systems for the treatment of human disease. ExRNA 1:24

Gao X, Zhang B, Zheng Y et al (2023) Neuroprotective effect of chlorogenic acid on Parkinson's disease like symptoms through boosting the autophagy in zebrafish. Eur J Pharmacol 956:175950

Geng J, Zhao H, Liu X et al (2021) MiR-101a-3p attenuated pilocarpine-induced epilepsy by downregulating c-FOS. Neurochem Res 46:1119–1128

Giralt-Steinhauer E, Jiménez-Balado J, Fernández-Pérez I et al (2022) Genetics and epigenetics of spontaneous intracerebral hemorrhage. Int J Mol Sci 23:6479

Gonzalez-Latapi P, Bustos B, Dong S et al (2024) Alterations in blood methylome as potential epigenetic biomarker in sporadic Parkinson's disease. Ann Neurol 95:1162–1172

Guedes JR, Custódia CM, Silva RJ et al (2014) Early miR-155 upregulation contributes to neuroinflammation in Alzheimer's disease triple transgenic mouse model. Hum Mol Genet 23:6286–6301

Guhathakurta S, Kim J, Adams L et al (2021) Targeted attenuation of elevated histone marks at SNCA alleviates α-synuclein in Parkinson's disease. EMBO Mol Med 13:e12188

Han D, Dong X, Zheng D et al (2020) MiR-124 and the underlying therapeutic promise of neurodegenerative disorders. Front Pharmacol 10:1555

He S, Qu Q, Chen X et al (2024) Downregulation of Ambra1 by altered DNA methylation exacerbates dopaminergic neuron damage in a fenpropathrin-induced Parkinson-like mouse model. Ecotoxicol Environ Saf 271:115995

Hoss AG, Kartha VK, Dong X et al (2014) MicroRNAs located in the Hox gene clusters are implicated in Huntington's disease pathogenesis. PLoS Genet 10:e1004188

Huang M, Malovic E, Ealy A et al (2023) Microglial immune regulation by epigenetic reprogramming through histone H3K27 acetylation in neuroinflammation. Front Immunol 14:1052925

Irwin AB, Bahabry R, Lubin FD (2021) A putative role for lncRNAs in epigenetic regulation of memory. Neurochem Int 150:105184

Juźwik CA, Drake SS, Zhang Y et al (2019) microRNA dysregulation in neurodegenerative diseases: a systematic review. Prog Neurobiol 182:101664

Kantor B, Tagliafierro L, Gu J et al (2018) Downregulation of SNCA expression by targeted editing of DNA methylation: a potential strategy for precision therapy in PD. Mol Ther 26:2638–2649

Kim T, Song S, Park Y et al (2019) HDAC inhibition by valproic acid induces neuroprotection and improvement of PD-like behaviors in LRRK2 R1441G transgenic mice. Exp Neurobiol 28:504–515

Klingl YE, Pakravan D, Van Den Bosch L (2021) Opportunities for histone deacetylase inhibition in amyotrophic lateral sclerosis. Br J Pharmacol 178(6):1353–1372

Komada M, Nishimura Y (2022) Epigenetics and neuroinflammation associated with neurodevelopmental disorders: a microglial perspective. Front Cell Dev Biol 12:852752

Kumar S, Goyal L, Singh S (2022) Tremor and rigidity in patients with Parkinson's disease: emphasis on epidemiology, pathophysiology and contributing factors. CNS Neurol Disord Drug Targets 21:596–609

Li Y (2021) Modern epigenetics methods in biological research. Methods 187:104–113

Li X, Liu T, Wu T-T et al (2021) SIRT1 deacetylates TET2 and promotes its ubiquitination degradation to achieve neuroprotection against Parkinson's disease. Front Neurol 12:652882

Lin HF, Hsi E, Huang LC et al (2017) Methylation in the matrix metalloproteinase-2 gene is associated with cerebral ischemic stroke. J Investig Med 65:794–799

Martínez-Iglesias O, Carrera I, Carril JC et al (2020) DNA methylation in neurodegenerative and cerebrovascular disorders. Int J Mol Sci 21:2220

Marzi SJ, Leung SK, Ribarska T et al (2018) A histone acetylome-wide association study of Alzheimer's disease identifies disease-associated H3K27ac differences in the entorhinal cortex. Nat Neurosci 21:1618–1627

Mastroeni D, Delvaux E, Nolz J et al (2015) Aberrant intracellular localization of H3k4me3 demonstrates an early epigenetic phenomenon in Alzheimer's disease. Neurobiol Aging 36:3121–3129

Mazzocchi M, Goulding SR, Wyatt SL et al (2021) LMK235, a small molecule inhibitor of HDAC4/5, protects dopaminergic neurons against neurotoxin- and α-synuclein-induced degeneration in cellular models of Parkinson's disease. Mol Cell Neurosci 115:103642

Meka ST, Bojja SL, Kumar G et al (2023) Novel HDAC inhibitors provide neuroprotection in MPTP-induced Parkinson's disease model of rats. Eur J Pharmacol 959:176067

Mishra T, Singh S, Singh TG (2024) Therapeutic implications and regulations of protein post-translational modifications in Parkinsons disease. Cell Mol Neurobiol 44:53

Morahan JM, Yu B, Trent RJ et al (2009) A genome-wide analysis of brain DNA methylation identifies new candidate genes for sporadic amyotrophic lateral sclerosis. Amyotroph Lateral Scler 10:418–429

Morris HR, Spillantini MG, Sue CM et al (2024) The pathogenesis of Parkinson's disease. Lancet 403:293–304

Mposhi A, Van der Wijst MG, Faber KN et al (2017) Regulation of mitochondrial gene expression, the epigenetic enigma. Front Biosci (Landmark Ed) 22:1099–1113

Mu M-D, Qian Z-M, Yang S-X et al (2020) Therapeutic effect of a histone demethylase inhibitor in Parkinson's disease. Cell Death Dis 11:927

Murugan M, Boison D (2020) Ketogenic diet, neuroprotection, and antiepileptogenesis. Epilepsy Res 167:106444

Myung NH, Zhu X, Kruman II et al (2008) Evidence of DNA damage in Alzheimer disease: phosphorylation of histone H2AX in astrocytes. Age (Dordr) 30:209–215

Nativio R, Donahue G, Berson A et al (2018) Dysregulation of the epigenetic landscape of normal aging in Alzheimer's disease. Nat Neurosci 21:497–505

Ng GYQ, Yun-An L, Sobey CG et al (2018) Epigenetic regulation of inflammation in stroke. Ther Adv Neurol Disord 11:1756286418771815

Noches V, Campos-Melo D, Droppelmann CA et al (2024) Epigenetics in the formation of pathological aggregates in amyotrophic lateral sclerosis. Front Mol Neurosci 17:1417961

Park JM, Kim YJ, Song MK et al (2018) Genome wide DNA methylation profiling in a rat model with vascular dementia. Mol Med Rep 18:123–130

Peng J, Ghosh D, Zhang F et al (2022) Advancement of epigenetics in stroke. Front Neurosci 16:981726

Ramazi S, Allahverdi A, Zahiri J (2020) Evaluation of post-translational modifications in histone proteins: A review on histone modification defects in developmental and neurological disorders. J Biosci 45:135

Rathore AS, Birla H, Singh SS et al (2021) Epigenetic modulation in Parkinson's disease and potential treatment therapies. Neurochem Res 46:1618–1626

Ratovitski T, Arbez N, Stewart JC et al (2015) PRMT5-mediated symmetric arginine dimethylation is attenuated by mutant huntingtin and is impaired in Huntington's disease (HD). Cell Cycle 14:1716–1729

Roser A-E, Caldi Gomes L, Halder R et al (2018) miR-182-5p and miR-183-5p act as GDNF mimics in dopaminergic midbrain neurons. Mol Ther Nucleic Acids 11:9–22

Sánchez-Jiménez P, Elizalde-Horcada M, Sanz-García A et al (2023) DNA methylation description of hippocampus, cortex, amygdala, and blood of drug-resistant temporal lobe epilepsy. Mol Neurobiol 60:2070–2085

Shukla S, Tekwani BL (2020) Histone deacetylases inhibitors in neurodegenerative diseases, neuroprotection and neuronal differentiation. Front Pharmacol 11:53

Słowikowski B, Owecki W, Jeske J et al (2024) Epigenetics and the neurodegenerative process. Epigenomics 16:473–491

Sogorb-Gonzalez M, Landles C, Caron NS et al (2024) Exon 1-targeting miRNA reduces the pathogenic exon 1 HTT protein in Huntington disease models. Brain 18:266

Song H, Chen J, Huang J et al (2023) Epigenetic modification in Parkinson's disease. Front Cell Dev Biol 11:1123621

Srivastava AK, Choudhury SR, Karmakar S (2021) Neuronal Bmi-1 is critical for melatonin induced ubiquitination and proteasomal degradation of α-synuclein in experimental Parkinson's disease models. Neuropharmacology 194:108372

Stanzione R, Cotugno M, Bianchi F et al (2020) Pathogenesis of ischemic stroke: role of epigenetic mechanisms. Genes 11:89

Sun JH, Tan L, Wang HF et al (2015) Genetics of vascular dementia: systematic review and meta-analysis. J Alzheimers Dis 46:611–629

Sun Z, Kantor B, Chiba-Falek O (2024) Neuronal-type-specific epigenome editing to decrease SNCA expression: implications for precision medicine in synucleinopathies. Mol Ther Nucleic Acids 35:102084

Surguchov A, Bernal L, Surguchev AA (2021) Phytochemicals as regulators of genes involved in synucleinopathies. Biomolecules 11(5):624

Tang J, Zhuang S (2019) Histone acetylation and DNA methylation in ischemia/reperfusion injury. Clin Sci 133:597–609

Tarr IS, McCann EP, Benyamin B et al (2019) Monozygotic twins and triplets discordant for amyotrophic lateral sclerosis display differential methylation and gene expression. Sci Rep 9:8254

Thomas B, Matson S, Chopra V et al (2023) A novel method for detecting 7-methyl guanine reveals aberrant methylation levels in Huntington disease. Anal Biochem 436:112–120

Toker L, Tran GT, Sundaresan J et al (2021) Genome-wide histone acetylation analysis reveals altered transcriptional regulation in the Parkinson's disease brain. Mol Neurodegener 16:31

Van Loo KMJ, Carvill GL, Becker AJ et al (2022) Epigenetic genes and epilepsy—emerging mechanisms and clinical applications. Nat Rev Neurol 18:530–543

Van Schependom J, D'haeseleer M (2023) Advances in neurodegenerative diseases. J Clin Med 12:1709

Xiao W, Liu C, Zhong K et al (2020) CpG methylation signature defines human temporal lobe epilepsy and predicts drug-resistant. CNS Neurosci Ther 26:1021–1030

Yang G, Song Y, Zhou X et al (2015) MicroRNA-29c targets β-site amyloid precursor protein-cleaving enzyme 1 and has a neuroprotective role in vitro and in vivo. Mol Med Rep 12:3081–3088

Yang X, Zhang Y, Chen Y et al (2021) LncRNA HOXA-AS2 regulates microglial polarization via recruitment of PRC2 and epigenetic modification of PGC-1α expression. J Neuroinflammation 18:197

Zhang SP, Zhang M, Tao H et al (2018) Dimethylation of Histone 3 Lysine 9 is sensitive to the epileptic activity, and affects the transcriptional regulation of the potassium channel Kcnj10 gene in epileptic rats. Mol Med Rep 17:1368–1374

Zhang Y, Zhang Z, Li L et al (2020) Selective loss of 5hmC promotes neurodegeneration in the mouse model of Alzheimer's disease. FASEB J 34:16364–16382

Zhang H-Q, Wang J-Y, Li Z-F et al (2021) DNA methyltransferase 1 is dysregulated in Parkinson's disease via mediation of miR-17. Mol Neurobiol 58:2620–2633

Zhang D, Zhang J, Wang Y et al (2023) Targeting epigenetic modifications in Parkinson's disease therapy. Med Res Rev 43:1748–1777

Zhang L-Y, Zhang S-Y, Wen R et al (2024) Role of histone deacetylases and their inhibitors in neurological diseases. Pharmacol Res 208:107410

Unraveling Epigenetic Modifications in Glaucoma: Biomarkers and Therapeutic Targets

Emanuele Tonti, Sofia Tonti, Fabiana D'Esposito, Caterina Gagliano, Leopoldo Spadea, and Marco Zeppieri

Contents

This chapter is based on the previously published Open Access review article Tonti, E.; Dell'Omo, R.; Filippelli, M.; Spadea, L.; Salati, C.; Gagliano, C.; Musa, M.; Zeppieri, M. Exploring Epigenetic Modifications as Potential Biomarkers and Therapeutic Targets in Glaucoma. Int. J. Mol. Sci. 2024, 25, 2822. https://doi.org/10.3390/ijms25052822

E. Tonti
UOC Ophthalmology, Cardarelli Hospital, Campobasso, Italy

S. Tonti
M.Sc. Biomedical Engineering (eHealth) at "Politecnico di Torino", Turin, Italy

F. D'Esposito
Imperial College Ophthalmic Research Group (ICORG) Unit, Imperial College, London, UK

Department of Neurosciences, Reproductive Sciences and Dentistry, University of Naples Federico II, Naples, Italy
e-mail: f.desposito@imperial.ac.uk

C. Gagliano
Department of Medicine and Surgery, University of Enna "Kore", Piazza dell'Università, Enna, Italy

Mediterranean Foundation "G.B. Morgagni", Catania, Italy
e-mail: caterina.gagliano@unikore.it

L. Spadea
Eye Clinic, Policlinico Umberto I, "Sapienza" University of Rome, Rome, Italy
e-mail: leopoldo.spadea@uniroma1.it

M. Zeppieri (✉)
Department of Ophthalmology, University Hospital of Udine, Udine, Italy

Department of Medicine, Surgery and Health Sciences, University of Trieste, Trieste, Italy

J. Barciszewski (ed.), *Decoding Aging and Neurodegeneration*,
RNA Technologies 16, https://doi.org/10.1007/978-3-031-91072-2_8

Abstract Epigenetic alterations have a crucial role in neurodegenerative disorders and aging, providing valuable information about the regulation of genes and the cellular mechanisms that influence the evolution of these diseases. They play a role in various areas of pathogenesis, such as inflammation, neuronal metabolism, proteostasis, and cell death, in neurodegenerative disorders like Alzheimer, Parkinson, amyotrophic lateral sclerosis, and glaucoma. Epigenetic modifications, which contribute to the susceptibility to neurodegenerative illnesses, are a manifestation of the aging process, which is a substantial risk factor for these diseases. Research emphasizes the identification of diagnostic biomarkers, pathways, and therapeutic targets, which can lead to the development of new techniques for prevention and treatment. Epigenetics plays a role in the natural aging process and provides insights for developing therapies to slow down or reverse aging. Glaucoma, a neurological illness with several factors, is a major cause of permanent blindness globally. Although there have been significant improvements in comprehending the causes and treatment of the condition, identifying it at an early stage and providing appropriate medical care still pose difficulties. Epigenetic alterations, such as DNA methylation, histone modifications, and noncoding RNAs, are important regulators of gene expression and cellular functions. This chapter focuses on the significance of epigenetics in understanding the genetic and molecular complexities of glaucoma. It provides an overview of the disease's pathogenesis, with a particular emphasis on the regulation of intraocular pressure, dysfunction of retinal ganglion cells, and intricacy of optic nerve injury. This study investigates the impact of epigenetic alterations on gene expression, specifically in relation to glaucomatous neurodegeneration and its contribution to the development of glaucoma. The text discusses evidence from both animal models and human investigations, providing insights into the epigenetic changes that are linked to the development and advancement of glaucoma. Furthermore, the discussion revolves around the possibility of using epigenetic modifications as diagnostic indicators and therapeutic targets for personalized and accurate glaucoma treatment.

Keywords Epigenetics · Glaucoma · Neurodegeneration · Neurodegenerative disorder · Retinal ganglion cell dysfunction · Glaucomatous neurodegeneration

1 Introduction

Glaucoma is a prominent cause of permanent blindness on a global scale. It is a neurodegenerative disease characterized by the gradual deterioration of retinal ganglion cells (RGCs) and the degeneration of the neuroretinal border of the optic nerve. This can result in significant loss of the visual field. Glaucoma has a worldwide prevalence of 3.54%, with the greatest rates observed in Africa. Global estimates suggest that by 2040, about 111.8 million individuals between the ages of 40 and 80 may experience the effects of glaucoma (Tham et al. 2014). Multiple risk factors lead to the initiation of glaucoma. The factors contributing to glaucoma are elevated intraocular pressure (IOP), a family history of the condition, genetic predisposition, advancing age, gender, race (specifically non-White ethnicity), environmental impacts, nearsightedness, pseudoexfoliation, disc hemorrhages, vasospasm, systemic hypotension/hypertension, corticosteroid usage, and smoking (McMonnies 2017). The precise mechanism by which the condition develops is still not fully understood, including factors such as mechanical or ischemia injury, oxidative stress, neuroinflammation, and others (Morrone et al. 2015).

Although there are several risk factors, currently the only one that can be specifically addressed through therapy is IOP. Medications, laser therapy, or surgery can be used to reduce intraocular pressure (IOP). Nevertheless, multiple investigations have demonstrated that even individuals with glaucoma who have intraocular pressure (IOP) levels within the expected range still experience a progressive loss of retinal ganglion cells (RGCs). Furthermore, there are emerging new strategies that promote neuroprotection (Gauthier and Liu 2016). In addition to intraocular pressure (IOP), genetics and environmental factors are significant factors in the progression of glaucoma (Wiggs 2012). Epigenetics is a branch of research that is seeing increasing attention. It focuses on the study of how environmental factors impact the way genes are expressed. The significance of epigenetics in the development and advancement of glaucoma should be recognized (Rezaei et al. 2023). Epigenetics, in conjunction with genetics and environmental circumstances, impacts the signaling pathways that are accountable for the advancement of diseases (Moore et al. 2013). Recently, there has been a surge in research focusing on the involvement of epigenetics in glaucoma, which is increasingly recognized as a significant contributing component. Research has established that hereditary variables account for 16–20% of the risk associated with primary open-angle glaucoma (POAG). Additionally, persons who are closely related to those affected by the condition, such as first- and second-degree relatives, are also at risk. The genes implicated in the development of early-onset glaucoma are OPTN, MYOC, CYP1BI, PAX6, PITX2, and FOXC1. However, mutations in these specific genes only explain a maximum of 10% of glaucoma cases globally. Environmental factors that influence intraocular pressure (IOP) include levels of physical activity, consumption of omega-3 and omega-6 fats, specific yoga poses, and engaging in weightlifting (Lee et al. 2021).

Epigenetics can control gene expression through different methods, with the most well-researched ones being DNA methylation, histone modification, and microRNAs (miRNAs). This review examines the impact of epigenetic changes on gene expression and their role in glaucomatous neurodegeneration and the development of glaucoma. Additionally, the discussion will focus on the prospect of utilizing epigenetic alterations as diagnostic indicators and therapeutic targets for a more individualized and precise approach to treating glaucoma. Gaining a more comprehensive comprehension of the mechanisms behind the development of glaucoma could potentially lead to the discovery of a highly efficient therapeutic approach to prevent or reverse the advancement of the illness.

2 Mechanisms

2.1 DNA Methylation

The discovery of DNA methylation in mammals coincided with the recognition of DNA as the hereditary material (Allmer 2023). In 1948, Rollin Hotchkiss first detected altered cytosine in calf thymus by the utilization of paper chromatography. DNA methylation is an epigenetic mechanism that entails the addition of a methyl group to the C5 position of cytosine, resulting in the formation of 5-methylcytosine. This alteration regulates gene expression by enlisting proteins that repress genes or impede the binding of transcription factors to DNA. DNA methyltransferases (DNMTs) catalyze the transfer of a methyl group from *S*-adenosylmethionine (SAM) to the 5′-position of the cytosine ring in DNA, leading to the formation of 5mC and completing the process of methylation.

Demethylation can happen either passively, when there are no mechanisms in place to maintain DNA methylation, or actively, through the oxidation of 5mC by ten-eleven translocation enzymes (TET). This process results in the production of 5-hydroxymethylcytosine (5hmC), 5-formylcytosine (5fC), and 5-carboxylcytosine (5caC). The levels of DNA methylation are strongly associated with the process of aging, which is currently being extensively investigated for its possible therapeutic uses. In addition, DNA methylation plays a vital role in genomic imprinting, X chromosome inactivation, regulation of tissue-specific gene expression, and the suppression of retroviral elements. The recognition of interactions between DNA methylation and other epigenetic processes is acknowledged. It is important to note that DNA methylation can have varying effects on gene activity depending on the specific genetic sequence in different areas of the genome (McDonnell et al. 2016).

Changes in DNA methylation caused by mutations might impact metabolic and neurodevelopmental processes, potentially playing a role in the development of different neurodegenerative disorders. Studies have demonstrated that in the specific setting of glaucoma, there are increased levels of DNA methylation in the trabecular meshwork affected by glaucoma, especially when exposed to low oxygen conditions. Glaucoma patients exhibit elevated levels of TGFβ1 and reduced levels of the antifibrotic factor RAS protein activator like 1 (RASAL1) (Roodnat et al. 2022). Increased levels of TGFβ1 contribute to the deposition of extracellular matrix in the trabecular meshwork (TM) and juxtacanalicular area, which affects the outcome of glaucoma filtration surgery.

TM fibrosis can be further exacerbated by the demethylation of the GDF7 promoter, which in turn stimulates the transcription of the growth differentiation factor 7 (GDF7) gene. GDF7 is a member of the TGFβ superfamily (Wan et al. 2021). Antibodies that specifically targeted GDF7 were able to prevent fibrosis in the TM of rhesus macaque models. Furthermore, Schlemm's canal (SC) cells affected by glaucoma display a unique methylation pattern in comparison to SC cells in the control group (Cai et al. 2020). Therefore, additional research is required to gain a better understanding of this phenomenon. The profibrotic alterations contribute to heightened resistance to the drainage of aqueous humor and raised intraocular pressure (IOP), indicating possible novel targets for pharmaceutical intervention.

2.2 *Histone Modification (Histone Acetylation, Histone Methylation)*

Histones are essential proteins that form part of the chromatin structure. They consist of protein octamers wrapped around DNA. Various histone modifications, such as acetylation, methylation, phosphorylation, and ubiquitination, have a major impact on gene expression by modifying the structure of chromatin. The primary focus of research on histone changes associated with glaucoma is on acetylation and methylation. Histone acetylation and deacetylation refer to the process of attaching or detaching acetyl groups from lysine residues located in the N-terminal tail and nucleosome core of histone proteins.

Acetylated and deacetylated histones act as epigenetic indicators in chromatin, causing relaxation (euchromatin) or condensation (heterochromatin) of chromatin structure, thus modulating gene transcription levels. Histone methylation has the ability to either suppress or stimulate transcription, depending on the specific locations it targets. Histone methyltransferases (HMTs) control DNA methylation by either repressing or activating transcription through chromatin-dependent mechanisms, which depend on the specific site of methylation (Xiao et al. 2021). Accurately measuring the activity and distribution of histone methyltransferases is essential for investigating the control of genes through epigenetic mechanisms and for identifying potential inhibitors. Both lysine and arginine side chains are impacted by histone methylation (McDonnell et al. 2016).

Studies on histone methylation and glaucoma have shown that the process of trimethylation of lysine 27 on histone H3 (H3K27me3), which is carried out by the enzyme histone methyltransferase zeste homolog 2 (Ezh2), is responsible for gene silencing (Ferrari et al. 2012). The presence of H3K27me3, which is produced by

Ezh2, hinders the activity of the PI3K/AKT pathway and diminishes the expression of Vgf, a factor that promotes neuroprotection. In a preclinical study, Zheng et al. (2018) showed that the inhibition of Ezh2 and H3K27me3 stimulates the PI3K-Akt pathway, resulting in a delay in retinal degeneration. Xiao recently demonstrated that DZNep, a selective inhibitor of EZH2 called 3-deazaneplanocin A, modifies EZH2 and H3K27me3, resulting in neuroprotective benefits against NMDA-induced retinal neurotoxicity (Xiao et al. 2021).

DZNep decreases the level of H3K27me3, increases the activity of PI3K-Akt, and boosts the expression of Vgf, thereby preventing the mortality of retinal ganglion cells triggered by NMDA. Tranylcypromine, which inhibits lysine-specific demethylase 1 (LSD1), also has an impact on p38 mitogen-activated protein kinase (MAPK) gamma, leading to the induction of stress and neurotoxicity (Bermudez et al. 2016; Ramaiah et al. 2021). In contrast, Tsutsumi provided evidence that tranylcypromine can decrease retinal ganglion cell (RGC) apoptosis by decreasing the harmful effects of glutamate and oxidative stress (Tsutsumi et al. 2016).

Recent research in the field of epigenetics have indicated that histone acetylation plays a vital role in the upregulation of TGFβ2 production, which in turn leads to damage in the trabecular meshwork in glaucoma (Wang et al. 2021a, b). Suppression of histone deacetylase (HDAC) promotes the survival of retinal ganglion cells (RGC) (Pan et al. 2022; Djebali et al. 2012). HDAC inhibitors are suggested for protecting and regenerating retinal ganglion cells (RGCs) in models of optic nerve crush, which simulate glutamate excitotoxicity and subsequent RGC mortality (Wei et al. 2017). The majority of these research are preclinical in nature.

METTL23, an enzyme that adds methyl groups to arginine residues on histone proteins, is found in retinal ganglion cells (RGCs) and is responsible for catalyzing the addition of two methyl groups to the arginine residue at position 17 on histone H3 in the retina. The development of normal tension glaucoma is associated with the mutated METTL23 mutation (c.A23G). Additional investigation is required. Pan Y. et al. linked the deletion of METTL23 to the loss of PS2, resulting in the activation of NF-κB-mediated TNF-α and IL-1β. Therefore, the methylation of arginine side chains on histones promotes the survival of retinal ganglion cells in this study (Li et al. 2022).

2.3 Noncoding RNAs

Approximately 60% of the human genome is made up of noncoding RNAs (ncRNAs) (Wang et al. 2021a, b). They control gene expression after transcription, once believed to only affect gene expression after transcription. Epigenetic modifications of noncoding RNA (ncRNA) have a significant impact on RNA metabolism, specifically affecting the stability of ncRNA, the processing of microRNA, the networks of competitive endogenous RNA (ceRNA), and the interactions between long noncoding RNAs (lncRNAs) and RNA binding proteins (Zhang et al. 2023).

Long noncoding RNAs (lncRNAs) are RNA molecules that are longer than 200 nucleotides. They have the ability to attach to certain proteins in order to assist in the process of chromatin remodeling, modification, and regulation of linear transcription factors. Multiple long noncoding RNAs (lncRNAs) exhibit differential expression in the aqueous humor, trabecular meshwork, iris and retinal cells, and venous blood of individuals with glaucoma. Long noncoding RNAs (lncRNAs) have been suggested as potential biomarkers for primary open-angle glaucoma (Dahbash et al. 2019), with a specific focus on competing endogenous RNA (ceRNA) mechanisms. Long noncoding RNAs (lncRNAs) control the process of retinal ganglion cell loss, the accumulation of extracellular matrix, and the programmed cell death of trabecular meshwork through mechanisms involving competing endogenous RNAs (ceRNAs). miRNAs disrupt mRNA translation in the field of glaucoma epigenetics (Lv et al. 2020).

MicroRNA-483-3p, a kind of miRNA, plays a protective role in glaucoma by reducing the occurrence of extracellular matrix fibrosis during periods of stress. In contrast, miRNAs that promote disease, such as microRNA-100, reduce nerve development, resulting in neuroinflammation, oxidative stress, and fibrosis. Possible therapeutic targets for microRNA (miRNA) in the fluid of the eye, known as aqueous humor, are miR-143-3p, miR-125b-5p, and miR-1260b. In preclinical investigations, the use of miRNA mimics (agomirs) or inhibitors (antagomirs) has shown the ability to modify the expression of specific miRNAs in patients with glaucoma, which might potentially slow down the progression of the condition (Dobrzycka et al. 2023).

2.4 *m6A Methylation*

N6-Methyladenosine (m6A) is the process of adding a methyl group to the nitrogen-6 position of adenosine. This modification is frequently observed in the genomes of prokaryotes and in the messenger RNA (mRNA) of eukaryotes. m6A alterations exhibit dynamic and reversible characteristics. Multiple elements, such as m6A erasers (FTO and ALKBH5), writers (METTL3, METTL14, METTL16, WTAP, RBM15, KIAA1429, and ZFP217), and readers (IGF2BP1–3, YTHDF1–3, eIF3, YTHDC1–2, HNRNPA2B1, FMR1, and LRPPRC), control the process of m6A methylation (Jian et al. 2020). METTL3 functions as an enzyme complex that catalyzes the methylation of m6A. FTO and ALKBH5 enzymes remove methyl groups from RNA, whereas reader proteins identify and bind to methylated RNA molecules.

M6A RNA methylation plays a crucial role in regulating gene expression, maintaining homeostasis, controlling cell differentiation and proliferation, and managing metabolism (Jian et al. 2020). Pathologically, m6A alterations have an impact on inflammation, angiogenesis, and fibrosis (Weng et al. 2018). Dysregulation could potentially be a contributing factor to both traumatic optic nerve injury (TON) and the process of optic nerve regeneration. Regulating m6A alteration has the potential to decrease inflammation and cell death in the optic nerve caused by injury. This can be achieved by regulating the MAPK and NF-kB signaling pathways to provide neuroprotection. Individuals diagnosed with pseudoexfoliative glaucoma (PXG) have increased expression of METTL3, indicating that METTL3 could be a promising target for suppressing scar formation after glaucoma surgery. This can be achieved by inhibiting the TGF-β1-induced proliferation of human Tenon's capsule fibroblasts and the deposition of extracellular matrix (Kennard 1940). The provided information is represented in Table 1.

Table 1 Glaucoma epigenetic

Glaucoma epigenetic modifications	Mechanisms	Main link to glaucoma
DNA methylation	DNA methylation controls gene expression by attracting proteins that are involved in gene repression or by preventing the binding of transcription factors to DNA	Elevated levels of DNA methylation have been detected in the trabecular meshwork of individuals with glaucoma In hypoxic conditions, methylation levels have a tendency to rise, leading to an increase in the production of fibrotic proteins
Histone modification	Histone methylation and histone demethylation are epigenetic changes that can modulate gene expression by modifying the structure of chromatin	The main text discusses a controversial role, but further studies are required to gain further understanding
Noncoding RNAs	Noncoding RNAs have the ability to modify epigenetics by altering the structure of chromatin. Noncoding RNAs (NcRNAs) have the ability to control the expression of genes either during the process of transcription or after transcription has occurred	Long noncoding RNA (lncRNA) has the potential to serve as a biomarker for primary open-angle glaucoma Long noncoding RNA (lncRNA) can regulate the loss of retinal ganglion cells, the deposition of extracellular matrix, and the death of human trabecular meshwork through the competing endogenous RNA (ceRNA) mechanism. "Protective" microRNAs, such as micro-RNA-483-3p, suppress the formation of fibrous tissue in the extracellular matrix in response to stress. On the other hand, disease-promoting microRNAs, such microRNA-100, have been found to inhibit nerve development, which can result in neuroinflammation, oxidative stress, and fibrosis
m6A methylation	m6A methylation can modulate the structure, stability, splicing, export, transcription, and decay of mRNA, miRNA, and lncRNA through the actions of methyltransferases, demethylases, and m6A binding proteins.	Regulating m6A alteration has the potential to decrease inflammatory responses and cell death in the visual nerve caused by damage. Pseudoexfoliative glaucoma patients have increased expression of METTL3 (methyltransferase-like 3).

Every epigenetic process, including DNA methylation, histone modification, noncoding RNAs, and m6A methylation, has a significant impact on the development of glaucoma and presents opportunities for therapeutic interventions and the creation of biomarkers.

3 Epigenetic Implications in Glaucoma Neurodegenerative Disease

The concept of "reprogramming" epigenetic information to regenerate tissues or biological functions is an innovative approach that has the potential to revolutionize the treatment of several disorders, such as vision loss. Utilizing genetic information restoration to recover visual function has the potential to offer more efficient and less disruptive treatments compared to the existing options available (Chen et al. 2020). Gaining insight into the acquisition, modification, and transmission of epigenetic information during cellular reprogramming is essential for comprehending the complex dynamics of epigenetics. Moreover, it is crucial to ensure the safety and minimize the potential dangers linked to epigenetic reprogramming to avoid any unforeseen outcomes or difficulties (Lu et al. 2020). Another notable obstacle pertains to the enduring efficacy of therapies relying on epigenetic reprogramming. It is crucial to determine the long-term durability of induced epigenetic modifications and whether vision restoration may be sustained without the need for ongoing supporting therapy.

Investigations into the epigenetics of glaucoma have yielded significant knowledge about the molecular processes that drive the illness, providing fresh insights into its development and prospective treatment options. By examining the findings of this research in relation to prior investigations, a thorough comprehension of the causes and development of glaucoma is attained (Zhang et al. 2019).

Initially, the identification of epigenetic modifications, such as alterations in DNA methylation, histone modifications, and noncoding RNA expression, has enhanced our understanding of the underlying mechanisms of glaucoma. The presence of these alterations, detected in different ocular tissues and body fluids of individuals with glaucoma, suggests a widespread disruption of gene expression networks that play a role in the development and advancement of the disease (Olova et al. 2019).

Prior research has emphasized the complex and varied factors that contribute to the development of glaucoma. Although genetic variables and intraocular pressure have been the main areas of study, epigenetic changes provide further levels of control that interact with genetic predispositions and environmental factors, influencing the intricate process of glaucomatous neurodegeneration (Sarkar et al. 2020). Furthermore, these discoveries have ramifications that go beyond glaucoma, encompassing other neurodegenerative conditions like Alzheimer's disease, Parkinson's disease, and multiple sclerosis. The presence of overlapping epigenetic fingerprints in these disorders indicates the existence of potential shared pathways and therapeutic targets that warrant additional exploration (Li 2020).

In summary, studies on glaucoma epigenetics highlight the interdependence of molecular pathways that play a role in the health and sickness of the eye. The statement emphasizes the significance of taking into account epigenetic factors when devising customized medicine strategies for managing glaucoma (Shukla and Tekwani 2020).

By selectively focusing on particular epigenetic alterations, it may be feasible to manipulate the patterns of gene expression and decrease harm to neurons, resulting in more efficient therapeutic treatments.

In a significant investigation conducted by Pan et al., a genetic foundation for normal-tension glaucoma (NTG) was discovered, elucidating the involvement of a mutant histone methyltransferase. The study examined a Japanese family that included three generations of individuals with normal tension glaucoma (NTG). All members of the family had a mutation in the METTL23 gene, which is responsible for encoding a histone arginine methyltransferase (Pan et al. 2022). The autosomal dominant inheritance pattern of the METTL23 c.A83G mutation was observed in all six affected family members (Jambhekar et al. 2019). The mutation led to incorrect splicing of METTL23 mRNA, causing a decrease in protein levels and improper distribution within the cell. Functionally, METTL23 facilitated the addition of a methyl group to H3R17 in the retina. This process was regulated by the estrogen receptor pS2, which acted as a downstream target to inhibit NF-κB activation. This information is supported by reference (Davydova et al. 2021).

The researchers utilized Mettl23-knockin and Mettl23-knockout mice, together with induced pluripotent stem cells (iPSCs) obtained from NTG sufferers, to establish the vital significance of METTL23 in the survival of retinal ganglion cells (RGCs) and the safeguarding of the optic nerve. Nevertheless, it is probable that there are other variables that impact the phenotype linked to METTL23 heterozygosity. For example, a specific genetic variation called METTL23 c.84+60delAT, which causes exon 2 to be skipped, was more common in patients with NTG (normal tension glaucoma) but was also detected in individuals without the condition. This suggests that the pathogenicity (ability to cause disease) of METTL23 splice variants can vary.

Epigenetics refers to modifications in gene expression or cellular phenotype that occur without changing the DNA sequence. These modifications are driven by environmental influences, lifestyle choices, and the aging process. Multiple variables can contribute to the correlation between cataract, which is characterized by the clouding of the lens owing to the aggregation of proteins, and glaucoma, which is characterized by damage to the optic nerve. Both situations are linked to oxidative stress, which results in epigenetic alterations such as DNA methylation and histone modifications. Furthermore, chronic inflammation, the process of aging, genetic predisposition, and environmental variables all have a role in the development of these disorders by modifying epigenetic regulation and gene expression.

4 Diagnostic Biomarkers

The study of glaucoma epigenetics is essential for early diagnosis and monitoring disease development since it helps identify diagnostic indicators. Recent research has emphasized the significance of epigenetic alterations, including DNA methylation, histone modifications, and noncoding RNA expression, in the progression of glaucoma. Analysis of DNA methylation across the entire genome has revealed noticeable differences in the methylation patterns seen in the fluid in the front part of the eye (aqueous humor) of persons with glaucoma compared to those without the disease. This indicates that these specific markers could potentially be valuable for detecting the disease at an early stage (Sharma 2017).

The disruption of histone-modifying enzymes such as histone deacetylases (HDACs) and histone methyltransferases (HMTs) has been associated with glaucoma, highlighting their potential as diagnostic biomarkers. Furthermore, the presence of atypical amounts of microRNAs (miRNAs) and long noncoding RNAs (lncRNAs) in glaucomatous tissues and biofluids provides additional evidence of their diagnostic significance (Patel 2017). Epigenome-wide association studies (EWAS) have identified particular epigenetic signatures that are unique to glaucoma. These findings provide valuable information about the underlying mechanisms of the disease and can help in the identification of possible targets for diagnostic purposes (Goldberg et al. 2002) (Table 2).

Recent research suggests that there is a connection between epigenetic changes and common risk factors for glaucoma, such as intraocular pressure (IOP) and the shape of the optic nerve head. This emphasizes the need of using epigenetic biomarkers to diagnose and categorize the risk of glaucoma. The pathogenesis of glaucoma is characterized by an intricate interaction between genetic predisposition, environmental variables, and epigenetic modifications. Epigenetic alterations induce changes in gene expression and cellular phenotypes without modifying the DNA sequence. Abnormal epigenetic regulation in glaucoma has been associated with retinal ganglion cell (RGC) death, optic nerve atrophy, and neuroinflammation, which contribute to the advancement of the disease (Cueto 2021).

Table 2 Epigenetic biomarkers in glaucoma

Biomarker	Diagnostic role
DNA methylation	Altered methylation patterns in ocular tissues Elevated DNA methylation levels Specific CpG island hypermethylation
Histone modifications	Aberrant histone acetylation levels Altered histone phosphorylation Dysregulated histone methylation
Noncoding RNAs	Abnormal levels of long noncoding RNAs Dysregulated expression of microRNAs Dysregulated expression of circRNAs

[a] Diagnostic biomarkers related to glaucoma epigenetics

Protein biomarkers associated with glaucoma have been detected in several parts of the eye. Nevertheless, the interpretation of these studies is difficult because of the varying methodologies, limited sample numbers, and inconclusive results. Although there are restrictions, excessive expression of specific biomarkers can have a harmful effect on the nervous system, while the disruption or decrease in expression of protective markers can result in degeneration of retinal ganglion cells through the TrkA receptor pathway. Biomarkers have the potential to be used for early screening, diagnosis, and prognosis of glaucoma. They may also be targeted for therapeutic purposes to improve visual stability or facilitate recovery. The buildup of biomarkers can cause disruption to the blood-aqueous barrier as a result of dysregulation of the extracellular matrix and inflammation (Jones et al. 2016).

Biomarkers also impact the autonomic regulation of the sympathetic nervous system, which in turn influences the structure of the trabecular meshwork and ciliary body. Despite the identification of more than 450 biomarkers, their validation in extensive patient cohorts or varied populations is still pending, highlighting the necessity for additional research. Biomarkers have been detected in several ocular tissues and fluids, including aqueous humor, optic nerve, retina, trabecular meshwork, tears, vitreous body, serum, and blood. The markers encompass indicators for immunological response, extracellular matrix, oxidative stress, apoptosis, inflammation, neuroprotection, and neurodegeneration (Kim and Park 2020).

Epigenetic enzymes, including DNA methyltransferases (DNMTs) and histone-modifying enzymes, control gene expression by altering DNA and histone proteins, therefore impacting chromatin structure and accessibility. DNMTs methylate cytosine residues in DNA, typically leading to gene silence by chromatin modification and inhibition of transcription factor binding. Abnormal DNMT activity in glaucoma has been associated with incorrect methylation patterns in ocular tissues, which contributes to the development of the disease (Fan and Wiggs 2010).

Histone-modifying enzymes carry out diverse posttranslational modifications (PTMs) on histone tails, such as acetylation, methylation, phosphorylation, and ubiquitination. These alterations impact the structure of chromatin and the expression of genes by influencing the accessibility of DNA to the machinery responsible for transcription. Neurodegeneration and dysfunction of retinal ganglion cells (RGCs) in glaucoma are linked to abnormal functioning of histone acetyltransferases (HATs), histone deacetylases (HDACs), histone methyltransferases (HMTs), and histone demethylases (HDMs).

Studying the connection between epigenetic enzymes and the development of glaucoma has great potential for discovering novel targets for therapy and creating individualized remedies. The use of epigenetic indicators into clinical practice has the potential to improve the early identification and individualized treatment of glaucoma.

5 Therapeutic Targets

Exploring therapeutic targets in the field of epigenetics has great promise for the development of innovative medicines that can help maintain eyesight and stop the progression of glaucoma. Epigenetic dysregulation, such as alterations in DNA methylation, histone modifications, and noncoding RNA expression, is a crucial factor in the development of glaucoma. This makes it a promising target for intervention.

- DNA methylation targets: DNMTs, enzymes responsible for DNA methylation, show promise as therapeutic targets. Blocking DNMTs could potentially correct atypical DNA methylation patterns linked to glaucomatous neurodegeneration, leading to the restoration of typical gene expression and a decrease in the severity of the disease (Quaranta et al. 2021).
- Enzymes that alter histones, such as histone acetyltransferases (HATs), deacetylases (HDACs), methyltransferases (HMTs), and demethylases (HDMs), offer other possibilities for therapy. By modulating these enzymes, it is possible to change the structure of chromatin and the expression of genes, which can have neuroprotective effects and help maintain retinal function.
- Noncoding RNAs (ncRNAs), such as microRNAs (miRNAs) and long noncoding RNAs (lncRNAs), have the potential to be targeted for therapeutic purposes because of their involvement in gene regulation. Modulating disease-related molecular pathways and reducing neuronal damage in glaucoma can be achieved by targeting certain noncoding RNAs (ncRNAs) (Lanza et al. 2019).

The TGF-β family plays a vital role in cellular activities such as the synthesis and buildup of extracellular matrix (ECM) in the trabecular meshwork (TM) and lamina cribrosa (LC). Imbalanced TGF-β signaling can cause irregular changes in the extracellular matrix (ECM), which can result in reduced drainage of aqueous fluid and injury to the optic nerve in glaucoma. A study discovered that a decrease in DNA methylation in the promoter region of the TGF-β1 gene resulted in an increase in its production in eyes affected by glaucoma. This finding emphasizes the significance of DNA methylation in the development of glaucoma and proposes potential novel therapeutic strategies.

Effectively addressing fibrosis and hypoxia is crucial in the treatment of glaucoma. The efficacy of the histone deacetylase inhibitor (HDACi) suberoylanilide hydroxamic acid (SAHA) in avoiding excessive wound healing and scarring was tested in a rabbit model of glaucoma filtration surgery (GFS). Despite advancements in minimally invasive procedures, GFS continues to be the preferred treatment for drug-refractory glaucoma. However, it is crucial to exercise caution in selecting and monitoring patients to reduce potential hazards. Scarring and wound healing after surgery are significant challenges for the effectiveness of GFS (glaucoma filtration surgery). Studies on rabbit models have demonstrated that SAHA (suberoylanilide hydroxamic acid) might decrease scarring, suggesting that histone acetylation may have a role in controlling chromatin and gene expression, and hence offer potential advantages in this regard (Huang et al. 2022).

By focusing on chromatin remodeling enzymes such SWI/SNF complexes and ATP-dependent remodelers, it is possible to influence the epigenetic characteristics of glaucoma. By integrating epigenetic regulators with current treatments, such as IOP-lowering medicines and neuroprotective medications, it is possible to enhance outcomes. Distinct epigenetic biomarkers are linked to the development and advancement of glaucoma, including DNA methylation patterns and histone changes. These indicate the potential for using these biomarkers in diagnosing and treating the disease.

By incorporating therapeutic biomarkers into clinical practice, it is possible to improve individualized glaucoma therapy by customizing medications based on patients' epigenetic patterns. Nevertheless, there are other obstacles that need to be overcome in this field, such as the need to establish uniform methods for conducting assays, verifying the reliability of biomarkers over time, and resolving ethical issues related to privacy and the sharing of data (Sharma et al. 2016). Subsequent investigations should prioritize examining the practical importance of revealed biomarkers and their potential for therapeutic applications. This can be achieved by employing multiomic methodologies and utilizing extensive biobanks to expedite the process of uncovering and implementing these findings in clinical settings.

Epigenetic profiling has the potential to enhance early detection, track the advancement of diseases, and create customized interventions to safeguard eyesight and enhance patient results (Springelkamp et al. 2014). By comprehending the interaction of genetic predisposition, environmental factors, and epigenetic alterations, we can uncover novel processes in the development of glaucoma. Conducting longitudinal studies to monitor epigenetic alterations in individuals with glaucoma could offer valuable information regarding the development of the disease and how patients respond to treatment. Additionally, the use of artificial intelligence could aid in identifying biomarkers that can predict disease progression and enhance treatment regimens.

It is essential to use advanced technology and collaborate across many disciplines to improve our knowledge of glaucoma epigenetics. This will help us apply our discoveries to clinical practice.

An individualized medicine strategy utilizing customized epigenetic profiles holds the potential to completely transform the management of glaucoma, providing superior results and enhancing quality of life. Ongoing research and innovation will play a crucial role in fully realizing the therapeutic possibilities of epigenetics in treating glaucoma and meeting the requirements of patients worldwide. In order to create targeted treatments for particular patients, it is necessary to thoroughly validate them in preclinical studies and conduct clinical trials to ensure their safety, effectiveness, and long-term advantages. Engaging in epigenetic-based treatments signifies a fundamental change in how glaucoma is managed, offering renewed optimism for individuals afflicted by this sight-endangering ailment.

6 Conclusions and Future Directions

Epigenetic modifications differ from genetic mutations as they can be reversed, providing new possibilities for potentially more efficient glaucoma therapies. Recent scientific progress has improved our comprehension of the epigenetic processes implicated in glaucoma, a condition frequently identified only after substantial and permanent deterioration of vision has taken place. Significant alterations in epigenetic markers, such as noncoding RNAs, m6a methylation, and DNA methylation, have been detected in different bodily fluids, including aqueous humor, tears, peripheral blood, and human trabecular fibroblasts (HTFs), in individuals with glaucoma and those undergoing glaucoma filtration surgery. These modifications indicate a wide range of possible areas to focus on for the early detection, prevention, and treatment of glaucoma.

Due to the intricate and multifaceted character of glaucoma, concentrating on a solitary epigenetic component may not comprehensively tackle the harm caused by the condition. Moreover, studies suggest that epigenetic pathways have the ability to interact with one other, highlighting the intricate nature of glaucoma development. Subsequent investigations should strive to elucidate the causal connections between alterations in epigenetics and the advancement of diseases. Longitudinal studies are necessary to monitor changes in epigenetic modifications over a period of time and in reaction to therapies, which will aid in the identification of biomarkers for the initiation and advancement of diseases. Additional investigation into the interplay between various epigenetic pathways and their correlation with environmental factors can enhance our comprehension of the progression of glaucoma. In summary, the investigation of glaucoma epigenetics represents a substantial change in our methodology toward the condition, opening up possibilities for early detection, specific therapies, and individualized medical approaches. To fully exploit the promise of epigenetics in increasing glaucoma therapy and enhancing patient outcomes, it is crucial to integrate various research findings and utilize emerging technology.

References

Allmer J (2023) Noncoding RNA databases. Curr Pharm Biotechnol 24:825–831

Bermudez JY, Webber HC, Patel GC et al (2016) HDAC inhibitor-mediated epigenetic regulation of glaucoma-associated TGFβ2 in the trabecular meshwork. Invest Ophthalmol Vis Sci 57:3698–3707

Cai J, Drewry MD, Perkumas K et al (2020) Differential DNA methylation patterns in human Schlemm's canal endothelial cells with glaucoma. Mol Vis 26:483–493

Chen HY, Ho YJ, Chou HC et al (2020) TGF-β1 signaling protects retinal ganglion cells from oxidative stress via modulation of the HO-1/Nrf2 pathway. Chem Biol Interact 331:109249

Cueto AF (2021) Candidate glaucoma biomarkers: from proteins to metabolites, and the pitfalls to clinical applications. Biology 10:763

Dahbash M, Sella R, Megiddo-Barnir E et al (2019) The histone deacetylase inhibitor AN7, attenuates choroidal neovascularization in a mouse model. Int J Mol Sci 20:714

Davydova E, Shimazu T, Schuhmacher MK et al (2021) The methyltransferase METTL9 mediates pervasive 1-methylhistidine modification in mammalian proteomes. Nat Commun 12:891

Djebali S, Davis CA, Merkel A et al (2012) Landscape of transcription in human cells. Nature 489:101–108

Dobrzycka M, Sulewska A, Biecek P et al (2023) miRNA studies in glaucoma: a comprehensive review of current knowledge and future perspectives. Int J Mol Sci 24:14699

Fan BJ, Wiggs JL (2010) Glaucoma: genes, phenotypes, and new directions for therapy. J Clin Invest 120:3064–3072

Ferrari G, Terushkin V, Wolff MJ et al (2012) TGF-β1 induces endothelial cell apoptosis by shifting VEGF activation of p38(MAPK) from the prosurvival p38β to proapoptotic p38α. Mol Cancer Res 10:605–614

Gauthier C, Liu J (2016) Neurodegeneration and neuroprotection in glaucoma. Yale J Biol Med 89:73–79

Goldberg JL, Klassen MP, Hua Y et al (2002) Amacrine-signaled loss of intrinsic axon growth ability by retinal ganglion cells. Science 296:1860–1864

Huang H, et al. Suberoylanilide hydroxamic acid (SAHA) treatment reveals crosstalk among proteome, phosphoproteome, and acetylome in nasopharyngeal carcinoma cells. Front Genet. 2022;13:873840.

Jambhekar A, Dhall A, Shi Y (2019) Roles and regulation of histone methylation in animal development. Nat Rev Mol Cell Biol 20:625–641

Jian D, Wang Y, Jian L et al (2020) METTL14 aggravates endothelial inflammation and atherosclerosis by increasing FOXO1 N6-methyladenosine modifications. Theranostics 10:8939–8956

Jones PA, Issa JP, Baylin S (2016) Targeting the cancer epigenome for therapy. Nat Rev Genet 17:630–641

Kennard MA (1940) Relation of age to motor impairment in man and in subhuman primates. Arch Neurol Psychiatr 44:377–397

Kim HJ, Park SJ (2020) Intravitreal injection of valproic acid regulates retinal ganglion cell survival in a mouse model of optic nerve crush. J Mol Neurosci 70:454–464

Lanza M, Benincasa G, Costa D et al (2019) Clinical role of epigenetics and network analysis in eye diseases: a translational science review. J Ophthalmol 2019:2424956

Lee DD, Komosa M, Sudhaman S et al (2021) Dual role of allele-specific DNA hypermethylation within the TERT promoter in cancer. J Clin Invest 131:e146915

Li H (2020) Single-cell transcriptomes reveal diverse regulatory strategies for olfactory receptor expression and axon targeting. Curr Biol 30:1189–1198

Li H, Ye Z, Li Z (2022) Identification of the potential biological target molecules related to primary open-angle glaucoma. BMC Ophthalmol 22:188

Lu Y, Brommer B, Tian X et al (2020) Reprogramming to recover youthful epigenetic information and restore vision. Nature 588:124–129

Lv Y, Zhang Z, Xing X et al (2020) lncRNA TGFbeta2-AS1 promotes ECM production via TGF-beta2 in human trabecular meshwork cells. Biochem Biophys Res Commun 527:888

McDonnell FS, McNally SA, Clark AF et al (2016) Increased global DNA methylation and decreased TGFβ1 promoter methylation in glaucomatous lamina cribrosa cells. J Glaucoma 25:e834–e842

McMonnies CW (2017) Glaucoma history and risk factors. J Optom 10:71–78

Moore LD, Le T, Fan G (2013) DNA methylation and its basic function. Neuropsychopharmacology 38:23–38

Morrone LA, Rombolà L, Corasaniti MT et al (2015) Natural compounds and retinal ganglion cell neuroprotection. Prog Brain Res 220:257–281

Olova N, Simpson DJ, Marioni RE et al (2019) Partial reprogramming induces a steady decline in epigenetic age before loss of somatic identity. Aging Cell 18:e12877

Pan Y, Suga A, Kimura I et al (2022) METTL23 mutation alters histone H3R17 methylation in normal-tension glaucoma. J Clin Invest 132:e153589
Patel NB (2017) Role of proteomics in the study of glaucoma. Proteom Clin Appl 11:1700048
Quaranta L, Bruttini C, Micheletti E et al (2021) Glaucoma and neuroinflammation: an overview. Surv Ophthalmol 66:693–713
Ramaiah MJ, Tangutur AD, Manyam RR (2021) Epigenetic modulation and understanding of HDAC inhibitors in cancer therapy. Life Sci 277:119504
Rezaei M, Faramarzpour M, Shobeiri P et al (2023) A systematic review, meta-analysis, and network analysis of diagnostic microRNAs in glaucoma. Eur J Med Res 28:137
Roodnat AW, Callaghan B, Doyle C et al (2022) Genome-wide RNA sequencing of human trabecular meshwork cells treated with TGF-β1: relevance to pseudoexfoliation glaucoma. Biomol Ther 12:1693
Sarkar TJ, Quarta M, Mukherjee S et al (2020) Transient non-integrative expression of nuclear reprogramming factors promotes multifaceted amelioration of aging in human cells. Nat Commun 11:1545
Sharma S (2017) Association of genetic variants in the TMCO1 gene with clinical parameters related to glaucoma and characterization of the protein in the eye. Invest Ophthalmol Vis Sci 58:6264–6274
Sharma A, Anumanthan G, Reyes M et al (2016) Epigenetic modification prevents excessive wound healing and scar formation after glaucoma filtration surgery. Invest Ophthalmol Vis Sci 57:3381–3389
Shukla S, Tekwani BL (2020) Histone deacetylases inhibitors in neurodegenerative diseases, neuroprotection and neuronal differentiation. Front Pharmacol 11:537
Springelkamp H, Höhn R, Mishra A et al (2014) Meta-analysis of genome-wide association studies identifies novel loci that influence cupping and the glaucomatous process. Nat Commun 5:4883
Tham YC, Li X, Wong TY et al (2014) Global prevalence of glaucoma and projections of glaucoma burden through 2040: a systematic review and meta-analysis. Ophthalmology 121:20812090
Tsutsumi T, Iwao K, Hayashi H et al (2016) Potential neuroprotective effects of an LSD1 inhibitor in retinal ganglion cells via p38 MAPK activity. Invest Ophthalmol Vis Sci 57:6461–6473
Wan P, Long E, Li Z et al (2021) TET-dependent GDF7 hypomethylation impairs aqueous humor outflow and serves as a potential therapeutic target in glaucoma. Mol Ther 29:1639–1657
Wang J, Struebing FL, Geisert EE (2021a) Commonalities of optic nerve injury and glaucoma-induced neurodegeneration: insights from transcriptome-wide studies. Exp Eye Res 207:108571
Wang L, Yu T, Zhang X et al (2021b) Network integration analysis and immune infiltration analysis reveal potential biomarkers for primary open-angle glaucoma. Front Cell Dev Biol 9:793638
Wei JW, Huang K, Yang C et al (2017) Non-coding RNAs as regulators in epigenetics (Review). Oncol Rep 37:3–9
Weng YL, Wang X, An R et al (2018) Epitranscriptomic m6A regulation of axon regeneration in the adult mammalian nervous system. Neuron 97:313–325
Wiggs JL (2012) The cell and molecular biology of complex forms of glaucoma: updates on genetic, environmental, and epigenetic risk factors. IOVS 53:2467–2469
Xiao L, Hou C, Cheng L et al (2021) DZNep protects against retinal ganglion cell death in an NMDA-induced mouse model of retinal degeneration. Exp Eye Res 212:108785
Zhang Y, Williams PR, Jacobi A et al (2019) Elevating growth factor responsiveness and axon regeneration by modulating presynaptic inputs. Neuron 103:39–51.e5
Zhang N, Cao W, He X et al (2023) Long non-coding RNAs in retinal ganglion cell apoptosis. Cell Mol Neurobiol 43:561–574
Zheng S, Xiao L, Liu Y et al (2018) DZNep inhibits H3K27me3 deposition and delays retinal degeneration in the rd1 mice. Cell Death Dis 9:310

Unveiling the Crucial Role of R-Loops in Genome Instability and Neurological Disorders

Yulin Jin and Peng Jin

Contents

Abstract R-loops, RNA:DNA hybrid structures formed during transcription, have emerged as critical regulators of gene expression and genome stability. Initially considered byproducts of transcription, R-loops are now recognized as key players in various cellular processes, including transcription regulation, chromatin remodeling, and DNA repair. However, abnormal accumulation of R-loops can lead to excessive DNA damage and genomic instability, contributing to the pathogenesis of various diseases. This chapter provides a comprehensive overview of recent advances in R-loop biology, emphasizing their formation, genomic distribution, and intricate interactions with epigenetic modifications. We discuss how R-loops influence chromatin architecture and gene expression and the mechanisms through which they contribute to genomic instability. Furthermore, we examine the involve-

---

Y. Jin · P. Jin (✉)
Department of Human Genetics, Emory University School of Medicine, Atlanta, GA, USA
e-mail: peng.jin@emory.edu

J. Barciszewski (ed.), *Decoding Aging and Neurodegeneration*,
RNA Technologies 16, https://doi.org/10.1007/978-3-031-91072-2_9

ment of R-loops in the pathogenesis of neurological diseases, including repeat expansion disorders and autism spectrum disorder (ASD). Finally, we discuss potential therapeutic strategies targeting pathological R-loops, emphasizing the need for precise compounds to selectively mitigate R-loop accumulation without causing off-target effects. This chapter aims to provide an updated perspective on R-loop biology and its significance in neurological disorder research.

Keywords R-loops · Transcription · Epigenetic modification · DNA damage and repair · Genome instability · Neurodegenerative disorders · ASD

1 Introduction

RNA:DNA hybrids form in a sequence-specific manner when nascent RNA transcripts anneal to their DNA template strand or homologous DNA regions. These hybrids are generated during essential cellular processes, including transcription, DNA replication, and DNA repair, where they function as critical intermediates (Petermann et al. 2022). For example, during DNA replication, short RNA primers (~11 nucleotides) synthesized by DNA primase are essential for lagging-strand synthesis at replication forks. Similarly, during transcription, an 8-bp RNA:DNA duplex forms transiently at the active site of RNA polymerase (RNAP). In contrast, R-loops represent longer, more stable three-stranded nucleic acid structures comprising an RNA:DNA hybrid and a displaced single-stranded DNA (ssDNA).

R-loops were first identified in 1976 through in vitro electron microscopy by the Davis laboratory. RNA was observed to hybridize with double-stranded DNA in the presence of 70% formamide, displacing the complementary DNA strand (Thomas et al. 1976). This discovery was initiated by R-loop mapping of the 18S and 28S rRNA sequences in *Drosophila melanogaster* (White and Hogness 1977). Early findings suggested that R-loops are stable structures that require RNA degradation for resolution. However, an important study in 1995 revealed that R-loop formation during transcription is dynamically regulated by DNA topoisomerases (TOPOs) and ribonuclease H (RNase H) (Drolet et al. 1995). In *Escherichia coli*, the absence of TOPO I led to growth defects caused by R-loop accumulation, which could be alleviated by Rnase H overexpression. This finding shifted the research toward understanding R-loop dynamics and their consequences across various organisms, including plants, bacteria, yeast, and mammals.

Earlier studies have shown that R-loops are naturally formed as key intermediates in various processes, such as initiating mitochondrial DNA replication, bacterial plasmid replication, and immunoglobulin (Ig) class-switch recombination (CSR) (Aguilera and García-Muse 2012). Over the past decade, accumulating evidence has shown that R-loops are pivotal for regulating gene expression and chromatin structure and serve as major sources of DNA damage and genomic instability. Technological advancements in R-loop mapping have provided detailed insights into their genome-wide distribution and dynamic behavior, particularly in disease-related contexts. This chapter explores the mechanisms driving R-loop formation and their distribution across the genome. We examine their impact on critical cellular functions, such as transcription regulation, chromatin architecture, DNA repair, and the detrimental consequences of R-loop accumulation on genome stability. Furthermore, we highlight recent discoveries linking R-loops to neurological disorders, including Fragile X-related disorders, C9orf72-mediated ALS/FTD, Friedreich's ataxia, and autism spectrum disorder (ASD). Additionally, we discuss the potential of targeting aberrant R-loops as therapeutic interventions. These advancements in understanding R-loop biology open promising avenues for novel therapeutic interventions to treat diseases linked to genomic instability.

2 The Biology of R-Loops

2.1 *R-Loop Formation and Factors Influencing Their Resolution*

R-loops typically form in *cis* near the transcription sites but can also arise in *trans*. The well-established "thread-back" model explains *cis* R-loop formation, where cotranscriptional R-loops occur when a nascent RNA transcript transiently reanneals back onto its template DNA strand behind the RNAP (Fig. 1a) (Kasahara et al.

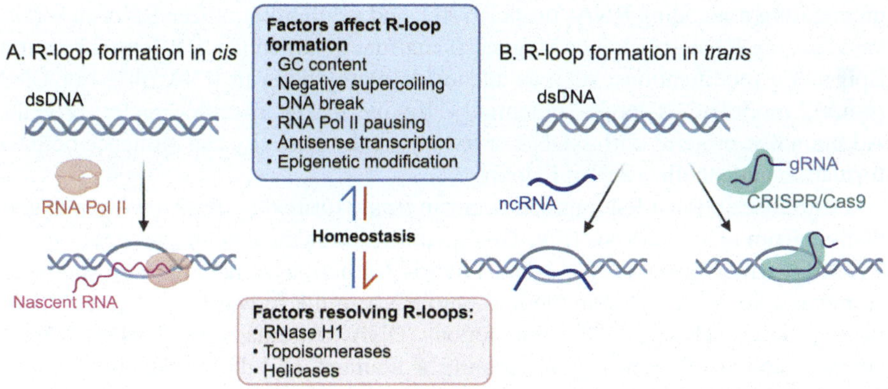

Fig. 1 R-loop formation and factors affect R-loop resolution. (**a**) *Cis* R-loops: R-loops can form cotranscriptionally when a nascent RNA transcript anneals to the DNA template strand behind the elongating RNA polymerase II (RNAPII). (**b**) *Trans* R-loops: R-loops can form in *trans* when a transcript can hybridize with a DNA strand at a distant location. R-loops occur when an RNA molecule hybridizes to a DNA strand at a distant genomic location. The latter is exemplified by the CRISPR/Cas9 system, where a guide RNA associates with the Cas9 protein to hybridize with its complementary DNA target sequence

2000). This model is supported by the discovery of the crystallographic structure of RNAP, which demonstrated that DNA and RNA exit from the RNAP complex at independent channels. Another proposed pathway for R-loop formation, discovered in bacteria, involves codirectional collisions between the replisome and backtracked RNAP (Dutta et al. 2011). In this model, a stable R-loop forms when the RNAP moves backward along the DNA without releasing the RNA it has already synthesized.

The efficient cotranscriptional R-loop formation depends on factors that make the RNA:DNA duplex more energetically favorable than the DNA:DNA duplex (Fig. 1). Key factors include the GC content of the nontemplate DNA strand, negative supercoiling that facilitates strand separation, and breaks within the nontemplate DNA strand. GC skew plays a critical role in initiating R-loop formation due to the thermodynamic stability of RNA:DNA hybrids, particularly in regions such as promoters and transcription termination sites with high GC content (Ginno et al. 2013; Sanz et al. 2016). After initiation, high guanine density, even without clustering, promotes R-loop stabilization and extension. Negative supercoiling from DNA unwinding destabilizes the DNA:DNA duplex, facilitating RNA invasion to form R-loops. This effect also extends R-loop length and reduces dependence on G-rich sequences (Massé and Drolet 1999; Roy et al. 2010). Further, DNA nicks downstream of promoters on the nontemplate strand drive R-loop formation by exposing single-stranded DNA (ssDNA) regions (Roy et al. 2010). Additional factors also play roles in R-loop dynamics. Certain ligands, such as peptide nucleic acids (PNAs), bind to the nontemplate strand, destabilizing the DNA structure and inducing R-loops that can block transcription (Belotserkovskii and Hanawalt 2015). Biological processes like RNAP II pausing, antisense transcription (frequently producing long noncoding RNAs or lncRNAs), and availability of free RNA ends (elevated during RNA splicing or biogenesis malfunctions) enhance R-loop formation. Epigenetic modifications, such as adenosine methylation in RNA (m^6A) or DNA (6mA), modulate R-loops, potentially by recruiting specific reader proteins. Additionally, proteins with ssDNA or R-loop binding activity can influence both the formation and stabilization of R-loops.

Posttranscriptional R-loops can occur in *trans*, formed by RNA invasion at sites distinct from their synthesis (Fig. 1b). Specific factors mediate this process, such as RAD51P, which promotes RNA-DNA hybrid formation through forward and inversion reactions. RAD51P polymerizes onto RNA in the forward reaction and facilitates strand exchange with homologous DNA. Alternatively, RAD51P binds single-stranded DNA in a gap, forming a filament on adjacent double-stranded DNA to promote RNA hybridization with homologous sequences (Wahba and Koshland 2013). In *Arabidopsis thaliana*, the long noncoding RNA APOLO recognizes its targets via short sequence complementarity and facilitates R-loop formation (Ariel et al. 2020). Upon invading target DNA, APOLO decoys components of the polycomb repressive complex, such as LHP1, thereby influencing local chromatin three-diemnsional structure. R-loops can also be artificially generated during

CRISPR-Cas-mediated genome editing (Xiao et al. 2017). Cas9 proteins guide small RNAs (gRNAs) to recognize target DNA, facilitating RNA invasion and R-loop formation that enables base or prime editing. Structural visualization of the Cas9-gRNA-DNA complex has shown that the R-loop is buried mainly within the Cas9 protein (Jiang et al. 2016). These insights underscore the diverse mechanisms and factors that drive R-loop formation, shaping their roles in biological processes and genome engineering applications.

2.2 *Factors Resolving R-Loops*

One conserved feature of R-loops is their transient nature, as they are quickly removed by cellular mechanisms (Fig. 1) designed to prevent the accumulation of unscheduled R-loops that can disrupt transcription and compromise genome integrity. The most well-characterized factor responsible for removing R-loops is RNase H, a conserved ribonuclease from bacteria to humans that specifically degrades the RNA strand of RNA hybrids. Studies across diverse organisms, including bacteria, yeast, and human cells, show that depletion or inactivation of RNase H leads to increased R-loop formation. Conversely, overexpression of RNase H alleviates R-loop-associated phenotypes and eliminates DNA:RNA hybrid signals detected by the S9.6 antibody (Drolet et al. 1995; Huertas and Aguilera 2003). In addition to RNase H, RNA-dependent ATPases with DNA-RNA hybrid unwinding activity represent another major class of R-loop-resolving factors, such as SETX, FANCM, AQR, DDX1, DDX5, and DDX19 (García-Muse and Aguilera 2019). For example, SETX is known to remove R-loops at transcription termination sites (Skourti-Stathaki et al. 2011), while AQR, a DEAD-like helicase, prevents R-loop accumulation and the associated DNA damage (Sollier et al. 2014). Similarly, DDX19, a helicase involved in mRNA export, limits R-loop formation and prevents R-loop-induced DSBs in proliferating cells (Hodroj et al. 2017). However, there are limitations to the current understanding of helicase activity in R-loop resolution. Many helicases have not been shown to specifically unwind RNA:DNA hybrids over double-stranded RNA, and their in vivo functionality remains unclear. Recent proteomic studies of DNA:RNA hybrid pull-downs have identified over 350 factors enriched in RNA:DNA hybrid-binding fractions, including more than 25 DEAD-box helicases (Cristini et al. 2018; Wang et al. 2018). While these studies highlight potential players in R-loop metabolism, there is limited overlap between the two datasets. Key factors like RNase H were not detected in the pull-down experiments. Further validation is necessary to confirm the roles of these identified proteins in R-loop dynamics.

2.3 Detection of R-Loops and Their Distribution on the Genome

Effective detection of RNA:DNA hybrids is essential for mapping their genomic distribution, tracking their dynamics, and understanding their regulatory roles. A range of methodologies has been developed for this purpose, spanning from classical techniques like electron microscopy (EM) to advanced immunoprecipitation (IP) and sequencing approaches. EM was one of the first techniques employed to visualize R-loops directly (Backert 2002). Another early method involved isolating and physical analysis of nucleic acids resistant to RNase A but susceptible to Rnase H digestion (Huertas and Aguilera 2003). Subsequently, DNA:RNA immunoprecipitation (DRIP) with the monoclonal S9.6 antibody, which specifically recognizes RNA:DNA hybrids, has become the primary tool for R-loop detection. This method has been adapted for various applications, including dot blot assays, immunofluorescence (IF), and sequencing of immunoprecipitated nucleic acids. Other methods include chromatin IP and/or IF using catalytically inactive Rnase H1 or Rnase H1 hybrid-binding domains fused to fluorescent proteins to visualize R-loops (Chen et al. 2017). In DRIP and IF assays, the inclusion of Rnase H1 treatment is critical to confirm specificity, as the S9.6 antibody also cross-reacts with double-stranded RNAs (dsRNAs), particularly in regions with high dsRNA abundance (Hartono et al. 2018). Genome-wide profiling of R-loops can be achieved using DRIP-Seq, which sequences the DNA strand involved in R-loops, or DRIPc-Seq, which targets the RNA strand via complementary DNA synthesis (Sanz et al. 2021). Enhanced techniques include S1-DRIP-Seq, which incorporates S1-nuclease treatment to degrade displaced single-stranded DNA before sequencing (Wahba et al. 2016), and bisDRIP-Seq, which combines sodium bisulfite treatment with the S9.6 antibody to detect promoter-associated R-loops (Dumelie and Jaffrey 2017). Recently, qDRIP-Seq has employed synthetic RNA:DNA hybrid standards as spike-ins to quantify R-loops (Crossley et al. 2020). The insights gained from these methods have established that R-loops are widely distributed across genomes and are particularly enriched in highly expressed genes (Fig. 2). These include loci associated with ribosomal RNA (rRNA) and transfer RNA (tRNA), transposable elements, centromeres, telomeres, ribosomal DNA (rDNA), antisense RNA, and noncoding RNA (ncRNA) regions (Niehrs and Luke 2020). In mammalian cells, transcriptional activity is the strongest predictor of R-loop formation, with enrichment occurring at promoter and terminator regions. R-loop-prone regions often correlate with high GC content, including GC-rich sequences, CpG islands, transcription termination sites with GC skew, and G-quadruplex (G4)-containing sequences. This characteristic distribution reflects the interplay between genomic sequence features and transcriptional activity in driving R-loop formation and stability.

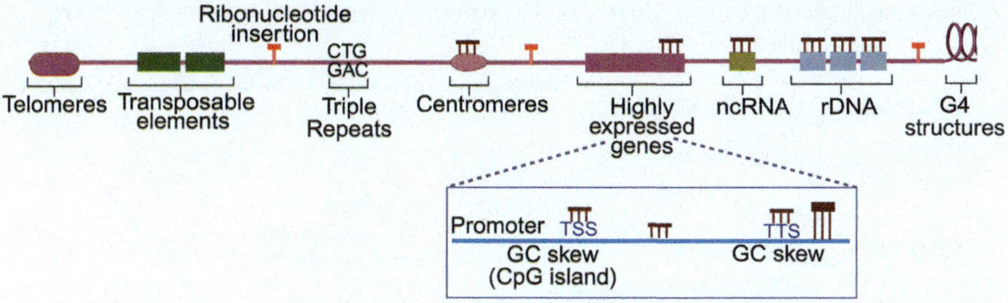

Fig. 2 R-loops across the genome. R-loops are widely distributed across genomes and are particularly enriched in highly expressed genes. These include loci associated with ribosomal RNA (rRNA) and transfer RNA (tRNA), transposable elements, centromeres, telomeres, ribosomal DNA (rDNA), antisense RNA, and noncoding RNA (ncRNA) regions (Niehrs and Luke 2020). In mammalian cells, transcriptional activity is the strongest predictor of R-loop formation, with enrichment occurring at promoter and terminator regions. R-loop-prone regions often correlate with high GC content, including GC-rich sequences, CpG islands, transcription termination sites with GC skew, and G-quadruplex (G4)-containing sequences

3 R-Loops in Transcriptional and Epigenetic Regulation

Once viewed as transcriptional byproducts, R-loops are now recognized as integral to gene expression and epigenetic regulation. Their enrichment at gene promoters, particularly at unmethylated CpG islands with a positive GC skew, highlights their regulatory roles in transcription. DRIP-Seq studies in human Ntera2 cells have shown that these R-loops protect against DNA methylation by DNMT3B1, a silencing enzyme, and are associated with active transcription markers such as H3K4me3, H3K79me2, and H4K20me1 (Ginno et al. 2012, 2013). These findings suggest that R-loops may facilitate transcriptional activation through their crosstalk with epigenetic signatures (Fig. 3a). In some instances, R-loop-dependent transcriptional regulation is mediated by antisense RNA (Fig. 3b). For example, the lncRNA ANRASSF1 forms R-loops that recruit PRC2 to silence the RASSF1A promoter (Beckedorff et al. 2013). In another expression outcome, TARID, an antisense transcript of TCF21, forms an R-loop that recruits GADD45A and TET1, promoting DNA demethylation and activating TCF21 transcription (Arab et al. 2019). Approximately 4% of TET1 binding sites in mouse embryonic stem cells coincide with R-loops, further supporting their role in epigenetic modulation. Additionally, R-loops at the promoter of the VIM gene in cancer cells enhance transcription by reducing nucleosome occupancy and facilitating transcription factor recruitment, such as NF-κB (Boque-Sastre et al. 2015). Beyond *cis*-acting regulation, R-loops also mediate *trans*-acting effects. For example, GATA3-AS1, a divergent lncRNA near GATA3, forms an R-loop loop approximately 2 kb upstream of its TSS, potentially recruiting the MLL methyltransferase to promote GATA3 expression (Gibbons et al. 2018). Interestingly, this effect can occur even when GATA3-AS1 is expressed from plasmids, suggesting trans-regulatory capabilities. Furthermore, R-loops also serve as

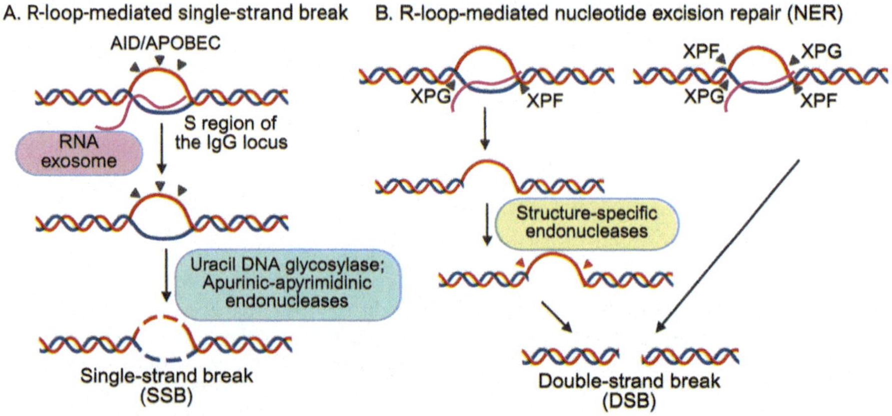

Fig. 3 Regulatory roles of R-loops in transcription. (**a**) R-loops can prevent DNA methylation by blocking the activity of DNA methyltransferases such as DNMT1. This function is associated with markers of active transcription, including H3K4me3, H3K79me2, and H4K20me1. R-loops can enhance gene expression by recruiting transcription factors or ten-eleven translocation (TET) DNA demethylases, removing methylation marks from promoter DNA, and activating transcription. (**b**) R-loops can act as platforms for recruiting repressive chromatin modifiers, leading to silencing of gene promoters and repression of gene expression. (**c**) Mechanisms by which R-loops lead to pausing of elongating RNA polymerase II. (1) Pol II Backtracking: Pol II oscillates between transcription elongation and backtracking. The presence of an R-loop may terminate the function of backtracked Pol II. (2) Torsional Stress: R-loop remains fixed relative to the DNA helix while elongating Pol II rotates along the helical path. The nascent transcript produced downstream of the R-loop must wrap around the DNA, ultimately causing Pol II arrest due to this unfavorable torsional stress. (3) Poll arrest by termination effectors: R-loops may promote dsRNA synthesis, which recruits RNase Dicer and the histone methyltransferase G9a. This creates a repressive chromatin environment, further pausing Pol II. Enzymes such as BRCA1, SETX, DHX9, DDX5, and XRN2 are critical in recognizing and resolving R-loops, ensuring proper transcription termination. Additionally, m6A modifications have been shown to bind R-loops, facilitating efficient transcriptional termination

promoters for antisense transcription, as shown by their ability to stimulate RNA polymerase II activity in plasmids. Removing R-loops across the human genome by RNase H1 overexpression selectively reduces antisense transcription (Tan-Wong et al. 2019).

In addition to their roles at promoters, R-loops accumulate at the 3′ ends of approximately 2000 human genes with GC-skewed sequences, suggesting a role in transcription termination (Ginno et al. 2012). These G-rich sequences stabilize R-loops, creating intrinsic pause sites for RNA polymerase II (Pol II) (Aguilera and García-Muse 2012). In gene-dense regions, R-loop-dependent transcription termination likely prevents transcriptional read-through, a process that could disrupt proper gene regulation and cellular function. Several mechanisms have been proposed to explain how R-loops-mediated transcription termination include backtracking, torsional stress, and heterochromatinization (Fig. 3c) (Niehrs and Luke

2020). In the backtracking termination model, Pol II oscillates between elongation and backtracking, and the formation of an R-loop may terminate the function of backtracked Pol II. The torsional stress termination suggests that the R-loop remains fixed relative to the DNA helix while elongating Pol II rotates along the helical path. The nascent transcript produced downstream of the R-loop must wrap around the DNA, ultimately causing Pol II arrest due to this unfavorable torsional stress. The heterochromatinization model proposes that R-loops induce antisense RNA production, forming dsRNA at termination sites, which recruits the RNA interference Dicer and the histone methyltransferase G9a, leading to heterochromatin formation and Pol II pausing (Skourti-Stathaki et al. 2011). Despite these roles, R-loop-enriched 3′ ends exhibit active transcription markers, implying they might attract specific termination factors (Ginno et al. 2012). Emerging evidence links R-loops to N6-methyladenosine (m6A) modifications. m6A can promote R-loop formation to ensure efficient transcription termination, while its reader, YTHDF2, prevents cotranscriptional R-loop accumulation during mitosis (Yang et al. 2019; Zhang et al. 2024). This dual role of m6A in promoting R-loops and their removal highlights the complexity of their regulatory functions. These studies demonstrate R-loops' multifaceted roles in transcriptional and epigenetic regulation.

4 R-Loops in Chromatin Architecture

R-loops play a significant role in shaping chromatin architecture, influencing both local chromatin states and the broader three-dimensional genome organization. Genome-wide mapping studies have revealed that R-loops are enriched at loci with reduced DNA methylation and increased DNase hypersensitivity, correlating with active chromatin states (Nadel 2015). Promoters prone to R-loop formation are marked by histone modifications associated with transcriptional activation, including H3K4me1, H3K4me3, H3K27ac, and H3K36me3. Conversely, R-loops have also been implicated in heterochromatin formation (Nakama et al. 2012) and chromatin compaction (Castellano-Pozo et al. 2013), demonstrating their dual role in chromatin regulation. In mouse embryonic stem cells (ESCs), R-loops influence promoter-proximal chromatin by modulating the binding of chromatin-regulatory complexes (Chen et al. 2015). R-loop-prone promoters are preferentially associated with the TIP60-p400 complex, linked to gene activation, rather than the repressive polycomb complex PRC2. Disruption of R-loops through RNase H overexpression reduces TIP60-p400 occupancy while enhancing PRC2 recruitment. However, at specific loci, R-loops can actively recruit PRC complexes to repress developmental regulator genes, indicating that the factors determining whether R-loops attract or repel PRC complexes remain poorly understood.

Beyond their impact on local chromatin, R-loops have emerged as pivotal regulators of three-dimensional genome architecture. They colocalize with key architectural proteins such as CTCF, the cohesin complex component RAD21, and ZNF143, particularly at gene terminator regions (Sanz et al. 2016). This association has recently been confirmed on a genome-wide scale (Zhang et al. 2023). In *trans* R-loop, formation by lncRNA HOTTIP at many genomic sites was observed to promote the binding of CTCF and the cohesin complex. This activity enhances TAD boundaries and strengthens promoter-enhancer contacts at oncogenes, collectively driving cancer-associated gene expression programs (Luo et al. 2022). A subsequent study has shown that R-loops removal by RNase H reduced CTCF occupancy at many genomic loci. At the same time, the stabilization of G4s enhanced CTCF binding and strengthened insulation of TAD boundaries, in addition to facilitating the formation of new chromatin loops (Wulfridge et al. 2023). Together, these studies highlight R-loops' broader role in shaping chromatin interactions and facilitating higher-order genome organization.

5 R-Loops in DNA Damage and Repair

R-loops are crucial in the cellular response to DNA damage, particularly in the repair of DSBs, where they play an essential role in promoting homologous recombination (HR). These R-loops form at DSB sites either through RNA-DNA hybridization on the ssDNA overhangs of DSBs or by RNA invasion into dsDNA (Petermann et al. 2022). The formation of R-loops at DSBs is often facilitated by de novo RNA synthesis. For instance, small RNAs generated by Dicer and Drosha have been shown to form hybrids around DSBs (Francia et al. 2012). These DSB-induced small RNAs (diRNAs) were reported to form a complex with AGO 2 and RAD51, facilitating RAD51 localization for repair (Gao et al. 2014). Similarly, DNA damage-induced long noncoding RNAs (dilncRNAs) can form hybrids at DSB sites, cooperating with γH2AX to drive liquid-liquid phase separation (LLPS) of DNA-damage response (DDR) factors into DDR foci (Pessina et al. 2019). DilncRNAs also aid HR repair by recruiting key repair proteins, including BRCA1, BRCA2, and RAD51 (D'Alessandro et al. 2018). Notably, BRCA1 has been identified as a sensor of R-loops, as its recruitment to DNA damage sites is reduced following RNase H treatment in S/G2-phase cells.

Genome-wide analyses of R-loops at DSB sites have shown their accumulation in actively transcribed genes, suggesting that preexisting RNA transcripts may contribute to damage-induced R-loop formation. Research has demonstrated that R-loops form specifically during transcriptional activity when DSBs or single-stranded breaks (SSBs) are induced by reactive oxygen species (ROS). These R-loops may lead to Pol II pausing before or after the break (Teng et al. 2018). Additionally, the exposure of ssDNA in R-loops at active genes may also trigger de novo RNA synthesis, further contributing to R-loop formation at DNA damage sites (Marnef and Legube 2021). Repair of ROS-induced DSBs requires a noncanonical HR pathway regulated by CSB, RAD52, and RAD51. Both CSB and RAD52 can directly recognize R-loops in vitro at DNA damage sites, including those near telomeres, to facilitate repair (Keskin et al. 2014; Teng et al. 2018). Furthermore, RAD51AP1 has been shown to drives R-loop formation and promotes HR only in the presence of active transcription (Ouyang et al. 2021). Overall, these studies demonstrate the critical role of R-loops in DSB repair, particularly in facilitating the recruitment of DNA repair factors and enhancing HR efficiency.

6 R-Loops Disrupt Genome Stability

The accumulation of unexpected R-loops is detrimental to cellular physiology and can contribute to genomic instability. If not properly resolved, misregulation of R-loops lead to an increased frequency of unscheduled recombination repair events and elevated numbers of repair foci containing recombination factors. One mechanistic explanation is that the ssDNA generated throng R-loop formation is more vulnerable to damage by nucleases and genotoxic agents (Fig. 4a). For example, AID/APOBEC enzymes can promote the conversion of cytosine (dC) to uracil (dU) residues specifically on ssDNA. This modification renders the DNA susceptible to uracil DNA glycosylase, which excises the uracil base, creating an abasic site and generating a DNA lesion (Pecori et al. 2022). Another structural vulnerability of R-loops lies in the DNA flaps that form at either end of the hybrid. These structures can be cleaved by structure-specific endonucleases, such as XPF and XPG, which are involved in nucleotide excision repair (NER) pathways (Fig. 4b) (Sollier et al. 2014). R-loops can also lead to the formation of DSBs by causing physical collisions with the replication machinery, which can disrupt replication fork progression and leads to genome instability (Fig. 4c) (Tuduri et al. 2010; Wahba et al. 2013). In human cells with the absence ASF or TOP1, R-loop-induced DSBs depend on replication and directly affect replication fork progression (Gan et al. 2011). R-loops can act as formidable physical barriers to replication forks, as their relatively stable secondary structures are associated with the transcriptional machinery and RNA-processing enzymes. These barriers can directly lead to replication fork collapse and DSB formation, even in the absence of nuclease-mediated processing by enzymes such as XPF and XPG. Additionally, mutations in genes involved in mRNP biogenesis, RNA export, and pre-mRNA splicing, have been linked to elevated R-loop formation and subsequent DNA damage, as well as transcription-associated recombination (Niehrs and Luke 2020). However, the precise mechanisms by which these RNA-related factors influence R-loop resolution and prevent genome instability remain poorly understood.

R-loops also disrupt chromatin reorganization, particularly during replication. The abnormal accumulation of R-loops can lead to replication stress, especially in regions of the genome with condensed chromatin (Fig. 4d). R-loops are associated with gene-repressive histone modifications, such as H3K9me2 and H3S10p, and with compacted chromatin. H3K9me2 can recruit HP1 to form heterochromatin. The compact chromatin induced by R-loops interferes with DNA replication and increases genome instability (Petermann et al. 2022). Fragile sites in the human genome are particularly prone to premature chromatin condensation under replicative stress, which increases susceptibility to DNA breakage and genomic fragility.

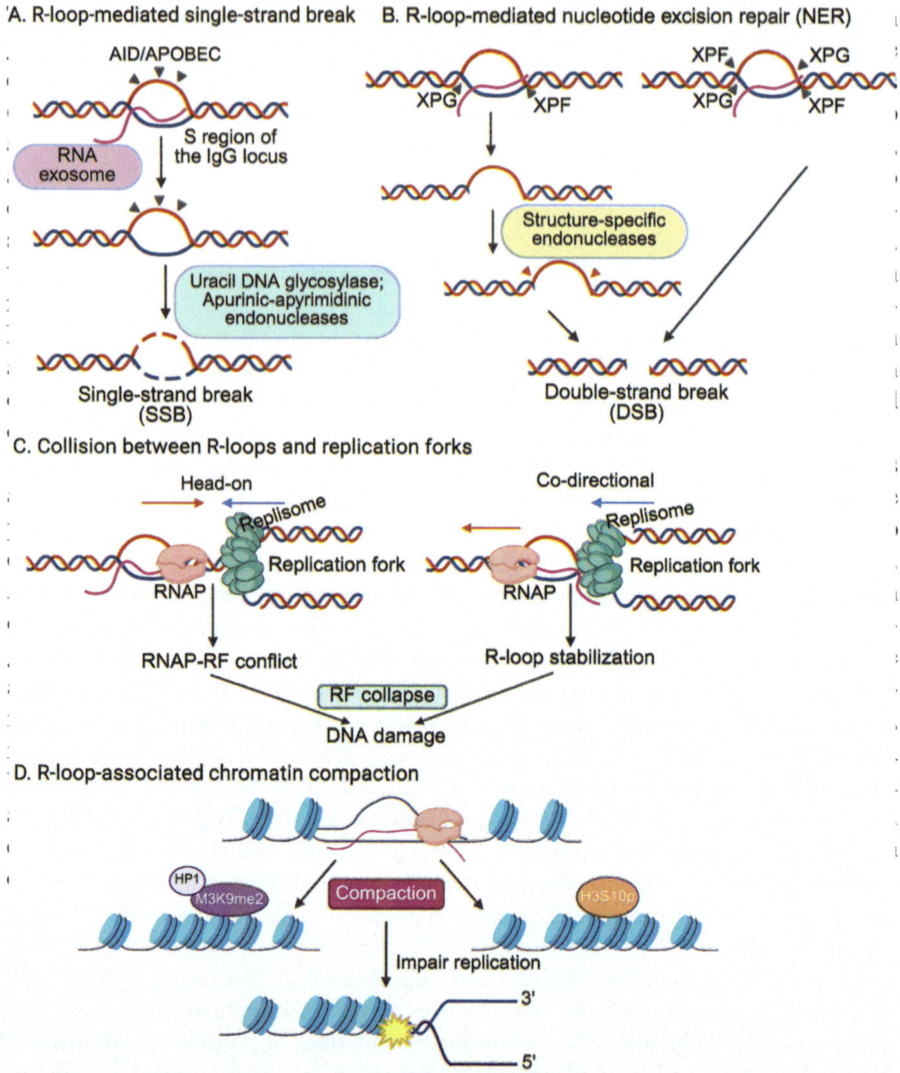

Fig. 4 R-loop-associated genome instability. (**a**) R-loop-induced single-strand breaks. R-loops can lead to single-strand DNA breaks through the activity of activation-induced cytidine deaminase (AID) or APOBEC enzymes, which deaminate cytosine residues on the exposed nontemplate DNA strand. Cytosine deamination on the template strand can also occur once the RNA exosome displaces RNA. The resulting uracil base is excised by uracil DNA glycosylase, creating an abasic site. This site is processed by apurinic/apyrimidinic (AP) endonucleases, leading to the formation of an SSB. (**b**) DNA flaps at the ends of the RNA:DNA hybrid structures may be cleaved by structure-specific endonucleases such as XPF and XPG, which participate in the NER pathway. This cleavage resolves R-loop-associated distortions but can also introduce genome instability if improperly repaired. (**c**) Collisions between R-loops and replication forks. R-loops can also lead to the formation of DSBs by causing physical collisions with the replication machinery (codirectional collision and head-on collision), which can disrupt replication fork progression and lead to genome instability. (**d**) R-loops are often associated with repressive chromatin modifications, such as histone H3 Lys9 di-methylation (H3K9me2) and histone H3 Ser10 phosphorylation (H3S10p). These modifications contribute to chromatin compaction and gene silencing. H3K9me2 can further recruit HP1, which enhances heterochromatin formation. The compacted chromatin environment interferes with DNA replication and exacerbates genome instability

7 R-Loops in Neurological Disorders

Neurons are particularly vulnerable to DNA damage due to their unique characteristics, including high energy demands, long lifespan, and elevated transcriptional activity, all of which make them prone to R-loop accumulation and dysregulation. Unlike proliferating cells, postmitotic neurons do not undergo cell division, which means that any accumulated R-loops cannot be easily resolved through cell division-based repair mechanisms. As a result, excessive R-loop formation in neurons can cause transcription-associated DNA damage and genomic instability, even without interference from DNA replication. These disruptions are implicated in various neurological diseases, particularly those related to repeat expansions and ASD. In these conditions, misregulated R-loops contribute to the pathogenesis by causing both transcriptional and genomic instabilities, which disrupts neuronal function.

7.1 *Fragile X Syndrome (FXS) and Fragile X-Associated Tremor/Ataxia Syndrome (FXTAS)*

FXS and FXTAS are caused by a trinucleotide repeat expansion (CGG) in the 5′UTR of the FMR1 gene on the X chromosome (Jin and Warren 2000). Under normal conditions, individuals typically have fewer than 50 CGG repeats. FXS is caused by expansions of more than 200 CGG repeats, leading to hypermethylation of the FMR1 gene and subsequent silencing of its expression. This results in a loss of the FMRP, a key player in synaptic plasticity and cognitive function, which is responsible for the intellectual disability and ASD associated with FXS. On the other hand, FXTAS is associated with expansions of CGG repeats in the premutation range (55–200 repeats). Unlike FXS, FXTAS does not involve a loss of FMRP expression but instead is marked by elevated FMR1 mRNA levels, which are thought to contribute to neurodegeneration through toxic RNA gain-of-function mechanisms.

R-loops could influence FMR1 expression. In cells derived from FXS, R-loops are transcription-dependent and stable, forming at the expanded CGG repeat region (Loomis et al. 2014; Colak et al. 2014). These R-loops colocalize with the repressive H3K9me2 chromatin mark, potentially impeding Pol II transcription of downstream sequences. Intriguingly, the formation of R-loops at the CGG repeat region appears to be independent of the repressive H3K9me2 mark, suggesting that R-loops may act as an initial trigger for FMR1 gene silencing. Furthermore, expansion of CGG repeats in the full mutation range may promote the formation of longer R-loops, which are prone to forming complex secondary structures. These structures could disrupt the protective role of R-loops against DNA methylation and trigger epigenetic silencing of FMR1 (Colak et al. 2014; Kraan et al. 2019). Efforts to restore FMR1 expression in FXS cells have linked between R-loops to repeat dynamics. Contraction of CGG repeats in patient-derived pluripotent stem cells and embryonic stem cells has been associated with FMR1 reactivation. For instance, targeting repeats in the FMR1 promoter with a dCas-Tet1 construct in the absence of the inhibitor treatment achieved a 90% restoration of FMR1 mRNA and 73% restoration of FMRP levels (Liu et al. 2018). Additionally, a partial restoration of FMR1 mRNA (45%) FXS-derived neurons restored functionality to levels comparable to wild-type neurons. Moreover, tethering RNase H to dCas9 to target the repeats in the FMR1 promoter abolishes R-loops within this region and leads to a loss of repeat contraction, demethylation, and reactivation of FMR1, even in the presence of inhibitor treatments. These findings indicate that even without catalytically active TET1/TET2, R-loops are essential for repeat contraction and subsequent FMR1 reactivation, likely by inducing DNA damage response at those sites to facilitate repeat contraction.

The expanded repeat is inherited due to repeat instability that occurs during germ cell generation and during early stages of embryogenesis (López Castel et al. 2010). While replication forks stall at CGG repeat regions of FMR1, the direction of the replication fork through the repeats differs at the endogenous FMR1 locus in FXS hESCs compared to control hESCs (Gerhardt 2017). This is attributed to the absence of replication initiation upstream of FMR1, which typically occurs in normal cells. Such altered replication origin usage, coupled with fork stalling, promotes repeat instability during early embryonic development of FXS. A pivotal study illustrated the role of FMRP in maintaining genomic stability by preventing R-loop formation and chromosome breakage. Chakraborty et al. (2020) demonstrated that FXS-derived cells exhibit over a twofold increase in DSBs compared to control cells, suggesting the genomes of FXS cells are inherently unstable (Chakraborty et al. 2020). Notably, these DSBs colocalize with R-loop-forming regions, where RNA transcripts hybridize with homologous DNA, forming RNA:DNA hybrids and displacing single-stranded DNA. The study also showed that exogenous FMRP reduces DSBs within genes prone to R-loop formation during transcription, particularly under conditions of programmed replication-transcription conflicts. However, the exact cause of the increased damage in FXS is still being investigated.

In FXTAS, hyperactive transcription of the FMR1 gene containing repeats is considered as the key driver of the disease pathogenesis, resulting in toxic RNA gain-of-function effects and RAN (repeat-associated non-AUG) protein toxicity. In addition, excessive FMR1 mRNA production due to CGG repeat expansions causes R-loop formation, which could offer transient protection against DNA methylation and silencing. The nontemplate strands of these R-loops are also prone to from secondary structures, which stabilize the R-loop after formation resulting in aberrant R-loop longevity at the FMR1 promoter. However, as R-loops become longer and form complex secondary structures, they might destabilize the genome and promote hypermethylation of the FMR1 locus (Loomis et al. 2014). This shift from protection to silencing is thought to exacerbate instability, contributing to the neurodegenerative features of FXTAS. Furthermore, these aberrant R-loops may trigger a DNA damage response, potentially leading to the activation of inflammatory pathways. Elevated DNA damage markers, such as γH2AX, are observed in FXTAS-derived cells, indicating a persistent DNA damage response, and these markers are found in the characteristic intranuclear inclusions in neurons of FXTAS patients (Iwahashi et al. 2006). However, the precise role of R-loop-mediated DNA damage and its potential dysregulation in FXTAS pathogenesis remains to be fully elucidated.

7.2 *Friedreich's Ataxia (FRDA)*

FRDA is a rare, neurodegenerative disorder that typically manifests in early childhood and is characterized by progressive degeneration of the cerebellum, spinal cord, and peripheral nerves (Koeppen 2011). This degeneration leads to impaired coordination, balance, gait, and sensory function. In healthy individuals, the FXN gene contains between 7 and 33 GAA repeats, while in individuals with FA, the repeats range from 66 to over 1000. These expanded repeats lead to a decrease in FXN mRNA levels, resulting in a deficiency of frataxin protein, which is essential for mitochondrial function and iron homeostasis (Campuzano et al. 1996). Similar as in FXS, the expanded GAA sequences in FRDA are prone to the formation of stable R-loops and are associated with the formation of heterochromatin and the transcriptional silencing of the FXN gene (Groh et al. 2014). Studies have shown that in cells derived from FA patients, there is a significant enrichment (approximately threefold) of R-loops over the regions of the FXN intron 1, which harbor the expanded GAA repeats. These R-loops can be removed by treatment with RNase H2. The levels of R-loops in FRDA patient-derived cells are found to correlate with the length of the GAA repeat expansion, with longer expansions leading to more abundant R-loops. Elevating R-loop levels through inhibition of DNA Topoisomerase I, which is involved in resolving DNA supercoiling during transcription, further increases the enrichment of repressive histone marks such as H3K9me2 at the FXN locus (Groh et al. 2014). This suggests that the accumulation of R-loops at the expanded GAA repeat regions contributes to the recruitment of repressive chromatin marks and silencing of FXN gene expression. This dysregulation of FXN expression might be a key factor in the pathogenesis of FRDA.

7.3 *Amyotrophic Lateral Sclerosis (ALS) and Frontotemporal Dementia (FTD)*

ALS and FTD are two closely related neurodegenerative disorders characterized by progressive cell degeneration in the brain. ALS predominantly involves the degeneration of motor neurons, leading to muscle weakness and paralysis, while FTD is associated with cognitive and behavioral impairments resulting from atrophy of the frontal and temporal lobes (Kiernan et al. 2011). Notably, 35–50% of ALS patients exhibit cognitive or behavioral changes, and approximately 20% progress to develop FTD (Grad et al. 2017). Both disorders share common pathological features, including abnormal accumulations of TDP-43 protein in nerve cells, with a smaller subset of cases also showing inclusions of the FUS protein. One of the most common genetic causes of familial ALS and inherited FTD is the expansion of a hexanucleotide repeat (GGGGCC) in the C9orf72 gene. The GGGGCC (G4C2) repeats occurs between two noncoding first exons upstream of the translation start site in the C9orf72 gene. While healthy individuals have fewer than 30 GGGGCC (G4C2) repeats, individuals with ALS can carry hundreds or thousands of repeats (Haeusler et al. 2014). These expansions are transcribed bidirectionally, producing sense and antisense transcripts that undergo repeat-associated non-AUG (RAN) translation, resulting in toxic dipeptide repeat proteins.

In ALS patient-derived motor neurons, increased levels of DNA damage markers (e.g., γH2A.X, p-ATM, 53BP1) suggest that R-loop accumulation and dysregulation may play a key role in pathogenesis (Farg et al. 2017). It has been found that the G4C2 DNA forms G-quadruplex structures in the displaced DNA strand (G4-DNA), while the G4C2 RNA forms G4-RNA that can bind to repeat DNA to facilitate the formation of R-loops (Haeusler et al. 2014). These R-loops may contribute to repeat instability and altered C9orf72 gene expression through mechanisms such as polymerase stalling, antisense transcription, and epigenetic modification. For example, R-loops might mediate reduced C9orf72 gene expression as observed in patient cells of ALS and the loss of full-length transcripts by introducing nucleolar stress within the repeats region. Trimethylation of H3K9, H3K27, and H4K20 has been linked to the reduced C9orf72 gene expression in ALS/FTD patient brain tissues and fibroblasts carrying the expanded repeat (Belzil et al. 2014), indicating that R-loops might trigger repressive histone modification marks to decrease C9orf72 gene expression. In addition, the C9orf72 hexanucleotide repeats can produce antisense transcripts which also exhibit histone methylation. The R-loop-dependent transcriptional silencing of the C9orf72 gene could be associated with these antisense transcripts.

Additionally, mutations in SETX and TDP-43 are among the most common contributors to ALS. SETX is a RNA/DNA helicase that has been shown to unwind RNA:DNA hybrids, suppress transcription-associated recombination, and prevent DSB accumulation (Kannan et al. 2022; Cohen et al. 2018). Loss-of-function mutations in SETX lead to AOA2, while gain-of-function mutations cause ALS4. Overexpression of SETX has been shown to reduce R-loop-driven DNA damage in cells with high expression of G4C2 RNA repeats, suggesting that SETX helps mitigate the toxic effects of R-loop accumulation. Although excessive R-loops are toxic, an abnormal loss of R-loops can also have severe consequences. R-loops are critical for various biological pathways, often serving as signals to recruit other proteins for normal cellular process. In ALS4, many R-loops are lost at gene promoter regions, which leads to increased DNA methylation and subsequent transcriptional silencing of key genes. For example, ALS4 patients experience R-loop loss at the BAMBI promoter (Grunseich et al. 2018); a regulator of TGFB signaling therefore leads to disruptions in downstream gene expression that exacerbate disease progression. This highlights the dual nature of R-loops in cellular homeostasis, as both excessive accumulation and depletion can contribute to neurodegenerative pathologies.

TDP-43 is a key protein implicated in ALS and FTD. TDP-43 protein aggregation is observed in 97% of all ALS cases, and mutation of TDP-43 are found in 3% of familial ALS (fALS) and 2% of sporadic ALS (sALS) cases (Grad et al. 2017). Loss of TDP-43 in human cells results in increased yH2A.X and DSB levels, as well as R-loop accumulation. These events are likely responsible for the replication fork stalling observed in TDP-43 knockdown models, as inhibiting transcription in these models reduces R-loop accumulation and alleviates replication stress (Wood et al. 2020). As an RNA-binding protein, TDP-43 potentially prevent RNA:DNA hybrid formation by binding to RNA during transcription, thereby allowing replication forks to proceed unhindered and therefore preventing aberrant DNA damage. Recent work has demonstrated that TDP-43 is involved in coordinating R-loop dynamics and regulating 5hmC in coding genes, distal regulatory elements, and transposable elements (Hou et al. 2024). These findings suggest that TDP-43 plays a central role in maintaining genomic integrity, and disruptions to its function contribute to the R-loop-associated genomic instability observed in ALS and FTD.

7.4 Autism Spectrum Disorder (ASD)

ASD is one of the most prevalent neurodevelopmental disorders, affecting approximately 1 in 36 children (Wormwood et al. 2023). It presents a wide range of phenotypes, including abnormalities in the gastrointestinal system, epilepsy, intellectual disability, sleep disturbances, and motor difficulties. The etiology of ASD involves complex interactions between genetic and environmental factors, with heritability estimates ranging from 40% to 80%. Functional clustering of over 350 autism-associated genes highlights significant enrichment for genes related to FMRP and β-catenin/chromatin remodeling networks (O'Roak et al. 2012; Iossifov et al. 2014). Vigilin, a ubiquitously expressed RNA-binding protein (RBP), plays critical roles in heterochromatin formation, chromosome segregation, and mRNA stability, showing notable structural similarities to FMR1 (Mushtaq et al. 2023). Hap loin sufficiency of Vigilin has been linked to ASD. Studies of neural progenitor cells (NPCs) derived from ASD-affected individuals have revealed elevated DNA DSBs, particularly in long genes involved in crucial cellular functions like cell migration and apical polarity (Wang et al. 2020). This genomic instability is linked to increased γ-H2AX signaling, a marker of DNA damage, and the improper recruitment of DSB repair proteins such as Rad51 and BRCA1 (Banday et al. 2021). DSB density in ASD-derived NPCs correlates with gene expression, suggesting that transcription-replication conflicts contribute to replication stress (Wang et al. 2020). This occurs when the replication machinery collides with the stalled transcriptional machinery, which is usually colocalized with R-loop accumulation. Therefore, maintaining proper Vigilin levels is crucial in preventing replication-transcription conflicts and ensuring genome stability in ASD.

Research by Sarma and colleagues has identified several chromatin regulators that are mutated in ASDs and regulate R-loop, such as ATRX and ADNP (Yan et al.

2022). Mutations in ATRX, a chromatin remodeler, have been shown to change histone binding and chromatin remodeling activities cluster in its PHD finger or the helicase domain, respectively (Bieluszewska et al. 2022). ATRX prevents R-loop formation by interacting with RNAs and inhibiting RNA:DNA hybrid structures. Similarly, ADNP, which encodes activity-dependent neuroprotective protein, dilutes R-loops both in vitro and in vivo. Deletion of the ADNP homeodomain results in R-loop accumulation at ADNP target sites and impairs neuronal differentiation. Moreover, patient-derived iPSCs carrying an ADNP syndrome-causing mutation exhibit R-loop and CTCF accumulation at ADNP targets, highlighting a unique role of ADNP-mediated R-loop resolution in neuronal function and developmental disorders (Bieluszewska et al. 2022). A recent genome-wide R-loop mapping study in the developing human brain revealed that R-loops are abundant in the progenitor-rich germinal matrix, particularly at promoters of genes poised for upregulated expression during later differentiation stages (LaMarca et al. 2023). Notably, a collection of R-loop primed genes is significantly enriched for risk loci associated with ASD, including nuclear protein AUTS2, multiple cell adhesion molecules (NRXN1, CDH7, and CDH8), and ion channels like CACNA1B. These findings suggest that R-loops may interact with ASD-associated risk loci, influencing gene expression programs that are critical for neuronal development and contributing to the pathogenesis of the disorder. Further investigations are essential to elucidate the broader impact of ASD-associated R-loop regulators on neurodevelopment and to determine how R-loop dysregulation affects neuronal gene expression programs.

8 Conclusion and Future Perspective

Advancements in biotechnology and analytical tools over the past decade have greatly enhanced researchers' ability to map R-loops across the genome, identify R-loop-associated proteins, and investigate their effects in various biological contexts. These findings have established that R-loops are not merely byproducts of transcription but also act as key regulators of cellular homeostasis, potential drivers of genome instability, and contributors to disease pathogenesis. However, their dual benefits in some contexts and detrimental in others remain a major challenge in understanding their precise biological roles. R-loops' effects depend on several factors, including their genomic context, the cell cycle stage at which they form, their size and subcellular localization, their dynamics, and how they influence chromatin status (Petermann et al. 2022). Additionally, current techniques are limited in determining whether detected R-loops are continuous or discontinuous and need more resolution to profile their spatial-temporal changes at the single-cell level. Future studies will aim to develop detection methods to achieve single-cell resolution and real-time profiling of R-loop dynamics. While numerous R-loop regulators have been identified, their roles in the brain still need to be better understood. Investigating why these regulators are essential and how they function in a context-specific manner is critical to uncovering their contribution to neurodevelopment and neuropathology.

R-loops have emerged as critical contributors to the pathogenesis of several neurological diseases, including neurodegenerative disorders and ASD. As discussed above, R-loops' involvement in repeat instability and genome instability suggests that they could be a promising target for new therapeutic approaches. Despite their potential, R-loops have yet to be extensively explored in drug design. Small molecules can nonspecifically bind to the exterior and specific grooves of nucleic acid structures. This presents an opportunity to design molecules that selectively target R-loops and mitigate their effects on genome instability. Several ligands, such as aminoglycosides (e.g., neomycin and paromomycin), ethidium bromide, and polyamides (e.g., distamycin and netropsin), have been identified to bind RNA:DNA hybrids. However, these ligands generally lack specificity, as they also interact with dsDNA and RNA. This nonspecific binding can lead to mutagenic effects or other unpredictable consequences (De Magis et al. 2019). A recent study has demonstrated the potential of inducing site-specific R-loop formation for treating fragile X syndrome (Lee et al. 2023). The study revealed that MEK and BRAF inhibitors can promote the R-loop formation between DNA and RNA, which cells recognize as DNA damage. This triggers DNA repair mechanisms that excise expanded CGG repeats, restoring them to normal levels and enabling the reexpression of the FMR1 gene. This study opens a new avenue for investigating therapeutic strategies targeting repeat expansion disorders. It will be important to establish clinically applicable assays to measure R-loop accumulation in patients, to elucidate how R-loops and associated genomic instability contribute to pathogenesis, and to develop new strategies to alleviate or exploit R-loop-associated defects in therapy. These efforts could provide transformative insights into R-loops' roles in human health and disease, ultimately leading to novel therapeutic interventions.

References

Aguilera A, García-Muse T (2012) R loops: from transcription byproducts to threats to genome stability. Mol Cell 46:115–124

Arab K, Karaulanov E, Musheev M et al (2019) GADD45A binds R-loops and recruits TET1 to CpG island promoters. Nat Genet 51:217–223

Ariel F, Lucero L, Christ A et al (2020) R-loop mediated trans action of the APOLO long noncoding RNA. Mol Cell 77:1055–1065

Backert S (2002) R-loop-dependent rolling-circle replication and a new model for DNA concatemer resolution by mitochondrial plasmid mp1. EMBO J 21:3128–3136

Banday S, Pandita RK, Mushtaq A et al (2021) Autism-associated vigilin depletion impairs DNA damage repair. Mol Cell Biol 41:e00082–e00021

Beckedorff FC, Ayupe AC, Crocci-Souza R et al (2013) The intronic long noncoding RNA ANRASSF1 recruits PRC2 to the RASSF1A promoter, reducing the expression of RASSF1A and increasing cell proliferation. PLoS Genet 9:e1003705

Belotserkovskii BP, Hanawalt PC (2015) PNA binding to the non-template DNA strand interferes with transcription, suggesting a blockage mechanism mediated by R-loop formation. Mol Carcinog 54:1508–1512

Belzil VV, Bauer PO, Gendron TF et al (2014) Characterization of DNA hypermethylation in the cerebellum of c9FTD/ALS patients. Brain Res 1584:15–21

Bieluszewska A, Wulfridge P, Doherty J et al (2022) ATRX histone binding and helicase activities have distinct roles in neuronal differentiation. Nucleic Acids Res 50:9162–9174

Boque-Sastre R, Soler M, Oliveira-Mateos C et al (2015) Head-to-head antisense transcription and R-loop formation promotes transcriptional activation. Proc Natl Acad Sci USA 112:5785–5790

Campuzano V, Montermini L, Molto MD et al (1996) Friedreich's ataxia: autosomal recessive disease caused by an intronic GAA triplet repeat expansion. Science 271:1423–1427

Castellano-Pozo M, Santos-Pereira JM, Rondón AG et al (2013) R loops are linked to histone H3 S10 phosphorylation and chromatin condensation. Mol Cell 52:583–590

Chakraborty A, Jenjaroenpun P, Li J et al (2020) Replication stress induces global chromosome breakage in the fragile X genome. Cell Rep 32:108179

Chen PB, Chen HV, Acharya D et al (2015) R loops regulate promoter-proximal chromatin architecture and cellular differentiation. Nat Struct Mol Biol 22:999–1007

Chen L, Chen J-Y, Zhang X et al (2017) R-ChIP using inactive RNase H reveals dynamic coupling of R-loops with transcriptional pausing at gene promoters. Mol Cell 68:745–757

Cohen S, Puget N, Lin Y-L et al (2018) Senataxin resolves RNA:DNA hybrids forming at DNA double-strand breaks to prevent translocations. Nat Commun 9:533

Colak D, Zaninovic N, Cohen MS et al (2014) Promoter-bound trinucleotide repeat mRNA drives epigenetic silencing in fragile X syndrome. Science 343:1002–1005

Cristini A, Groh M, Kristiansen MS et al (2018) RNA/DNA hybrid interactome identifies DXH9 as a molecular player in transcriptional termination and R-loop-associated DNA damage. Cell Rep 23:1891–1905

Crossley MP, Bocek MJ, Hamperl S et al (2020) qDRIP: a method to quantitatively assess RNA–DNA hybrid formation genome-wide. Nucleic Acids Res 48:e84

D'Alessandro G, Whelan DR, Howard SM et al (2018) BRCA2 controls DNA:RNA hybrid level at DSBs by mediating RNase H2 recruitment. Nat Commun 9:5376

De Magis A, Manzo SG, Russo M et al (2019) DNA damage and genome instability by G-quadruplex ligands are mediated by R loops in human cancer cells. Proc Natl Acad Sci USA 116:816–825

Drolet M, Phoenix P, Menzel R et al (1995) Overexpression of RNase H partially complements the growth defect of an Escherichia coli delta topA mutant: R-loop formation is a major problem in the absence of DNA topoisomerase I. Proc Natl Acad Sci USA 92:3526–3530

Dumelie JG, Jaffrey SR (2017) Defining the location of promoter-associated R-loops at near-nucleotide resolution using bisDRIP-seq. elife 6:e28306

Dutta D, Shatalin K, Epshtein V et al (2011) Linking RNA polymerase backtracking to genome instability in E. coli. Cell 146:533–543

Farg MA, Konopka A, Soo KY et al (2017) The DNA damage response (DDR) is induced by the C9orf72 repeat expansion in amyotrophic lateral sclerosis. Hum Mol Genet 26:2882–2896

Francia S, Michelini F, Saxena A et al (2012) Site-specific DICER and DROSHA RNA products control the DNA-damage response. Nature 488:231–235

Gan W, Guan Z, Liu J et al (2011) R-loop-mediated genomic instability is caused by impairment of replication fork progression. Genes Dev 25:2041–2056

Gao M, Wei W, Li M-M et al (2014) Ago2 facilitates Rad51 recruitment and DNA double-strand break repair by homologous recombination. Cell Res 24:532–541

García-Muse T, Aguilera A (2019) R loops: from physiological to pathological roles. Cell 179:604–618

Gerhardt J (2017) Epigenetic modifications in human fragile X pluripotent stem cells; Implications in fragile X syndrome modeling. Brain Res 1656:55–62

Gibbons HR, Shaginurova G, Kim LC et al (2018) Divergent lncRNA GATA3-AS1 regulates GATA3 transcription in T-helper 2 cells. Front Immunol 9:2512

Ginno PA, Lott PL, Christensen HC et al (2012) R-loop formation is a distinctive characteristic of unmethylated human CpG island promoters. Mol Cell 45:814–825

Ginno PA, Lim YW, Lott PL et al (2013) GC skew at the 5′ and 3′ ends of human genes links R-loop formation to epigenetic regulation and transcription termination. Genome Res 23:1590–1600

Grad LI, Rouleau GA, Ravits J et al (2017) Clinical spectrum of amyotrophic lateral sclerosis (ALS). Cold Spring Harb Perspect Med 7:a024117

Groh M, Lufino MM, Wade-Martins R et al (2014) R-loops associated with triplet repeat expansions promote gene silencing in Friedreich ataxia and fragile X syndrome. PLoS Genet 10:e1004318

Grunseich C, Wang IX, Watts JA et al (2018) Senataxin mutation reveals how R-loops promote transcription by blocking DNA methylation at gene promoters. Mol Cell 69:426–437

Haeusler AR, Donnelly CJ, Periz G et al (2014) C9orf72 nucleotide repeat structures initiate molecular cascades of disease. Nature 507:195–200

Hartono SR, Malapert A, Legros P et al (2018) The affinity of the S9.6 antibody for double-stranded RNAs impacts the accurate mapping of R-loops in fission yeast. J Mol Biol 430:272–284

Hatchi E, Skourti-Stathaki K, Ventz S et al (2015) BRCA1 recruitment to transcriptional pause sites is required for R-loop-driven DNA damage repair. Mol Cell 57:636–647

Herrera-Moyano E, Mergui X, García-Rubio ML et al (2014) The yeast and human FACT chromatin-reorganizing complexes solve R-loop-mediated transcription–replication conflicts. Genes Dev 28:735–748

Hodroj D, Recolin B, Serhal K et al (2017) An ATR-dependent function for the Ddx19 RNA helicase in nuclear R-loop metabolism. EMBO J 36:1182–1198

Hou Y, Li Y, Xiang J-F et al (2024) TDP-43 chronic deficiency leads to dysregulation of transposable elements and gene expression by affecting R-loop and 5hmC crosstalk. Cell Rep 43:113662

Huertas P, Aguilera A (2003) Cotranscriptionally formed DNA:RNA hybrids mediate transcription elongation impairment and transcription-associated recombination. Mol Cell 12:711–721

Iossifov I, O'Roak BJ, Sanders SJ et al (2014) The contribution of de novo coding mutations to autism spectrum disorder. Nature 515:216–221

Iwahashi C, Yasui D, An H-J et al (2006) Protein composition of the intranuclear inclusions of FXTAS. Brain 129:256–271

Jiang F, Taylor DW, Chen JS et al (2016) Structures of a CRISPR-Cas9 R-loop complex primed for DNA cleavage. Science 351:867–871

Jin P, Warren ST (2000) Understanding the molecular basis of fragile X syndrome. Hum Mol Genet 9:901–908

Kannan A, Cuartas J, Gangwani P et al (2022) Mutation in senataxin alters the mechanism of R-loop resolution in amyotrophic lateral sclerosis 4. Brain 145:3072–3094

Kasahara M, Clikeman JA, Bates DB et al (2000) RecA protein-dependent R-loop formation in vitro. Genes Dev 14:360–365

Keskin H, Shen Y, Huang F et al (2014) Transcript-RNA-templated DNA recombination and repair. Nature 515:436–439

Kiernan MC, Vucic S, Cheah BC et al (2011) Amyotrophic lateral sclerosis. Lancet 377:942–955

Koeppen AH (2011) Friedreich's ataxia: pathology, pathogenesis, and molecular genetics. J Neurol Sci 303:1–12

Kraan CM, Godler DE, Amor DJ (2019) Epigenetics of fragile X syndrome and fragile X-related disorders. Dev Med Child Neurol 61:121–127

LaMarca EA, Saito A, Plaza-Jennings A et al (2023) R-loop landscapes in the developing human brain are linked to neural differentiation and cell-type specific transcription. bioRxiv

Lee H-G, Imaichi S, Kraeutler E et al (2023) Site-specific R-loops induce CGG repeat contraction and fragile X gene reactivation. Cell 186:2593–2609

Liu XS, Wu H, Krzisch M et al (2018) Rescue of fragile X syndrome neurons by DNA methylation editing of the FMR1 gene. Cell 172:979–992
Loomis EW, Sanz LA, Chédin F et al (2014) Transcription-associated R-loop formation across the human FMR1 CGG-repeat region. PLoS Genet 10:e1004294
López Castel A, Cleary JD, Pearson CE (2010) Repeat instability as the basis for human diseases and as a potential target for therapy. Nat Rev Mol Cell Biol 11:165–170
Luo H, Zhu G, Eshelman MA et al (2022) HOTTIP-dependent R-loop formation regulates CTCF boundary activity and TAD integrity in leukemia. Mol Cell 82:833–851
Marnef A, Legube G (2021) R-loops as Janus-faced modulators of DNA repair. Nat Cell Biol 23:305–313
Massé E, Drolet M (1999) Escherichia coli DNA topoisomerase I inhibits R-loop formation by relaxing transcription-induced negative supercoiling. J Biol Chem 274:16659–16664
Mushtaq A, Mir US, Altaf M (2023) Multifaceted functions of RNA-binding protein vigilin in gene silencing, genome stability, and autism-related disorders. J Biol Chem 299:102988
Nadel JA (2015) The influence of RNA:DNA hybrids on chromatin composition, genome structure, and transcription. Dissertation, Yeshiva University
Nakama M, Kawakami K, Kajitani T et al (2012) DNA–RNA hybrid formation mediates RNAi-directed heterochromatin formation. Genes Cells 17:218–233
Niehrs C, Luke B (2020) Regulatory R-loops as facilitators of gene expression and genome stability. Nat Rev Mol Cell Biol 21:167–178
O'Roak BJ, Vives L, Fu W et al (2012) Multiplex targeted sequencing identifies recurrently mutated genes in autism spectrum disorders. Science 338:1619–1622
Ouyang J, Yadav T, Zhang J-M et al (2021) RNA transcripts stimulate homologous recombination by forming DR-loops. Nature 594:283–288
Pecori R, Di Giorgio S, Paulo Lorenzo J et al (2022) Functions and consequences of AID/APOBEC-mediated DNA and RNA deamination. Nat Rev Genet 23:505–518
Pessina F, Giavazzi F, Yin Y et al (2019) Functional transcription promoters at DNA double-strand breaks mediate RNA-driven phase separation of damage-response factors. Nat Cell Biol 21:1286–1299
Petermann E, Lan L, Zou L (2022) Sources, resolution and physiological relevance of R-loops and RNA–DNA hybrids. Nat Rev Mol Cell Biol 23:521–540
Roy D, Zhang Z, Lu Z et al (2010) Competition between the RNA transcript and the nontemplate DNA strand during R-loop formation in vitro: a nick can serve as a strong R-loop initiation site. Mol Cell Biol 30:146–159
Santos-Pereira JM, Aguilera A (2015) R loops: new modulators of genome dynamics and function. Nat Rev Genet 16:583–597
Sanz LA, Hartono SR, Lim YW et al (2016) Prevalent, dynamic, and conserved R-loop structures associate with specific epigenomic signatures in mammals. Mol Cell 63:167–178
Sanz LA, Castillo-Guzman D, Chédin F (2021) Mapping R-loops and RNA:DNA hybrids with S9.6-based immunoprecipitation methods. J Vis Exp 10:3791
Skourti-Stathaki K, Proudfoot NJ (2014) A double-edged sword: R loops as threats to genome integrity and powerful regulators of gene expression. Genes Dev 28:1384–1396
Skourti-Stathaki K, Proudfoot NJ, Gromak N (2011) Human senataxin resolves RNA/DNA hybrids formed at transcriptional pause sites to promote Xrn2-dependent termination. Mol Cell 42:794–805
Sollier J, Stork CT, García-Rubio ML et al (2014) Transcription-coupled nucleotide excision repair factors promote R-loop-induced genome instability. Mol Cell 56:777–785
Tan-Wong SM, Dhir S, Proudfoot NJ (2019) R-loops promote antisense transcription across the mammalian genome. Mol Cell 76:600–616
Teng Y, Yadav T, Duan M et al (2018) ROS-induced R loops trigger a transcription-coupled but BRCA1/2-independent homologous recombination pathway through CSB. Nat Commun 9:4115
Thomas M, White RL, Davis RW (1976) Hybridization of RNA to double-stranded DNA: formation of R-loops. Proc Natl Acad Sci USA 73:2294–2298

Tuduri S, Crabbe L, Tourrière H et al (2010) Does interference between replication and transcription contribute to genomic instability in cancer cells? Cell Cycle 9:1886–1892
Wahba L, Koshland D (2013) The Rs of biology: R-loops and the regulation of regulators. Mol Cell 50:611–612
Wahba L, Gore SK, Koshland D (2013) The homologous recombination machinery modulates the formation of RNA–DNA hybrids and associated chromosome instability. elife 2:e00505
Wahba L, Costantino L, Tan FJ et al (2016) S1-DRIP-seq identifies high expression and polyA tracts as major contributors to R-loop formation. Genes Dev 30:1327–1338
Wang IX, Grunseich C, Fox J et al (2018) Human proteins that interact with RNA/DNA hybrids. Genome Res 28:1405–1414
Wang M, Wei P-C, Lim CK et al (2020) Increased neural progenitor proliferation in a hiPSC model of autism induces replication stress-associated genome instability. Cell Stem Cell 26:221–233
White RL, Hogness DS (1977) R loop mapping of the 18S and 28S sequences in the long and short repeating units of Drosophila melanogaster rDNA. Cell 10:177–192
Wood M, Quinet A, Lin Y-L et al (2020) TDP-43 dysfunction results in R-loop accumulation and DNA replication defects. J Cell Sci 133:jcs244129
Wormwood KL, Charette L, Ryan JP et al (2023) A proteomics investigation of salivary profiles as potential biomarkers for autism spectrum disorder (ASD). Protein J 42:607–620
Wulfridge P, Yan Q, Rell N et al (2023) G-quadruplexes associated with R-loops promote CTCF binding. Mol Cell 83:3064–3079
Xiao Y, Luo M, Hayes RP et al (2017) Structure basis for directional R-loop formation and substrate handover mechanisms in type I CRISPR-Cas system. Cell 170:48–60
Yan Q, Wulfridge P, Doherty J et al (2022) Proximity labeling identifies a repertoire of site-specific R-loop modulators. Nat Commun 13:53
Yang X, Liu Q-L, Xu W et al (2019) m6A promotes R-loop formation to facilitate transcription termination. Cell Res 29:1035–1038
Zhang Z, Zhang H, Hu B et al (2023) R-loop defines neural stem/progenitor cells during mouse neurodevelopment. Stem Cells Dev 32:719–730
Zhang J, Chen F, Tang M et al (2024) The ARID1A-METTL3-m6A axis ensures effective RNase H1-mediated resolution of R-loops and genome stability. Cell Rep 43:113779

Noncoding RNAs and RNA Epigenetics in Brain Aging

Aya Sugai-Munson, Christopher Simon, Occam Kelly Graves, Tina B. McKay, and Haobo Li

Contents

Abstract As an individual ages, the brain undergoes complex structural and functional changes that ultimately lead to cognitive dysfunction. With the aging population fast expanding, there is an unmet clinical need for interventions to prevent age-related cognitive decline. Noncoding RNAs have emerged as crucial regulators of cellular and biochemical processes in age-related cognitive dysfunction and neurodegenerative disease in general. This paper provides an overview of the current knowledge on the role of noncoding RNAs in brain aging, emphasizing the mechanism of RNA epigenetics and modifications that regulate noncoding RNAs' function.

A. Sugai-Munson · O. K. Graves
Department of Anesthesia, Critical Care and Pain Medicine, Massachusetts General Hospital, Boston, MA, USA

C. Simon · T. B. McKay · H. Li (✉)
Department of Anesthesia, Critical Care and Pain Medicine, Massachusetts General Hospital, Boston, MA, USA

Harvard Medical School, Boston, MA, USA
e-mail: haobo.li@mgh.harvard.edu

J. Barciszewski (ed.), *Decoding Aging and Neurodegeneration*,
RNA Technologies 16, https://doi.org/10.1007/978-3-031-91072-2_10

Keywords Noncoding RNA · RNA epigenetics · Aging · Brain · RNA modification

1 Introduction

Older age is a vital risk factor for developing most chronic diseases. The increased susceptibility to disease with age is primarily attributed to genetic, epigenetic, and metabolic changes acquired over time, leading to a loss of normal cellular function. Recent studies have focused on distinguishing chronological and biological age, defining the former as the time from the day of birth and the latter as the cellular and genetic changes associated with higher chronological age. Biological age and the overall healthspan of an organism are at least partially determined by the body's ability to manage endogenous and exogenous stress at the cellular level. With time, the accumulation of oxidative damage to DNA, proteins, and lipids may lead to morphological changes in tissue structure, cellular senescence, and death.

Like all tissues in the body, the brain is susceptible to the effects of biological aging. Maintaining optimal brain health through curated dietary, behavioral, and environmental features has emerged as a targeted approach to combat cognitive decline associated with aging (Santisteban and Iadecola 2025; Sun et al. 2025). Studies in mice have shown that serum from young animals can promote tissue rejuvenation in aged animals, providing evidence that factors found in circulation are important in aging and longevity (Katsimpardi et al. 2014; Sinha et al. 2014).

Alzheimer's disease (AD) and AD-related dementias (ADRD) are the most common neurodegenerative diseases, affecting over 6.9 million Americans, with the number expected to double in the next 30 years (Chen et al. 2024; Jia et al. 2024). Transcriptomic studies of postmortem brain tissue have revealed age-related signatures involved in cognitive function during development and aging (Chen et al. 2017; Kang et al. 2011). AD has been associated with differential RNA expression and mislocalization, suggesting that the mechanisms underlying gene expression may also be involved in its pathogenesis (Dube et al. 2019; Smukowski et al. 2024). Further studies in RNA dynamics may reveal their role in neurodegeneration as biomarkers of disease and potential drug targets. In this chapter, we provide an overview of the mechanisms underlying brain aging, the role of RNA regulation in these processes, and novel opportunities for therapeutic intervention.

2 Brain Aging

Aging in the brain has been associated with neurodegenerative processes that influence normal cellular function, involving transcription and translation, protein degradation, and recycling. Working memory and recall deficits are the most common changes associated with older age, while long-term memory and implicit memory are generally retained throughout life. Moreover, some parts of brain function may improve with age, including emotional responses and maturity, which have been attributed to changes in the prefrontal cortex. There are well-defined hallmarks of aging that contribute to neuronal and glial cell loss in the brain (López-Otín et al. 2013). Genomic instability, mitochondrial dysfunction, telomere shortening, and epigenetic modifications are common denominators in the pathological processes underlying age-associated neurodegeneration (Fig. 1) (Fraser et al. 2022; López-Otín et al. 2023; Vijg and Suh 2013).

Genomic integrity during biological aging is susceptible to aberrant DNA replication, defects in chromosome segregation, oxidative processes, and spontaneous hydrolysis, which collectively contribute to altered gene and protein expression (Vermeij et al. 2016). Neurons are terminally differentiated cells, with long-lasting regenerative mechanisms for maintaining cell integrity. When these processes go awry, genomic instability occurs, leading to translocation and deletion errors, single- to double-strand breaks, chromosomal rearrangements, and interferences caused by viral and transposable elements (Vijg and Dong 2020). These changes may lead to cell loss and progressive thinning of the gray and white matter in the cortex (Giorgio et al. 2010), contributing to decreased cognitive performance with age (Bennett and Madden 2014). Cells from aged tissues may gather somatic mutations, chromosomal aneuploidy, and copy number variations in the nuclear DNA (Blokzijl et al. 2016; Huang et al. 2022). These alterations disrupt transcriptional pathways and organismal homeostasis, making dysfunctional cells vulnerable to cell death (Miller et al. 2022). This mutational burden may be moderately tolerated because of the high energetic cost associated with genomic repair (Martincorena et al. 2018). As cells favor survival over genomic integrity, mutations alone may not be sufficient to accelerate aging, and, in general, a permissive microenvironment is required for these changes to lead to permanent tissue damage (Balmain 2020; Nik-Zainal and Hall 2019). Alterations in DNA repair mechanisms have been shown to accelerate aging in mice and underlie human progeroid syndromes (Hennekam 2020). Transgenic mice overexpressing BubR1, a mitotic checkpoint kinase, exhibit extended longevity (North et al. 2014), while overexpression of Sirtuin 6 has been found to enhance double-strand break repairs in mice (Tian et al. 2019).

Regarding neurodegeneration, mutations in genes associated with the production and clearance of amyloid beta (Aβ), including amyloid precursor protein (APP), ApoE, and lysosomal proteins, are related to AD development. Recent genome-wide association studies have revealed several genetic risk factors for AD involving APP expression, activity, and microglial activation (Bellenguez et al. 2022). Other AD risk factors include elevated systolic blood pressure and lipid levels, highlighting the importance of cardiovascular health in long-term cognitive outcomes (Alzheimer's and Collaboration 2023). Overexpression of the protein tyrosine kinase, Fyn, has also been associated with aberrant phosphorylation of tau isoforms and ERK activation in AD, leading to pathological accumulation and aggregation of Aβ to form insoluble plaques (Li and Götz 2017). In addition to neurodegeneration attributed to aberrant Aβ and tau accumulation, defects in neurogenesis in the dentate gyrus of the hippocampus are linked to impaired memory and cognition in AD (Tobin et al. 2019).

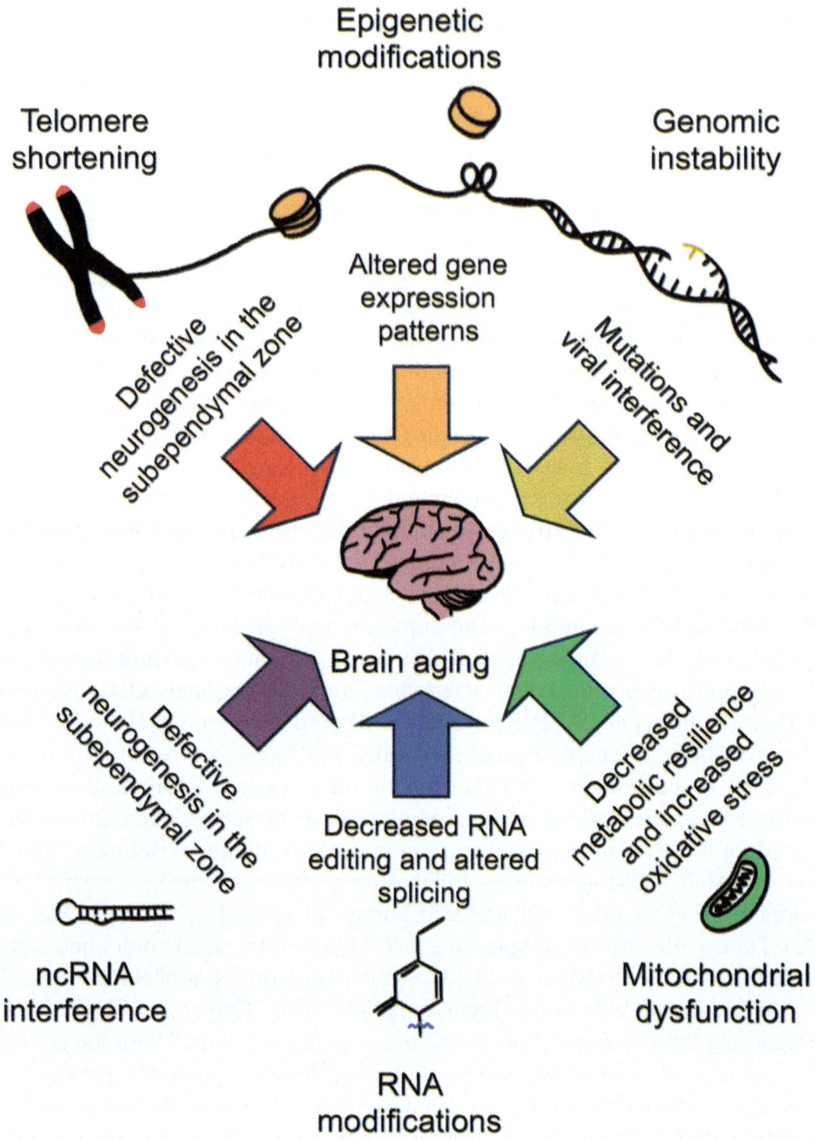

Fig. 1 Major determinants of biological aging in the brain. Brain aging is influenced by multiple molecular and cellular mechanisms, including telomere shortening, epigenetic modifications, genomic instability, mitochondrial dysfunction, RNA modifications, and noncoding RNA (ncRNA) interference. These factors contribute to various downstream effects, such as altered gene expression patterns, mutations and viral interference, metabolic decline, oxidative stress, defective neurogenesis, and decreased RNA editing and splicing. Together, these processes lead to the progressive decline in brain function associated with aging.

Defects in mitochondrial structure and function are also significant contributors to neurodegenerative processes in the brain (Chinnery et al. 2002). Mitochondrial DNA (mtDNA) is highly susceptible to age-associated mutations due to its limited repair mechanisms, low replication fidelity, high oxidative microenvironment, and lack of protective histones in the mitochondria compared to the nucleus (Corral-Debrinski et al. 1992; Cortopassi et al. 1992; Sanchez-Contreras and Kennedy 2022; Soong et al. 1992). However, the causal effects of mitochondrial mutations may be masked by simultaneous changes occurring in genomic DNA as a function of aging that broadly impacts gene expression patterns and metabolic processes (Stewart and Chinnery 2015). Deep sequencing of aged cells has revealed that mtDNA mutational load may accelerate with age during clonal expansion (Greaves et al. 2014), even though most mtDNA mutations seem to arise from replication errors caused by the mtDNA polymerase (Graziewicz et al. 2006). Supporting this mechanism, mice deficient in DNA polymerase exhibit accelerated aging and reduced lifespan, possibly associated with deletions rather than point mutations in mtDNA (Lujan et al. 2020). AD has been associated with aberrant mtDNA localization, indicative of increased oxidative stress in hippocampal regions, supporting a role for mitochondrial dysfunction in age-associated neurodegeneration (Hirai et al. 2001; Scheff et al. 2016).

Another fundamental cause of aging is telomere shortening. DNA polymerases generally cannot complete the copy of telomere regions of eukaryotic DNA. Consequently, DNA damage at telomeres is a significant factor that contributes to biological aging (Blackburn et al. 2015). Telomere shortening is observed during normal aging in humans and mice, with an attrition rate influenced by genetic variants and proliferation rates, and predicts lifespan (Whittemore et al. 2019). Studies of genetically engineered mice have revealed causal associations between telomere attrition and organismal aging, with shortened or lengthened telomeres exhibiting decreased or increased lifespan (Muñoz-Lorente et al. 2019; Tomas-Loba et al. 2008). Telomeres undergo a substantial shortening that induces genomic instability (Chakravarti et al. 2021), which can be prevented by telomerase (Blasco 2005). Since telomerase is not expressed in most normal human cells, progressive shortening of telomeric sequences from chromosomal ends occurs as a function of DNA replication and cell division (Chakravarti et al. 2021). In terms of disease, telomerase deficiency in humans has been associated with the premature development of diseases that hinder normal tissue regeneration and maintenance (Alder and Armanios 2022; Whittemore et al. 2019). In telomerase-deficient mice, premature aging can be prevented with reactivation of telomerase (Jaskelioff et al. 2011), while mice engineered to maintain telomerase's physiological levels generally show neuron preservation and survival (Shim et al. 2021). Since AD pathology is associated with aged microglia cells, microglial proliferation linked to telomere shortening also constitutes a risk factor for neurodegeneration. In AD mice, early-stage inhibition of microglial proliferation has been shown to impair the onset of microglia senescence and reduce amyloid-induced pathology (Hu et al. 2021; Rolyan et al. 2011).

Genetic alterations of DNA methylation and histone posttranslational modifications may additionally influence gene expression, resulting in the development and progression of neurodegeneration. DNA methylation changes over time, and further analyses have revealed that specific loci, including those of polycomb target genes, are hypermethylated with age (Bejaoui et al. 2022). Cells from progeroid patients and mice exhibit DNA methylation changes within the intronic and intergenic regions (Horvath and Raj 2018). Indeed, it has been proposed that DNA methylation-based clocks may extend the human lifespan, since protocols aimed at thymus regeneration have resulted in an epigenetic age of 1.5 years less than the baseline. In contrast, predictions of human mortality show a 2-year decrease in epigenetic age (Fahy et al. 2019). Nevertheless, further studies will be necessary to determine whether defective maintenance of DNA methylation produces accelerated aging.

Histone modifications can also lead to transcriptional alterations, loss of cellular homeostasis, and age-associated decline in metabolism (Oh and Petronis 2021). There is ample evidence that histone demethylases modulate lifespan by targeting insulin/insulin growth factor-1 (IGF-1) signaling and the SIRT family of protein deacetylases and ADP-ribosyl transferases (Bhatt and Tiwari 2023). Transgenic overexpression of SIRT1 appears to improve genomic stability in aged mice, while overexpression of mitochondrial SIRT3 reverses stem cell regenerative capacity (Brown et al. 2013). Additionally, SIRT6-deleted mice may accelerate aging, whereas SIRT6 overexpression extends lifespan (Mostoslavsky et al. 2006). Of note, histone acetyltransferase inhibitors extend progeroid mice's lifespan, while histone deacetylase activators promote longevity by upregulating SIRT1 activity (Bonkowski and Sinclair 2016). Together, these studies prove that the aging processes may be blunted with therapeutic interventions and possibly reversed to prolong a healthy lifespan.

RNA epigenetic modifications mediated primarily by methylation and acetylation of select nucleotides likewise influence gene expression patterns that contribute to the development and progression of neurodegeneration. These posttranscriptional modifications include m^6A (N6-methyladenosine), m^5C (5-methylcytosine), m^1A (N1-methyladenosine), m^7G (7-methylguanosine), and N4-acetylcytidine (ac^4C). Further studies in RNA dynamics may reveal their role in neurodegeneration as biomarkers of disease and potential drug targets.

3 Noncoding RNAs in Brain Aging

Noncoding RNAs (ncRNAs) lack protein-coding potential and constitute most of the human transcriptome (Park et al. 2022). Based on their sequence length, ncRNAs can be classified as either small ncRNAs (shorter than 200 nucleotides) or long ncRNAs (longer than 200 nucleotides). Emerging evidence has indicated the critical roles of ncRNAs in the development and progression of age-related cognitive dysfunction.

3.1 Small Noncoding RNAs in Brain Aging

3.1.1 MicroRNAs in Brain Aging

MicroRNAs (miRNA) are shorter variations of small ncRNAs, typically 18–22 nucleotides in length, regulating gene expression at the posttranscriptional level (Bushati and Cohen 2007). In most cases, miRNAs bind to complementary sequences on the 3′ untranslated regions of messenger RNAs (mRNA) and either repress transcription or initiate mRNA degradation to decrease protein synthesis (Bushati and Cohen 2007). Sixty percent of all human mRNAs are regulated by miRNA, with miRNA dictating nearly all cellular and molecular functions (Friedman et al. 2009). Various miRNAs have been identified and shown to be potential biomarkers and therapeutic targets for age-related cognitive dysfunction. miRNAs are the most extensively studied noncoding RNAs in brain aging and age-related cognitive decline (Table 1).

Deep sequencing of mouse hippocampus found that miR-34c was highly expressed in the hippocampus (Zovoilis et al. 2011), where its expression was upregulated and correlated with impaired memory function in aged mice (24 months vs. 3 months old) and in APP/PS1-21 AD mice, as well as in AD patients compared with age-matched controls (Zovoilis et al. 2011). Direct injection of a miR-34c-mimic to the hippocampus led to impaired learning function, while inhibition of miR-34c improved memory performance in 12-month-old APP/PS1-21 AD mice (Zovoilis et al. 2011). Mechanistically, miR-34c induced age- and AD-related cognitive impairment through binding to the 3′ UTR of SIRT1 and the subsequent repression of SIRT1 translation (Zovoilis et al. 2011). Interestingly, in Drosophila, miR-34 was upregulated in the brain with age (Liu et al. 2012; Srinivasan et al. 2022). In these animals, loss of miR-34 led to a shorter lifespan. It triggered late-onset neurodegeneration, where its upregulation extended the median lifespan and mitigated human pathogenic polyglutamine protein-induced neurodegeneration (Liu et al. 2012). These age-associated effects of miR-34 were partially mediated by suppressing Eip74EF, an essential ETS domain transcription factor in steroid hormone pathways (Liu et al. 2012). Similarly, in Drosophila, animals with a miR-34 mutation exhibited early aging phenotypes, including reduced lifespan and neurodegeneration, mediated via Lst8, a subunit of Tor Complex 1, leading to an increase in gene translation activity and accumulation of protein aggregation (Srinivasan et al. 2022). These controversial findings describing the role of miR-34 in brain aging between vertebrates and Drosophila may be due to species differences. Still, they may also suggest that miR-34 functions differently in diverse brain regions (hippocampus in vertebrates vs. whole brain in Drosophila). In addition to miR-34, various miRs have been shown to participate in brain aging and age-related neurodegeneration. miR-144 and miR-451 expression decreased in the hippocampal dentate gyrus in 24-month-old mice compared with 6-month-old mice (Wang et al. 2022). Compared with age-matched (24-month-old) wild-type mice, 6-month-old mice with miR-144 and miR-451 knockout exhibited reduced radial glia cells in the hippocampal dentate gyrus that were associated with increased oxidative stress in

Table 1 MiRNAs in brain aging

miRNA	Organism	Age of animal	Effect	Mechanism
miR-34c (Zovoilis et al. 2011)	Mouse, human	24-month-old	Overexpression of miR-34c improves memory performance in 12-month-old APP/PS1 AD mice	miR-34c binds to the 3′ UTR of SIRT1 and reduces SIRT1 translation
miR-34 (Liu et al. 2012)	Drosophila	30 and 60 days	Overexpression of miR-34 increases lifespan and prevents neurodegeneration	miR-34 suppress Eip74EF expression
miR-34 (Srinivasan et al. 2022)	Drosophila	20 and 50 days	miR-34 mutation leads to early aging phenotypes, including reduced lifespan and neurodegeneration	miR-34 directly binds to Lst8 and inhibit its expression
miR-144 and miR-451 (Wang et al. 2022)	Mouse	24-month-old	Knockout of miR-144 and miR-451 reduces radial glia cells in the hippocampal dentate gyrus and increases oxidative stress in the hippocampal dentate gyrus	miR-144 and miR-451 knockout inactivates of the antioxidant and the Akt/GSK3β/β-catenin pathway
miR-206 (Zhang et al. 2022)	Mouse	17-week-old	Downregulation of miR-206 improves cognitive function in D-galactose-induced accelerated brain aging	Downregulation of miR-206 increases BDNF, Egr-1, and PSD-95
miR-130a (Shen et al. 2021)	Rat	21-month-old	Reduced miR-130a in the hippocampus is associated with impaired spatial learning and memory capacity in aged rats	miR-130a restores autophagy flux
miR-331-3p and miR-9-5p (Chen et al. 2021)	Mouse	12-month-old	Inhibition of miR-331-5p and miR-9-5p or their combination improve long-term memory and motor function in 12-month-old APPswe/PS1dE9 AD mice	Inhibition of miR-331-5p and miR-9-5p increases Sqstm1 and Optn, two autophagy receptors, and the subsequent upregulation of autophagy
miR-195 (Cao et al. 2021)	Mouse and human	With ApoE4+/−	Overexpression of miR-195 improves cognitive function, reduces amyloid plaque deposition, decreases Tau hyper-phosphorylation, and inhibits lysosomal defects	miR-195 directly binds to synaptojanin 1 and inhibits its expression

the hippocampal dentate gyrus. These effects of miR-144 and miR-451 deficiency were likely mediated through inactivation of the antioxidant and the Akt/GSK3β/β--catenin pathway (Wang et al. 2022). Seventeen-week-old male mice injected with D-galactose for 8 weeks led to accelerated brain aging and declining cognitive function. Transplantation of human umbilical cord mesenchymal stem cells (hUCMSCs) with downregulated miR-206 showed improved learning and spatial memory abilities in these mice. hUCMSCs are thought to improve cognitive function by increasing BDNF, Egr-1, and PSD-95 (Zhang et al. 2022). Compared to 12-week-old young rats, 21-month-old rats developed impaired spatial learning and memory capacity associated with reduced miR-130a in the hippocampus (Shen et al. 2021). These changes were reversed by 12 weeks of voluntary wheel running, likely through the miR-130a-mediated restoration of autophagy flux (Shen et al. 2021). This was supported by in vitro results showing that a miR-130a-mimic prevented D-galactose-induced cell senescence and autophagy impairment in SH-SY5Y cells (Shen et al. 2021). However, the causal role of miR-130a and autophagy in age-related cognitive decline and the neuroprotective effects of exercise need further in vivo validation. Interestingly, in APPswe/PS1dE9 AD mice, hippocampal miR-331-3p and miR-9-5p were downregulated in 6-month-old AD mice while upregulated at the age of 12 months in these mice (late stage of AD) (Chen et al. 2021), indicating a biphasic change in miRNAs in AD pathogenesis. At the late stage of AD, inhibition of miR-331-3p, miR-9-5p, or their combination improved long-term memory and motor function in AD mice (Chen et al. 2021). Mechanistically, inhibition of miR-331-3p and mR-9-5p conferred neuroprotective effects through Sqstm1 and Optn, two autophagy receptors, and the subsequent upregulation of autophagy (Chen et al. 2021). These results reinforce the importance of miRNAs in AD and the role of autophagy in this process. In brain tissues from patients with ApoE4+/– with clinical diagnosis of mild cognitive impairment or early AD, miR-195 was decreased. A similar reduction of miR-195 was observed in ApoE4+/+ mice hippocampal brain tissue and cultured neurons (Cao et al. 2021). Overexpression of miR-195, through reducing synaptojanin 1 (a brain phosphoinositol biphosphate-degrading enzyme), improved cognitive function, reduced amyloid plaque deposition, decreased Tau hyper-phosphorylation, and inhibited lysosomal defects in ApoE4+/+ mice (Cao et al. 2021). Interestingly, CSF miR-195 levels were positively correlated with cognitive performance while negatively correlated with phosphorylated Tau levels in patients with mild cognitive impairment (Cao et al. 2021). These suggest that miR-195 may not only serve as a therapeutic target for AD but may also serve as a biomarker for the development and progression of AD.

Indeed, the promise of miRNAs as potential biomarkers for brain aging and age-related cognitive dysfunction has been shown in numerous studies (Islam et al. 2021; Melas et al. 2024; Shokhirev and Johnson 2022). In blood samples from a large patient population (2689 participants), differential expression of 21 circulating miRNAs was associated with cognitive domains (Melas et al. 2024). Among them, seven correlated with cortical thickness and two associated with hippocampal volume (Melas et al. 2024). Multiomic analysis suggests these miRNAs regulated pathways linked to neurogenesis and synapse assembly (Melas et al. 2024), and thus supports the potential of these miRNAs as biomarkers of brain aging and age-related cognitive dysfunction. In addition to circulating miRNAs, brain tissue-derived miRNAs have also exhibited unique signatures correlated with brain aging and age-related cognitive dysfunction. Coexpression network analysis of miRNA expression levels in the brain tissues of the Religious Orders Study and Rush Memory and Aging Project (ROS/MAP, 702 participants) and an independent dataset from the Gene Expression Omnibus (GEO, 16 participants) identified five hub miRNAs (miR-129-5p, miR-433, miR-1260, miR-200a, and miR-221) that were associated with AD neuropathology markers and cognition, with miR-129-5p showing the strongest correlation (Han et al. 2024). Although these discovery-focused studies provide valuable results identifying miRNAs as biomarkers for brain aging, they will be enhanced by further integration with functional validation. By relating circulating miRNAs in patients and animals with age-associated memory decline, three miRNAs (miR-181a-5p, miR-146a-5p, and miR148a-3p) were identified to be highly associated with age-associated memory decline (Islam et al. 2021). Importantly, these findings were further validated in cell culture and an animal model. Overexpression of these three miRNAs impaired neural plasticity in primary hippocampal mouse cultures while inhibiting these three miRNAs by injecting antisense oligonucleotides directly into the hippocampus improved learning and memory function in 16.5-month-old mice (Islam et al. 2021). These findings support using these three miRNAs as a signature to predict the onset and as a treatment strategy for brain aging and age-related cognitive function.

3.1.2 Other Small Noncoding RNAs in Brain Aging

Other than miRNAs, different types of small noncoding RNAs have been suggested to participate in the development of brain aging and age-related cognitive dysfunction. PiRNA is another type of sncRNA, typically 26–31 nucleotides long, associated with PIWI proteins, and is involved in gene regulation and genomic stability (Czech et al. 2018). PiRNA forms complexes with PIWI proteins primarily expressed in germ cells, targeting RNA sequences via transcriptional and posttranscriptional gene silencing. piRNAs recruit DNA methyltransferases to the target genomic DNA in germline cells to suppress transposons and silence the gene (Czech et al. 2018). A key link between DNA methylation and aging is using DNA methylation patterns to measure aging and as an indicator of overall mortality risk (Czech et al. 2013; Lenart et al. 2018). However, piRNAs have been found to act beyond germline cells, playing a role in aging genetic factors in the human brain (Mao et al. 2019). piRNAs hold potential as biomarkers for AD, with 149 of the 9453 human brain piRNAs displaying differential expression in brains with AD (Roy et al. 2017; Sato et al. 2023). In exosomes derived from human cerebrospinal fluid (CSF), the expression levels of three piRNAs were altered: *piR_019324*, *piR_019949*, and *piR_020364*. The former displayed decreased expression, and the latter two had increased expression in the CSF of patients with AD (Jain et al. 2019). Combined with three miRNAs associated with memory and neurodegeneration, a sncRNA signature was formed, distinguishing AD patients from non-AD patients (Jain et al. 2019). When combined with existing AD markers, the Aβ42/40 ratio, and pTau levels, diagnostic tests demonstrated high accuracy (Jain et al. 2019), highlighting piRNA's potential as a sensitive and specific biomarker for AD. On the other hand, the presence of commonly dysregulated piRNAs suggests their potential role as biomarkers. Jain et al. (2019) identified a miRNA/piRNA signature comprised of three miRNAs, miR-27-a-3p, miR-30a-5p, and miR-34c, and three piRNAs, piR_019324, piR_019949, and piR_020364, that demonstrated high diagnostic accuracy. The distinction of AD dementia patients from the control yielded an AUC (area under the curve) of 0.83 and 0.86 for predicting the advancement of mild cognitive impairment to AD (Jain et al. 2019). When combined with the widely used method of measuring phosphorylated tau (pTau) and the amyloid-beta 42/40 (Aβ42/40) ratio, the AUC increased to 0.96. This evidence suggests using a miRNA/piRNA signature as a sensitive diagnostic biomarker for AD.

SiRNAs are short, double-stranded RNA molecules involved in the RNA interference (RNAi) pathway. SiRNAs downregulate molecular targets posttranscription with high accuracy and potency, making them a potential treatment for neurodegenerative disease (Amiri et al. 2021; Khvorova 2023). In the treatment of AD, siRNAs have the capability of targeting PS1 and reducing amyloid plaque formation, potentially mitigating AD progression (Kandimalla et al. 2012). Amyloid plaques, or Aβ42 accumulation, are a diagnostic feature of AD. PS1 is part of the γ-secretase enzyme complex, which completes the final cleavage of the amyloid precursor protein. This, in turn, releases Aβ42, with mutations in PS1 connected to increased Aβ42 production, leading to early-onset familial AD. However, targeting PS1 with siRNA led to a dose-dependent reaction in IMR-32 cells, a human neuroblastoma cell line (Kandimalla et al. 2012). This consequently led to a decrease in Aβ42 production and silenced PS1 gene expression in the cell (Kandimalla et al. 2012), suggesting that siRNAs may serve as targeted therapeutics to reduce the amyloid load associated with AD progression.

3.2 Long Noncoding RNAs in Brain Aging

Long noncoding RNAs (lncRNAs) are RNA molecules longer than 200 nucleotides that, unlike mRNA, do not code for protein (Wang and Chang 2011). LncRNAs are crucial in gene regulation, protein metabolism control, and epigenetic regulation. They interact with chromatin-modifying complexes to activate or suppress specific gene expression (Wang and Chang 2011). Due to this significant role, lncRNA mutations and dysregulation can significantly impact protein synthesis and lead to neurodegenerative diseases (Table 2).

In brain aging, multiple lncRNAs have been identified and suggested to play roles in brain aging and age-related cognitive dysfunction. However, only a few of these have been functionally tested due to the low conservation of lncRNAs across

Table 2 lncRNAs in brain aging

lncRNA	Organism	Age of animal	Effect	Mechanism
Silc1 (Ben-Tov Perry et al. 2023)	Mouse	18-month-old	Silc1 deficiency delays spatial learning without affecting long-term memory. In both CA3 and DG, Silc1 expressions are reduced during aging and further reduced in 5xFAD AD mice	Silc1 knockout reduces Sox11 expression and reduces chromatin binding at Sox family binding sites
GAS5 (Patel et al. 2023)	Mouse and human	20-month-old	GAS5 gene knockdown increases Tau phosphorylation and induces neuroinflammation while its overexpression decreases neuroinflammation in aged mice	N/A
BACE1-AS (Faghihi et al. 2008)	Mouse	3-month-old	BACE1-AS increases Aβ production and increasing the risk of amyloid plaques in the brain	*BACE1-AS* upregulates the transcription of *BACE1* mRNA by forming a stabilizing RNA duplex
Lamr1-ps1 (Wu et al. 2025)	Mouse	3-, 6-, and 12-month-old	Overexpression of Lamr1-ps1 in the hippocampus exacerbates early spatial learning and memory deficits in APP/PS1 AD mice	Lamr1-ps1 binds to miR-29c and prevents miR-29c from binding to BACE1, resulting in increased BACE1 levels

different species, the complexity of their functions (Chen and Kim 2024; Kung et al. 2013; Ponting et al. 2009; Wang and Chang 2011), and their region-specific and sex-specific alterations during brain aging (Cao et al. 2019; Zhou et al. 2019). Silc1 (Sciatic Injury Induced LincRNA Upregulator Of SOX11) is a sequence conserved lncRNA broadly expressed in the central nervous system. Under physiological conditions, Silc1 was expressed higher in the CA3 region of mice following exposure to a novel environment, while Silc1 expression was increased in both CA3 and DG. In Silc1 knockout mice, Silc1 deficiency delayed spatial learning without affecting long-term memory (Ben-Tov Perry et al. 2023). Interestingly, in both CA3 and DG, Silc1 expressions were reduced during aging and further reduced in a 5xFAD AD mouse model (Ben-Tov Perry et al. 2023). Together, these results point to a possibility that the reduction of Silc1 in the brain during aging and neurodegeneration may lead to age-related cognitive dysfunction, which needs further investigation. Similar to the change in Silc1 in the brain, lncRNA GAS5 (growth arrest-specific transcript 5) levels were reduced in the brains of AD patients and aged mice (20-month-old vs. 6-month-old) (Patel et al. 2023). Depletion of GAS5 by siRNA decreased insulin receptor levels, impaired insulin signaling, increased Tau phosphorylation, and induced proinflammatory gene expression in mouse neurons in vitro (Patel et al. 2023). In contrast, activation of GAS5 by the small molecule NPC86 that disrupts UPF1 binding to a GAS5 premature termination codon improved neuronal insulin signaling and decreased neuroinflammation in 20-month-old mice (Patel et al. 2023). These indicate the antineuroinflammatory property of GAS5 and its potential in treating age-related cognitive dysfunction. *BACE1-AS* is the conserved antisense transcript of the gene coding for beta-secretase 1 (*BACE1*), a key enzyme in the amyloid pathway that cleaves amyloid precursor proteins (Faghihi et al. 2008). *BACE1* overexpression increased the production of misfolded proteins, making BACE1 levels a valuable blood plasma marker for detecting brain amyloidosis in AD patients. Indeed, BACE1-AS expression was elevated in various brain regions (cerebellum, hippocampus, entorhinal cortex, and superior frontal gyrus) of AD patients and Tg19959 mice (Faghihi et al. 2008), an AD mouse model expressing doubly mutated human APP. *BACE1-AS* expression upregulates the transcription of *BACE1* mRNA by forming a stabilizing RNA duplex. This binds to *BACE1*'s open reading frame, blocking the miRNA-485-5p binding site and elevating *BACE1* mRNA and protein levels, boosting Aβ production, and increasing the risk of amyloid plaques in the brain (Faghihi et al. 2008). In another study, profiling lncRNA expression in the hippocampus of APP/PS1 mice at different ages (3-, 6-, and 12 months), Lamr1-ps1 (a pseudogene of the laminin receptor (Lamr1)) was identified where Lamr1-ps1 expression was increased in the hippocampus of APP/PS1 AD mice (Wu et al. 2025). Overexpression of Lamr1-ps1 in the hippocampus exacerbated early spatial learning and memory deficits in APP/PS1 AD mice (Wu et al. 2025). Interestingly, this study found that Lamr1-ps1 regulated cognitive function through the miR-29c/BACE1 pathway (Wu et al. 2025), suggesting that there are multiple regulatory mechanisms on BACE1 in AD pathology.

In addition to lncRNAs' function in neurons, there are also lncRNAs expressed in other cell types in the brain that are functionally important for age-related cognitive decline. PRDM16-DT is a lncRNA highly enriched in the human brain and almost exclusively expressed in astrocytes (Schroder et al. 2024). PRDM16-DT and its murine homolog Prdm16os expressions were reduced in the brains of AD patients and mouse primary astrocytes after proinflammatory cytokines or Abeta42 simulations (Schroder et al. 2024). Knockdown of Prdm16os in astrocytes led to impaired glutamate update, lactate secretion, and impaired neuronal activity and spine density. While its overexpression protected astrocytes against cytokine-induced functional decline (Schroder et al. 2024), Prdm16os's neuroprotective role in AD requires further validation in AD animal models.

3.3 *Circular RNAs in Brain Aging*

Circular RNAs (circRNA) are a class of noncoding RNAs that form a covalent loop with no 5′ or 3′ end and have been found to play diverse regulatory roles in parental gene transcription, miRNA activity modulators, and RNA-binding proteins (Chen and Yang 2015). Various circRNAs have been identified and proven functionally important in age-related cognitive dysfunction (Lu et al. 2019; Wang et al. 2024; Xu et al. 2020).

Genome-wide screening of circRNAs in the cerebral cortex in 5xFAD AD mice between 5 and 7 months of age identified a highly conserved circRNA, circGigyf2 (Wang et al. 2024). Cerebral circGigyf2 expression was reduced in 7-month-old 5xFAD mice and negatively correlated with dementia severity (Wang et al. 2024). Functionally, circGigyf2 worked as a sponge binding to various miRNAs (e.g., miR-335-5p) and enhanced the polyadenylation site processing efficiency of CPSF6 targets. In Neuro-2a cells in vitro, circGigyf2 gene knockdown disrupted circGigyf2 binding to miRNAs and CPSF6, leading to cell apoptosis (Wang et al. 2024), suggesting a role for circGigyf2 in regulating neurodegeneration. In the rhesus macaque prefrontal cortex, circGRIA1 was increased during aging (Xu et al. 2020). circGRIA1 was a conserved and male-specific circRNA derived from the Gria1 gene that negatively regulated Gria1 expression by binding to its promoter. Injection of AAV expressing siRNA targeting circGRIA1 in the hippocampus in 20-year-old macaques reduced circGRIA1 while increasing Gria1. This reduction of circGRIA1 improved synaptogenesis and GluR1 activity-dependent synaptic plasticity in these aged macaques (Xu et al. 2020), indicating its protective effect against brain aging. MicroRNA-7 (miR-7), a highly abundant miRNA in the human brain, is associated with ciRS-7, a circRNA, which acts as a sponge to reduce miR-7 activity (Zhao et al. 2016). In patients with AD, the miR-7-circRNA interaction is disrupted with decreased ciRS-7 levels and increased miR-7 levels. This leads to the downregulation of AD-associated genes such as UBE2A (Zhao et al. 2016), a ubiquitin-protein ligase and phagocytic protein crucial for amyloid clearance in the AD brain. Inhibitions of "miRNA sponging systems" and the subsequent increase in inducible miRNAs may be a player in sporadic AD. In contrast, in in vitro in SH-SY5Y cells, it has been shown that ciRS-7, by inhibiting NF-κB nuclear translocation, promoted APP and BACE1 protein degradation, thereby reducing Abeta generation, indicating a potential neuroprotective effect of ciRS-7 (Shi et al. 2017). The controversy of these results is not apparent but may be due to the use of the model system (in vivo vs. in vitro). circCwc27 is another circRNA that has recently been identified, with an expression that was enriched and abundantly expressed in neurons (Song et al. 2022). circCwc27 was also upregulated in APP/PS1 AD mice and patients with AD. Injection of lentivirus expressing circCwc27 shRNA into the lateral ventricle in 9-month-old APP/PS1 mice decreased circCwc27 and impaired spatial learning and memory function (Song et al. 2022). circCwc27 does this by directly binding to the Pur-α protein, thereby preventing Abeta deposition (Song et al. 2022). In addition to their functional roles in the brain, owing to their closed covalent loop that prevents degradation in circulation, circRNAs also potentially serve as biomarkers for age-related cognitive dysfunction. Indeed, a signature consisting of six circRNAs in the circulation was identified and could distinguish patients with AD from patients with other types of dementia (Ren et al. 2022). Of note, the sample size used in this study was relatively small and requires validation in an independent patient cohort with a larger sample size. Nevertheless, the results support the potential use of circRNAs as biomarkers for age-related cognitive dysfunction. These findings, of course, will be beneficial by integrating multiomic results from screening circRNAs in the brains of AD patients and functional studies to establish circRNAs as promising biomarkers for age-related cognitive dysfunction.

4 RNA Modifications in Brain Aging

Eukaryotic RNA modifications are chemical changes that occur on the nucleotides of RNA molecules. RNA modifications are initially identified on tRNAs and rRNAs but are now increasingly observed on other RNA species, including mRNAs, miRNAs, and lncRNAs (Roundtree et al. 2017). Among the identified RNA modifications, N6-methyladenosine (m^6A) is the most extensively studied. m^6A methylation is a dynamic process regulated by "writers," "erasers," and "readers." Writers are methyltransferase proteins that deposit specific RNA m^6A modifications, which can be removed by erasers (i.e., demethylases) (Zaccara et al. 2019). Readers are a large class of proteins or domains that specifically identify different types of RNA methylation and link RNA methylation to specific biological functions (Zaccara et al. 2019). m^6A has been shown to regulate many aspects of RNA metabolism, such as RNA stability and translation, and has been implicated in neurodevelopment, disease, and brain aging.

Profiling of m^6A changes across five different time points (2-, 4-, 6-, 26-, and 52-week-old) in mice suggests temporal and spatial dynamics of m^6A during neurodevelopment and aging (Shafik et al. 2021). During aging, m^6A sites were increased, where these sites were enriched in alternative untranslated regions of genes that affect aging-related pathways (Shafik et al. 2021). Similar m^6A site changes were observed in humans when comparing m^6A in the brain from old postmortem to adolescence. However, a decrease in m^6A was observed in AD-related transcripts in 5xFAD mice (Shafik et al. 2021). In contrast, another study showed that m^6A levels were consistently decreased in the brain during aging in mice (16-month-old vs. 3-month-old) and in patients with AD (Castro-Hernandez et al. 2023). Moreover, transcripts that had reduced m^6A during aging and AD were linked to synaptic function, including CAMKII and GLUA1, which were validated by proximity ligation assays showing that reduced m^6A levels led to reduced CAMKII and GLUA1-induced synaptic protein synthesis and synaptic function (Castro-Hernandez et al. 2023). The difference in findings may be due to the ages of the animals used, while the exact reason needs further investigation. Nevertheless, these results point to a potential role of m^6A in brain aging and age-related neurodegeneration. Indeed, specific m^6A enzymes have been shown to regulate brain aging and neurodegeneration. Hu and colleagues (2023) showed that differentially methylated transcripts in the age mice (64-week-old vs. 11-week-old) hippocampus were enriched in microglia. Further, hippocampus METTL3, a m^6A writer, was reduced in aged compared to young adult mice. Lentiviral expression of METTL3 shRNA in mice led to cognitive deficit (Hu et al. 2023), suggesting a neurotoxic effect of METTL3 knockdown, likely through activation of the PD-1/PD-L1 pathway. NOP2/Sun RNA methyltransferase 2 (NSun2) is a brain-enriched methyltransferase shown to regulate noncoding RNAs in the brain (Kim et al. 2023). Autosomal-recessive mutation mediated loss of function of NSun2 was associated with neurological dysfunction in patients (Kim et al. 2023). In AD patients, NSun2 protein expression was reduced in the hippocampal and prefrontal cortex associated with increased pTau level. In a *Drosophila melanogaster* model of tau-induced toxicity, inhibition of NSun2 was exacerbated while its overexpression attenuated tau toxicity (Kim et al. 2023). Functionally, NSun2 deficiency decreased miR-125b and induced pTau (Kim et al. 2023). In addition to m^6A, other RNA modifications have been suggested to be associated with brain aging and neurodegeneration. Monomethylation of adenosine in the N1 position, N1-methyl adenosine (m^1A), has been linked to dysregulated mitochondrial tRNA expression and mitochondrial function in a Drosophila tau model (Shafik et al. 2022). Moreover, adenosine-to-inosine (A-to-I) editing is a prevalent posttranscriptional RNA modification within the brain, and changes in A-to-I editing were observed when comparing A-to-I editing between postmortem and living prefrontal cortical tissues (Rodriguez de Los Santos et al. 2024), suggesting a role of A-to-I editing in brain aging. These results highlight the critical role of m^6A, or RNA modifications, in age-related cognitive dysfunction. Additional studies are needed to uncover the full picture of RNA modifications in brain aging.

5 Conclusion and Future Directions

The rising prevalence of age-related cognitive decline in the United States and globally highlights the urgent need for innovative therapeutic approaches. Emerging evidence identifies noncoding RNAs as valuable diagnostic biomarkers and promising therapeutic targets for addressing age-related cognitive decline. RNA targeting strategies, particularly antisense oligonucleotides and small interfering RNAs that promote RNA degradation or cleavage, have demonstrated potential, although in diseases other than cognitive dysfunction, in preclinical models and are currently being investigated in multiple clinical trials. Unique challenges need to be considered when targeting noncoding RNAs as therapeutic strategies. First, noncoding RNAs are usually expressed at relatively low levels compared to protein-coding genes, which may require highly sensitive detection techniques (i.e., droplet digital PCR) when adopting noncoding RNAs as biomarkers for disease diagnosis. The low abundance of noncoding RNAs may also limit the effect of therapies that target noncoding RNA inhibition and degradation. However, noncoding RNAs have been shown to exhibit tissue-specific and cell type-specific expression patterns, especially under disease conditions. Second, noncoding RNAs, such as lncRNAs, have low species conservation, requiring extra effort when selecting homologous lncRNAs as targets for functional studies and further clinical trials. Third, RNA modifications are common in RNA molecules, and usually, multiple RNA modification sites exist in one RNA or for the same modification site present on different RNAs under certain disease conditions. Attempts to target selective modification sites for a specific RNA modification type may lead to off-target effects. Thus, optimization is needed when targeting RNA modifications for therapeutic benefit. This is also true for lncRNAs that confer their functions through diverse mechanisms that cover a wide range of downstream effectors. Nevertheless, noncoding RNAs are increasingly recognized as pivotal regulators of age-related cognitive dysfunction. Targeting them at the expression level or through posttranscriptional mechanisms offers innovative therapeutic opportunities for addressing brain aging and age-related cognitive dysfunction.

References

Alder JK, Armanios M (2022) Telomere-mediated lung disease. Physiol Rev 102:1703–1720

Amiri A, Barreto G, Sathyapalan T et al (2021) siRNA therapeutics: future promise for neurodegenerative diseases. Curr Neuropharmacol 19:1896–1911

Balmain A (2020) The critical roles of somatic mutations and environmental tumor-promoting agents in cancer risk. Nat Genet 52:1139–1143

Bejaoui Y, Razzaq A, Yousri NA et al (2022) DNA methylation signatures in Blood DNA of Hutchinson–Gilford Progeria syndrome. Aging Cell 21:e13555

Bellenguez C, Küçükali F, Jansen IE et al (2022) New insights into the genetic etiology of Alzheimer's disease and related dementias. Nat Genet 54:412–436

Bennett IJ, Madden DJ (2014) Disconnected aging: cerebral white matter integrity and age-related differences in cognition. Neuroscience 276:187–205

Ben-Tov Perry R, Tsoory M, Tolmasov M et al (2023) Silc1 long noncoding RNA is an immediate-early gene promoting efficient memory formation. Cell Rep 42:113168

Bhatt V, Tiwari AK (2023) Sirtuins, a key regulator of ageing and age-related neurodegenerative diseases. Int J Neurosci 133:1167–1192

Blackburn EH, Epel ES, Lin J (2015) Human telomere biology: a contributory and interactive factor in aging, disease risks, and protection. Science 350:1193–1198

Blasco MA (2005) Telomeres and human disease: ageing, cancer and beyond. Nat Rev Genet 6:611–622

Blokzijl F, De Ligt J, Jager M et al (2016) Tissue-specific mutation accumulation in human adult stem cells during life. Nature 538:260–264

Bonkowski MS, Sinclair DA (2016) Slowing ageing by design: the rise of NAD+ and sirtuin-activating compounds. Nat Rev Mol Cell Biol 17:679–690

Brown K, Xie S, Qiu X et al (2013) SIRT3 reverses aging-associated degeneration. Cell Rep 3:319–327

Bushati N, Cohen SM (2007) microRNA functions. Annu Rev Cell Dev Biol 23:175–205

Cao M, Li H, Zhao J et al (2019) Identification of age- and gender-associated long noncoding RNAs in the human brain with Alzheimer's disease. Neurobiol Aging 81:116–126

Cao J, Huang M, Guo L et al (2021) MicroRNA-195 rescues ApoE4-induced cognitive deficits and lysosomal defects in Alzheimer's disease pathogenesis. Mol Psychiatry 26:4687–4701

Castro-Hernandez R, Berulava T, Metelova M et al (2023) Conserved reduction of m(6)A RNA modifications during aging and neurodegeneration is linked to changes in synaptic transcripts. Proc Natl Acad Sci USA 120:e2204933120

Chakravarti D, LaBella KA, DePinho RA (2021) Telomeres: history, health, and hallmarks of aging. Cell 184:306–322

Chen LL, Kim VN (2024) Small and long non-coding RNAs: past, present, and future. Cell 187:6451–6485

Chen LL, Yang L (2015) Regulation of circRNA biogenesis. RNA Biol 12:381–388

Chen BJ, Ueberham U, Mills JD et al (2017) RNA sequencing reveals pronounced changes in the noncoding transcriptome of aging synaptosomes. Neurobiol Aging 56:67–77
Chen ML, Hong CG, Yue T et al (2021) Inhibition of miR-331-3p and miR-9-5p ameliorates Alzheimer's disease by enhancing autophagy. Theranostics 11:2395–2409
Chen S, Cao Z, Nandi A et al (2024) The global macroeconomic burden of Alzheimer's disease and other dementias: estimates and projections for 152 countries or territories. Lancet Glob Health 12:e1534–e1543
Chinnery PF, Samuels DC, Elson J et al (2002) Accumulation of mitochondrial DNA mutations in ageing, cancer, and mitochondrial disease: is there a common mechanism? Lancet (London, England) 360:1323–1325
Corral-Debrinski M, Horton T, Lott MT et al (1992) Mitochondrial DNA deletions in human brain: regional variability and increase with advanced age. Nat Genet 2:324–329
Cortopassi GA, Shibata D, Soong NW et al (1992) A pattern of accumulation of a somatic deletion of mitochondrial DNA in aging human tissues. Proc Natl Acad Sci USA 89:7370–7374
Czech B, Preall JB, McGinn J et al (2013) A transcriptome-wide RNAi screen in the Drosophila ovary reveals factors of the germline piRNA pathway. Mol Cell 50:749–761
Czech B, Munafo M, Ciabrelli F et al (2018) piRNA-guided genome defense: from biogenesis to silencing. Annu Rev Genet 52:131–157
Dube U, Del-Aguila JL, Li Z et al (2019) An atlas of cortical circular RNA expression in Alzheimer disease brains demonstrates clinical and pathological associations. Nat Neurosci 22:1903–1912
European Alzheimer's and DBMR Collaboration (2023) Genetic associations between modifiable risk factors and Alzheimer disease. JAMA Netw Open 6:e2313734
Faghihi MA, Modarresi F, Khalil AM et al (2008) Expression of a noncoding RNA is elevated in Alzheimer's disease and drives rapid feed-forward regulation of beta-secretase. Nat Med 14:723–730
Fahy GM, Brooke RT, Watson JP et al (2019) Reversal of epigenetic aging and immunosenescent trends in humans. Aging Cell 18:e13028
Fraser HC, Kuan V, Johnen R et al (2022) Biological mechanisms of aging predict age-related disease co-occurrence in patients. Aging Cell 21:e13524
Friedman RC, Farh KK, Burge CB et al (2009) Most mammalian mRNAs are conserved targets of microRNAs. Genome Res 19:92–105
Giorgio A, Santelli L, Tomassini V et al (2010) Age-related changes in grey and white matter structure throughout adulthood. Neuroimage 51:943–951
Graziewicz MA, Longley MJ, Copeland WC (2006) DNA polymerase γ in mitochondrial DNA replication and repair. Chem Rev 106:383–405
Greaves LC, Nooteboom M, Elson JL et al (2014) Clonal expansion of early to mid-life mitochondrial DNA point mutations drives mitochondrial dysfunction during human ageing. PLoS Genet 10:e1004620
Han SW, Pyun JM, Bice PJ et al (2024) miR-129-5p as a biomarker for pathology and cognitive decline in Alzheimer's disease. Alzheimers Res Ther 16:5
Hennekam RC (2020) Pathophysiology of premature aging characteristics in Mendelian progeroid disorders. Eur J Med Genet 63:104028
Hirai K, Aliev G, Nunomura A et al (2001) Mitochondrial abnormalities in Alzheimer's disease. J Neurosci 21:3017–3023
Horvath S, Raj K (2018) DNA methylation-based biomarkers and the epigenetic clock theory of ageing. Nat Rev Genet 19:371–384
Hu Y, Fryatt GL, Ghorbani M et al (2021) Replicative senescence dictates the emergence of disease-associated microglia and contributes to Aβ pathology. Cell Rep 35:109228
Hu W, Xie H, Zeng Y et al (2023) N6-methyladenosine participates in mouse hippocampus neurodegeneration via PD-1/PD-L1 pathway. Front Neurosci 17:1145092
Huang Z, Sun S, Lee M et al (2022) Single-cell analysis of somatic mutations in human bronchial epithelial cells in relation to aging and smoking. Nat Genet 54:492–498

Islam MR, Kaurani L, Berulava T et al (2021) A microRNA signature that correlates with cognition and is a target against cognitive decline. EMBO Mol Med 13:e13659

Jain G, Stuendl A, Rao P et al (2019) A combined miRNA-piRNA signature to detect Alzheimer's disease. Transl Psychiatry 9:250

Jaskelioff M, Muller FL, Paik J-H et al (2011) Telomerase reactivation reverses tissue degeneration in aged telomerase-deficient mice. Nature 469:102–106

Jia J, Ning Y, Chen M et al (2024) Biomarker changes during 20 years preceding Alzheimer's disease. N Engl J Med 390:712–722

Kandimalla RJ, Wani WY, Binukumar BK et al (2012) siRNA against presenilin 1 (PS1) down regulates amyloid beta42 production in IMR-32 cells. J Biomed Sci 19:2

Kang HJ, Kawasawa YI, Cheng F et al (2011) Spatio-temporal transcriptome of the human brain. Nature 478:483–489

Katsimpardi L, Litterman NK, Schein PA et al (2014) Vascular and neurogenic rejuvenation of the aging mouse brain by young systemic factors. Science 344:630–634

Khvorova A (2023) siRNAs—a new class of medicines. JAMA 329:2185–2186

Kim YA, Siddiqui T, Blaze J et al (2023) RNA methyltransferase NSun2 deficiency promotes neurodegeneration through epitranscriptomic regulation of tau phosphorylation. Acta Neuropathol 145:29–48

Kung JT, Colognori D, Lee JT (2013) Long noncoding RNAs: past, present, and future. Genetics 193:651–669

Lenart P, Novak J, Bienertova-Vasku J (2018) PIWI-piRNA pathway: setting the pace of aging by reducing DNA damage. Mech Ageing Dev 173:29–38

Li C, Götz J (2017) Somatodendritic accumulation of Tau in Alzheimer's disease is promoted by Fyn-mediated local protein translation. EMBO J 36:3120–3138

Liu N, Landreh M, Cao K et al (2012) The microRNA miR-34 modulates ageing and neurodegeneration in Drosophila. Nature 482:519–523

López-Otín C, Blasco MA, Partridge L et al (2013) The hallmarks of aging. Cell 153:1194–1217

López-Otín C, Blasco MA, Partridge L et al (2023) Hallmarks of aging: an expanding universe. Cell 186:243–278

Lu Y, Tan L, Wang X (2019) Circular HDAC9/microRNA-138/Sirtuin-1 pathway mediates synaptic and amyloid precursor protein processing deficits in Alzheimer's disease. Neurosci Bull 35:877–888

Lujan SA, Longley MJ, Humble MH et al (2020) Ultrasensitive deletion detection links mitochondrial DNA replication, disease, and aging. Genome Biol 21:1–34

Mao Q, Fan L, Wang X et al (2019) Transcriptome-wide piRNA profiling in human brains for aging genetic factors. Jacobs J Genet 4:014

Martincorena I, Fowler JC, Wabik A et al (2018) Somatic mutant clones colonize the human esophagus with age. Science 362:911–917

Melas K, Talevi V, Imtiaz MA et al (2024) Blood-derived microRNAs are related to cognitive domains in the general population. Alzheimers Dement 20:7138–7159

Miller MB, Huang AY, Kim J et al (2022) Somatic genomic changes in single Alzheimer's disease neurons. Nature 604:714–722

Mostoslavsky R, Chua KF, Lombard DB et al (2006) Genomic instability and aging-like phenotype in the absence of mammalian SIRT6. Cell 124:315–329

Muñoz-Lorente MA, Cano-Martin AC, Blasco MA (2019) Mice with hyper-long telomeres show less metabolic aging and longer lifespans. Nat Commun 10:4723

Nik-Zainal S, Hall BA (2019) Cellular survival over genomic perfection. Science 366:802–803

North BJ, Rosenberg MA, Jeganathan KB et al (2014) SIRT 2 induces the checkpoint kinase BubR1 to increase lifespan. EMBO J 33:1438–1453

Oh ES, Petronis A (2021) Origins of human disease: the chrono-epigenetic perspective. Nat Rev Genet 22:533–546

Park EG, Ha H, Lee DH et al (2022) Genomic analyses of non-coding RNAs overlapping transposable elements and its implication to human diseases. Int J Mol Sci 23:8950

Patel RS, Lui A, Hudson C et al (2023) Small molecule targeting long noncoding RNA GAS5 administered intranasally improves neuronal insulin signaling and decreases neuroinflammation in an aged mouse model. Sci Rep 13:317

Ponting CP, Oliver PL, Reik W (2009) Evolution and functions of long noncoding RNAs. Cell 136:629–641

Ren Z, Chu C, Pang Y et al (2022) A circular RNA blood panel that differentiates Alzheimer's disease from other dementia types. Biomark Res 10:63

Rodriguez de Los Santos M, Kopell BH, Buxbaum Grice A et al (2024) Divergent landscapes of A-to-I editing in postmortem and living human brain. Nat Commun 15:5366

Rolyan H, Scheffold A, Heinrich A et al (2011) Telomere shortening reduces Alzheimer's disease amyloid pathology in mice. Brain 134:2044–2056

Roundtree IA, Evans ME, Pan T et al (2017) Dynamic RNA modifications in gene expression regulation. Cell 169:1187–1200

Roy J, Sarkar A, Parida S et al (2017) Small RNA sequencing revealed dysregulated piRNAs in Alzheimer's disease and their probable role in pathogenesis. Mol BioSyst 13:565–576

Sanchez-Contreras M, Kennedy SR (2022) The complicated nature of somatic mtDNA mutations in aging. Front Aging 2:805126

Santisteban MM, Iadecola C (2025) The pathobiology of neurovascular aging. Neuron 113:49–70

Sato K, Takayama KI, Inoue S (2023) Role of piRNA biogenesis and its neuronal function in the development of neurodegenerative diseases. Front Aging Neurosci 15:1157818

Scheff SW, Ansari MA, Mufson EJ (2016) Oxidative stress and hippocampal synaptic protein levels in elderly cognitively intact individuals with Alzheimer's disease pathology. Neurobiol Aging 42:1–12

Schroder S, Fuchs U, Gisa V et al (2024) PRDM16-DT is a novel lncRNA that regulates astrocyte function in Alzheimer's disease. Acta Neuropathol 148:32

Shafik AM, Zhang F, Guo Z et al (2021) N6-methyladenosine dynamics in neurodevelopment and aging, and its potential role in Alzheimer's disease. Genome Biol 22:17

Shafik AM, Zhou H, Lim J et al (2022) Dysregulated mitochondrial and cytosolic tRNA m1A methylation in Alzheimer's disease. Hum Mol Genet 31:1673–1680

Shen K, Liu X, Chen D et al (2021) Voluntary wheel-running exercise attenuates brain aging of rats through activating miR-130a-mediated autophagy. Brain Res Bull 172:203–211

Shi Z, Chen T, Yao Q et al (2017) The circular RNA ciRS-7 promotes APP and BACE1 degradation in an NF-kappaB-dependent manner. FEBS J 284:1096–1109

Shim HS, Horner JW, Wu C-J et al (2021) Telomerase reverse transcriptase preserves neuron survival and cognition in Alzheimer's disease models. Nat Aging 1:1162–1174

Shokhirev MN, Johnson AA (2022) An integrative machine-learning meta-analysis of high-throughput omics data identifies age-specific hallmarks of Alzheimer's disease. Ageing Res Rev 81:101721

Sinha M, Jang YC, Oh J et al (2014) Restoring systemic GDF11 levels reverses age-related dysfunction in mouse skeletal muscle. Science 344:649–652

Smukowski SN, Danyko C, Somberg J et al (2024) mRNA and circRNA mislocalization to synapses are key features of Alzheimer's disease. PLoS Genet 20:e1011359

Song C, Zhang Y, Huang W et al (2022) Circular RNA Cwc27 contributes to Alzheimer's disease pathogenesis by repressing Pur-alpha activity. Cell Death Differ 29:393–406

Soong NW, Hinton DR, Cortopassi G et al (1992) Mosaicism for a specific somatic mitochondrial DNA mutation in adult human brain. Nat Genet 2:318–323

Srinivasan AR, Tran TT, Bonini NM (2022) Loss of miR-34 in Drosophila dysregulates protein translation and protein turnover in the aging brain. Aging Cell 21:e13559

Stewart JB, Chinnery PF (2015) The dynamics of mitochondrial DNA heteroplasmy: implications for human health and disease. Nat Rev Genet 16:530–542

Sun ED, Nagvekar R, Pogson AN et al (2025) Brain aging and rejuvenation at single-cell resolution. Neuron 113:82–108

Tian X, Firsanov D, Zhang Z et al (2019) SIRT6 is responsible for more efficient DNA double-strand break repair in long-lived species. Cell 177:622–638.e622

Tobin MK, Musaraca K, Disouky A et al (2019) Human hippocampal neurogenesis persists in aged adults and Alzheimer's disease patients cell stem. Cell 24:974–982.e973

Tomas-Loba A, Flores I, Fernandez-Marcos PJ et al (2008) Telomerase reverse transcriptase delays aging in cancer-resistant mice. Cell 135:609–622

Vermeij WP, Hoeijmakers JH, Pothof J (2016) Genome integrity in aging: human syndromes, mouse models, and therapeutic options. Annu Rev Pharmacol Toxicol 56:427–445

Vijg J, Dong X (2020) Pathogenic mechanisms of somatic mutation and genome mosaicism in aging. Cell 182:12–23

Vijg J, Suh Y (2013) Genome instability and aging. Annu Rev Physiol 75:645–668

Wang KC, Chang HY (2011) Molecular mechanisms of long noncoding RNAs. Mol Cell 43:904–914

Wang J, Xia Z, Sheng P et al (2022) Targeting MicroRNA-144/451-AKT-GSK3beta axis affects the proliferation and differentiation of radial glial cells in the mouse hippocampal dentate gyrus. ACS Chem Neurosci 13:897–909

Wang F, Li Y, Shen H et al (2024) Identification of pathological pathways centered on circRNA dysregulation in association with irreversible progression of Alzheimer's disease. Genome Med 16:129

Whittemore K, Vera E, Martínez-Nevado E et al (2019) Telomere shortening rate predicts species life span. Proc Natl Acad Sci USA 116:15122–15127

Wu Z, Liu X, Wang Y et al (2025) Pseudogene Lamr1-ps1 aggravates early spatial learning memory deficits in Alzheimer's disease model mice. Neurosci Bull 41:600. https://doi.org/10.1007/s12264-024-01336-6

Xu K, Zhang Y, Xiong W et al (2020) CircGRIA1 shows an age-related increase in male macaque brain and regulates synaptic plasticity and synaptogenesis. Nat Commun 11:3594

Zaccara S, Ries RJ, Jaffrey SR (2019) Reading, writing and erasing mRNA methylation. Nat Rev Mol Cell Biol 20:608–624

Zhang Y, Deng W, Wang W et al (2022) MicroRNA-206 down-regulated human umbilical cord mesenchymal stem cells alleviate cognitive decline in D-galactose-induced aging mice. Cell Death Discov 8:304

Zhao Y, Alexandrov PN, Jaber V et al (2016) Deficiency in the ubiquitin conjugating enzyme UBE2A in Alzheimer's disease (AD) is linked to deficits in a natural circular miRNA-7 sponge (circRNA; ciRS-7). Genes (Basel) 7:116

Zhou M, Zhao H, Wang X et al (2019) Analysis of long noncoding RNAs highlights region-specific altered expression patterns and diagnostic roles in Alzheimer's disease. Brief Bioinform 20:598–608

Zovoilis A, Agbemenyah HY, Agis-Balboa RC et al (2011) microRNA-34c is a novel target to treat dementias. EMBO J 30:4299–4308

Long Noncoding RNAs: Emerging Players in Neurodegenerative Diseases

Prarthana Guha, Ashcharya Rishi, Nagashree Bhat, and Subhrangsu S. Mandal

Contents

Abstract Growth and integrity of neurons are critical in proper physiological development and functions. Malfunction or abnormal development of neurons causes neurological and neurodegenerative diseases. This could be linked to various genetic and environmental factors. Neurodegenerative diseases, marked by irreversible neuronal damage, myelin loss, and gliosis, present substantial challenges for the aging global population. This is becoming more challenging in terms of the cost of health care and social responsibility. Diseases such as Alzheimer's disease,

P. Guha · A. Rishi · N. Bhat · S. S. Mandal (✉)
Gene Regulation and Epigenetics Research Laboratory, Department of Chemistry and Biochemistry, The University of Texas at Arlington, Arlington, TX, USA
e-mail: smandal@uta.edu

J. Barciszewski (ed.), *Decoding Aging and Neurodegeneration*,
RNA Technologies 16, https://doi.org/10.1007/978-3-031-91072-2_11

Parkinson's diseases, amyotrophic lateral sclerosis, and multiple sclerosis are associated with neuronal dysfunction, chronic neuroinflammation, and eventual cell death. At present, there are very limited solutions in terms of diagnosis and therapy. Hence, there is an urgent need to develop novel biomarkers for early diagnosis and discover novel therapeutics. While peptide and protein-based pathological mechanism have been extensively studied, emerging research emphasizes the pivotal role of noncoding RNAs, in disease pathogenesis. Here, we emphasize the functions and implications of long noncoding RNAs (lncRNAs) that are associated with neurological and neurodegenerative diseases, delving into their involvement in mitochondrial dysfunction, genetic factors, and neurovascular interactions. This chapter aims to bridge the gap between basic research and clinical applications, highlighting the potential of lncRNAs as therapeutic targets and biomarkers.

Keywords Noncoding RNAs · LncRNAs · Neurodegenerative diseases · Neuroinflammation · Neurological disorders · Alzheimer's disease · Parkinson's diseases · Multiple sclerosis

1 Introduction

The neuronal system, comprising the central and peripheral nervous systems, is crucial for regulating physiological functions, cognition, and behavior. Neuronal growth and integrity are vital for maintaining neurological health, with disruptions often leading to various neurological disorders such as Alzheimer's disease (AD), Parkinson's diseases (PD), and multiple sclerosis (MS). These conditions place a significant burden on individuals and healthcare systems globally, with direct medical costs exceeding billions of dollars annually. According to recent studies, the economic impact of neurological disorders in the United States alone reaches over $800 billion each year (Gooch et al. 2017), underscoring the urgent need for effective treatments, prevention strategies, and ongoing research to improve outcomes and reduce costs.

Neurodegenerative diseases (NDDs) are a heterogeneous group of disorders resulting from irreversible neuronal damage, loss of myelin sheath, and gliosis, posing a challenge for the world population (Azam et al. 2021). Recent reports from the World Health Organization suggest that the global population of individuals aged 60 and older is projected to double, reaching 22% by 2050 (https://www.who.int/news-room/fact-sheets/detail/ageing-and-health). A significant increase in the morbidity and mortality of the aging world population can be correlated with several diseases related to the progressive degeneration of the structure and function of the central nervous system (CNS) and peripheral nervous system, including AD, PD, Huntington's disease (HD), amyotrophic lateral sclerosis (ALS), MS, etc. (Reddy and Abeygunaratne 2022). These diseases are associated with neuronal dysfunction, chronic neuroinflammatory processes, and eventually death.

Current therapeutic strategies remain largely symptomatic, offering only temporary relief without halting or reversing the underlying neurodegenerative processes. One of the hallmarks of neurodegenerative diseases is neuroinflammation, a complex response involving resident immune cells of the CNS, such as microglia and astrocytes (Adamu et al. 2024). These cells play dual roles, contributing to both protective and pathological outcomes. Under normal conditions, microglia and astrocytes support neuronal health by maintaining homeostasis and clearing cellular debris. However, in NDDs, these cells often shift to a pro-inflammatory phenotype, releasing cytokines, chemokines, and reactive oxygen species (ROS) that exacerbate neuronal damage. Chronic neuroinflammation disrupts the delicate balance between protective and destructive immune responses, accelerating disease progression. Understanding the molecular mechanisms underlying neuroinflammation is crucial for identifying novel therapeutic targets.

Numerous studies have been performed to understand the pathological mechanism of neurological and neurodegenerative diseases, including genetic and epigenetic factors, lifestyles, food habits, and so on (Popa-Wagner et al. 2020). The studies centered around finding various proteins and peptide-based factors associated with neurodegenerative and neurological diseases. For example, in AD, the accumulation of misfolded proteins such as amyloid-beta (Aβ) plaques and tau neurofibrillary tangles is considered central to its pathology (Hampel et al. 2023). Research has identified pathways involving amyloid precursor protein (APP) processing and tau phosphorylation, emphasizing their roles in neuronal damage and synaptic dysfunction. In PD, alpha (α)-synuclein (SNCA) aggregation forming Lewy bodies is a hallmark of the disorder (Balaj and Kaku 2024). The role of this protein in synaptic transmission and its pathological aggregation due to genetic mutations, such as SNCA or LRRK2, has been extensively studied. Additionally, mitochondrial dysfunction and oxidative stress contribute significantly to dopaminergic neuron degeneration in the substantia nigra (SN). For HD, an autosomal-dominant disorder, the expanded cytosine, adenine, and guanine (CAG) repeats in the huntingtin (HTT) gene, leads to the production of mutant huntingtin protein, which aggregates and disrupts cellular processes (Tabrizi et al. 2019; Tong et al. 2024). Epigenetic mechanisms such as histone modification and transcriptional dysregulation play critical roles in disease pathology. Emerging research in ALS focuses on the role of RNA-binding proteins like TAR DNA-binding protein of 43 kDa (TDP-43) and fused in sarcoma (FUS), as well as the involvement of C9ORF72 hexanucleotide repeat expansions (Barmada 2015).

While there has been a significant enhancement in the understanding of the genetic and protein-based factors associated with NDDs, in recent years, research has increasingly focused on the role of noncoding RNAs, especially lncRNAs, in regulating neuroinflammatory pathways and contributing to the pathology of NDDs. LncRNAs are a subclass of noncoding RNAs, longer than 200 nucleotides, that mostly do not encode proteins but potentially contribute toward the regulation of chromosomal dynamics, gene expression, cell signaling, and physiological functions (Cuevas-Diaz Duran et al. 2019). Despite not being translated into proteins, the lncRNA, being large stretch of nucleic acids, can fold into various three-dimensional structures, can interact with various proteins and nucleic acids, and hence may modulate target protein functions and physiological responses. LncRNAs exhibit also tissue specificity and temporal regulation, making them particularly significant in the CNS, where they contribute to processes such as neural differentiation, synaptic plasticity, and neuronal survival (Zimmer-Bensch 2019). These findings underscore the functional diversity of lncRNAs in neurodegenerative pathways, highlighting their potential for developing targeted interventions. Despite their promise, leveraging lncRNAs as therapeutic targets faces numerous challenges. The selective permeability of the blood-brain barrier (BBB) complicates the delivery of lncRNA-based therapeutics to the CNS (Mathew and Sivasubbu 2022). Additionally, the complex and redundant functions of lncRNAs necessitate precise targeting to avoid off-target effects and unintended consequences. Nevertheless, advances in molecular biology, such as high-throughput sequencing, CRISPR-based gene editing, and nanoparticle-based delivery systems, are gradually overcoming these obstacles, bringing lncRNA-based therapies closer to clinical application.

In this article, we aim to explore and review the multifaceted roles of lncRNAs in the pathogenesis of various neurological and neurodegenerative diseases and discuss their potential implications as biomarkers and therapeutic targets. It will also address the challenges associated with translating basic research on lncRNAs into clinical practice and outline future directions for developing effective treatments. Understanding the intricate regulatory networks involving lncRNAs may pave the way for novel interventions, offering hope for slowing or reversing the devastating effects of neurodegenerative diseases.

2 Long Noncoding RNAs

Long noncoding (lncRNAs) have garnered considerable attention over the past decade due to their roles in regulating gene expression and modulating various cellular and physiological processes (Bhan et al. 2013, 2014). Initially dismissed as transcriptional noise due to their low expression levels and lack of conservation, lncRNAs have now been shown to play pivotal roles in nearly all biological processes, including chromatin remodeling, transcriptional and post-transcriptional regulation, and signal transduction pathways. Recent studies have identified nearly 27,919 lncRNAs in the human genome, underscoring their potential significance in cellular function and disease pathology (Hon et al. 2017). Understanding the structure and function of lncRNAs presents a significant challenge due to their poor sequence conservation across species, functional redundancy, and extensive cross-regulation among different lncRNA molecules. Unlike protein-coding genes, lncRNAs often lack conserved sequences that could be used to predict their function, complicating efforts to study them using traditional comparative genomics approaches. LncRNAs display a broad range of biogenesis pathways and structural complexities, with some lncRNAs transcribed from enhancer regions (eRNAs) or as antisense transcripts from the opposite strand of protein-coding genes, while others originate from intronic or intergenic regions (Fig. 1a) (Bhan et al. 2017; Wang and Chang 2011). The biogenesis and processing of lncRNAs are tightly regulated, involving RNA-binding proteins (RBPs) and ribonucleoprotein complexes. These factors can influence the stability, localization, and function of lncRNAs, further contributing to their diverse regulatory roles. For example, lncRNAs such as X-inactive specific transcript (XIST) and Hox transcript antisense intergenic RNA (HOTAIR) interact with polycomb repressive complex 2 (PRC2) to guide chromatin modifications and gene silencing at specific genomic loci, demonstrating their involvement in epigenetic regulation (Li et al. 2022; Price et al. 2021).

LncRNAs exert their regulatory functions through various mechanisms, including acting as molecular scaffolds, decoys, guides, and enhancers (Fig. 1b) (Harries 2012). They can serve as molecular scaffolds to assemble multiple protein complexes, thereby coordinating chromatin remodeling and transcriptional activation or repression. For example, the lncRNA-HOTAIR acts as a scaffold to bring together PRC2 and lysine-specific demethylase 1 (LSD1) complexes, leading to the silencing of HOX genes and influencing chromatin dynamics (Bhan and Mandal 2015; Price et al. 2021). Additionally, lncRNAs can function as decoys by sequestering transcription factors or RBPs away from their target DNA or RNA sequences. The lncRNA PANDA sequesters the transcription factor NF-YA, preventing it from activating pro-apoptotic genes, thereby regulating cell survival under stress conditions (Hung et al. 2011). Moreover, some lncRNAs act as guides, directing chromatin-modifying complexes to specific genomic loci, as demonstrated by the lncRNA XIST in X-chromosome inactivation (Li et al. 2022). Furthermore, lncRNAs can act as enhancer RNAs (eRNAs) that promote the transcription of nearby genes by interacting with transcription factors and coactivators. The lncRNA CCAT1-L, for instance, enhances c-Myc (MYC) expression by looping into the MYC promoter region, demonstrating a role in chromatin organization and long-range gene regulation (Xiang et al. 2014).

High-throughput deep sequencing technologies have significantly expanded our knowledge of the human transcriptome, revealing that ncRNAs comprise about 50% of the human transcriptome (Yin and Shen 2023). This discovery underscores the vast and largely uncharted landscape of noncoding RNA biology, positioning lncRNAs as a critical area of research for understanding gene regulation and its implications in health and disease. As research continues to uncover the complexity and versatility of lncRNAs, it becomes increasingly evident that these molecules are not mere transcriptional byproducts but are central to the regulation of gene expression and the maintenance of cellular homeostasis. Despite the understanding of LncRNA mechanisms in a plethora of immune and metabolic disorders, the role of these RNA molecules in brain and neural function is largely unknown. The next sections of this chapter aim to summarize the role of lncRNAs in neural diversity and disorders.

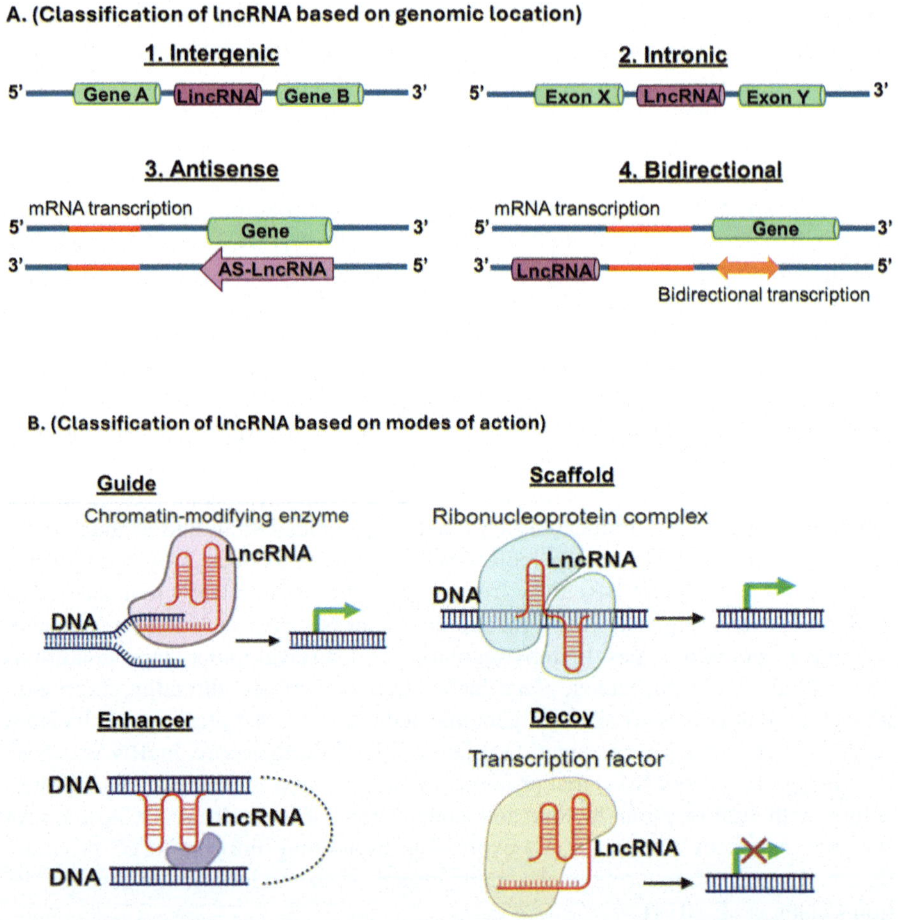

Fig. 1 Classification and functional mechanisms of lncRNAs. (**a**) Classification of lncRNA based on genomic location. Based on their genomic location, lncRNAs are categorized into intergenic, intronic, antisense, bidirectional, and enhancer RNAs. Intergenic lncRNAs are transcribed from regions between two protein-coding genes, while intronic lncRNAs originate from the introns of protein-coding genes. Antisense lncRNAs are synthesized from the complementary strand of a protein-coding gene. Bidirectional lncRNAs are produced from both strands. (**b**) Classification of lncRNA based on modes of action. Mechanisms of lncRNA functions include molecular guides, scaffolds, signals, and decoys. As guides, they recruit chromatin modifiers to regulate gene expression and chromatin structure. As scaffolds, they facilitate the assembly of protein complexes at target sites. Acting as signals, lncRNAs are transcribed in response to stimuli and collaborate with transcription factors to influence downstream gene transcription. As decoys, lncRNAs inhibit transcriptional regulators by sequestering them away from their targets, whereas enhancer RNAs are transcribed from enhancer regions

3 LncRNAs in the Central Nervous System

Studies show that lncRNAs are highly abundant in the brain, with approximately 40% of them being specifically expressed in the brain tissue, greatly exceeding the number of protein-coding genes (Navandar et al. 2024). Compared to other tissues, lncRNAs in the brain show higher levels of conservation, suggesting a critical role in the evolution and function of the nervous system (Navandar et al. 2024). Furthermore, lncRNAs in the brain exhibit greater temporal and spatial specificity than mRNAs, indicating that their expression is tightly regulated according to developmental stages and specific brain regions. For instance, the expression patterns of lncRNAs differ significantly between various brain regions, such as the cerebral cortex and hippocampus, and vary with age, reflecting their specialized roles in brain development and function. One notable example is the human accelerated region 1 (HAR1), a cis-antisense locus targeted by RE1-silencing transcription (REST) factor (Johnson et al. 2010). It is part of the overlapping lncRNA gene HAR1F (also known as HAR1A). HAR1F is specifically expressed in Cajal-Retzius neurons within the developing human cerebral cortex, indicating a role in early brain development. The HAR1F gene, through its lncRNA product, is thought to contribute to the unique aspects of human brain evolution and cortical development. A pioneering study of the HAR1F variant revealed that human embryonic brain exhibits a strong HAR1F expression alongside cortical patterning protein reelin. Additionally, lincRNA1230 plays a critical role in repressing neural commitment in mouse embryonic stem (ES) cells (Wang et al. 2016). It modulates H3K4me3 accumulation on the promoters of key transcription factors, Pax6 and Sox1, by interacting with WDR5, a trithorax complex component. Overexpression of lincRNA1230 reduces WDR5 occupancy and H3K4me3 marks, inhibiting neuronal differentiation, highlighting its pivotal function in preventing neural induction. LncRNAs are increasingly recognized as key regulators of neural differentiation. In proliferative neural stem and progenitor cells, lncRNAs help orchestrate the sequential activation of gene regulatory programs specific to different cell types (Zimmer-Bensch 2019). This regulation is essential for the progression from pluripotent cells in early embryonic development to the fully differentiated neuronal and glial cell types present in the mature mammalian brain. A specific example is lncRNA_DALI (deleted in AML1/ETO leukemia 1 intergenic), which promotes neural differentiation by modulating the expression of genes necessary for neuronal development in neuroblastoma cells through multiple mechanisms (Chalei et al. 2014). DALI regulates the expression of the neuronal differentiation gene Pou3f3 in cis and forms a trans-acting regulatory complex that further modulates the expression of neural differentiation genes. Another lncRNA, lncND, expressed from a locus linked to neurodevelopmental disorders, increases during neurogenesis and decreases with neuronal differentiation (Rani et al. 2016). Predicted to contain 16 miR-143-3p binding sites, lncND interacts with AGO2, acting as a miRNA sponge. By stabilizing miR-143-3p, lncND regulates Notch signaling, maintaining neural progenitors. Knockdown reduces Notch activation, while overexpression expands radial glial cells, preventing premature differentiation. These interactions exemplify the multifaceted roles of lncRNAs in guiding the complex process of neural development and differentiation.

LncRNAs are also involved in critical processes such as neurite outgrowth, synaptogenesis, and synaptic plasticity, all of which require intricate regulation of gene expression and signaling pathways (Zimmer-Bensch 2019). These processes are fundamental to the establishment and refinement of neural circuits during development and for the adaptability of the brain throughout life. A key player in this regard is the brain-derived neurotrophic factor (BDNF), which is crucial for synaptogenesis and neurite outgrowth (Bohnsack et al. 2019). The antisense lncRNA of BDNF, known as BDNF-AS, regulates the expression of the BDNF gene by recruiting the PRC2 to the BDNF promoter. This recruitment suppresses the expression of BDNF, thereby affecting BDNF-mediated processes such as axonal growth, neuronal proliferation, and apoptosis. By influencing these key neurotrophic pathways, lncRNAs like BDNF-AS play essential roles in neurodevelopment and plasticity, impacting learning, memory, and recovery from neural injuries. Another example of lncRNAs involved in neurite outgrowth and synaptic plasticity is metastasis-associated lung adenocarcinoma transcript 1 (MALAT1), a highly expressed lncRNA in neurons, which is enriched in nuclear speckles and regulates neurite elaboration (Briggs et al. 2015). While MALAT1 knockout mice show no overt phenotype, in vitro studies reveal its role in synaptic density regulation by recruiting serine/arginine family splicing proteins to synaptogenesis-related gene sites. Knockdown reduces synaptic density, while overexpression increases it, highlighting MALAT1's critical regulatory function.

Given their critical roles in brain development, differentiation, and plasticity, lncRNAs are increasingly recognized as potential therapeutic targets for neurological disorders. Dysregulation of lncRNA expression has been associated with various neurodevelopmental and neurodegenerative diseases, including autism spectrum disorders (ASD), schizophrenia, Alzheimer's disease, and Parkinson's disease. Understanding these mechanisms will pave the way for developing lncRNA-based biomarkers and therapeutic approaches, potentially revolutionizing the treatment of neurological and psychiatric disorders.

4 LncRNAs in Neurodegenerative Diseases

4.1 Pathological Mechanisms of Neurodegenerative Diseases

Neurodegenerative diseases, being mostly incurable, are prominent causes of most issues related to memory loss, cognitive dysfunctions, and ataxias. The aging of neurons in CNS decreases their survival rate leading the classic neurodegenerative diseases such as AD, PD, HD, ALS, MS, etc. The pathophysiology of these diseases remains elusive despite current research indicating several factors including neuroinflammation and metabolic stress.

4.1.1 Neuroinflammation

Neuroinflammation has recently garnered attention as a common and crucial pathological mechanism across a spectrum of NDDs, including AD, PD, ALS, MS, etc. (Gao et al. 2023). The term refers to the inflammatory response within the CNS involving various cellular components including microglia, the resident immune cells of the brain, astrocytes, and, occasionally, infiltrating immune cells from the peripheral system. CNS is a site of immune abundance where homeostasis is largely dependent on the innate immune response (Guo and Schluesener 2007). Neuroinflammation is a complicated response to the malfunction of the innate immune system and involves the alteration of reactive species in the CNS. Recent evidence surrounding the suppressed innate immune response in CNS parenchyma indicates the anti-inflammatory activities of the resident immune cells and shortage of dendritic cells in the tissue (Guo and Schluesener 2007). Inflammation is a biological process mediated by chemokines, cytokines, pro-inflammatory factors, ROS, etc. Inflammation, when unchecked, may have irreversible effects such as severe tissue damage, low general immunity, and even death. Elevation and activation of interleukin-1β (IL-1β), IL-6, tumor necrosis factor-α (TNF-α), granulocyte-macrophage colony-stimulating factor, and nuclear factor kappa-B (NF-κB) released by glial cells such as activated microglia, astrocytes, oligodendrocytes, neurons, T lymphocytes, and mast cells mediate neuroinflammation promoting toward neurodegenerative diseases. Therefore, deciphering the intricate mechanisms of neuroinflammation and identifying its critical regulators may provide novel strategies toward biomarker and drug development.

Microglia, the brain resident macrophages, regulate CNS development and neuronal networks and protect against infection and injury (Lull and Block 2010). Microglia are the first subsets of glial cells to be activated and the earliest responders to any insult or invasion. Activated microglia can have two different phenotypes such as M1 and M2 phenotypes. M1, the pro-inflammatory microglia, defends the brain against microbial infection and prevents acute inflammation. On the contrary, M2 microglia are anti-inflammatory in nature and contribute toward maintaining tissue homeostasis and resolving tissue injury. The neuroprotective activities of M2 microglia are owed to the release of anti-inflammatory factors such as IL-10, transformation growth factor-β, etc. However, overactive M1 microglia may release pro-inflammatory factors leading to a significant damage of the neighboring neurons. In the AD brain, Aβ plaque formation is associated with microglial activation and chronic inflammation, mediated via generation of ROS and nitric oxide (NO). Along with known cytokines and chemokines, emerging studies suggest that lncRNAs play a significant role in microglial activation. For instance, small nucleolar RNA host genes (SNHGs) are lncRNAs involved in tumor development and microglial polarization (Gao et al. 2022b; Zhang et al. 2020). SNHG4 sponges miR-449c-5p to activate STAT6, promoting M2 polarization, while SNHG15 sponges miR-302a-3p to activate STAT1/NF-κB, driving M1 polarization. Other lncRNAs, such as TUG1, HOTAIR, and GAS5, regulate M1 polarization through distinct miRNA interactions (Briggs et al. 2015). SNHG14 suppresses miR-145-5p to facilitate M1 polarization (Zhou et al. 2017), and KCNQ1OT1 downregulates miR-873-5p to promote TRAF6-mediated pathways (Liu et al. 2021). Exosomal lncGm37494 enhances M2 polarization via the PPARγ pathway (Shao et al. 2020). Interestingly, lncRNA H19 exhibits dual roles, modulating STAT3 activation under hypoxia or promoting M2 polarization via miR-29b-3p inhibition, showcasing complex regulatory mechanisms (Busscher et al. 2022).

Astrocytes, another major subset of the glial cells, significantly influence the occurrence of neuroinflammation and progression of neurodegenerative diseases (Stoklund Dittlau and Freude 2024). In normal physiological conditions, astrocytes are the protectors of neurons, their functions extending from regulation of blood flow, neurotransmitter balance, synaptic activity toward the maintenance of BBB. However, in neurodegenerative diseases, astrocytes exhibit pro-inflammatory characteristics leading to the secretion of neurotoxic elements associated with disease progression (Gudkov et al. 2023). They rapidly proliferate and cause hypertrophy, eventually introducing a glial scar that can evade neuronal damage repair. In AD, astrocytes promote Aβ accumulation and synaptic dysfunction, thereby exacerbating neuroinflammation and associated neuronal loss (Gudkov et al. 2023). The initiation and propagation of neuroinflammation in neurodegenerative diseases can be driven by various factors such as protein aggregates, neuronal injury, genetic factors, etc. For example, in AD, misfolded Aβ peptides accumulate to form plaques, which can activate microglia and astrocytes (Ahmad et al. 2019). Similarly, in PD, misfolded α-synuclein accumulates in neurons and can activate glial cells (Kwon and Koh 2020). Apart from that, damage to neurons and axons can release signals that activate glial cells. These signals include adenosine triphosphate, DNA, and damaged cell components that can be recognized by pattern recognition receptors on microglia and astrocytes. Variants in genes like TREM2, CD33, and ApoE can modulate the immune response in the brain (Krasemann et al. 2017). For instance, ApoE4, a risk factor for AD, is associated with an altered inflammatory response by glial cells. Glaucoma, a leading cause of irreversible blindness, is now recognized as a neurodegenerative disease with a strong inflammatory component (Feng et al. 2023). In glaucoma, microglia and astrocytes are activated in response to elevated intraocular pressure, leading to optic nerve head damage and retinal ganglion cell degeneration. The inflammatory response can trigger further damage to the optic nerve and exacerbate the progression of the disease. Chronic neuroinflammation can lead to several detrimental outcomes, including synaptic dysfunction and loss, neuronal death, and barrier dysfunction.

Recent studies from our laboratory also demonstrate that lncRNAs are critically associated with expression, inflammation, and macrophage activation (Chini et al. 2023; Obaid et al. 2018). In particular, we discovered a series of novel human lncRNAs (termed as hLinfRNAs—long noncoding inflammation-associated RNAs) involved in inflammation and macrophage activation (Chini et al. 2023). Several hLinfRNAs (hLinfRNA1 and others) regulate cytokine expression via regulation of NF-κB activation and other signaling pathways. However, the structure and functions of these hLinfRNAs in inflammation and human diseases remain mostly elusive. Additionally, we demonstrated that lncRNA such as HOTAIR is overexpressed in macrophages and regulates inflammation-associated glucose metabolism in macrophages (Chini et al. 2024; Obaid et al. 2021). Furthermore, we also found that HOTAIR is an overexpression in Aβ-mediated inflammation in mouse microglia and AD mouse brain (unpublished data). Our observation suggests that lncRNA are integral to inflammation and macrophage activation including neuroinflammation and neurodegenerative diseases.

4.1.2 Mitochondrial Dysfunction, Genetic Factors, and Neovascular Interactions

Beyond neuroinflammation, mitochondrial dysfunction, genetic factors, and neovascular interactions are also implicated in neurodegenerative diseases. For example, mitochondrial dynamics play a central role in PD where mutations in mitochondrial genes lead to neuronal loss (Van Acker et al. 2025). During mitophagy, the process of mitochondrial removal, there are several gene mutations that are involved in maintaining mitochondrial quality, such as PINK1 and Parkin (Imai and Lu 2011). Impaired mitophagy caused by these genes results in the accumulation of dysfunctional mitochondria and eventual neuronal death. This process is mostly associated with the loss of dopaminergic neurons, mostly reliant on mitochondrial function for dopamine synthesis and neurotransmission. Another example of mitochondrial dysfunction can be found in AD where it is considered as a hallmark for diagnosis. In AD, mitochondrial dynamics, disrupted by accumulation of Aβ plaques and tau fibrils, leads to the disruption of energy renewal (Rajmohan and Reddy 2017). This disruption affects neuronal function and contributes to synaptic loss and cognitive decline. Furthermore, increased ROS production from dysfunctional mitochondria can exacerbate oxidative stress, leading to further neuronal damage.

Genome-wide association studies (GWASs) have identified several risk alleles associated with neurodegenerative diseases, providing insights into the genetic factors that contribute to disease pathogenesis. In AD, the APOE gene plays a crucial role in determining disease risk and progression (Hunsberger et al. 2019). The APOE4 allele is associated with a higher risk of developing AD and an earlier onset of symptoms. The mechanism by which APOE4 contributes to AD involves its impact on amyloid-beta clearance and neuroinflammatory responses. APOE4 disrupts the normal function of glial cells, leading to increased Aβ accumulation and neuroinflammation. Conversely, the APOE2 allele is associated with a lower risk of AD and may have a protective effect (Li et al. 2020). Other genetic factors in AD include mutations in the APP gene, leading to increased production of Aβ, and mutations in the presenilin genes (PSEN1 and PSEN2), which are involved in the cleavage of APP to produce Aβ (Lanoiselee et al. 2017). These mutations are associated with early-onset AD and significantly increase the risk of developing the disease. In PD, genetic mutations in the SNCA gene, which encodes α-synuclein, contribute to the formation of Lewy bodies and neuronal death (Jeong and Lee 2020). Additionally, mutations in the LRRK2 gene, linked to PD, affect mitochondrial function and cellular signaling pathways. These genetic variations influence the pathogenesis and progression of PD and offer potential targets for therapeutic intervention. Understanding these genetic factors can lead to personalized medicine approaches, where treatments are tailored to an individual's genetic profile. Genetic testing can identify individuals at higher risk for neurodegenerative diseases, allowing for early intervention and monitoring. Additionally, gene therapy offers the potential to correct or mitigate the effects of harmful genetic mutations, providing a new avenue for treatment.

The integrity of the neurovascular unit, comprising blood vessels, endothelial cells, pericytes, and glial cells, is crucial for maintaining CNS homeostasis. The BBB and blood-retina barrier (BRB) regulate the flow of nutrients and neurotoxic substances, ensuring a stable environment for neuronal function (Carvey et al. 2009). Disruptions in these barriers are implicated in neurodegenerative disease progression. Therapeutic strategies to address neurovascular interactions focus on restoring the integrity of the BBB and BRB, improving blood flow, and reducing vascular damage. Drugs that promote endothelial cell health and support pericyte function could help maintain the structural integrity of the neurovascular unit. Additionally, therapies targeting neurovascular coupling, the coordination between neuronal activity and blood flow, could improve CNS function and reduce neurodegenerative processes.

4.2 LncRNAs in Alzheimer's Disease

AD is the most common form of dementia, characterized by progressive neurodegeneration leading to severe cognitive decline, accounting for 60% of all dementia cases in people over 65 years of age (DeTure and Dickson 2019). According to data collected in 2018 on 50 million dementia cases, more than 50% were found to exhibit AD (Xiao et al. 2021). Statistically, in 2015, the global cost of dementia amounted to $957.56 billion and is projected to escalate to $9.12 trillion by 2050 (Wimo et al. 2017). Histopathologically, AD is defined by the extracellular accumulation of Aβ plaques and intracellular neurofibrillary tangles (NFTs) composed of hyperphosphorylated tau protein (Hampel et al. 2023). Aβ is derived from the cleavage of APP by β- and γ-secretases, resulting in neurotoxic Aβ oligomers. These oligomers aggregate into plaques, triggering synaptic dysfunction, oxidative stress, and neuroinflammation through microglial activation. Additionally, tau, a microtubule-associated protein, becomes abnormally hyperphosphorylated in AD, detaching from microtubules and forming NFTs. This disrupts axonal transport and cellular homeostasis. Aβ accumulation is thought to initiate tau pathology, as Aβ promotes kinases such as glycogen synthase kinase (GSK)-3β, leading to tau

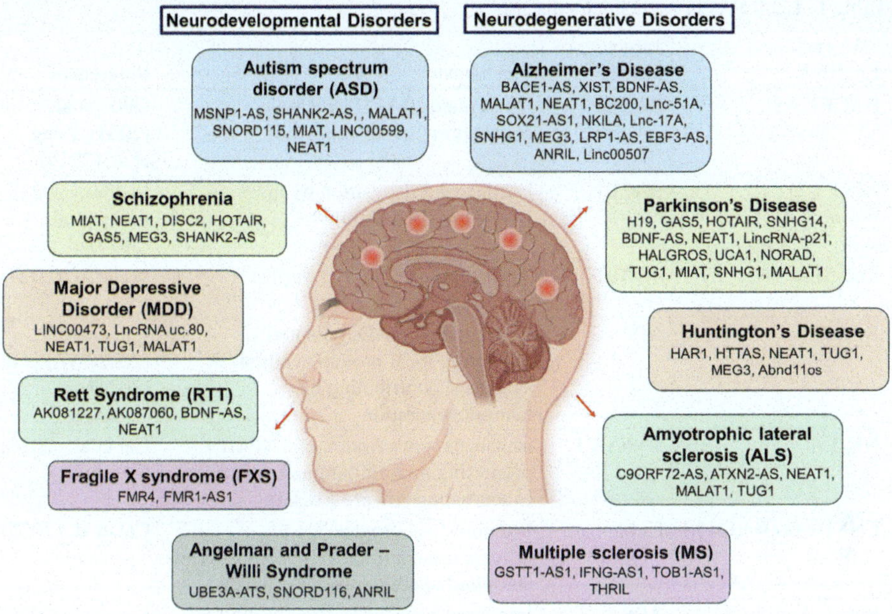

Fig. 2 List of lncRNAs in neurodegenerative and neurodevelopmental diseases

hyperphosphorylation (Lauretti et al. 2020). Furthermore, tau pathology correlates more closely with cognitive decline, emphasizing its role in neuronal death and synaptic loss. In the context of lncRNAs, several have been identified to play pivotal roles in regulating the pathways involved in AD pathology, and they are summarized below. The imbalance of lncRNA expression has been detected in AD, principally associated with Aβ accumulation, hyperphosphorylation of tau protein, neuroinflammation, mitochondrial dysfunction, autophagy regulation, and other processes (Fig. 2, Table 1).

4.2.1 BACE1-AS

The antisense gene of the β-secretase β-site APP-cleaving enzyme (BACE1)-AS is a lncRNA that has gained significant attention due to its role in enhancing the stability of BACE1 mRNA (Cai et al. 2025). BACE1 is an enzyme critical for the production of amyloid-beta, the main component of the amyloid plaques found in AD brains. BACE1-AS binds to BACE1 mRNA, preventing its degradation and thus promoting amyloid-beta production. Studies have demonstrated that the expression of BACE1-AS is upregulated in the brains of AD patients, correlating with increased BACE1 mRNA and protein levels. The loss of BACE1 in animal models had shown a deleterious effect on growth and behavior including memory dysfunction, emotional loss, and decreased learning ability (Ohno et al. 2004). Another study using

Table 1 LncRNAs in Alzheimer's disease

LncRNA	Target mRNA	Mechanism	References
BACE1-AS	BACE1	Upregulates BACE1 mRNA contributing in Aβ deposition	Ohno et al. (2004), Zeng et al. (2019)
XIST	BACE1	Leads to Aβ deposition by inducing BACE1 mRNA stability	Li et al. (2022), Liang et al. (2017)
Lnc-51A	SORL1	Downregulates SORL1 variant A causing Aβ deposition	Wu et al. (2024)
BC200	APP	Upregulated in AD, potentially disrupting local protein synthesis at synapses, contributing to neurodegeneration	Lan et al. (2022), Wu et al. (2024)
NKILA	TNFAIP1	Enhances transcription of TNFAIP1, promoting neuroinflammation and exacerbating AD progression	Cai et al. (2025)
ENST00000440246.1	Unknown	Promotes AD progression by influencing neuronal cell apoptosis and autophagy pathways	Cai et al. (2025)
Lnc-17A	APP	Increases APP expression, leading to higher β-amyloid levels and plaque formation	Cai et al. (2025)
SOX21-AS1	SOX21	Modulates neuronal differentiation and may influence AD pathology through SOX21 regulation	Lu et al. (2023), Zhang et al. (2019)
NEAT1	FZD3	Alters the FZD3/GSK3β/p-tau signaling leading to tau hyperphosphorylation	Ghafouri-Fard and Taheri (2019), Zhao et al. (2020)
BDNF-AS	BACE1	Induces BACE1 mRNA expression through competitive binding of miR-9-5p, causing amyloid deposition	Asadi et al. (2021), Gao et al. (2022a)
ANRIL	Multiple	ANRIL knockdown reduces apoptosis and pro-inflammatory cytokine levels while promoting neurite outgrowth by interacting with microRNA-125a	Wang et al. (2024)
Linc00507	MAPT/TTBK1	Regulates MAPT and TTBK1 causing tau hyperphosphorylation	
MALAT1	IL-10/SOC1	Regulates IL-10 and SOC1 expression and induces neuroinflammation	Cai et al. (2025)
EBF3-AS	EBF3	Responsible for neuronal apoptosis in AD	Cai et al. (2025)

(continued)

Table 1 (continued)

LncRNA	Target mRNA	Mechanism	References
SNHG1	Multiple	Targets miR-137 to inhibit KREMEN1 and modulates ZNF217 through miR-361-3p sponging ultimately altering neuronal damage	Cai et al. (2025)
MEG3	PI3/Akt pathway molecules	MEG3 downregulation promotes neuroinflammation via PI3/Akt pathway	Lan et al. (2022)
LRP1-AS	LRP1	LRP1-AS expression deregulates the expression of LRP1 via direct binding with to HMGB2 affecting AD progression	von Einem et al. (2010)
NDM29	APP	NDM29 expression increases APP mRNA, thus enhancing the production of Aβ and increment of Aβ x-42/Aβ x-40 ratio	Lan et al. (2022)
GDNFOS3	GDNF	GDNFOS3 increases GDNF, and exon 2 decreases the mature GDNF peptide in early AD brain	Lan et al. (2022), Wu et al. (2024)
NAT-Rad18	Rad18	NAT-Rad18 expression increases in AD leading to disruption in DNA repair	Lan et al. (2022), Wu et al. (2024)
LoNA	rRNA	LoNA decreases 28s rRNA transcription contributing to AD related ribosomal deficiency	Lan et al. (2022), Wu et al. (2024)

siRNA-based knockdown of BACE1-AS in the mouse brain showed that BACE1 mRNA levels are significantly decreased in cortex, central, and dorsal parts of the mouse brain upon BACE1-AS silencing (Zhang et al. 2018). Stress response could be one of the contributing factors in the induction of BACE1-AS, promoting the development of AD. BACE1-AS also interferes with microRNA (miRNA) regulation; specifically, it can sequester miRNAs that would otherwise target BACE1 mRNA for degradation (Zeng et al. 2019). By preventing these miRNAs from binding to BACE1 mRNA, BACE1-AS further enhances BACE1 expression. This suggests that BACE-AS is a potential biomarker and targeting BACE1-AS could be a potential strategy for reducing amyloid-beta levels and treatment of AD.

4.2.2 XIST

LncRNA XIST is a crucial RNA molecule responsible for X-chromosome inactivation in female mammals, ensuring dosage compensation between sexes (Li et al. 2022). It coats one of the two X chromosomes in female cells, recruiting chromatin modifiers to establish a transcriptionally silent heterochromatin state. It can act as an oncogenic RNA and induce growth in pancreatic and bladder tumors by interacting and inhibiting a microRNA-miR-124 (Liang et al. 2017). Recent study suggests that XIST is significantly upregulated in AD model mice and N2a cells as a result of H_2O_2-mediated oxidative stress (Zhou et al. 2021). It has been reported that miR-124 regulates the expression of BACE1 and is decreased in AD, implying that XIST might play a role in AD via modulation of miR-124 (Wu et al. 2024). Bioinformatics studies have revealed a potential binding site between XIST and miR-124 which may have implications in the negative correlation between XIST and the miRNA. Additionally, silencing of XIST in N2a cells resulted in an elevated level of miR-124 contrary to the decrease in BACE1 expression. Interplay between XIST, miR-124, and BACE1 was further validated by a luciferase reporter assay. Overall, current research confirms that XIST knockdown attenuates alteration of BACE1 via miR-124/BACE1 signaling, indicating a major role of XIST in the progression and development of AD.

4.2.3 Lnc-51A

Among the lncRNAs related to the deposition of Aβ plaques, lnc-51A has emerged as a significant factor (Lan et al. 2022). It is an antisense transcript located within the SORL1 gene, which encodes the sortilin-related receptor 1 involved in APP trafficking. Upregulation of lnc-51A was observed in the plasma of patients with sporadic AD. Moreover, lnc-51A expression was negatively correlated with disease progression, as measured by the Mini-Mental State Examination score, likely due to its upregulation altering the splicing pattern of SORL1. SORL1 encodes SorLA, a key protein involved in the processing of the amyloid-β precursor protein and the secretion of Aβ peptide. SORL1 interacts with APP and functions as a receptor for APP. Downregulation of SORL1 expression shifts APP to the β-secretase cleavage pathway, promoting Aβ formation. These findings suggest that lnc-51A could serve as a reliable diagnostic biomarker for AD.

4.2.4 NEAT1

LncRNA NEAT1 (nuclear-enriched abundant transcript 1) is a lncRNA that plays a significant role in the pathogenesis of AD. NEAT1 is known for its involvement in the formation of nuclear paraspeckles, which are subnuclear structures associated with gene regulation, RNA metabolism, and stress responses (An 2018; Ghafouri-Fard and Taheri 2019). Recent studies have shown that NEAT1 is dysregulated in neurodegenerative diseases, including AD, where it is implicated in various molecular mechanisms that contribute to the disease's progression.

In AD, NEAT1 expression is notably upregulated, particularly in brain regions that are vulnerable to AD pathology, such as the hippocampus and the cortex (An 2018; Ghafouri-Fard and Taheri 2019). This upregulation of NEAT1 has been associated with several pathological features of AD, including Aβ plaque formation, neuroinflammation, oxidative stress, and neuronal apoptosis. NEAT1 can interact with various proteins and RNA-binding factors that influence the alternative splicing and expression of key genes involved in Aβ generation, such as APP. Dysregulation of NEAT1 can enhance the aberrant splicing of APP or alter its cleavage, promoting increased production of Aβ peptides that aggregate to form toxic plaques in the brain. In addition to Aβ pathology, NEAT1 has been linked to tau protein hyperphosphorylation (Zhao et al. 2020). NEAT1 can regulate the expression of tau kinases, such as GSK-3β and CDK5, which are responsible for tau phosphorylation. Elevated levels of NEAT1 may exacerbate tau hyperphosphorylation, leading to the formation of neurofibrillary tangles. NEAT1 is also a crucial regulator of neuroinflammation, which is increasingly recognized as a driving force in AD progression. NEAT1 upregulation has been associated with the activation of microglia and astrocytes. It modulates the expression of pro-inflammatory cytokines such as IL-1β, TNF-α, and IL-6 through interactions with NF-κB signaling pathways. Chronic inflammation induced by NEAT1 can exacerbate neuronal damage, contributing to the cognitive decline in AD patients.

In summary, NEAT1 is a multifaceted lncRNA that plays a pivotal role in the pathogenesis of Alzheimer's disease through its involvement in amyloid-beta production, tau pathology, neuroinflammation, and epigenetic regulation. Its upregulation and complex interactions in AD make it a promising biomarker and potential therapeutic target. Further research into the precise mechanisms and regulatory networks of NEAT1 in AD could pave the way for novel interventions that target this lncRNA to slow or halt the progression of Alzheimer's disease (Fig. 2, Table 2).

4.2.5 HOTAIR

HOTAIR is a lncRNA located on chromosome 12q13.13, transcribed from the antisense strand of the HOXC gene cluster (Bhan and Mandal 2015; Guha et al. 2024). Initially identified as a crucial regulator of gene expression through chromatin remodeling, HOTAIR exerts its effects by interacting with chromatin-modifying complexes, such as the PRC2 and LSD1 complex. Through these interactions, HOTAIR plays a pivotal role in epigenetic modifications, contributing to the repression of gene expression. Although extensively studied in cancer, emerging evidence suggests that HOTAIR also plays a significant role in AD, by influencing neuroinflammation, Aβ accumulation, and neuronal apoptosis (Lan et al. 2022). Recent studies have indicated that HOTAIR contributes to the pathological processes underlying AD. One of the primary mechanisms by which HOTAIR is implicated in AD is through its interaction with epigenetic regulators, leading to aberrant gene expression patterns. Furthermore, HOTAIR has been linked to the regulation of neuroinflammation, a key factor in AD pathology. Studies from our lab also demonstrated that HOTAIR is a critical player in inflammation and macrophage activation (Chini et al. 2024; Obaid et al. 2018, 2021). Understanding the precise mechanisms by which HOTAIR contributes to AD will provide further insights into disease progression and potential interventions.

Table 2 LncRNAs in Parkinson's disease

LncRNA	Target mRNA	Mechanism	References
H19	HPRT1	Induces HPRT1 mRNA stability leading to dopaminergic neuron loss	Jiang et al. (2020), Matouk et al. (2015)
GAS5	NLRP3	Binds to miR-223-3p and affects NLRP3 activation altering PD pathogenesis	Chen et al. (2018), Ji et al. (2019)
HOTAIR	ELAVL1	Mediates miR-326 suppression and upregulation of ELAVl1 and inflammatory pathways via NLRP3 leading to PD progression	Cantile et al. (2021), Wang et al. (2017)
SNHG1	SIAH1	Ubiquitinates α-synuclein affecting its aggregation	Zhou et al. (2021)
SNHG14	3′ UTR of α-synuclein	Induces PD pathogenesis via targeting miR-133b	Zhou et al. (2021)
NEAT1	LRRK2	Induces dopaminergic neuron loss via regulation of oxidative stress	Jeong and Lee (2020)
LincRNA-p21	α-Synuclein	Acts as a transcription repressor affecting aggregation of α-synuclein	Zhou et al. (2021)
BDNF-AS	Multiple	Regulates autophagy and apoptosis via targeting miR-125b-5p	Gao et al. (2022b)
NORAD	α-Synuclein	Rescues cellular destruction and apoptosis	
TUG1	PTEN	Binds to miR-152-3p and regulates PTEN expression	Zhou et al. (2021)
MIAT	SYT1	Upregulates the expression of synaptotagmin-1 (SYT1) through sponging of miR-34-5p leading to reduced PD progression	Zhou et al. (2021)
MALAT1	NRF2 mRNA	MALAT1 epigenetically inhibits NRF2 by regulating EZH2, thus activating NLRP3 in MPTP induced PD mice	Ma et al. (2019)

4.2.6 MALAT1

MALAT1 is expressed abundantly in neurons and is involved in synaptic density. It is said to have a neuroprotective role in AD and contribute toward neuroinflammation associated with AD anti-inflammatory effects. In studies involving AD, the stimulation of Sprague-Dawley rat embryonic PC12 cells and primary cortical neurons with nerve growth factor was shown to inhibit apoptosis in AD models mediated by lncMALAT1 (Ma et al. 2019). This treatment also promoted neurite outgrowth in the primary neuronal AD model, suggesting that MALAT1 plays a protective role in neurodegeneration. These results underscore MALAT1's role in modulating inflammatory responses associated with AD. Further investigations revealed that MALAT1 likely exerts its anti-inflammatory effects by inhibiting specific inflammation-associated microRNAs, such as miR-125b and miR-155. These miRNAs are involved in targeting genes within key signaling pathways in inflammation. By interfering with these pathways, MALAT1 mitigates neuroinflammation, offering insights into its contribution to AD pathogenesis and its potential as a target for therapeutic interventions.

4.2.7 SOX21-AS1

SOX21-AS1 is a recently discovered lncRNA that has come up as a biomarker for squamous cell carcinoma marked by its low expression (Lu et al. 2023). It is also shown to be abnormally expressed in the nervous system and therefore possibly plays a significant role in neurodegenerative diseases. Studies suggest that silencing SOX21-AS1 upregulates the expression of frizzled 3/5 (FZD3/5) receptor which is required for the activation of Wnt signaling pathway and thus reduced apoptosis of hippocampal neuron cells in AD mice (Zhang et al. 2019) (Fig. 2, Table 2).

4.2.8 BDNF-AS

BDNF-AS is involved in neurogenesis and synaptic plasticity, and its decrease in the brain leads to memory and learning damage (Asadi et al. 2021; Gao et al. 2022a). Levels of BDNF-AS are significantly decreased in advanced and early AD patients. BDNF-AS is an antisense transcript to BDNF and therefore could negatively regulate BDNF. Silencing the BDNF-AS lncRNA leads to an increase in cell viability and inhibits oxidative stress.

4.2.9 MEG3

The lncRNA maternally expressed gene 3 (MEG3) was suggested to be involved in oxidative stress and inflammatory injury in subjects with AD (Xie et al. 2021). The level of MEG3 is decreased in the tissues of AD rats. MEG3 expression is known to inhibit the activation of the PI3/Akt signaling pathway in the hippocampal tissues of AD rats. The upregulation of MEG3 contributes to decreased amyloid plaque deposition, as well as inhibition of oxidative stress and inflammatory injury and the improvement of cognitive impairment in AD rats.

4.2.10 LRP1-AS

LRP1-AS inhibits the expression of LRP1; LRP1 is a receptor that is involved in many physiological processes such as intracellular signaling, lipid homeostasis, and APP formation (von Einem et al. 2010). LRP1-AS directly binds to high mobility group box 2 (HMGB2) and inhibits its activity, which in turn reduces LRP1 mRNA-dependent transcription of SREBP1A. Short oligonucleotides targeting LRP1-AS can disrupt this interaction with HMGB2, allowing SREBP1A to resume its role in driving LRP1 gene transcription. Interestingly, LRP1-AS is upregulated in the brains of AD patients, suggesting that inhibiting LRP1-AS could be a potentially novel therapeutic strategy for AD by restoring LRP1 expression (Fig. 2, Table 1).

In addition to these, there are several other lncRNAs that are potentially involved in Alzheimer's disease pathogenesis, but the exact mechanism of many of these lncRNAs is not known. Some of these lncRNAs have been considered as potential biomarkers and therapeutic targets for AD, although the effect of lncRNAs is complex and relative research on lncRNAs in AD is still limited.

4.3 *LncRNAs in Parkinson's Disease*

PD is the second most common neurodegenerative disorder, primarily affecting motor function due to the loss of dopaminergic neurons in the substantia nigra (Balaj and Kaku 2024; Srinivasan et al. 2021). Other typical features of PD include the accumulation of α-synuclein and neuroinflammation. The current understanding of the pathophysiology of PD is still preliminary, which makes developing efficient strategies to treat the disease difficult although dopamine replacement therapy and surgical interventions can provide temporary relief. LncRNAs have been found to be involved in several key processes of PD pathology, including dopaminergic neuronal loss, α-synuclein aggregation, and neuroinflammation (Fig. 2, Table 2).

4.3.1 LncRNA H19

The LncRNA H19 is located in the imprinted gene cluster H19-IGF2, while the IGF2-proinsulin precursor (INS)-TH gene cluster, located at the telomeric end of chromosome 11, has been identified as encoding various proteins critical for dopamine neuron homeostasis (Matouk et al. 2015). This gene cluster has also been linked to the risk of PD. In a PD mouse model induced by 6-hydroxydopa (6-OHDA), H19 expression was significantly reduced (Jiang et al. 2020). Notably, overexpression of H19 prevented dopaminergic neuron loss in this model by activating the HPRT1-mediated Wnt/β-catenin signaling pathway through inhibition of miR-301b-3p-mediated repression of HPRT1 transcription. Furthermore, the significant upregulation of miR-301b in the pars compacta of the substantia nigra in PD models leads to the downregulation of HPRT1 expression. Acting as a sponge for miR-301b-3p, H19 mitigates dopaminergic neuron loss and helps slow the degeneration of the SN.

4.3.2 UCHL1-AS1

This lncRNA regulates the expression of UCHL1, a gene important for the ubiquitin-proteasome system, which is crucial for protein degradation (Carrieri et al. 2015). Dysregulation of this system contributes to the accumulation of alpha-synuclein, a hallmark of PD. UCHL1-AS1 is believed to modulate UCHL1 expression at both the transcriptional and post-transcriptional levels, thus influencing the degradation of misfolded proteins (Carrieri et al. 2015). Targeting UCHL1-AS1 could help enhance the clearance of alpha-synuclein aggregates, offering a potential therapeutic strategy for PD.

4.3.3 LncRNA GAS5

GAS5, a lncRNA located on chromosome 1 and comprising 4087 base pairs, has emerged as a critical player in tumor and inflammatory diseases (Ji et al. 2019). GAS5 demonstrates dual roles in cancer, acting as a tumor suppressor in renal cell carcinoma, cervical cancer, and osteosarcoma, while functioning as an oncogene in esophageal and liver cancers by inhibiting apoptosis. Beyond cancer, GAS5 contributes to inflammatory diseases by elevating pro-inflammatory mediators in airway epithelial and smooth muscle cells. In neurological disorders, GAS5 exacerbates ischemic stroke by targeting miR-137 and activating the Notch1 signaling pathway (Chen et al. 2018). In a recent study, GAS5 was upregulated in PD mouse models and LPS-induced microglia, activating microglia and increasing inflammatory cytokine levels, implicating its role in PD progression (Xu et al. 2020). MiR-223-3p has also been implicated in various diseases. Interestingly, it not only serves as a biomarker for Alzheimer's and Parkinson's diseases but also aids in nerve repair and ameliorates autoimmune myocarditis, liver injuries, and retinal damage by targeting the NLRP3 inflammasome. This study observed miR-223-3p downregulation in PD models and LPS-stimulated microglia. Overexpression of miR-223-3p suppressed inflammatory cytokines and NLRP3 activation, revealing its anti-inflammatory role in PD. These findings underscore the GAS5/miR-223-3p/NLRP3 axis as a crucial regulator of PD pathogenesis. By targeting this pathway, therapeutic strategies could mitigate excessive inflammation, offering a promising avenue for PD treatment.

4.3.4 HOTAIR

In PD mouse models, the expression of HOTAIR was found to be significantly upregulated. This upregulation is closely linked to the progression of PD, as knockdown of HOTAIR effectively mitigated neuronal damage, suggesting its involvement in PD-related pathological changes (Wang et al. 2017). Mechanistically, lncRNA-HOTAIR acts as a competing endogenous RNA (ceRNA), modulating the expression of miR-326 by sponging it (Cantile et al. 2021). This regulatory interaction establishes a lncRNA-HOTAIR/miR-326/ELAVL1 axis, wherein ELAVL1 (embryonic lethal abnormal vision-like RNA-binding protein 1), a direct target of miR-326, is upregulated upon HOTAIR overexpression. The downregulation of miR-326 by HOTAIR or overexpression of ELAVL1 leads to enhanced expression of the NLRP3 inflammasome, a central mediator of the inflammatory response. NLRP3 upregulation exacerbates neuroinflammatory injury by promoting the maturation and secretion of pro-inflammatory cytokines, such as IL-1β and IL-18, contributing to excessive inflammation and neuronal death in PD. Thus, the lncRNA-HOTAIR/miR-326/ELAVL1 axis plays a crucial role in amplifying inflammatory responses and aggravating PD pathology. Furthermore, experimental evidence shows that silencing HOTAIR significantly suppresses NLRP3 activation, reduces the production of inflammatory cytokines, and alleviates neuroinflammatory injury, underscoring its potential as a therapeutic target. Taken together, these findings highlight HOTAIR's role in PD pathogenesis by mediating miR-326 suppression and driving the upregulation of inflammatory pathways via NLRP3. Targeting the HOTAIR-miR-326/ELAVL1 axis could offer a promising strategy to reduce neuroinflammation and protect neuronal function in PD patients.

Further research is required to fully elucidate the roles of lncRNAs in PD, as many more lncRNAs remain to be characterized with respect to their expression and functional relevance in the disease context.

4.4 LncRNAs in Huntington's Disease

HD is a rare autosomal-dominant neurodegenerative disorder characterized by neuronal loss in the striatum and cerebral cortex (Tabrizi et al. 2019; Tong et al. 2024). The disease arises from an abnormal expansion of CAG repeats in the HTT gene, leading to a mutant HTT protein with an extended polyglutamine region. This mutant protein disrupts various cellular processes, including transcription, mitochondrial function, and protein homeostasis. LncRNAs have emerged as significant players in HD pathogenesis, mediating transcriptional dysregulation and contributing to the neurodegenerative cascade (Fig. 2, Table 3).

4.4.1 REST-Related LncRNAs in HD

The REST factor plays a central role in HD by abnormally translocating to the nucleus and repressing target genes, including LncRNAs (Buckley et al. 2010). For instance, HAR1 is significantly downregulated in the striatum of HD patients due to REST binding at its regulatory elements (Bhattacharyya et al. 2021). MEG3, another lncRNA containing REST binding sites near its transcription start site, is similarly downregulated in HD brains, suggesting REST-mediated suppression. In contrast, NEAT1 is upregulated in HD and interacts with REST, potentially modulating HD pathogenesis through different mechanisms.

Table 3 LncRNAs in Huntington's disease, amyotrophic lateral sclerosis, and multiple sclerosis

Disease	LncRNA	Target mRNA	Mechanism	References
Huntington's disease	HAR1	REST	Downregulated in HD models and related to p53 expression	Bhattacharyya et al. (2021), Buckley et al. (2010)
	HTTAS	HTT	Upregulated in HD patients and controls HTT expression	Buckley et al. (2010), Chung et al. (2011)
	NEAT1	TP53/HTT	Reduces the expression of endogenous HTT and TP53	Bhattacharyya et al. (2021), Chanda et al. (2018)
	MEG3	HTT	Downregulated in HD patients and regulates HTT expression	Chanda et al. (2018)
	Abhd11os	HTT	Downregulated in HD mice model and targets HTT expression	Francelle et al. (2015)
Amyotrophic lateral sclerosis	MEG3	Hox	Induces the formation of PRC2-Jarid2 complexes	
	NEAT1 and MALAT1	TDP-43	Promotes TDP-43 toxicity	Sekar et al. (2022), Zeng et al. (2024)
	NEAT1_2	RBP	Upregulated in early ALS and alters ALS related RBP	An et al. (2019), Ruffo et al. (2021)
	C9ORF72-AS	C9ORF72	Regulates repetition of (G4C2) n in C9ORF72	Suzuki et al. (2023), Van Es (2024)
	ATXN2-AS	ATXN2/ TDP-43	Its neurotoxic CUG repeat expansion disrupts RNA metabolism by forming hairpin and sequesters RNA-binding proteins	Sekar et al. (2022), Suzuki et al. (2023)
Multiple sclerosis	GSTT1-AS1	TNF and INFG	Downregulated in MS patients. Suppresses TNF and INFG through the recruitment of PRC2-EZH2 complex	Ruberto et al. (2024)
	IFNG-AS1	INFG	Checkpoint for IFNG in T helper cells	Hosseini et al. (2019), Ruberto et al. (2024)
	TOB1-AS1	TOB1, TSG, and SKP2	Increases the expression of the TOB1 and TSG genes via suppressing IL-2 and positive feedback between TOB1 and SKP2	Hosseini et al. (2019)
	PANDA	p53	In oligodendrocytes and neurons, activates apoptosis	Jalaiei et al. (2021)
	MIAT	Unknown	Downregulated in MS patients	Hosseini et al. (2019)

4.4.2 Dysregulated LncRNAs in HD Pathogenesis

The expression profiles of various lncRNAs are altered in HD. LINC0341 and RPS20P22 are upregulated, while LINC00342 is downregulated, reflecting disease-specific changes. In HD models, overexpression of HTTAS, an antisense transcript of HTT, decreases HTT transcription, whereas knockdown of HTTAS v1 increases HTT levels, demonstrating its regulatory role (Chung et al. 2011).

4.4.3 Neuroprotective Roles of Specific LncRNAs

Studies in HD mouse models highlight the neuroprotective functions of certain lncRNAs. Abhd11os, significantly reduced in HD, exhibits neuroprotective effects against mutant HTT (Francelle et al. 2015). Its loss increases striatal vulnerability, suggesting its potential therapeutic relevance.

The pathological mechanisms of lncRNAs in HD include transcriptional dysregulation, modulation of protein homeostasis, and interference with cellular stress responses (Chanda et al. 2018). NEAT1 may influence proteasome degradation, a process impaired in HD. Additionally, lncRNAs like MEG3 and NEAT1 regulate mutant HTT aggregation by interacting with key proteins and signaling pathways, including the p53 pathway. Given the neuroprotective and regulatory roles of certain lncRNAs, they present potential therapeutic targets. Modulating the expression of lncRNAs like NEAT1 and Abhd11os could mitigate HD-associated neurodegeneration. Moreover, targeting lncRNA-miRNA interactions may restore miRNA levels and reduce the pathological effects of mutant HTT. Understanding these lncRNAs' functions and interactions provides valuable insights into HD mechanisms and highlights their potential as therapeutic targets to mitigate disease progression.

4.5 Amyotrophic Lateral Sclerosis and Multiple Sclerosis

ALS is a fatal neurodegenerative disorder primarily affecting motor neurons in the spinal cord, brainstem, and motor cortex (Van Es 2024). The disease manifests through progressive limb weakness, muscle atrophy, and pyramidal tract signs, often accompanied by bulbar palsy, dysphagia, and respiratory muscle dysfunction. ALS can be sporadic (SALS) or familial (FALS), with 90% of cases being sporadic. While the genetic basis of ALS remains incompletely characterized, mutations in over 20 genes, including C9ORF72 and TARDBP (encoding TDP-43), have been implicated, illustrating a complex interplay between genetic and molecular factors. LncRNAs have emerged as critical players in ALS pathogenesis, particularly through interactions with key proteins like TDP-43 and FUS (Fig. 2, Table 3). The lncRNA NEAT1, significantly upregulated in early ALS motor neurons, is integral to nuclear paraspeckle function (An et al. 2019). NEAT1 binds FUS, a protein frequently mutated in ALS, and mediates paraspeckle formation. In ALS, FUS mislocalizes to the cytoplasm, disrupting paraspeckle integrity. This disruption contributes to disease progression, as studies show that NEAT1_2, typically minimal in healthy motor neurons, is highly expressed in ALS, suggesting its potential as a biomarker for early diagnosis. NEAT1 also serves as a scaffold for RBPs, regulating their expression and illustrating the central role of lncRNAs in ALS. TDP-43, a multifunctional RBP implicated in 95% of SALS cases, also interacts with lncRNAs. Transcriptomic studies reveal that 293 lncRNAs are differentially expressed in ALS, with many exhibiting antisense functions or being categorized as long intergenic noncoding RNAs (lincRNAs). Notably, NEAT1 binds to TDP-43 in spinal motoneurons, forming paraspeckles, which are aberrantly abundant in ALS motor neurons. This suggests that lncRNA dysregulation, particularly NEAT1, is a hallmark of ALS pathogenesis. Additionally, mutations in C9ORF72, the most common genetic factor in ALS, lead to the expansion of (GGGGCC)n repeats, affecting the expression of the antisense transcript lnc-C9ORF72-AS (DeJesus-Hernandez et al. 2011). These expansions disrupt RNA stability and transcription, potentially contributing to ALS via toxic gain-of-function mechanisms. Antisense oligonucleotide (ASOs) targeting these repeats have shown a promise in rescuing transcriptional abnormalities in ALS models. Other lncRNAs, such as MALAT1 and lnc-BDNFOS, also interact with TDP-43 and FUS, further emphasizing the intricate network of lncRNA-mediated regulatory mechanisms in ALS (Ruffo et al. 2021). These findings underscore the potential of lncRNAs as diagnostic markers and therapeutic targets, providing new avenues for understanding and addressing ALS pathogenesis.

MS, a chronic inflammatory and neurodegenerative disease, is characterized by the dysregulation of immune responses (Ruberto et al. 2024). Emerging evidence highlights the critical role of lncRNAs in MS pathogenesis by modulating immune pathways, inflammation, and neural function (Fig. 2, Table 3). Several lncRNAs, such as GSTT1-AS1 and IFNG-AS1, exhibit dysregulated expression in MS (Jalaiei et al. 2021). GSTT1-AS1, which suppresses pro-inflammatory cytokines like TNF and IFNG, is downregulated in MS, potentially enhancing inflammatory responses. IFNG-AS1 regulates IFNG expression in Th1 cells, contributing to MS pathogenesis. LncRNAs like AC007278.2 and TOB1-AS1 regulate Th1/Th17 pathways, crucial for MS progression (Hosseini et al. 2019). AC007278.2 influences IL18R1 expression, promoting pro-inflammatory cytokine release. Similarly, TOB1-AS1 modulates immune tolerance by enhancing TOB1 expression, potentially mitigating MS inflammation. Their dysregulation underscores their potential as biomarkers and therapeutic targets in MS.

5 LncRNAs in Other Neurodegenerative Diseases

NDDs are one of the major challenges in modern medicine. In addition to AD, PD, HD, ALS, and MS, several other NDDs are associated with dysregulation of lncRNAs. These include frontotemporal dementia (FTD), spinal muscular atrophy

(SMA), ataxias, etc. The involvement of lncRNAs in these disorders highlights their broad role in regulating neuronal function and pathogenesis.

5.1 LncRNAs in Frontotemporal Dementia

FTD is characterized by progressive degeneration of the frontal and temporal lobes, leading to behavioral changes and cognitive decline (Zeng et al. 2024). LncRNAs play important roles in the molecular mechanisms underlying FTD. TDP-43 aggregation is a hallmark of FTD (Zeng et al. 2024). Several lncRNAs interact with TDP-43, affecting its nuclear and cytoplasmic localization: Upregulated in FTD, NEAT1 is involved in paraspeckle formation, which regulates gene expression and cellular stress responses (An 2018). The hexanucleotide repeat expansion in C9ORF72, a major genetic cause of FTD, generates antisense transcripts (C9ORF72-AS) that form RNA foci. These foci sequester RNA-binding proteins, impairing RNA metabolism and contributing to neurodegeneration. The dysregulation of NEAT1 and C9ORF72-AS makes them potential biomarkers for early FTD diagnosis. Targeting RNA foci with antisense oligonucleotides (ASOs) has shown a promise in preclinical models.

5.2 LncRNAs in Spinal Muscular Atrophy

SMA is a genetic disorder characterized by motor neuron degeneration due to reduced levels of survival motor neuron (SMN) protein (d'Ydewalle et al. 2017). LncRNAs regulate SMN expression and neuronal survival in SMA. SMN-AS1 is an antisense lncRNA that modulates SMN2 splicing by recruiting splicing factors, enhancing the inclusion of exon 7, and increasing functional SMN protein levels. NEAT1 is upregulated in SMA models and promotes motor neuron survival by regulating cellular stress responses. LncRNAs like SMN-AS1 present promising therapeutic targets for modulating SMN2 splicing. Small molecules and ASOs targeting these lncRNAs have shown efficacy in preclinical SMA models.

5.3 LncRNAs in Ataxias

Ataxias, such as spinocerebellar ataxias (SCAs) and Friedreich's ataxia (FRDA), are characterized by progressive cerebellar degeneration (D'Oria et al. 2013). LncRNAs influence gene expression and stress responses in these disorders. In FRDA, FXN-AS1 regulates the expression of the FXN gene, which encodes frataxin. Dysregulation of FXN-AS1 contributes to mitochondrial dysfunction and oxidative stress. TUG1 is upregulated in SCA models and influences apoptosis by regulating p53 signaling (Koka et al. 2024). Modulating FXN-AS1 and TUG1 levels could restore frataxin expression and reduce neuronal apoptosis, offering therapeutic potential in ataxias.

Dysregulated lncRNAs such as NEAT1 and SMN-AS1 are implicated in diverse NDDs, highlighting their broad relevance. Their potential as biomarkers and therapeutic targets underscores the importance of continued research into lncRNA biology and offers hope for addressing the unmet needs of NDD patients.

6 LncRNAs in Other Neurological Disorders

Since LncRNAs are essentially involved in the development and functions of nervous systems, it is quite likely that they play a role in defining the progression of several neurological disorders apart from known NDDs. Comparative transcriptomic studies have revealed the importance of lncRNAs in conditions like ASD, schizophrenia, Rett syndrome (RTT), Asperger's syndrome, depression, bipolar disorder, Angelman's syndrome (AS), fragile X syndrome, etc. (Table 4, Fig. 2) (Tang et al. 2017). Although the mechanism behind each lncRNA function in these diseases is still being explored, we can definitely say lncRNAs are emerging as the main players in neurological disorders alongside protein-coding genes.

ASD is a neurodevelopmental condition characterized by cognitive disorientation and repetitive behavior (Kerin et al. 2012). GWASs of ASD have identified variants in the 5p14.1 region, adjacent to a lncRNA-designated MSNP1AS, which shares 94% homology with the MSN gene (Kerin et al. 2012). Because MSN is a key regulator of synapse development, these variants may disrupt its function. Evidence indicates that MSNP1AS expression correlates with MSN mRNA and protein levels and is elevated in the ASD brain tissue. In vitro analyses demonstrate that MSNP1AS can bind and stabilize MSN mRNA, thereby enhancing MSN protein levels. Based on these findings, scientists have hypothesized that variants within MSNP1AS contribute to ASD pathogenesis by disrupting MSN regulation and synaptic function. Additionally, the lncRNA SHANK2-AS, transcribed antisense to the SHANK2 gene, is upregulated in ASD patients, negatively correlating with SHANK2 mRNA and protein levels (Luo et al. 2019). SHANK2-AS binds SHANK2 mRNA, reducing its expression, which decreases neurite length and number, suggesting its abnormal expression contributes to ASD and schizophrenia by affecting neuronal structure and growth.

Another example of behavioral, cognitive, and emotional impairment is the disorder schizophrenia (McCutcheon et al. 2023). The lack of clarity in diagnosis and the high heritability (almost 80%) set this disease apart from other forms of psychosis. Several molecular machineries have been predicted to be involved in the pathophysiology of this disease. For example, altered splice isoform distributions of ERRB4 and DISC1 are implicated in schizophrenia pathogenesis, with mutations in these genes sometimes causing splicing defects (Nakata et al. 2009). Recent evidence suggests that the lncRNA GOMAFU may drive aberrant splicing of these

Table 4 NcRNAs involved in neurodevelopmental and neuroimmune disorders

NcRNA name	Target gene(s)	Disease associated	Mechanism	References
MSNP1AS	MSN	Autism spectrum disorder	Correlates with MSN mRNA and protein and upregulated in ASD brain	Kerin et al. (2012), Tang et al. (2017)
SHANK2AS	SHANK2	Autism spectrum disorder and schizophrenia	Binds SHANK2 mRNA, reduces its expression, which decreases neurite length and number, suggesting its abnormal expression contributes to ASD and schizophrenia	Luo et al. (2019), Tang et al. (2017)
AK081227 and AK087060	GABA receptor Rho 2	Rett syndrome	Downregulates GABA receptor Rho 2, important for inhibitory neurotransmission	Reble et al. (2018)
UBE3A-ATS	UBE3A	Angelman's syndrome	Silencing of UBE3A allele in neurons	Srinivas et al. (2023)
BC1 RNA	Various mRNAs	Fragile X syndrome	Impairs synaptic plasticity and local protein translation regulation at synapses, affecting cognitive and developmental processes	Chalei et al. (2014)
MALAT1	Synapse-related genes	Autism spectrum disorder	Regulates synapse formation and dendritic spines through modulation of synaptic genes (e.g., Nlgn1, SynCAM1)	Tang et al. (2017)
MEG3	SOX2, p53	Schizophrenia	Enhances p53 activity and modulates neural differentiation and synaptic plasticity	McCutcheon et al. (2023), Nakata et al. (2009)
NEAT1	Multiple	Rett syndrome, autism	Regulates synaptic gene expression and alternative splicing via interaction with splicing factors (e.g., MALAT1)	Reble et al. (2018)
SNORD115	Serotonin receptor 2C	Autism spectrum disorders	Modulates alternative splicing of serotonin receptor 2C pre-mRNA, influencing serotonergic signaling in ASD	Policarpo et al. (2021)
SNORD116	Imprinting genes	Prader-Willi syndrome	Deletion results in dysregulation of imprinted gene networks, leading to developmental and behavioral abnormalities	Policarpo et al. (2021)

(continued)

Table 4 (continued)

NcRNA name	Target gene(s)	Disease associated	Mechanism	References
ANRIL	CDKN2A/B (predicted)	Neurovascular disorders	Modulates expression of cell cycle inhibitors (CDKN2A/B), potentially influencing neurovascular development and immune responses	Wang et al. (2024)
PTCHD1AS1, PTCHD1AS2, PTCHD1AS3	Multiple	Autism spectrum disorders	Deletion found in males with ASD	Tang et al. (2017)
GOMAFU	mRNAs in synapse formation	Schizophrenia, ASD	Regulates synaptic plasticity and neural connectivity via interaction with RNA-binding proteins	Srinivas et al. (2023), Zakutansky and Feng (2022)

genes, as its dysregulation correlates with schizophrenia risk (Zakutansky and Feng 2022). GOMAFU forms a ribonucleoprotein complex with splicing proteins, controlling splicing in neurons. Chronic GOMAFU downregulation in vitro induces DISC1 and ERRB4 splicing defects, mirroring schizophrenia patient samples and implicating GOMAFU in disease pathogenesis. Additionally, this lncRNA can act as a ceRNA for miR-150-5p, miR-24, miR-22-3p, or miR-150, thereby promoting cell proliferation, apoptosis, and migration.

The lncRNAs AK081227 and AK087060 are upregulated in RTT mouse brains, correlating with the downregulation of GABA receptor Rho 2, crucial for inhibitory neurotransmission (Srinivas et al. 2023). CircRNAs, another noncoding RNA class, influence NDDs by regulating brain proteins and AMPA receptor biogenesis in RTT models (Srinivas et al. 2023). Reduced BDNF and its antisense transcript, BDNF-AS lncRNA, are potential RTT therapeutic targets. Despite technical challenges, dysregulated lncRNAs offer diagnostic insights into NDD etiologies, including RTT.

The severity of symptoms in major depressive disorder (MDD) may correlate with elevated plasma inflammatory factors caused by microglia activation (Rahimian et al. 2022). LncRNAs, linked to synaptic plasticity and microglia regulation, may regulate MDD. Genome-wide analyses identified differentially expressed lncRNAs, such as LINC00473, which synergize with mRNAs in MDD (Gu et al. 2020; Rahimian et al. 2022). LncRNA uc.80 promotes M2 polarization, improving depression (Gu et al. 2020). Other lncRNAs and their regulatory targets may contribute to fluoxetine's antidepressant effects, highlighting the need for further exploration of lncRNA roles in MDD.

UBE3A-ATS is a nuclear lncRNA associated with AS, a severe neurodevelopmental disorder resulting from maternal deficiency of the imprinted UBE3A gene (Meng et al. 2013). Although patients possess a functional paternal UBE3A allele, it is silenced in neurons by UBE3A-ATS. The mechanism of silencing, whether due to the lncRNA's presence or its transcription, remains uncertain. Restoring UBE3A protein expression by targeting and reducing UBE3A-ATS has been proposed as a potential therapeutic approach for AS. This strategy aims to overcome the silencing of the paternal allele to compensate for maternal deficiency.

Apart from the neurological diseases mentioned above, lncRNAs play key roles in traumatic brain injury (TBI), cerebral hemorrhage, and glioma development by regulating neuroinflammation, cell death, angiogenesis, and tissue repair (Li et al. 2019). In TBI, lncRNAs like MALAT1, NEAT1, and H19 affect neuronal apoptosis and blood-brain barrier integrity, while MIAT and PVT1 are implicated in vascular damage and neuronal survival in hemorrhage (Li et al. 2019). In gliomas, lncRNAs such as MEG3, H19, and HOTAIR regulate cell proliferation, apoptosis, invasion, and epigenetic modifications, influencing tumor progression (Kciuk et al. 2023). Despite their potential as biomarkers and therapeutic targets, further research is needed to fully understand their mechanistic roles in disease pathology and treatment.

7 Therapeutic Potential of LncRNAs in Treating Neurodegenerative Diseases

Neurodegenerative diseases pose a significant global health challenge due to their progressive nature, leading to motor, cognitive, and behavioral impairments. Neuroinflammation is increasingly recognized as a key contributor to their pathophysiology, driving interest in therapeutic strategies aimed at modulating inflammation to preserve neuronal function and slow disease progression. Unlike traditional drug targets that primarily focus on proteins, lncRNAs can modulate gene expression at the transcriptional, post-transcriptional, and epigenetic levels. This versatility makes them attractive therapeutic candidates, especially in diseases where conventional approaches have shown limited success. Emerging therapeutic strategies, including RNA interference, ASOs, and gene editing, are being explored to harness lncRNAs for clinical applications in NDDs (Chery 2016).

One of the most promising therapeutic strategies involves silencing disease-promoting lncRNAs using ASOs. These short, synthetic RNA molecules are designed to bind to specific lncRNAs, leading to their degradation or preventing their interaction with other cellular components. For example, BACE1-AS, a lncRNA that stabilizes the BACE1 mRNA involved in amyloid-β production, has been targeted by ASOs in preclinical models of AD. Reducing BACE1-AS levels led to a significant decrease in amyloid-β deposition, a hallmark of AD pathology, demonstrating the potential for lncRNA-targeted therapies to halt or slow disease progression. The US Food and Drug Administration (FDA) has approved three ASO-based drugs for treating muscular dystrophy. Nusinersen has been authorized for various forms of spinal muscular atrophy, eteplirsen for Duchenne muscular dystrophy, and inotersen for familial amyloid polyneuropathy (Amanat et al. 2022). These approvals highlight the potential of ASOs as a new generation of innovative treatments for neurological disorders. Currently, ASOs are primarily being explored for conditions such as PD, ALS, HD, and certain cancers. Additionally, numerous ASO-based therapies are in various stages of clinical trials. Moreover, lncRNAs such as NEAT1 and MALAT1, which regulate cellular stress responses, present additional therapeutic targets. Gene editing technologies, including CRISPR-Cas systems, offer another promising approach to lncRNA-based therapies (Zhen and Li 2019). By precisely editing lncRNA genes or their regulatory elements, it is possible to either upregulate protective lncRNAs or downregulate pathogenic ones. In HD, CRISPR-mediated knockout of lncRNA HTT-AS reduced mutant huntingtin protein levels and improved neuronal function in vitro, providing proof of concept for gene editing approaches targeting lncRNAs. This precision targeting minimizes the risk of off-target effects, which is a critical consideration in developing lncRNA-based therapies.

Small molecules offer an alternative approach to targeting lncRNAs alongside oligonucleotide and gene therapies for the CNS (Chen et al. 2021). These compounds, typically under 1 kDa, are orally administered, allowing systemic distribution. Their small size, minimal charge, and hydrophobic nature enable cell membrane penetration and intracellular binding. For example, ibuprofen can cross the blood-brain barrier (Parepally et al. 2006). Small molecules often exhibit favorable pharmacokinetics and pharmacodynamics, are easier and cheaper to synthesize than oligonucleotides or viral vectors, and can be repurposed from the existing drugs, reducing development costs. While protein-targeting drugs dominate, RNA-targeting, especially for ncRNAs, is emerging, with challenges in binding due to limited chemical diversity and structural complexity. However, secondary and tertiary RNA structures offer potential binding sites. Bacterial riboswitches illustrate RNA's regulatory potential, and in 2020, risdiplam became the first FDA-approved RNA-splicing modifier for spinal muscular atrophy by enhancing SMN2 mRNA processing (Ratni et al. 2021). Furthermore, small molecule drugs targeting lncRNA-protein interactions are gaining attention (Chen et al. 2021). These molecules can disrupt pathological interactions between lncRNAs and their binding partners, thereby altering disease-related pathways. Despite these promising developments, several challenges must be addressed for the clinical translation of lncRNA-based therapies. Advances in nanoparticle-based delivery systems and viral vectors are being explored to overcome this obstacle and achieve precise targeting of affected brain regions. Additionally, long-term safety and efficacy need to be rigorously evaluated through clinical trials to ensure that lncRNA-targeted therapies do not induce off-target effects or unanticipated toxicity. In conclusion, the therapeutic potential of lncRNAs in treating neurodegenerative diseases is immense, offering novel approaches to modulate disease pathways that are inaccessible to traditional therapies. Through RNA-based interventions, gene editing, and small molecule drugs, lncRNA-based therapeutics have the potential to revolutionize the treatment landscape of NDDs.

8 Conclusion

Neurodegenerative diseases, characterized by progressive neuronal loss and declining function, pose a significant medical challenge, especially with an aging population (Adamu et al. 2024). While traditional research has centered on protein misfolding and synaptic dysfunction, emerging evidence highlights the crucial role of lncRNAs in these disorders. LncRNAs influence neuroinflammation, a key driver of neurodegeneration in conditions such as AD, PD, HD, ALS, and MS. They regulate gene expression, protein stability, and inflammatory pathways, offering deeper insights into disease mechanisms and potential therapeutic targets. In AD, lncRNAs like BACE1-AS and XIST contribute to amyloid-beta plaque accumulation and tau hyperphosphorylation. In PD, lncRNAs such as GAS5 and H19 play a role in maintaining neuronal homeostasis and survival. Beyond neurodegenerative diseases, lncRNAs also play a role in various neurological disorders, including ASD, schizophrenia, RTT depression, bipolar disorder, AS, and fragile X syndrome. Comparative transcriptomic studies have identified lncRNAs as critical regulators in these conditions. For example, in ASD, lncRNA MSNP1AS interacts with the MSN gene, which is vital for synapse development, while SHANK2-AS disrupts neuronal growth by suppressing SHANK2 mRNA expression. Additionally, circRNAs, another class of noncoding RNAs, regulate key brain proteins and AMPA receptor biogenesis, further influencing neurodevelopmental disorders. LncRNAs also regulate neuroinflammation, cell death, angiogenesis, and tissue repair in TBI, cerebral hemorrhage, and gliomas. Despite their potential as biomarkers and therapeutic targets, clinical applications of lncRNAs face challenges. The BBB limits drug delivery to the central nervous system, while specificity and off-target effects pose risks. Additionally, the redundancy and poor sequence conservation of lncRNAs complicate therapeutic targeting. However, advancements in high-throughput sequencing and CRISPR-based gene editing are improving the understanding of lncRNA regulatory networks. Nanotechnology-based delivery systems offer potential solutions for crossing the BBB, while artificial intelligence and machine learning accelerate the identification of lncRNA targets for drug development. Stem cell therapies and gene editing techniques provide new opportunities for correcting dysregulated lncRNA expressions and repairing neural damage. Moving forward, a multidisciplinary approach integrating molecular biology, genome editing, neurology, bioinformatics, and clinical research will be essential to translating preclinical findings into effective treatments. The development of comprehensive lncRNA databases and advanced bioinformatics tools will aid in identifying disease-specific lncRNAs and their functional roles, while clinical trials will be crucial for evaluating the safety and efficacy of lncRNA-targeted interventions. LncRNAs represent a promising frontier in neurodegenerative and neurological disease research, offering new insights into disease mechanisms and potential therapeutic solutions. Overcoming current challenges and leveraging technological advancements could revolutionize the management of these complex disorders. As research continues to uncover the intricate roles of lncRNAs, their integration into diagnostic and therapeutic strategies may pave the way for personalized medicine, offering hope to millions of patients worldwide.

Acknowledgments We thank all the Mandal lab members for their helpful discussions. Research in Mandal laboratory is supported by grants from the National Institute of Health (1 R15 HL170257-01), National Science Foundation (NSF AGEP Award-2243017), and the Schwartzberg Companies.

References

Adamu A, Li S, Gao F et al (2024) The role of neuroinflammation in neurodegenerative diseases: current understanding and future therapeutic targets. Front Aging Neurosci 16:1347987

Ahmad MH, Fatima M, Mondal AC (2019) Influence of microglia and astrocyte activation in the neuroinflammatory pathogenesis of Alzheimer's disease: rational insights for the therapeutic approaches. J Clin Neurosci 59:6–11

Amanat M, Nemeth CL, Fine AS et al (2022) Antisense oligonucleotide therapy for the nervous system: from bench to bedside with emphasis on pediatric neurology. Pharmaceutics 14:2389

An H (2018) NEAT1 and paraspeckles in neurodegenerative diseases: a missing lnc found? Noncoding RNA Res 3:243–252

An HY, Skelt L, Notaro A et al (2019) ALS-linked FUS mutations confer loss and gain of function in the nucleus by promoting excessive formation of dysfunctional paraspeckles. Acta Neuropathol Commun 7:7

Asadi MR, Hassani M, Kiani S et al (2021) The perspective of dysregulated LncRNAs in Alzheimer's disease: a systematic scoping review. Front Aging Neurosci 13:709568

Azam S, Haque ME, Balakrishnan R et al (2021) The ageing brain: molecular and cellular basis of neurodegeneration. Front Cell Dev Biol 9:683459

Balaj AR, Kaku H (2024) A novel approach to Parkinson's disease treatment with a potentially dual-acting therapeutic agent that targets alpha-synuclein aggregation and neuron death. Neural Regen Res 19:2577–2578

Barmada SJ (2015) Linking RNA dysfunction and neurodegeneration in amyotrophic lateral sclerosis. Neurotherapeutics 12:340–351

Bhan A, Mandal SS (2015) LncRNA HOTAIR: a master regulator of chromatin dynamics and cancer. Biochim Biophys Acta 1856:151–164

Bhan A, Hussain I, Ansari KI et al (2013) Antisense transcript long noncoding RNA (lncRNA) HOTAIR is transcriptionally induced by estradiol. J Mol Biol 425:3707–3722

Bhan A, Hussain I, Ansari KI et al (2014) Bisphenol-A and diethylstilbestrol exposure induces the expression of breast cancer associated long noncoding RNA HOTAIR in vitro and in vivo. J Steroid Biochem Mol Biol 141:160–170

Bhan A, Soleimani M, Mandal SS (2017) Long noncoding RNA and cancer: a new paradigm. Cancer Res 77:3965–3981

Bhattacharyya N, Pandey V, Bhattacharyya M et al (2021) Regulatory role of long non coding RNAs (lncRNAs) in neurological disorders: from novel biomarkers to promising therapeutic strategies. Asian J Pharm Sci 16:533–550

Bohnsack JP, Teppen T, Kyzar EJ et al (2019) The lncRNA BDNF-AS is an epigenetic regulator in the human amygdala in early onset alcohol use disorders. Transl Psychiatry 9:34

Briggs JA, Wolvetang EJ, Mattick JS et al (2015) Mechanisms of long non-coding RNAs in mammalian nervous system development, plasticity, disease, and evolution. Neuron 88:861–877

Buckley NJ, Johnson R, Zuccato C et al (2010) The role of REST in transcriptional and epigenetic dysregulation in Huntington's disease. Neurobiol Dis 39:28–39

Busscher D, Boon RA, Juni RP (2022) The multifaceted actions of the lncRNA H19 in cardiovascular biology and diseases. Clin Sci 136:1157–1178

Cai J, Ni YQ, Liu YS (2025) Exploring the frontier: antisense long non-coding RNAs as key regulators in Alzheimer's disease. Aging Dis. https://doi.org/10.14336/AD.2024.0762

Cantile M, Di Bonito M, De Bellis MT et al (2021) Functional interaction among lncRNA HOTAIR and microRNAs in cancer and other human diseases. Cancers (Basel) 13:570

Carrieri C, Forrest AR, Santoro C et al (2015) Expression analysis of the long non-coding RNA antisense to Uchl1 (AS Uchl1) during dopaminergic cells' differentiation in vitro and in neurochemical models of Parkinson's disease. Front Cell Neurosci 9:114

Carvey PM, Hendey B, Monahan AJ (2009) The blood-brain barrier in neurodegenerative disease: a rhetorical perspective. J Neurochem 111:291–314

Chalei V, Sansom SN, Kong LS et al (2014) The long non-coding RNA is an epigenetic regulator of neural differentiation. Elife 3:e04530

Chanda K, Das S, Chakraborty J et al (2018) Altered levels of long NcRNAs Meg3 and Neat1 in cell and animal models of Huntington's disease. RNA Biol 15:1348–1363

Chen FH, Zhang LX, Wang EW et al (2018) LncRNA GAS5 regulates ischemic stroke as a competing endogenous RNA for miR-137 to regulate the Notch1 signaling pathway. Biochem Biophys Res Commun 496:184–190

Chen Y, Li Z, Chen X et al (2021) Long non-coding RNAs: from disease code to drug role. Acta Pharm Sin B 11:340–354

Chery J (2016) RNA therapeutics: RNAi and antisense mechanisms and clinical applications. Postdoc J 4:35–50

Chini A, Guha P, Malladi VS et al (2023) Novel long non-coding RNAs associated with inflammation and macrophage activation in human. Sci Rep 13:4036

Chini A, Guha P, Rishi A et al (2024) Discovery and functional characterization of LncRNAs associated with inflammation and macrophage activation. Methods 227:1–16

Chung DW, Rudnicki DD, Yu L et al (2011) A natural antisense transcript at the Huntington's disease repeat locus regulates expression. Hum Mol Genet 20:3467–3477

Cuevas-Diaz Duran R, Wei H, Kim DH et al (2019) Invited Review: Long non-coding RNAs: important regulators in the development, function and disorders of the central nervous system. Neuropathol Appl Neurobiol 45:538–556

D'Oria V, Petrini S, Travaglini L et al (2013) Frataxin deficiency leads to reduced expression and impaired translocation of NF-E2-related factor (Nrf2) in cultured motor neurons. Int J Mol Sci 14:7853–7865

d'Ydewalle C, Ramos DM, Pyles NJ et al (2017) The antisense transcript regulates SMN expression and is a novel therapeutic target for spinal muscular atrophy. Neuron 93:66–79

DeJesus-Hernandez M, Mackenzie IR, Boeve BF et al (2011) Expanded GGGGCC hexanucleotide repeat in noncoding region of C9ORF72 causes chromosome 9p-linked FTD and ALS. Neuron 72:245–256

DeTure MA, Dickson DW (2019) The neuropathological diagnosis of Alzheimer's disease. Mol Neurodegener 14:32

Feng KM, Tsung TH, Chen YH et al (2023) The role of retinal ganglion cell structure and function in glaucoma. Cells 12:2797

Francelle L, Galvan L, Gaillard MC et al (2015) The striatal long noncoding RNA Abhd11os is neuroprotective against an N-terminal fragment of mutant huntingtin in vivo. Neurobiol Aging 36:1601.e7–1601.16

Gao L, Zhang Y, Sterling K et al (2022a) Brain-derived neurotrophic factor in Alzheimer's disease and its pharmaceutical potential. Transl Neurodegener 11:4

Gao X, Cao Z, Tan H et al (2022b) LncRNA, an emerging approach for neurological diseases treatment by regulating microglia polarization. Front Neurosci 16:903472

Gao C, Jiang J, Tan Y et al (2023) Microglia in neurodegenerative diseases: mechanism and potential therapeutic targets. Signal Transduct Target Ther 8:359

Ghafouri-Fard S, Taheri M (2019) Nuclear Enriched Abundant Transcript 1 (NEAT1): a long non-coding RNA with diverse functions in tumorigenesis. Biomed Pharmacother 111:51–59

Gooch CL, Pracht E, Borenstein AR (2017) The burden of neurological disease in the United States: a summary report and call to action. Ann Neurol 81:479–484

Gu XH, Xu LJ, Zheng LL et al (2020) Long non-coding RNA uc.80-overexpression promotes M2 polarization of microglias to ameliorate depression in rats. IUBMB Life 72:2194–2203

Gudkov SV, Burmistrov DE, Kondakova EV et al (2023) An emerging role of astrocytes in aging/neuroinflammation and gut-brain axis with consequences on sleep and sleep disorders. Ageing Res Rev 83:101775

Guha P, Chini A, Rishi A et al (2024) Long noncoding RNAs in ubiquitination, protein degradation, and human diseases. Biochim Biophys Acta Gene Regul Mech 1867:195061

Guo LH, Schluesener HJ (2007) The innate immunity of the central nervous system in chronic pain: the role of Toll-like receptors. Cell Mol Life Sci 64:1128–1136

Hampel H, Hu Y, Hardy J et al (2023) The amyloid-beta pathway in Alzheimer's disease: a plain language summary. Neurodegener Dis Manag 13:141–149

Harries LW (2012) Long non-coding RNAs and human disease. Biochem Soc Trans 40:902–906

Hon CC, Ramilowski JA, Harshbarger J et al (2017) An atlas of human long non-coding RNAs with accurate 5′ ends. Nature 543:199–204

Hosseini A, Teimuri S, Ehsani M et al (2019) LncRNAs associated with multiple sclerosis expressed in the Th1 cell lineage. J Cell Physiol 234:22153–22162

Hung T, Wang YL, Lin MF et al (2011) Extensive and coordinated transcription of noncoding RNAs within cell-cycle promoters. Nat Genet 43:621–629

Hunsberger HC, Pinky PD, Smith W et al (2019) The role of APOE4 in Alzheimer's disease: strategies for future therapeutic interventions. Neuronal Signal 3:NS20180203

Imai Y, Lu B (2011) Mitochondrial dynamics and mitophagy in Parkinson's disease: disordered cellular power plant becomes a big deal in a major movement disorder. Curr Opin Neurobiol 21:935–941

Jalaiei A, Asadi MR, Sabaie H et al (2021) Long non-coding RNAs, novel offenders or guardians in multiple sclerosis: a scoping review. Front Immunol 12:774002

Jeong GR, Lee BD (2020) Pathological functions of LRRK2 in Parkinson's disease. Cells 9:2565

Ji J, Dai X, Yeung SJ et al (2019) The role of long non-coding RNA GAS5 in cancers. Cancer Manag Res 11:2729–2737

Jiang J, Piao X, Hu S et al (2020) LncRNA H19 diminishes dopaminergic neuron loss by mediating microRNA-301b-3p in Parkinson's disease via the HPRT1-mediated Wnt/beta-catenin signaling pathway. Aging (Albany NY) 12:8820–8836

Johnson R, Richter N, Jauch R et al (2010) Human accelerated region 1 noncoding RNA is repressed by REST in Huntington's disease. Physiol Genomics 41:269–274

Kciuk M, Yahya EB, Mohamed MMI et al (2023) Insights into the role of LncRNAs and miRNAs in glioma progression and their potential as novel therapeutic targets. Cancers (Basel) 15:3298

Kerin T, Ramanathan A, Rivas K et al (2012) A noncoding RNA antisense to moesin at 5p14.1 in autism. Sci Transl Med 4:128ra40

Koka M, Li H, Akther R et al (2024) Long non-coding RNA is downregulated in Friedreich's ataxia. Brain Commun 6:fcae170

Krasemann S, Madore C, Cialic R et al (2017) The TREM2-APOE pathway drives the transcriptional phenotype of dysfunctional microglia in neurodegenerative diseases. Immunity 47:566–581.e569

Kwon HS, Koh SH (2020) Neuroinflammation in neurodegenerative disorders: the roles of microglia and astrocytes. Transl Neurodegener 9:42

Lan Z, Chen YT, Jin JL et al (2022) Long non-coding RNA: insight into mechanisms of Alzheimer's disease. Front Mol Neurosci 14:821002

Lanoiselee HM, Nicolas G, Wallon D et al (2017) APP, PSEN1, and PSEN2 mutations in early-onset Alzheimer disease: a genetic screening study of familial and sporadic cases. PLoS Med 14:e1002270

Lauretti E, Dincer O, Pratico D (2020) Glycogen synthase kinase-3 signaling in Alzheimer's disease. Biochim Biophys Acta Mol Cell Res 1867:118664
Li ZX, Han KW, Zhang DF et al (2019) The role of long noncoding RNA in traumatic brain injury. Neuropsychiatr Dis Treat 15:1671–1677
Li Z, Shue F, Zhao N et al (2020) APOE2: protective mechanism and therapeutic implications for Alzheimer's disease. Mol Neurodegener 15:63
Li J, Ming Z, Yang L et al (2022) Long noncoding RNA XIST: mechanisms for X chromosome inactivation, roles in sex-biased diseases, and therapeutic opportunities. Genes Dis 9:1478–1492
Liang S, Gong X, Zhang G et al (2017) The lncRNA XIST interacts with miR-140/miR-124/iASPP axis to promote pancreatic carcinoma growth. Oncotarget 8:113701–113718
Liu N, Sun H, Li X et al (2021) Downregulation of lncRNA KCNQ1OT1 relieves traumatic brain injury induced neurological deficits via promoting "M2" microglia polarization. Brain Res Bull 171:91–102
Lu Z, Xu J, Cao B et al (2023) Long non-coding RNA SOX21-AS1: a potential tumor oncogene in human cancers. Pathol Res Pract 249:154774
Lull ME, Block ML (2010) Microglial activation and chronic neurodegeneration. Neurotherapeutics 7:354–365
Luo T, Liu P, Wang XY et al (2019) Effect of the autism-associated lncRNA Shank2-AS on architecture and growth of neurons. J Cell Biochem 120:1754–1762
Ma PZ, Li YL, Zhang W et al (2019) Long non-coding RNA MALAT1 inhibits neuron apoptosis and neuroinflammation while stimulates neurite outgrowth and its correlation with MiR-125b mediates PTGS2, CDK5 and FOXQ1 in Alzheimer's disease. Curr Alzheimer Res 16:596–612
Mathew S, Sivasubbu S (2022) Long non coding RNA based regulation of cerebrovascular endothelium. Front Genet 13:834367
Matouk IJ, Halle D, Gilon M et al (2015) The non-coding RNAs of the H19-IGF2 imprinted loci: a focus on biological roles and therapeutic potential in Lung Cancer. J Transl Med 13:113
McCutcheon RA, Keefe RSE, McGuire PK (2023) Cognitive impairment in schizophrenia: aetiology, pathophysiology, and treatment. Mol Psychiatry 28:1902–1918
Meng LY, Person RE, Huang W et al (2013) Truncation of unsilences paternal and ameliorates behavioral defects in the Angelman syndrome mouse model. PLoS Genet 9:e1004039
Nakata K, Lipska BK, Hyde TM et al (2009) Splice variants are upregulated in schizophrenia and associated with risk polymorphisms. Proc Natl Acad Sci USA 106:15873–15878
Navandar M, Vennin C, Lutz B et al (2024) Long non-coding RNAs expression and regulation across different brain regions in primates. Sci Data 11:545
Obaid M, Udden SMN, Deb P et al (2018) LncRNA HOTAIR regulates lipopolysaccharide-induced cytokine expression and inflammatory response in macrophages. Sci Rep 8:15670
Obaid M, Udden SMN, Alluri P et al (2021) LncRNA HOTAIR regulates glucose transporter Glut1 expression and glucose uptake in macrophages during inflammation. Sci Rep 11:232
Ohno M, Sametsky EA, Younkin LH et al (2004) BACE1 deficiency rescues memory deficits and cholinergic dysfunction in a mouse model of Alzheimer's disease. Neuron 41:27–33
Parepally JM, Mandula H, Smith QR (2006) Brain uptake of nonsteroidal anti-inflammatory drugs: ibuprofen, flurbiprofen, and indomethacin. Pharm Res 23:873–881
Policarpo R, Sierksma A, De Strooper B et al (2021) From junk to function: LncRNAs in CNS health and disease. Front Mol Neurosci 14:714768
Popa-Wagner A, Dumitrascu DI, Capitanescu B et al (2020) Dietary habits, lifestyle factors and neurodegenerative diseases. Neural Regen Res 15:394–400
Price RL, Bhan A, Mandal SS (2021) HOTAIR beyond repression: in protein degradation, inflammation, DNA damage response, and cell signaling. DNA Repair (Amst) 105:103141
Rahimian R, Belliveau C, Chen R et al (2022) Microglial inflammatory-metabolic pathways and their potential therapeutic implication in major depressive disorder. Front Psych 13:871997
Rajmohan R, Reddy PH (2017) Amyloid-beta and phosphorylated Tau accumulations cause abnormalities at synapses of Alzheimer's disease neurons. J Alzheimers Dis 57:975–999

Rani N, Nowakowski TJ, Zhou H et al (2016) A primate lncRNA mediates notch signaling during neuronal development by sequestering miRNA. Neuron 90:1174–1188

Ratni H, Scalco RS, Stephan AH (2021) Risdiplam, the first approved small molecule splicing modifier drug as a blueprint for future transformative medicines. ACS Med Chem Lett 12:874–877

Reble E, Dineen A, Bar CL (2018) The contribution of alternative splicing to genetic risk for psychiatric disorders. Genes Brain Behav 17:e12430

Reddy DS, Abeygunaratne HN (2022) Experimental and clinical biomarkers for progressive evaluation of neuropathology and therapeutic interventions for acute and chronic neurological disorders. Int J Mol Sci 23:11734

Ruberto S, Dominguez-Mozo MI, Garcia-Martinez MA et al (2024) Immune response profiling of HERV-W envelope proteins in multiple sclerosis: potential biomarkers for disease progression. Front Immunol 15:1505239

Ruffo P, Strafella C, Cascella R et al (2021) Deregulation of ncRNA in neurodegenerative disease: focus on circRNA, lncRNA and miRNA in amyotrophic lateral sclerosis. Front Genet 12:784996

Sekar D, Tusubira D, Ross K (2022) TDP-43 and NEAT long non-coding RNA: roles in neurodegenerative disease. Front Cell Neurosci 16:954912

Shao MH, Jin MM, Xu S et al (2020) Exosomes from long noncoding RNA-Gm37494-ADSCs repair spinal cord injury shifting microglial M1/M2 polarization. Inflammation 43:1536–1547

Srinivas T, Mathias C, Oliveira-Mateos C et al (2023) Roles of lncRNAs in brain development and pathogenesis: emerging therapeutic opportunities. Mol Therapy 31:1550–1561

Srinivasan E, Chandrasekhar G, Chandrasekar P et al (2021) Alpha-synuclein aggregation in Parkinson's disease. Front Med (Lausanne) 8:736978

Stoklund Dittlau K, Freude K (2024) Astrocytes: the stars in neurodegeneration? Biomolecules 14:289

Suzuki N, Nishiyama A, Warita H et al (2023) Genetics of amyotrophic lateral sclerosis: seeking therapeutic targets in the era of gene therapy. J Hum Genet 68:131–152

Tabrizi SJ, Ghosh R, Leavitt BR (2019) Huntingtin lowering strategies for disease modification in Huntington's disease. Neuron 102:899

Tang J, Yu Y, Yang W (2017) Long noncoding RNA and its contribution to autism spectrum disorders. CNS Neurosci Ther 23:645–656

Tong H, Yang T, Xu S et al (2024) Huntington's disease: complex pathogenesis and therapeutic strategies. Int J Mol Sci 25:3845

Van Acker ZP, Leroy T, Annaert W (2025) Mitochondrial dysfunction, cause or consequence in neurodegenerative diseases? BioEssays 47:e2400023

Van Es MA (2024) Amyotrophic lateral sclerosis; clinical features, differential diagnosis and pathology. Int Rev Neurobiol 176:1–47

von Einem B, Schwanzar D, Rehn F et al (2010) The role of low-density receptor-related protein 1 (LRP1) as a competitive substrate of the amyloid precursor protein (APP) for BACE1. Exp Neurol 225:85–93

Wang KC, Chang HY (2011) Molecular mechanisms of long noncoding RNAs. Mol Cell 43:904–914

Wang C, Li G, Wu Y et al (2016) LincRNA1230 inhibits the differentiation of mouse ES cells towards neural progenitors. Sci China Life Sci 59:443–454

Wang S, Zhang X, Guo Y et al (2017) The long noncoding RNA HOTAIR promotes Parkinson's disease by upregulating LRRK2 expression. Oncotarget 8:24449–24456

Wang R, Yuan Q, Wen Y et al (2024) ANRIL: a long noncoding RNA in age-related diseases. Mini-Rev Med Chem 24:1930–1939

Wimo A, Guerchet M, Ali GC et al (2017) The worldwide costs of dementia 2015 and comparisons with 2010. Alzheimers Dement 13:1–7

Wu XB, Xia PC, Yang L et al (2024) The roles of long non-coding RNAs in Alzheimer's disease diagnosis, treatment, and their involvement in Alzheimer's disease immune responses. Non-Coding RNA Res 9:659–666

Xiang JF, Yin QF, Chen T et al (2014) Human colorectal cancer-specific lncRNA regulates long-range chromatin interactions at the locus. Cell Res 24:513–531

Xiao S, Chan P, Wang T et al (2021) A 36-week multicenter, randomized, double-blind, placebo-controlled, parallel-group, phase 3 clinical trial of sodium oligomannate for mild-to-moderate Alzheimer's dementia. Alzheimers Res Ther 13:62

Xie B, Qiao M, Xuan J (2021) lncRNA MEG3 downregulation relieves intracerebral hemorrhage by inhibiting oxidative stress and inflammation in an miR-181b-dependent manner. Med Sci Monit 27:e929435

Xu W, Zhang L, Geng Y et al (2020) Long noncoding RNA GAS5 promotes microglial inflammatory response in Parkinson's disease by regulating NLRP3 pathway through sponging miR-223-3p. Int Immunopharmacol 85:106614

Yin Y, Shen X (2023) Noncoding RNA-chromatin association: functions and mechanisms. Fundam Res 3:665–675

Zakutansky PM, Feng Y (2022) The long non-coding RNA GOMAFU in schizophrenia: function, disease risk, and beyond. Cells 11:1949

Zeng T, Ni HT, Yu Y et al (2019) BACE1-AS prevents BACE1 mRNA degradation through the sequestration of BACE1-targeting miRNAs. J Chem Neuroanat 98:87–96

Zeng J, Luo C, Jiang Y et al (2024) Decoding TDP-43: the molecular chameleon of neurodegenerative diseases. Acta Neuropathol Commun 12:205

Zhang W, Zhao H, Wu Q et al (2018) Knockdown of BACE1-AS by siRNA improves memory and learning behaviors in Alzheimer's disease animal model. Exp Ther Med 16:2080–2086

Zhang L, Fang Y, Cheng X et al (2019) Silencing of long noncoding RNA SOX21-AS1 relieves neuronal oxidative stress injury in mice with Alzheimer's disease by upregulating FZD3/5 via the Wnt signaling pathway. Mol Neurobiol 56:3522–3537

Zhang S, Sun WC, Liang ZD et al (2020) LncRNA SNHG4 attenuates inflammatory responses by sponging miR-449c-5p and up-regulating STAT6 in microglial during cerebral ischemia-reperfusion injury. Drug Des Devel Ther 14:3683–3695

Zhao YW, Wang ZQ, Mao YH et al (2020) NEAT1 regulates microtubule stabilization via FZD3/GSK3β/P-tau pathway in SH-SY5Y cells and APP/PS1 mice. Aging 12:23233–23250

Zhen S, Li X (2019) Application of CRISPR-Cas9 for long noncoding RNA genes in cancer research. Hum Gene Ther 30:3–9

Zhou D, Yang K, Chen L et al (2017) Promising landscape for regulating macrophage polarization: epigenetic viewpoint. Oncotarget 8:57693–57706

Zhou SY, Yu X, Wang M et al (2021) Long non-coding RNAs in pathogenesis of neurodegenerative diseases. Front Cell Dev Biol 9:719247

Zimmer-Bensch G (2019) Emerging roles of long non-coding RNAs as drivers of brain evolution. Cells 8:1399

Genetic-Epigenetic Regulatory Mechanisms of Extrachromosomal Circular DNA: Evidence for a Role in Aging and Age-Related Neurodegenerative Disorders

Marcos J. Araúzo-Bravo, Daniela Gerovska, Matthias Schwab, and Alexandra Kretz

Contents

M. J. Araúzo-Bravo (✉)
Biogipuzkoa Health Research Institute, Paseo Dr. Begiristain, San Sebastián, Spain

IKERBASQUE, Basque Foundation for Science, Bilbao, Spain

Department of Cell Biology and Histology, Faculty of Medicine and Nursing, University of Basque Country (UPV/EHU), Leioa, Spain

D. Gerovska
Biogipuzkoa Health Research Institute, Paseo Dr. Begiristain, San Sebastián, Spain

M. Schwab · A. Kretz (✉)
Department of Neurology, Jena University Hospital, Jena, Germany
e-mail: alexandra.kretz@med.uni-jena.de

J. Barciszewski (ed.), *Decoding Aging and Neurodegeneration*,
RNA Technologies 16, https://doi.org/10.1007/978-3-031-91072-2_12

Abstract Aging is a prime risk factor for neurodegeneration and central nervous system (CNS) disorders. Connecting causalities reside, among others, in accumulating genomic damage, altered genomic plasticity, and modified epigenetic control. One product of genome plasticity involved in gene and epigenetic regulation is extrachromosomal circular DNA (eccDNA). Though pervasive in the eukaryotic genome, its importance in health and disease and in aging and neurodegeneration is poorly understood. Recent data, propagated by advanced sequencing and computational technologies, have revealed an intriguing complexity of mechanisms by which eccDNA influences the copy number, expression, and transcription of pathogenicity genes hosted on the circular DNA and throughout the linear genome.

This chapter discusses established concepts of eccDNA generation and inherent gene including epigenetic regulatory capabilities, conferred by particle-autonomous properties, or episomal-episomal and episomal-linear modes of chromatin interactions. Methodically, we delineate how our new computational approach, which links merged eccDNAs to individual genes to form a measure called Produced per Gene Circles and Common Produced per Gene Circles, has the capability to complement length- or size-determined analysis of eccDNA marker detection in the CNS tissue. In conclusion, we propose eccDNA as an unconventional episomal genome regulator that should be considered in the context of epigenetic modifications, offering a novel perspective for the diagnostic evaluation of age-related CNS disorders.

Keywords Amyotrophic lateral sclerosis · Aging · Differentially Produced per Gene Circles · Disease biomarkers · Epigenomics · Extrachromosomal circular DNA · Multiple sclerosis · Neurodegeneration · Produced per Gene Circles · Systemic lupus erythematosus

1 Introduction: Neurodegeneration and Its Liaison with Aging

1.1 Chronobiological and Epigenetic Dimensions of Aging

Visible aging in humans emerges at early mid-ages and continues to progress thereafter. Although older age inversely correlates with the number of detrimental hits required to induce age-related pathology, as it has been modeled for several neurodegenerative diseases (Gerovska and Araúzo-Bravo, 2022), the aging process is not linear, and there are no accepted timelines for the onset of aging. Some concepts propose aging to commence before birth or after sexual maturity, when the organism disengages from its evolutionary pressure to regenerate. More recent molecular modeling emphasizes the role of accumulating genomic damage and of epigenetic alterations throughout lifetime in the process of aging (Kinzina et al. 2019). Epigenomic studies further disclose that different organs can age at different times, eliciting discrepancies between the chronological age of the body and the epigenetic age of individual organs (Horvath 2013). Non-linear changes in epigenetic patterns implicate genes relevant to nervous system maintenance and function and therefore might contribute to neurodegenerative phenotypes (Olecka et al. 2024). Thus, aging is multifaceted, with the chronological axis being one of several dimensions that contribute to the accumulation of deleterious events. On such intricate background complexity, neurodegeneration occurs early and late in life but is more prevalent in later life.

1.2 Aging as a Risk Factor for Neurodegeneration

Aging is a major risk factor for several chronic and incurable disorders, including those of the central nervous system (CNS). As differentiated neurons remain unreplaced, they accumulate interacting forms of demise over time, culminating in dysfunction, senescence, and neuronal death. The types of cellular injury involved are largely not neuron-specific, but result from the wear and tear of a cell over time, including physiological limitations of cell renewal, compromised debris clearance, altered anabolic-catabolic turnover, loss of proteostasis, genomic instability, and epigenetic alterations, among other factors that are well understood as cellular and molecular "hallmarks of aging" (López-Otín et al. 2013; Schmauck-Medina et al. 2022). The still-evolving list of such factors (Schmauck-Medina et al. 2022) and their interdependent influence is critical in explaining the aging process, including brain aging. They also set the framework conditions for the co-manifestation of age-related neurodegenerative pathologies such as Alzheimer's dementia (AD), Parkinson's disease (PD), amyotrophic lateral sclerosis (ALS), and vision disorders (VD) (Fraser et al. 2022). Although each of these disorders is associated with a distinct set of pathophysiological events that can be linked in a common genealogical tree (Gerovska et al. 2020; Gerovska and Araúzo-Bravo 2022), explaining, for example, cortical brain atrophy, altered vasculature, neurotransmitter disturbance, formation of cytotoxic protein aggregates, or population-specific vulnerability to neuronal death, they might converge on impairing somatic homeostatic control or repair capacity. Compromised cellular homeostasis due to diminished adaptive response repertoires and repair quality control may play a common, crucial role in the transformation of aging into pathology (Kirkwood and Austad 2000) and be further challenged by comorbidities and environmental exposures.

Given such age-related risk basis, specific neurodegenerative presentations implicate to 5–10% additional familial hereditary traits that can be broadly explored by molecular genetic testing. A comprehensive assessment of non-Mendelian somatic mutations, however, has quite lately become approached, requiring methodological advances. Only recently, refined single-cell genomic analyses revealed an almost linear increase in the frequency of somatic mutations with age in human neurons, involving aging-associated mutational signatures such as signs of oxidative damage and defective DNA repair (Lodato et al. 2018).

1.3 Early-Onset Neurodegeneration

In addition to chronological aging, non-linear biological aging and senescence-like neuronal states are proposed to play a role in neurodegeneration, with an assumed impact for disorders manifesting in young adults. One example is autoimmune-mediated neurodegenerative multiple sclerosis (MS). Apart from the deleterious cellular consequences of acute neuroinflammatory attacks and the accumulating lesion load brought about over time, biological aging contributes to explain the longitudinal, relapse-independent deterioration of the disease course seen in many patients and the waning efficacy of disease-modifying therapies (DMT) with age (Weideman et al. 2017).

Although biological aging causes a decline in structural and functional plasticity, penetrance of neurodegeneration in the young is again linked to additional catalyzing hits, such as hereditary or sporadic genetic mutations. Causal genetic determinants are now being discovered for several early-onset neurodegenerative diseases, such as juvenile PD, early-onset ALS, presenile dementia, and juvenile stroke. In neuro-immunological MS, where the knowledge of hereditary genetic risk factors has long been limited to *HLA* haplotypes, and in neuromyelitis optica (NMO) spectrum disorders (NMOSD), the importance of epigenetic and somatically acquired genome alterations is increasingly being discussed. Among, epigenetic events related to the distribution of X chromosomal inactivation required for dosage compensation of X-linked genes are proposed to contribute to disease vulnerability and gynecotropism in MS (Brooks and Renaudineau 2015). Further, a recent genome-wide association study addressing copy number variations (CNV) in an Asian cohort identified 5–50 kb large deletions at specific T cell receptor (TCR) γ and α loci in 16 and 17% of MS patients, respectively, and at the TCR α locus in 13% of NMO/NMOSD patients, involving the entire J gene and parts of the V gene segments of the TCR γ locus (Sato et al. 2015). Such deletions may be disparate from the somatic rearrangements that occur physiologically in TCR regions but arise from de novo mutations in certain T cell populations. Their somatic rather than hereditary origin is supported by the finding that these CNV were restricted to peripheral blood lymphocytes (Sato et al. 2015).

The relevance of such somatic mutations, in parallel to the programmed somatic rearrangements required to expand both the TCR and B cell receptor (BCR) repertoire, in disease manifestation is not yet understood. Adaptive immunity mediated by B cell and T cell populations and their receptors is critical in both MS and NMOSD and a central target of established treatment regimens for these diseases. Thus, exploring the genetic history of somatic recombination in TCR and BCR genes may provide valuable new insights into the etiology of MS and NMOSD.

Further, somatic genetic variation and damage are the major sources of pathology in genomic instability syndromes (GIS) manifesting in the young. Strikingly, GIS regularly present with immunologic deficits, due to unrepaired DNA injury in genes that control immune system development, and with neurological symptoms and signs of premature aging. Examples of GIS with such features include ataxia telangiectasia, Fanconi anemia, the Nijmegen breakage syndrome, Werner's syndrome, Bloom's syndrome, and others (Taylor et al. 2019).

In summary, genomic instability and both hereditary and somatic mutations play a critical role in aging and neurodegeneration.

2 Extrachromosomal Circular DNA and Genome Variations

2.1 Mendel's Laws and Selection of Somatic Variants

When exploring pathologies, disease classifications have traditionally separated hereditary genetic from non-genetic disorders. Inheritance by Mendel's laws provides the key framework for understanding monogenic disease causalities. However, based on new approaches in molecular genetics, evidence has emerged for parallel engagement of more than a single gene in several classical neurodegenerative disorders. Moreover, it is now understood that genomic signatures causal in hereditary forms are also present in sporadic courses of the same disease, rendering the dissection of hereditary and sporadic classifications unsharp. The inheritance principles predicated on Gregor Mendel in 1865, and modeling of oligogenic traits has meanwhile been superseded by an ever-rising intricacy of concepts that aim at understanding complex genetic including epigenetic characteristics in a disease. In the dynamics of this elaborateness, the combination of classical genetics, high-throughput genomics, epigenomics, and exciting new predictive tools based on artificial intelligence, as well as the discovery of somatic genetic variants, has become increasingly important.

Regarding the latter, somatic variants can disseminate in a tissue over time and contribute not only to cancer by the propagation of driver mutations but also to the pathogenesis of non-neoplastic disorders. Such concept is supported by the knowledge that the individual landscape of somatic drivers is subject to selection pressure, favoring the parallel selection of different variants in the same genes or pathways. In other words, though inter-individually different, the set of somatic variants found in each individual is not random but seems shaped by a conserved process of positive selection and parallel evolution of variants that convey a selective advantage on a cell in health or under disease conditions. Parallel evolution is particularly strong in aging, where large fractions of wild-type cells can be replaced by mutant cell clones, a process that can be further accentuated by exogenous factors (Olafsson and Anderson 2021).

In the CNS tissue, where clonal cell expansion is not carried on beyond cell differentiation, a mosaic of somatic variants might be acquired early in development or accumulate due to conditions of genomic instability (Mustjoki and Young 2021). The propensity to acquire strand breaks is particularly high for certain gene loci called chromosomal fragile sites (CFS), which might carry a so far underestimated role in aging and age-related neurodegenerative pathologies (Araúzo-Bravo et al. 2025). Moreover, germline haplotypes and hereditary mutations known to cause a neurodegenerative pathology might be in linkage disequilibrium with such CFS and other somatic mutational hotspots, meaning that their coincidence shows a non-random association. Beyond the category of rare GIS aforementioned, a role of somatic variants in the genesis of non-tumorous diseases is currently suggested, e.g., for epileptic syndromes as well as vascular and autoimmune disorders (Mustjoki and Young 2021).

Although the expansion and selection of individual cell clones carrying a specific variant rule out a mechanism of mutational spread in differentiated CNS neurons, clonality may be an indirect risk factor to elicit neurodegenerative diseases. Some examples are the epidemiological association of clonal hematopoiesis with the manifestation of AD and age-related VD, although the details of the causal relationship are not yet understood. A linkage of lymphoproliferative disorders with ALS has also been suggested (Gordon et al. 1997). In addition, atherosclerosis as a dominant public disease has been proposed to arise from somatic variants that undergo monoclonal mutational selection (Lenormand and Lipsker 2021).

A better understanding of somatic variants might also help explain the population-specific incidence of several neurodegenerative disorders, like the higher age-adjusted incidence of ALS and MS in European populations, or of specific mutations linked with clonal hematopoiesis and T cell and B cell malignancies (Terao et al. 2020). While the contribution of environmental versus germline processes to these differences is unclear, they indicate the requirement to understand the evolution of somatic mutations in diverse populations.

In summary, the observations of accumulating autosomal mutations in neurons of the aging brain, the association with neurological disorders, a putative linkage disequilibrium with CFS, and the unresolved epidemiological finding of population-specific disease incidence highlight the need for a better comprehension of somatic mosaicism and evolution in neurodegenerative disorders. Overall, the combined integration of somatic genomic, epigenomic, and germline variants based on appropriate methodical tools for high-throughput and high-precision variant calling seems promising to achieve a new complexity and resolution of the genetic origin of health and disease.

2.2 *Categorization, Molecular, and Structural Features of eccDNA*

A novel approach for the study of somatic genomic variants is the examination of extrachromosomal circular DNA (eccDNA). This primarily double-stranded circular DNA structure is spatially separated from the linear chromosomes and resides episomally in the nucleus. Depending on the size and type, it can possess self-replicating and autonomous regulatory features or interact with the linear genome. By this, it contributes to both somatic genetic variation and epigenetic control. Following the initial discovery of eccDNA in boar sperm and wheat embryos by Hotta and Bassel (1965), research in this field remained for long stagnant, mostly due to methodical limitations. In recent years, however, this area of research has grown enormously, driven by the availability of modern high-throughput and omics technologies, coupled with computational biology algorithms that allow precise annotation, mapping, progeny identification, and functional characterization of eccDNA sequences (Yang et al. 2022) down to a capture with single-cell resolution (Chamorro González et al. 2023).

One persistent obstacle to the dissection and understanding of the various forms of eccDNA has been the inconsistencies in nomenclature. Therefore, some aspects on the history of terminology and classification of eccDNA will be given before addressing the scope of scientific importance.

2.2.1 Nomenclature: Classification According to Size

Although the terminology remains heterogeneous, there is a growing practice of using eccDNA for the broad indication of circular DNA of any size and genomic origin in eukaryotes. In the context of this chapter, we will relate to eccDNA as circular DNA of limited size (up to kb range), as found in healthy, diseased, and aging CNS tissue. Conversely, ecDNA is reserved for specifically denoting cancer-associated circular DNAs of large size (50 kb–5 Mbp) that carry and amplify oncogenes or drug resistance genes (Ilić et al. 2022). As earlier cytogenetic studies had observed small/minute paired chromatin bodies in metaphase spread preparations of cancer cells by light microscopy (Cox et al. 1965; Spriggs et al. 1962), ecDNA has historically been named double minutes (DMs). However, since DMs are only one entity of large circular DNA, this term is successively replaced by ecDNA. Unlike supernumerary neochromosomes formed in tumors, DMs lack telomeres and centromeres and are randomly segregated into daughter cells following cell division (Ilić et al. 2022; Yi et al. 2022b). DMs can be the conversion products of self-replicating precursor episomes of ~250 kb in size, which are circular structures with replication origins that arise from chromosomal deletions and that enlarge up to DMs (Carroll et al. 1988).

In contrast, small circular DNAs that carry non-repetitive unique sequences are known as microDNAs (Shibata et al. 2012). Given their limited size (in most cases

100–400 bp), microDNAs typically encompass gene fragments rather than complete genes. The term spcDNA (small polydisperse circular DNA) has been more previously used to refer to small-size circular DNA (100 bp–10 kb), highlighting the fact that size and sequence characteristics can be broadly heterogeneous (Smith and Vinograd 1972).

Beyond, the class of eccDNA includes extragenomic telomeric circles present either as double-stranded t-circles (Tomaska et al. 2000) or as partially single-stranded c-circles (Henson et al. 2009). When deriving from the terminal genomic regions in yeast with linear mitochondrial genome, t-circles were identified to possess a size of integral multiples of 738 bp (Tomaska et al. 2000, 2009). In cell lines that maintain telomere length independently of telomerase activity, a process called alternative lengthening of telomeres (ALT), t-circles reach a size of up to 56.8 kb (Cesare and Griffith 2004). In this process, t-circles convey the important capacity to maintain telomere length by acting as template structures for rolling circle nascency of telomere repeats, while the c-circle counterparts can be used as diagnostic markers for ALT (Henson et al. 2009). Apart, telomeric circles are excised from internal telomere loops (i-loops) that emerge after damage to telomere repeats, resulting in telomere length attrition (Mazzucco et al. 2020).

The scope of eccDNA further comprises ribosomal DNA (rDNA; Cohen et al. 2010; Sinclair and Guarente 1997). In yeast, where extrachromosomal rDNA circles (ERCs) accumulate over time in the mother cell due to asymmetric segregation during cell budding and other retention factors, they propagate aging of the mother cell (Cohen et al. 2010; Sinclair and Guarente 1997; Denoth-Lippuner et al. 2014).

The presence of eccDNA is not restricted to the nuclear cell compartment. In analogy to linear DNA fragments (Lo et al. 2021), circular DNA is released into the systemic circulation where it can be assessed as cell-free eccDNA (cf-eccDNA) from bodily fluids (Sin et al. 2020). In plasma, eccDNA shows periodicity with peaks at 202 bp and 338 bp that well correspond to the length of one or two nucleosome cores and two linkers, respectively. Overall, eccDNA is larger compared to linear DNA fragments in human plasma (Sin et al. 2020). However, size dimensions of cf-eccDNA vary substantially and can reach up to 20 kb (Zhu et al. 2017).

Further entities of circular DNA are bacterial plasmid DNA and chloroplast (or plastid) DNA in plants, however these forms, as well as the circular mitochondrial genome, will not specifically be addressed here. A summary of the different forms of circular DNA can be found in Table 1.

2.2.2 Classification According to Genomic Structural Content

Another proposal for classifying eccDNA is based on the structural elements incorporated: full gene eccDNA, exon eccDNA, intron eccDNA, repeat eccDNA, repeat-intergenic eccDNA, intergenic eccDNA, transposable element (TE) eccDNA, and promoter/enhancer eccDNA (Ling et al. 2021).

Table 1 Circular DNA entities, size categories, and nomenclature

Form of circular DNA	Size	Features	References
spcDNA	<100 bp–10 kb	Byproducts of genomic rearrangements, contribute to genomic instability	Smith and Vinograd (1972)
microDNA	200 bp–400 bp, up to 3 kb (5%)	Unique, non-repetitive sequences; can produce functionally active small regulatory RNAs, miRNAs, and si-like RNAs	Shibata et al. (2012); Paulsen et al. (2019)
eccDNA	<100 bp up to >10 kb	Circular DNAs in non-tumor tissue; carry repeat or unique sequences including gene fragments; rarely embed full genes	Møller et al. (2018); Gerovska et al. (2023b)
ecDNA	Up to several Mb	Large circular DNAs; can carry amplicons of oncogenes and drug resistance genes in tumor cells; devoid of telomeres and centromeres	Nathanson et al. (2014); Turner et al. (2017)
DM	100 kb up to several Mb	Large circular DNAs; can amplify oncogenes in tumors; can reinsert into chromosomes to form HSR	Spriggs et al. (1962); Cox et al. (1965); Ilić et al. (2022)
Episome	≥100 kb–Mb	Give rise to DMs by multimerization; can carry full genes; circularize from chromosomal deletions	Carroll et al. (1987, 1988)
Telomeric circle	Multiples of 738 bp, up to 56.8 kb repeat	Partially single-stranded c-circles or double-stranded t-circles; composed of telomeric arrays; can serve as rolling circle amplification templates of telomere DNA; play a role in ALT	Cesare and Griffith (2004); Tomaska et al. (2009); Henson et al. (2009)
ERC	<2 kb up to >20 kb repeat	Contain replication origins; contribute to replicative aging in yeast	Sinclair and Guarente (1997); Cohen et al. (2010)
cf-eccDNA	200 bp–400 bp (plasma)	Are released in body fluids; larger than linear DNA fragments in plasma	Sin et al. (2020); Lo et al. (2021)

2.3 *Genome Plasticity Involving ecDNA-Mediated Gene Amplifications*

Following the initial discovery by Hotta and Bassel (1965), circular DNA has been identified in all organisms studied so far. To date, it has been most extensively characterized in tumors of different types and aggressiveness (Yi et al. 2022a). A considerable portion of tumor entities has been shown to integrate ecDNA-producing cell subsets that increase somatic oncogene copies by focal amplifications and the unequal segregation of circles into daughter cells following cell division. Likewise, up to 47 and 60% of newly diagnosed cases of sarcoma and glioblastoma amplify ecDNA, respectively (Yi et al. 2022a). By means of gene copy number alterations and other DNA adaptions (delineated in the further), circles contribute to the diversification of the tumor genome. The presence of ecDNA has been associated with malignant cell transformation, acquired drug resistance and a worse prognosis (Luebeck et al. 2023; Helmsauer et al. 2020; Kim et al. 2020). Different mechanisms are discussed based on which ecDNA mediates drug resistance: First, drug efficacy can be impeded by the focal amplification of driver oncogenes on ecDNA in parallel with other forms of genomic plasticity (Song et al. 2022). By regulating specific ecDNA amounts, tumor cells gain the capability to dynamically adjust the dosage of genes relevant for tumor growth and survival. For example, loss of ecDNAs that express mutant *EGFR*, a gene that dually promotes tumor cell proliferation and drug-related cell death, explains drug resistance of glioblastoma cells over time. By contrast, drug pause can rescue the level of such oncogene carrying ecDNAs and re-sensitize the tumor (Nathanson et al. 2014). Notably, it is now evidenced that mutant gene variants are primarily amplified extrachromosomally, highlighting the importance of ecDNAs in tumor growth and adaption (Nikolaev et al. 2014). Complementary to this paradigm, amplification of drug resistance genes like *DHFR* on ecDNA contributes to longitudinally compromised drug responses. Thus, ecDNA generation is a clinically relevant mechanism by which the tumor cell selects the expression of genes that are favorable for its growth and evasion from drug-induced death. Overall, these observations illustrate that ecDNA is not merely a passive genomic component but that it plays a critical role in tumor genome evolution, adaption, and response to therapy.

2.4 Molecular Concepts of How ecDNA Regulates the Genome and Epigenome

In addition to amplifying oncogenes and driver mutations (Nikolaev et al. 2014), thereby efficiently increasing intratumor heterogeneity (Turner et al. 2017), ecDNA can influence gene transcription by several means. First, rearrangements subsequent to chromosomal reintegration of circles activate proto-oncogenes or silence tumor suppressor genes (Koche et al. 2020). Moreover, oncogene transcription on the circles is boosted by improved chromatin accessibility, as evaluated by increased ATAC-seq signals from circular as compared to linear amplicons, and likely a reduction in the compactness of the nucleosomes that comprise the circular DNA structure (Wu et al. 2019). Due to rewired amplicon architecture, involving chimeric elements, and new three-dimensional topologies, cargo genes on circles can build unconventional connections with co-amplified local (proximal and distant) enhancers or ectopic regulatory units (Morton et al. 2019; Helmsauer et al. 2020). In addition, as being mobile vehicles, ecDNAs can physically approximate and cluster in interactive hubs that permit intermolecular usage of regulatory elements, including enhancer hijacking by those circles that harbor cargo genes devoid of regulatory units. Hence, circle interaction further amplifies cargo gene transcription above the level expected from the gene copy numbers alone (Zhu et al. 2021; Hung et al. 2021; Yi et al. 2022a). Alternatively, ecDNAs can interact with transcriptionally active regions of the linear genome and operate as mobile transcriptional enhancers of genes in proximity to their contact sites (Zhu et al. 2021). In summary, such intricate epigenetic mechanisms render ecDNA potent flexible regulators of transcription, implicating the capability to operate throughout the genome and to cause synthetic aneuploidy comparable to structural chromosomal anomalies.

As ecDNA can feature a complex composite structure incorporating DNA segments from different genomic locations, and with variable transcriptional orientation, it is postulated that ecDNAs are participating in the processes of chromothripsis, breakage-fusion-bridge cycles and other forms of DNA damage (Shoshani et al. 2021), each of which can further propagate genome remodeling. Finally, as ecDNA can back-integrate into the linear genome to form homogeneously staining regions (HSR) (Song et al. 2022; Nikolaev et al. 2014), or at sites of DNA repair, it might influence gene expression beyond oncogenes.

2.5 *Molecular Concepts of How Short eccDNA Regulates the Genome and Epigenome*

The function of short eccDNAs, which commonly do not carry complete genes and may not self-replicate, remains poorly understood. Recent investigations have shown that small eccDNA can regulate gene expression through the promoter-independent generation of si-like RNAs and miRNAs containing exon sequences that repress the endogenous gene of eccDNA origin (Paulsen et al. 2019). Such repressor function can principally target any gene in the genome. Apart, short eccDNAs are assumed to be reunified in larger ecDNAs, thereby contributing to the genome evolution.

An alternative, though indirect mechanism by which eccDNA may regulate gene expression is through its relocation to extranuclear compartments, and its capacity to activate innate immunity. In case eccDNA is extruded from the nuclear space and enters the cytosol, it might be sensed by the cyclic GMP-AMP synthase (cGAS), ensued by the activation of the stimulator of interferon genes, STING. Finally, these steps trigger the activation of interferon- and pro-inflammatory cytokine cascades. In support of this scenario, Wang and colleague demonstrated that eccDNAs that are released from cultured apoptotic cells into the cell-free supernatant are sensed by co-cultured immune cells, which elicit STING-dependent type 1 interferon expression (Wang et al. 2021b). Targets of the cGAS-STING axis can either reinforce apoptosis or, in tumor cells, activate pathways that influence the process of ALT (Chen et al. 2017). Therefore, players in the cGAS-STING cascade represent potential candidates for therapeutic interference in tumors and other disease conditions.

2.6 Translational Approaches on ecDNA, eccDNA, and cf-eccDNA Epigenetics

The term "epigenetic mechanisms" traditionally implicates processes that affect gene expression and silencing without altering the DNA sequence. This can be achieved, e.g., through modifications of the chromosomal superstructure, the nucleosome formation and chromatin accessibility, and by alterations in the methylation status of the DNA. Moreover, non-coding RNA sequences can exert gene regulatory function. Epigenetic regulation is dynamic and reversible over time but can persist through multiple cycles of DNA replication.

As outlined before, cell- and tissue-derived ecDNA/eccDNAs exhibit several aspects that render these circular DNA structures intriguing participants in epigenetic regulation. Although the detailed molecular mechanisms remain to be elucidated, several established features allow for the consideration of ecc/ecDNAs as potent nuclear bodies of epigenetic control:

- ecDNA amplicons are reorganized in that new spatial and topological interaction sites between the cargo gene promoters and co-amplified local regulatory elements are established (Helmsauer et al. 2020; Morton et al. 2019).
- Chimerically circularized, multi-fragment ecDNA amplicons can create new topologies of diverse genomic provenience, with the influence on gene regulation and expression by involving ectopic enhancers and insulators (Helmsauer et al. 2020).
- ecDNA amplicons incorporate enhancer and cell lineage-specific transcriptional super-enhancers with active histone marks such as H3K27ac and H3K4me1 (Morton et al. 2019; Zhu et al. 2021; Helmsauer et al. 2020).
- ecDNAs that are spatially tethered in hubs allow intermolecular enhancer–cargo gene interactions and may provide enhancers for ecDNA rings that harbor a pathogenicity gene or drug-resistant gene only and those that carry the corresponding enhancer domain (Hung et al. 2021; Yi et al. 2022a).
- ecDNAs provide ectopic mobile enhancers that can establish genome-wide contact sites with the linear chromosomes, conveying trans-regulatory impact (Hung et al. 2021; Zhu et al. 2021).
- ecDNA interact with the linear genome particularly at transcriptionally active sites, as indicated by the enrichment of RNAPII and H3K27ac in the ecDNA–chromosome interactomes (Zhu et al. 2021).
- ecDNA is enriched in accessible chromatin, suggesting new nucleosomal organization and less chromatin compaction (Wu et al. 2019).
- Telomeric circles can continuously reconstitute telomeres in ALT cells and thus influence cell proliferation and survival (Tomaska et al. 2009).
- eccDNAs can play a role in gene regulation and silencing by producing miRNAs and si-like RNAs that interact with the linear genome (Paulsen et al. 2019).
- Similar to ecDNAs, small eccDNAs might serve as mobile trans-regulators of chromosomal genes.

- Small eccDNAs are proposed to operate as a molecular sponge that captures other regulatory factors.

In conclusion, there is ample evidence that tumors containing ecDNA alter their transcriptional patterns through a complex, interactive battery of genetic-epigenetic mechanisms. These altered patterns of non-classical control of gene expression and epigenetic regulation are believed to cooperatively involve both the cell's circular and chromosomal genomes (Lin et al. 2024). Some of these mechanisms might be applicable to non-tumor pathologies including CNS disorders.

For targeted interference with such genome control capacities in clinical practice, access to circular DNA needs to be established. One route amenable for clinical translation of eccDNA research is the study of cf-eccDNA retrieved from biological fluids by liquid biopsies. While linear plasma cf-DNA is well-characterized as for DNA methylation variants and fragmentation in open chromatin regions that can reveal the tissue and cell type of origin (Sun et al. 2015, 2019), such information is still rare for circular cf-eccDNA (Sin et al. 2020). The following advances in linear plasma fragmentomics involving epigenetic signatures of cf-DNA have disclosed enormous diagnostic power in multiple clinical dimensions and thus might become a blueprint for circular DNA research:

I. When analyzing cf-DNA in plasma of pregnant women, methylation differences of the maternal allele and the maternally inherited fetal allele can distinguish between either origin, as proven for imprinted, maternally expressed transcript regions (Poon et al. 2002). Moreover, aberrant, short fragment moieties in maternal blood reveal strong sensitivity and specificity in detecting fetal chromosomal aneuploidies (Yu et al. 2014). Accordingly, cf-DNA marker analysis has been adopted worldwide with high profitability for non-invasive prenatal diagnostics with the possibility to omit invasive amniocentesis. In expansion of such achievements, signatures of placental DNA fragments are under investigation for the diagnostics of gestation-related disorders. Overall, detection algorithms that combine the parameters of cf-DNA counts, size, and methylation profiles appear highly promising in providing sensitive information for diagnostic applications.
II. A second forefront of cf-DNA-associated marker employment is the field of cancer detection and surveillance. Plasma cf-DNAs derived from tumor cells recover tumor-inherent signatures of methylation and other post-translational histone modifications, and this information is useful to detect early tumor stages and events of recurrence. Notably, cf-DNA is also proposed to dissect types of primary CNS tumors and metastatic processes that commonly require tissue surgery for histological affirmation (Nassiri et al. 2020). Therapeutically, cf-DNA-derived methylation information, e.g., related to the promoter regions of tumor suppressor genes or genes relevant in DNA repair, such as O6-methylguanine-DNA-methyltransferase, can support tailored treatment choices.
III. A third evolving field of application is the monitoring of donor-derived cf-DNA in the plasma of patients after organ transplantation, which can give supportive

information about the recipient's graft function or risk of rejection (Sun et al. 2015; Schütz et al. 2017).

Considering these exciting properties, the concept of "liquid biopsy" assessment based on cf-DNA might be broadened to several other disorders including CNS pathologies, in which local biopsies are grossly impracticable. In a similar manner, circular cf-DNA research may identify tissue- and pathology-specific signatures and open up a new class of biomarkers for diagnostic approval, longitudinal monitoring, and putatively prognostic considerations in CNS disorders. Still, such exciting functions need confirmation for circular cf-eccDNA and in large-scale clinical trials linked to appropriate biocomputational platforms.

2.7 Compartments of eccDNA Assessment and Biomarker Function

Extrachromosomal circular DNA exists in different cell compartments. Though its location in cell nuclei of cell lines or tissues is most frequently addressed, diluted amounts can be detected and assayed from bodily fluids such as plasma, serum, cerebrospinal fluid (CSF) and urine (Kumar et al. 2017; Sin et al. 2020; Lv et al. 2022; Zhu et al. 2022). While the analysis of cf-eccDNA in fluid compartments suggests to be a new field of research, experimental routes date back into the late 1940s when linear cf-DNA was first discovered in blood (Mandel and Metais 1948). As aforementioned, linear fragmentomics have meanwhile attained prominent attention as non-invasive biomarkers in the field of prenatal diagnostics and in several disease and medical conditions (Lo et al. 1997, 2021).

Currently, the routes through which eccDNA is shuttled from intracellular organelles into the systemic circulation is not well understood. Similarly unclear is whether the moiety of cf-eccDNA isolated from liquid compartments conveys the same informational spectrum as tissue eccDNA. In the frame of neurological disorders, cf-eccDNA might have enormous advantages since brain or spinal cord biopsies are unavailable in most cases, with non-invasive plasma and CSF sampling being the only routes to access eccDNA. Moreover, due to their circular structure, eccDNA might be more resistant to exonuclease digestion than linear DNA fragments (Sin et al. 2020), increasing the time window to track and monitor a disease.

As assessed in mouse and human samples, the size of cf-eccDNA categorized as microDNA exceeds that of linear fragmented DNA in blood (Kumar et al. 2017). In malignant tumor disorders, the size of cf-eccDNA in serum decreases after tumor resection (Kumar et al. 2017), suggesting that size categories of eccDNA could, similar to that of linear cf-DNA, serve as a monitoring tool for disease states. Thus, in addition to quantitative and sequence-specific information, size dynamics may be meaningful in longitudinal disease assessments. Moreover, recent data from Zhu and colleagues collected from a limited number of pediatric patients with medulloblastoma suggest that eccDNA extracted from CSF may disclose the chromosomal origin, overall abundance, and length spectrum of eccDNA in a similar integrity than eccDNAs isolated directly from the tumor tissue (Zhu et al. 2022). Hence, the information profile of eccDNA in CSF may be promising to match that of the CNS tissue.

The initial success of using numerical and length parameters as eccDNA markers has led to a broad use of these metrics in several diseases, while underutilizing the rich sequence-inherent information provided by eccDNAs. Overall, further methodologically advanced studies based on larger patient cohorts are promising in revealing new exploitation levels.

3 Role of Extrachromosomal Circular DNA in Aging and Neurodegeneration

3.1 eccDNA in Physiological and Pathological Aging

The delineated concept that the accumulation of somatic variants over time is a central feature of genome aging and a driver of organismal aging and neurodegeneration has been proposed for over 60 years. Only recently, this notion has become scientifically consolidated in a milestone work by Lodato and colleagues (Lodato et al. 2018). The dimensions of genome aging may be of particular consequence for CNS neurons due to their long lifespan and reliance on error-prone DNA repair mechanisms like replication-independent canonical non-homologous end-joining (c-NHEJ). Moreover, repair efficiency of c-NHEJ declines with aging, a factor that augments the propensity for enlarged deletions and mutagenic repair (Seluanov et al. 2004). The question of whether neurons are inherently susceptible to genosenescence, aging, and cell degeneration remains unresolved and represents a persistent challenge for researchers.

Despite the perceived importance of this topic, a comprehensive analysis of the mutational burden in CNS neurons has yet to be conducted. Using single-cell whole genome sequencing (WGS) linked to a bioinformatics pipeline called linked-read analysis (LiRA), Lodato and colleagues could demonstrate that somatic single nucleotide variants (sSNV) in neurons increase almost linearly with age. This conclusion was inferred from the analysis of ~90 single neurons from the prefrontal cortex and ~30 neurons derived from the hippocampal dentate gyrus of 15 and 6 individuals with no history of neurological disorders at ages between 4 months and 82 years. The accumulation of sSNV was accelerated in cases of early-onset Cockayne syndrome and xeroderma pigmentosum, both of which implicate defects in nucleotide excision repair (NER), as assessed from nine individuals. A compromised or age-related decline in NER mechanisms may therefore explain the accumulation of sSNV and be linked to accelerated aging (Lodato et al. 2018). Notably, it is anticipated that the LiRA pipeline is effectively differentiating between amplification artifacts and biologically derived sSNV and identifying singletons in WGS datasets with highest call precision (Bohrson et al. 2019).

Beyond SNV, other structural genome alterations like small insertions/deletions (indels) up to larger chromosomal structural variants such as CNV, inversions, translocations, and numerical chromosomal imbalances might be a relevant contributor to genome variations and brain aging (McConnell et al. 2017; Luquette et al. 2022). Notably, in humans, variable regions by CNV account for ~12% of the genome (Redon et al. 2006), with a putative impact on gene dosages. In an innovative, computation-assisted approach using PhaseDel and high-coverage single-cell WGS (scWGS), Kim and colleagues could demonstrate from ~110 neurons isolated from 18 individuals that the frequency of somatic deletions in neurons also increases with age, with the genomic scars being identified particularly in highly expressed genes (Kim et al. 2022). Deletions were also increased in neurons from individuals diagnosed with Cockayne syndrome, xeroderma pigmentosum, or ataxia telangiectasia, as compared to age-matched controls, with a particular origin from NHEJ-based deletions in either syndrome.

Notably, somatic mutations that accumulate in adult stem cells over lifetime exhibit a tissue-specific pattern with striking inter-individual similarities as for frequency and variation type and can correspond to driver mutations in cancers. Such patterns might be traceable in neurons as well and overlap with mutations that are causal in neurodegeneration. As the mutations are succeeded by epigenetic events, such as of spontaneous deamination of methylated cytosine residues (Blokzijl et al. 2016), structural genome and epigenetic alterations might be interdependent beyond a stochastic association.

3.1.1 eccDNA as a Result of Lesion Load and DNA Repair

Understanding the load and frequency of somatic mutations in the aging brain is instructive in getting a general idea of the importance of episomal research in age-related CNS disorders. Somatic mutations and repair processes that entail focal deletions are of particular interest, as the excised fragments can form eccDNA by head-to-tail ligation of the broken ends. When illuminating the repair processes involved, genes of alternative double-strand break (DSB) repair cascades, such as of microhomology-mediated end-joining (MM-EJ), also known as alternative end-joining, appear to be stronger stimulated in ecDNA containing tumors than those repair genes engaged in c-NHEJ (Lin et al. 2024). Such observations are supported by the identification of microhomologies at the breakpoint junctions of ecDNA (Kim et al. 2020) and by studies addressing the repair-related provenience of small circles. Mechanistically, using cell lines with a functional depletion of key repair enzymes, Paulsen and colleagues proposed that end resection may determine whether eccDNAs are generated during the repair of DSB or not. While inhibition of c-NHEJ, capable of repairing DSB devoid of extended end resection, increased microDNA production, blocking of resection-dependent MM-EJ decreased eccDNA abundance, possibly by a shift to less eccDNA-genic repair pathways. Inhibition of single-strand annealing (SSA) and homologous recombination (HR) had no or variable effects on microDNA abundance (Paulsen et al. 2021). Genesis of microDNAs might also result from mismatch repair, involving the excision of chromatin loops that arise from annealing of short homology sequences (Paulsen et al. 2021). Other computational studies characterizing the junction sequences of eccDNAs isolated from human plasma suggested that blunt end-joining by c-NHEJ is also involved in circle production (Zhu et al. 2021); however this event may require two DSB on the same chromosome (Paulsen et al. 2021). Alternative predictive models exist (Eugen-Olsen et al. 2025), leaving the contribution of individual repair pathways to circle generation incompletely understood.

In summary, chromosomal fragments left behind by genomic stress and related repair processes are expected to be a relevant source of non-self-replicating eccDNA in the aging CNS and in benign pathologies of post-replicative organs such as the brain. Notably, the key features of somatic mutations—(I) presence in the form of a cellular mosaicism, (II) but with elements of patterned occurrence, (III) a positive association with transcription, (IV) gene length, and (V) the dependence on DNA repair mechanisms—mirror the features that are central to eccDNA genesis and origin (as also discussed later). Molecularly, the excision scar grossly indicates the gap corresponding to the excised DNA flap, which is circularized in the form of an eccDNA body. In conclusion, the detection of somatic mutations and eccDNA appear to be two sides of the same coin. The use of eccDNA could therefore serve as a marker tool to elegantly localize and characterize the origin of somatic mutations and better understand brain aging. Since the excision scar in the linear genome and the corresponding ligation site in the circular DNA can show sequence differences depending on the repair mechanisms involved (Rose et al. 2025), the analysis of eccDNA has the potential to provide additional valuable information about a cell's repair choices and those that favor circle generation. Bioinformatics in silico analyses could therefore be valuable to complement information on whether, in addition to the excision process, alterations in strand break repair may also be implicated. Overall, the combination of ultra-deep mutational characterization with breakpoint-specific eccDNA analysis could add a further level of complexity and understanding of the variants relevant to brain aging and neurodegeneration.

3.1.2 eccDNA, Nuclear Pore Complex Structures, and CNS Aging

Currently, the functional role of eccDNA in the process of physiological and pathological aging of mammalians is poorly understood. Although several concepts of how eccDNA might influence human aging have been proposed, they still require experimental validation. Extrachromosomal DNA has long been discussed as a driver of cellular aging in yeast (Sinclair and Guarente 1997). Extending such knowledge, a model has been proposed, in which ERCs excised by HR from highly repetitive rDNA loci tether to the nuclear pore complexes (NPCs) of yeast mother cells. The asymmetric segregation of these ERCs and their exponential accumulation in the mother cell disrupts the distribution of NPCs in the nuclear membrane and promotes replicative aging (Denoth-Lippuner et al. 2014). The model implicates that ERCs promote aging independent of their DNA sequence and not as a function of their presence alone, but as a consequence of ERC anchoring to NPC via the multifunctional Spt-Ada-Gcn5 acetyltransferase, or SAGA complex, and due to modified NPC abundance and an overall altered nuclear organization (Denoth-Lippuner et al. 2014). These data are supported by findings in mutant yeast, in which the lifespan negatively correlates with the amount of ERCs they retain (Defossez et al. 1999), and by studies in which artificial introduction of circular DNA propagates replicative aging in yeast (Falcón and Aris 2003).

Interestingly, a follow-up study showed that the tethering of ERCs to NPCs induces the remodeling of NPC components, implicating the secession of the nuclear baskets and cytoplasmic structures from the NPC upon ERC recruitment. Undocking of basket components impaired mRNA export shuttling mechanisms, and the altered functional state induced by ERC retention may contribute to replicative aging in yeast (Meinema et al. 2022). Finally, the authors raised the question whether NPC rearrangements could play a similar role in human aging pathologies, e.g., by influencing mRNA availability. Likewise, in Hutchinson-Gilford progeria, nuclear expression of progerin, a truncated lamin A precursor protein causal in the disease, is associated with disturbed topological arrangement of NPC basket structures (Cobb et al. 2016).

Whether eccDNA adherence to NPC structures could also be a scenario in human cells and neurons is worth to be explored. Notably, loss of WRN and BLM proteins in Werner progeria and Bloom's syndrome, respectively, shifts DSB repair choices toward those that require end resection such as MM-EJ. Unscheduled resection precipitates extended deletions (>200 bp), causes telomere fusions, and propagates genomic instability (Shamanna et al. 2016; Grabarz et al. 2013). Therefore, one may speculate increased eccDNA abundance in Bloom's and Werner's syndromes that could interact with NPC structures (Denoth-Lippuner et al. 2014).

In conclusion, understanding the interplay between genomic stability, eccDNA formation, and NPC structural and functional states appears to be an exciting domain that may further elucidate the aging process in eukaryotic mammalian cells. Contextualization in terms of CNS aging seems particularly interesting, as some scaffolding NPC structures are not renewed during the lifetime of a neuron, rendering them prone for deterioration of structural integrity and functionality over time (Savas et al. 2012). Accordingly, the selectivity of NPC substrate shuttling drastically declines in aging mammalian neurons. Disturbed and "leaky" NPC barrier function causes the relocation of cytoplasmic proteins into the nucleus (D'Angelo et al. 2009) and impairs the nuclear export and import machinery, involving compromised mRNA export.

Given that there is a multiplication of genotoxic and mutational events, paralleled by a reduced efficiency of DNA protection and repair systems and NPC dysfunction in neurons over time, two possible scenarios emerge how eccDNA could interfere with CNS aging (Araúzo-Bravo et al. 2025): First, if eccDNAs are retained in the nucleus over time, they could associate with basket structures of the NPC and promote cellular aging, analogous to the models established in yeast and speculated for progeria syndromes. Alternatively, eccDNA may exit the nucleus through NPCs that become "leaky" with age, be deposited in the cytosol, and, as recognized as misplaced DNA, trigger an innate immune response via cGAS-STING that causes cellular damage or stimulates aging cascades, senescence, and neurodegeneration (Gulen et al. 2023; Glück et al. 2017). The extent to which these scenarios contribute to mammalian aging remains to be determined. The two events might not necessarily be mutually exclusive, particularly if the nucleus produces sufficient numbers of eccDNA.

3.1.3 Quantitative and Sequence Features of eccDNA in the Young Versus Aged Brain

Contrary to the first scenario, a recent study exploring eccDNA in seven young versus old murine brains failed to confirm an increased eccDNA tissue load with aging. When the Circle-seq pipeline used for eccDNA assessment was applied to six anatomical brain regions, no significant change of the mean number of eccDNAs was detected between the 3-month-old and 19-month-old groups, regardless of the brain region analyzed (Hong et al. 2024).

We wondered whether the applied univariate marker analysis might have masked signatures and create uniformity of results between the groups. Several of our own studies have illustrated that the use of metrics that capture the number or length of eccDNAs underscores the complex information carried by eccDNA sequences (Gerovska and Araúzo-Bravo 2023a). For this reason, we have developed a new computational tool, DifCir, which merges eccDNA split reads related to each individual gene (Gerovska and Araúzo-Bravo 2023a) and implements a new metric to quantify them by the concept of "Produced per Gene Circles" (PpGC). This approach further facilitates quantitative differential analysis between samples of different categories by yielding the "Differentially Produced per Gene Circles" (DPpGC) or by designating the "Common Produced per Gene Circles" (CPpGC) for samples of the same category.

As elements from coding genes were, apart from TE, the most frequently annotated eccDNA structure, regardless of age or anatomical brain region (Hong et al. 2024), we re-assessed the raw DNA circulomics data generated by Hong et al. (2024) based on the DifCir pipeline (Gerovska and Araúzo-Bravo 2023a). Our objective was to see whether a sequence-based quantitative and qualitative eccDNA analysis can disclose aging-relevant genomic fingerprints. Applying a Wilcoxon rank-sum test, our analysis indicated that the number of unique eccDNAs was slightly but significantly higher in young as compared to the aged samples at the 1% significance level (p-value of 0.00795; Fig. 1a, b). The average absolute numbers were 38,089.6 ± 2247.9 (SEM) eccDNAs in the young versus 29,442.1 ± 2211.8 (SEM) eccDNAs in the aged group, representing a 1.29-fold increase (Fig. 1a, b). When assessing length distributions, we found that the maxima of eccDNA frequencies corresponded to multiples of the nucleosome length, as described in other

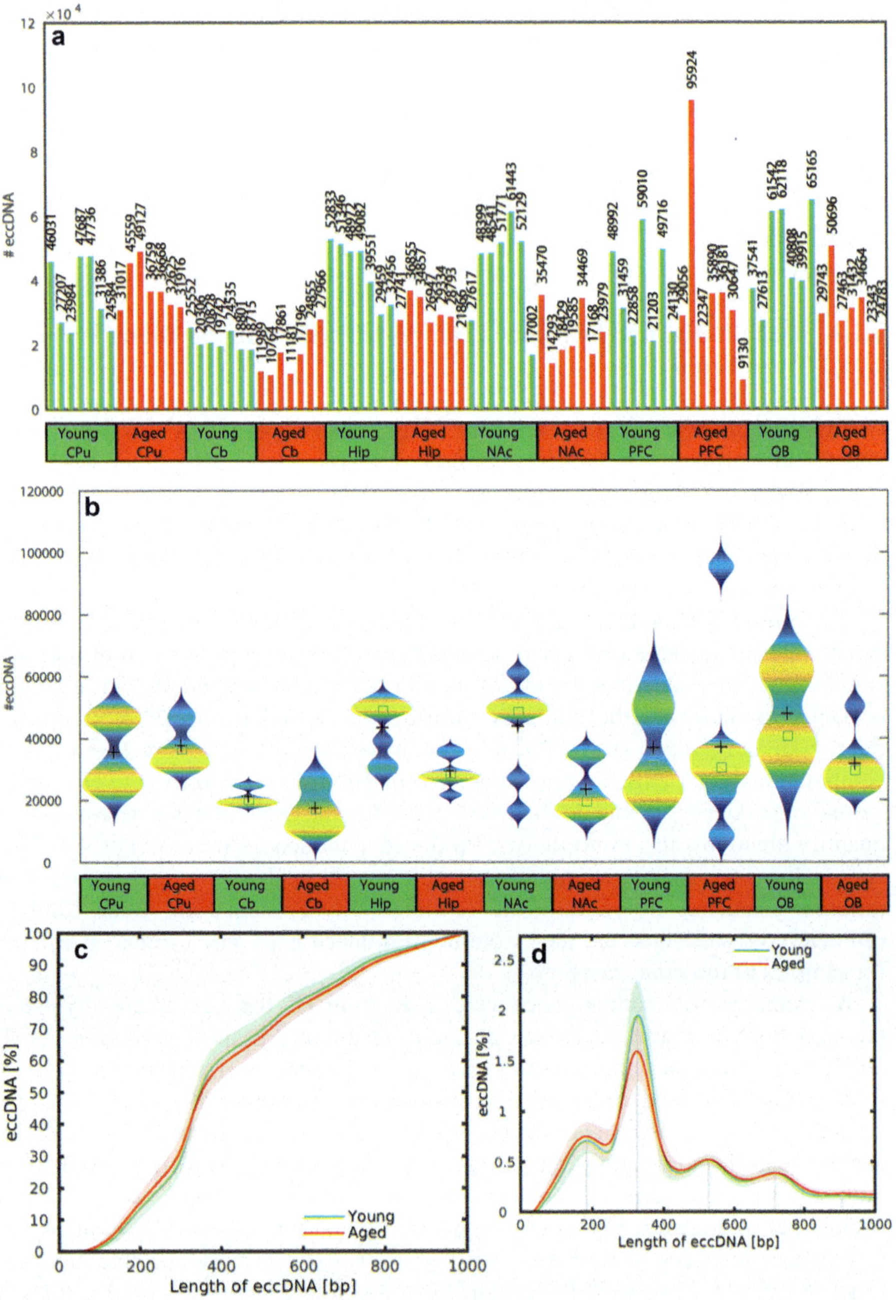

Fig. 1 Number and length distribution of unique eccDNAs in young versus aged brains, mapped separately for six anatomic regions. (**a**) Number (#) of unique eccDNAs mapped to the mm10 reference genome with at least two split reads. Included were eccDNAs ≤10 kb in length. (**b**) Violin plots showing the distribution of # eccDNAs in the young and aged groups. The black "+"

studies, with a higher amount of longer eccDNAs in young than in old brains (Fig. 1c, d).

We further performed a differential analysis for the genic eccDNAs produced per gene, defined as PpGC, in each age category. Selecting a FC of 4 (in $\log_2$ scale) and a significance level $\alpha = 0.01$ as threshold parameters for the p-values used to identify the DPpGC, we detected 124 genes that up-produced eccDNA in young versus aged brains (Fig. 2), and almost none in aged versus young brains (only two gene models). These observations suggest that eccDNA might be associated with additional factors other than DNA damage and repair in the CNS tissue such as, e.g., gene length and transcription.

Differential analyses of the genic eccDNAs were further specified for the six murine brain regions discriminated in the source data of Hong et al. (2024), including olfactory bulb (OB), prefrontal cortex (PFC), caudate putamen (CPu), nucleus accumbens (NAc), and cerebellum (Cb). We identified specific genic eccDNA profiles in both directions of comparison, suggesting a topological association with CFS (Fig. 3). As in the genic eccDNA analyses of the combined sample data from all six anatomical brain locations, the number of region-specific DPpGCs was significantly higher in the young than in the aged group. The list of the top 12 statistically significant up-DPpGCs per region and group, and their association with CFS, is shown in Fig. 3.

For conditions of FC 2 on $\log_2$ scale and significance level $\alpha = 0.01$, we found that the number of genes up-producing eccDNA accounted for 1267 genes in young versus old brains and 43 genes in the aged versus young comparison (not shown here). Thus, by relaxing FC stringency, up-DPpGCs were ~10-fold in young versus aged (1267/124 = 10.2) brains and ~20-fold in aged versus young (43/2 = 21.5) brains. In conclusion, we found a much higher number of genes that were up-producing eccDNA in young than in aged brains. Apparently, this difference is better contrasted by our DPpGC metric than by the number of eccDNAs alone, regardless of whether the number of detected eccDNAs was normalized to the total number of reads per sample (Hong et al. 2024) or not (Fig. 1a).

When considering the entirety of DPpGCs that were either up- or down-produced as a function of age in one of the six brain regions investigated, we identified a large number of DPpGCs for each direction of regulation, as illustrated in the volcano plots in Fig. 4. A comprehensive list of the number of up-/down-DPpGCs identified per anatomic region is presented in Table 2. Notably, when applying gene ontology (GO)-based gene set enrichment analysis (GSEA), we found a downregulation of

Fig. 1 (continued) represents the position of the mean, and the green square the median of each distribution. (**c**) Cumulative length distribution of eccDNA ≤1000 bp. (**d**) Length distribution of eccDNA ≤1000 bp. The shaded areas in (**c**) and (**d**) represent the standard deviation across the groups. Figures were generated after mapping of eccDNA with in-house software (Gerovska and Araúzo-Bravo 2023a; Gerovska et al. 2023b) using accessible raw DNA circulomics data from Hong et al. (2024)

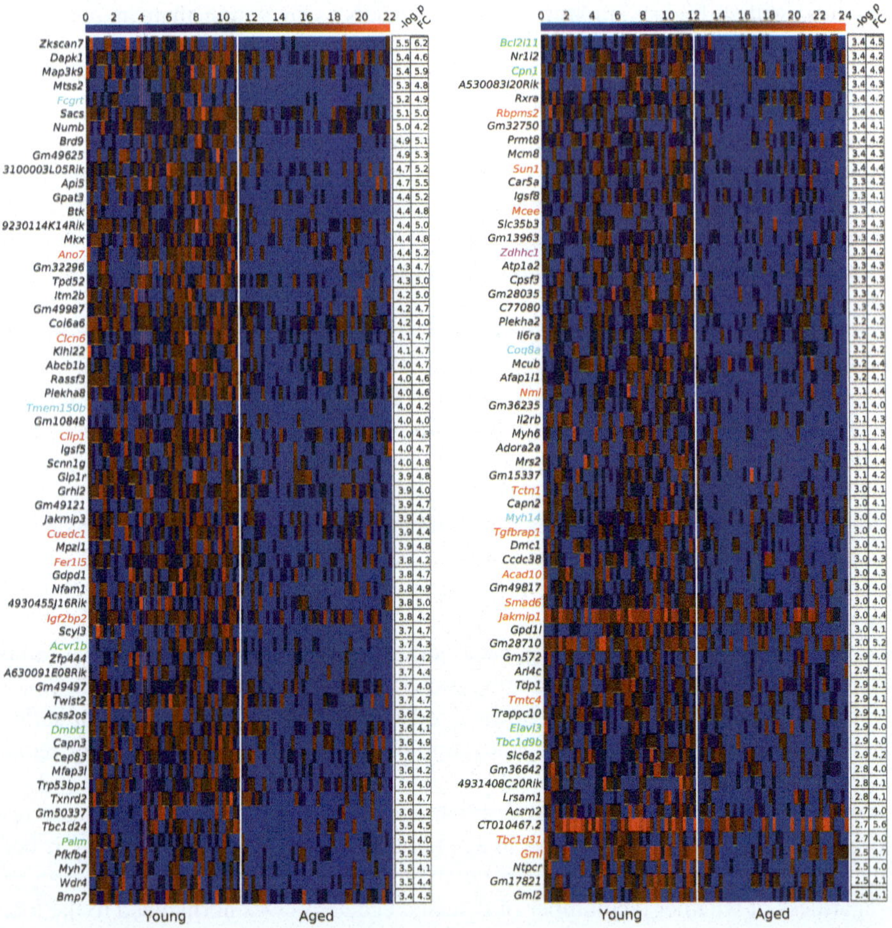

Fig. 2 Heatmaps of the up-DPpGCs in young versus aged brains. Shown are the 124 genes in descending order of statistical significance that produced more eccDNA in 3-month-old compared to 19-month-old murine brains assessed with the measure called PpGC. Color bars indicate the value of PpGCs on the $\log_2$ scale. High values correspond to red, and lower values to blue colors. The $-\log_{10}$(p-value) and the absolute $\log_2$ of the FC for each DPpGC are presented in a table to the right of the heatmaps. Gene names on the left are colored in red, green, blue, cyan, and magenta for genes associated with CFS induced by five different agents, aphidicolin, folic acid, BrdU, 5-azacytidine, and distamycin A, respectively. The figures illustrate the differential analysis DifCir performed on the publicly available raw DNA circulomics data from Hong et al. (2024). *DPpGC* Differentially Produced per Gene Circle; *PpGC* Produced per Gene Circle

eccDNAs derived from genes that are functionally involved (I) in DNA recombination and DSB repair via HR and c-NHEJ, (II) in nuclear pore organization and NPC assembly, (III) in histone modification, and (IV) in inhibition of cytokine-mediated immune functions (Table 2). Up-produced eccDNAs were functionally associated with genes engaged in the maintenance of molecule and protein homeostasis, e.g.,

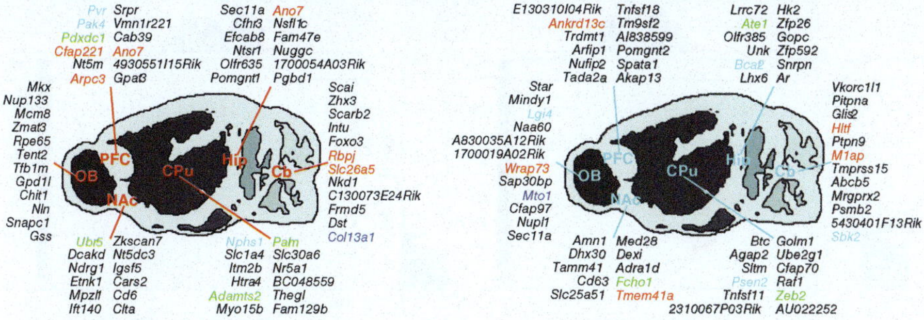

Fig. 3 Top 12 up-DPpGCs in young versus aged brains (left panel) and in aged versus young brains (right panel), according to six anatomical locations, and their association with CFS. OB, olfactory bulb; PFC, prefrontal cortex; Hip, hippocampus; CPu, caudate putamen; NAc, nucleus accumbens; Cb, cerebellum. Red-, green-, blue-, cyan-, and magenta-colored gene names indicate CFS induction by different chemicals, i.e., aphidicolin, folic acid, BrdU, 5-azacytidine, and distamycin A, respectively. Figures were generated after DifCir analysis and applying our lab-internal MATLAB functions on the publicly available raw DNA circulomics data from Hong et al. (2024). *DPpGC* Differentially Produced per Gene Circle; *CFS* chromosomal fragile site

through nuclear export shuttling and endo- and transcytosis, respectively, as well as in immune responses. Though these GO associations are a selection among others, their emergence from an unbiased eccDNA sequence mapping approach supports a crucial functional link with DNA repair, NPC transport, and immunity.

Following our DifCir algorithm (Gerovska and Araúzo-Bravo 2023a), we further revisited the data of Hong et al. (2024) to find the genes that produced eccDNA in the young (Fig. 5a) or aged (Fig. 5b) brains independently of an intergroup reference, termed CPpGC. Since a minimum threshold of 30 votes produced a high number of CPpGCs in the young category, we increased the selection stringency to a number of 34 votes in this group while maintaining a threshold of 30 votes for the aged samples.

Based on these parameters, we identified 80 and 88 CPpGCs, i.e., genes that consistently produced eccDNA under young (Fig. 5a) and aged (Fig. 5b) conditions, respectively. A comparative examination of the CPpGCs revealed that, although a greater number of genes excised genic eccDNA in the young brains than in the aged brains, there were "active" genes common to both groups that shed eccDNA at high rates in the brain in general, albeit at different rates.

Studying CPpGCs opens the possibility of tissue-, organ-, and organ region-specific genic eccDNA profiling in health and disease and may serve to create an atlas of genic eccDNA in mice or humans. Such an approach, prospectively complemented by ATAC-seq-based eccDNA data at a cell type-specific level, could be profitable for broad applications, given the public availability of DNA circulomics data generated in frame of different studies, their integrative alignment, and continuous curation for updated information.

The second study that has so far addressed eccDNA as a function of age evaluated eccDNA production between developmental stage E17.5 up to 3 and 22 months

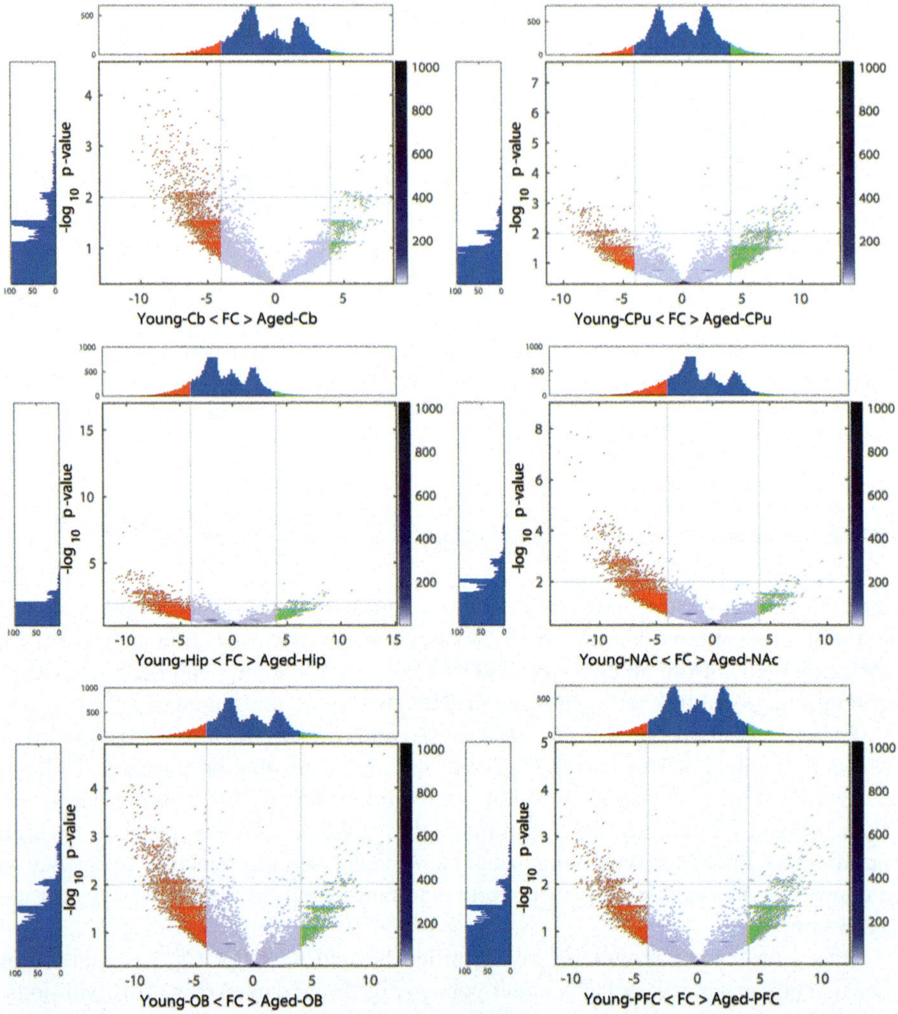

Fig. 4 Volcano plots illustrating the cloud of DPpGC that are up- (green) or down- (red) regulated with age, according to the anatomic brain region of detection. *Cb* cerebellum; *CPu* caudate putamen; *Hip* hippocampus; *NAc* nucleus accumbens; *OB* olfactory bulb; *PFC* prefrontal cortex

of age in seven healthy murine tissues (Liang et al. 2025). Consistent with the observations of Hong et al. (2024) and our own re-assessments, the authors did not find an absolute increase in eccDNA abundance with age in any of the tissues analyzed, including cerebral sources. Notably, healthy cortex and hippocampus produced similar amounts of unique eccDNAs as the liver, muscle, pancreas, and two types of adipose tissue, as demonstrated by normalization strategies that factor the number of reads and nuclei per sample at either age.

In accordance with the concept that endogenous DNA damage, e.g. by transcription stress, affects genes proportional to their length, the study further

Table 2 DPpGCs up- and down-produced with aging in a specific brain region and a selection of related, GO term-based functional categories

Brain region	DPpGC up in aging	DPpGC down in aging	GO function up	GO function down
Cb	34	361	• Protein localization • Cell division, cell cycle • CNS development	• DNA DSB repair via HR and c-NHEJ • DNA recombination • Isotype switching to IgA isotypes
CPu	166	171	• Nuclear membrane function • Nuclear export	• Amino acid catabolic process • Negative regulation of IL10 production
Hip	49	492	• Activation of innate immune response • Response to leukemia inhibitory factor • Maintenance of protein location • Developmental induction	• Nuclear pore organization • NPC assembly • Nucleus organization • Histone h3k27 demethylation
NAc	16	907	• Signaling • Receptor-mediated endocytosis	• Myeloid leukocyte-mediated immunity • Monocyte chemotactic protein-1 production • Neutrophil migration • Microtubule and cytoskeletal motor activity • Dendrite development
OB	57	513	• No significant terms	• Chromatin binding • K63-linked polyubiquitin modification-dependent protein binding • Neurotransmitter reuptake • Neural stem cell population maintenance
PFC	99	177	• Regulation of endocytic recycling • Transcytosis	• Deoxyribonucleoside monophosphate (dNMP) metabolic process • Myeloid dendritic cell activation • Type 2 immune response • Microtubule nucleation • Regulation of cell shape • Negative regulation of IL-8 production

complemented that shedding of eccDNA in healthy tissues is positively correlated with the length of the source gene, thereby disclosing an overrepresentation of long genes connected to neuronal functions and neurodevelopment. In compliance with the proposition of a transcription-related provenience of genic eccDNA in young

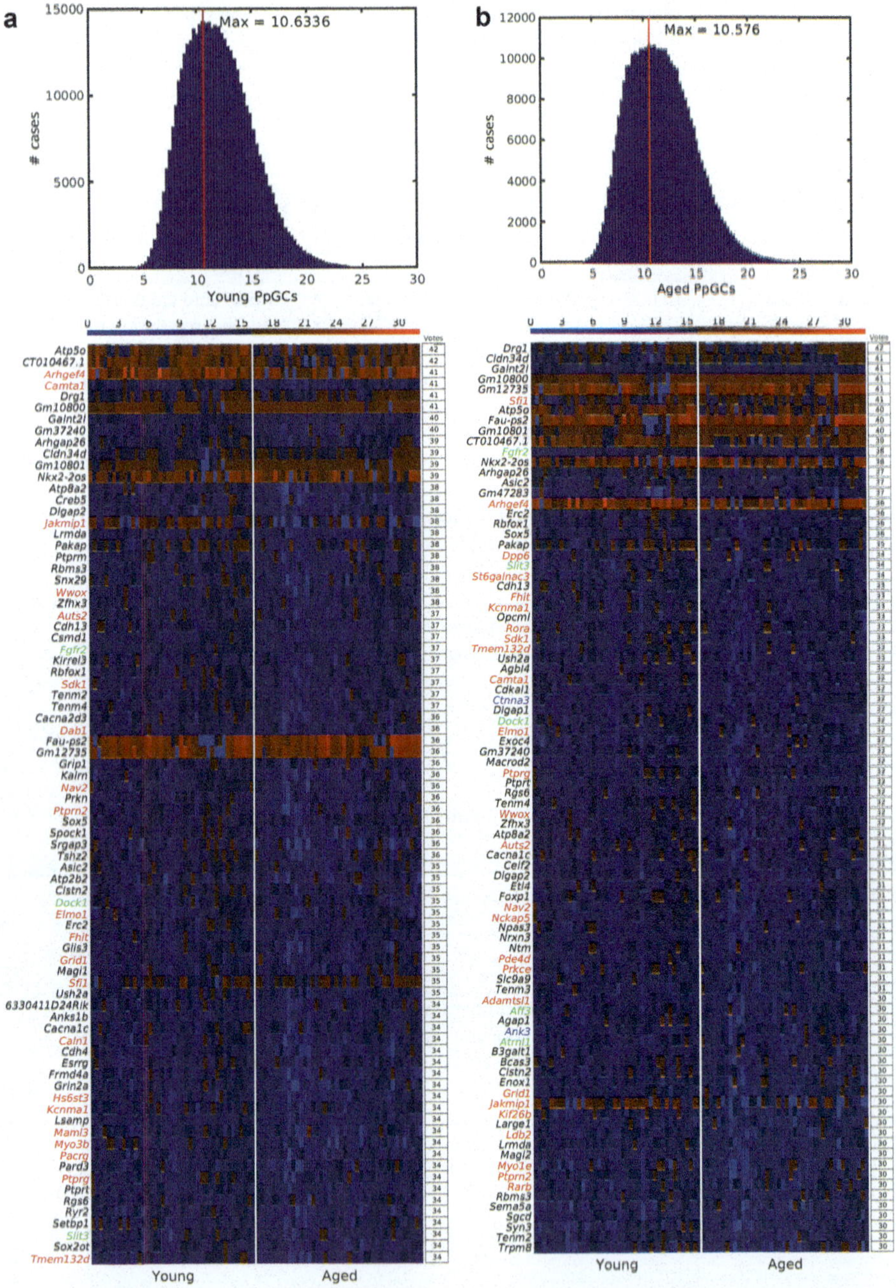

Fig. 5 CPpGCs in young and aged murine brains. Although the heatmaps illustrate the set of PpGCs for both age categories, CPpGC identification in (**a**) was based on young samples only, and CPpGC detection in (**b**) was based on aged samples only. The minimum threshold value for a PpGC to be considered as a positive vote in each age category was 11. The FC threshold for

brains, as depicted when re-assessing the dataset of Hong et al. (2024), the authors found a positive nexus between eccDNA genesis and transcription. Actually, a logarithmic relationship between eccDNA amounts and the transcription rate of a gene was found if a certain threshold level of transcriptional activity per individual gene was exceeded (Liang et al. 2025). In most except brain tissues, this threshold decreased with age, suggesting an age-dependent exhaustion of endogenous repair capacities that can neutralize genomic mutations induced by transcriptional stress. In support of a strong impact by transcription, eccDNA was enriched from genomic regions with open chromatin marks, as indicated by H3K27ac and H3K4me1 at younger and H3K36me3 across all ages (Liang et al. 2025). Epigenomic associations however were found to be reverted in the aged samples. Loss of epigenomic marks is proposed as a strong driver of aging and assumed to be elicited by DNA repair processes (Yang et al. 2023). Whether epigenomic "erosion" occurs on eccDNA similar as on the linear chromosomes is a question of future research.

Interestingly, the work also provides evidence that some genes, such as those with high intron density, are shielded from eccDNA release. Since many neural genes have particularly long introns, structural features may provide a partial explanation for the stable amount of eccDNA produced in the healthy but aging brain. In light of transcription-dependent eccDNA release, it will be interesting to explore in how far the phenomenon of gene length-dependent transcription decline (GLTD) recently identified as a key feature of genome aging (Soheili-Nezhad et al. 2024) can restrict eccDNA genesis in later life. Given that GLTD is triggered by DNA injuries, increased DNA damage load coincident with exhausting DNA repair capacities might stall transcription and thereby downscale rather than propagate eccDNA accumulation in the aging brain.

Overall, a limited number of studies that currently address eccDNA in the context of CNS aging raise arguments for the following considerations:

- Total eccDNA abundance in healthy brain tissues appears stable across different ages and is suggested to have a major source from genic regions.
- Certain genes appear to consistently shed eccDNA in brain of either age.
- The genic regions that up-produced eccDNA in the form of DPpGC in the young compared to aged CNS tissue might be associated with the transcriptional activity of the source genes and involve long neuronal genes.
- GLTD triggered by DNA injury and potentiated by declining DNA repair capacities might mitigate eccDNA excess in spite of crescent DNA damage load in the aging CNS.

Fig. 5 (continued) selection of PpGCs was 0.5 on the log2 scale, regardless of age. The minimum number of votes required for a PpGC to be considered as a CPpGC was 34 for the young group and 30 for the aged group. A total of 80 CPpGCs were identified in the young group and 88 in the aged group. *PpGC* Produced per Gene Circle; *CPpGC* Common Produced per Gene Circle

- Logarithmic amplification of eccDNA occurring beyond a threshold of transcriptional activity emphasizes the role of efficient DNA repair in limiting eccDNA.
- The contribution of CFS to amplifying eccDNA release on the one, and of long introns to shielding from eccDNA excision on the other hand, is not sufficiently understood.
- Currently unclear is whether the real-time frequency of DNA damaging events in young and aged brains is similar or different, as the cumulative quantity of somatic mutations might grossly reflect the moiety of unrepaired or wrongly repaired lesions over time (with many being cleared by cell division cycles or cell death). Further studies will elucidate the relationship between eccDNA genesis and efficient or erroneous DNA repair.
- To date, it is unknown which cell type—neurons or glia populations—strongest contributes to eccDNA genesis in the CNS. Enhanced replicative activity in young relative to aged glia cells might add to the abundance of eccDNA in young brains.
- It has to be explored in how far clearing mechanisms (e.g., coincident with NPC leakiness) are relevant to a washing-out of eccDNAs from aging neurons.
- Accumulation of eccDNA might occur in some susceptible but not all neurons, meaning that single-cell analyses are required to elucidate eccDNA in aging.

In synopsis, under healthy conditions, transcription appears to be a strong amplification factor of eccDNA, while the effect of aging on eccDNA shedding remains insufficiently understood. Further studies are needed to close the current caveats and balance putative discrepancies between theoretical considerations and evidence from new, still limited data. Investigations based on larger cohorts and covering even higher biological age categories are to be conducted. Such studies will help disclose whether, in brain tissues, the release of eccDNA as byproducts of transcription and maybe replication are sufficient to outweigh the expected rise in eccDNA production with aging due to cumulative DNA injuries and altered DNA repair efficiencies, thereby explaining the stable or slightly higher netto amount of eccDNA currently evidenced for young ages.

3.2 eccDNA in Neurodegeneration

The genesis of eccDNA in age-related neurodegenerative disorders is currently broadly unexplored. Source considerations suggest that eccDNA formation in the human CNS may be associated with sites of high transcriptional activity (Liang et al. 2025) and preferred sites of DNA injury and repair including CFS (Gerovska et al. 2023a; Araúzo-Bravo et al. 2025). CFS often intersect with genes that regulate neurogenesis and neurodevelopment (Smith et al. 2006). This raises the question of whether neurodegenerative disorders with known pathogenic germline mutations might implicate a second hit of somatic variants occurring early in neurodevelopment, e.g., at loci that regulate neuronal differentiation, maturation, and network integration and that are susceptible to acquire strand breaks. During brain development, neural progenitor cells traverse multiple rounds of cell division, with subsequent processes of DNA replication and transcription being assumed to provoke DNA injuries that exceed the endogenous repair capacities and remain unresolved. Persistence of such somatic mutations in neuronal subsets may help rationalize the increased susceptibility of certain cell populations to age-related degeneration such as motor neurons in ALS, cholinergic neurons in AD, and dopaminergic neurons in PD. Somatic mutations acquired during neurogenesis can be inherited clonally and become shared by many neurons. If the event occurs early enough, the mutant neuronal population might be effectually large to cause a neurological or neuropsychiatric phenotype (Cai et al. 2014). Likewise, the prefrontal cortex of patients with autism spectrum disorders (ASD) showed more somatic mutations in neural enhancer sequences relative to control individuals as assessed by ultra-deep WGS, suggesting that mosaic enhancer mutations may contribute to the risk of ASD. Projected frequencies accounted for 2–3 SNV per cell generation, ending up with 80 SNV present in $\geq$2% of the cells, a number comparable to de novo germline mutations (Rodin et al. 2021). Elucidating somatic mosaicism and patterned signatures will improve our understanding of aging and age-related neurodegenerative disorders.

3.3 eccDNA in Amyotrophic Lateral Sclerosis

Although for long suspected, it has required technical innovations to demonstrate that somatic variants are enriched in diseased neurons. Likewise, in 2018, Lodato and colleagues proved by scWGS that sSNV in premature, early-onset hereditary neurodegenerative disorders accumulate even stronger than in healthy aging, implicating region- and disease-associated peculiarities (Lodato et al. 2018). Proceeding with an innovative primary template-directed amplification technique, the authors confirmed an increase in somatic variants in aging neurons, which were particularly attributable to transcribed genes and regulatory elements. To assess the pure effect of aging on long-lived neurons, clonally expanded variants likely to have origin in development were filtered out (Luquette et al. 2022). Thus, there might be cumulative but partly independent variant signatures in aging and neurodegeneration.

Mutations that are somatically mosaic but found with distinct molecular patterns may therefore be relevant in late-onset hereditary and sporadic degenerative disorders of the CNS such as those manifesting as dementia, extrapyramidal movement disorders, or motor neuron spectrum disorders (Lodato and Walsh 2019). Likewise in AD, the frequency of somatic variants is enriched for C > A nucleotide exchanges, occurs preferentially on transcribed strands, affects exons, is associated with gene length, and can progressively impair function (Miller et al. 2022). Overall, the mosaicism of somatic variants, with SNV being one of different entities, may trace the history of genomic scars acquired by neurons from development to postmitotic functions. In this sense, somatic variants may be particularly informative for understanding age-related

ALS, where DNA integrity is threatened over time by variants in genes that have a primary or additional function in genome protection, DNA damage response (DDR) signaling, and DNA repair. Encompassed are alterations in the major pathogenicity genes, *SOD1*, *TDP-43*, *FUS*, and *C9ORF72*, which account for ~70% of familial (fALS) and a percentage of sporadic ALS (sALS) cases, as well as by variants in *C21ORF2*, *NEK1*, and *VCP,* among others (Wang et al. 2021a). Pathogenic variants in the *SOD1* gene, responsible for ~20% of human fALS cases, cause oxidative genomic damage, compromise SOD1-dependent transcription of DDR-relevant and antiapoptotic genes in the nucleus, and are proposed to propagate R-loop formation (Tsang et al. 2014; Salvi and Mekhail 2015). Similarly, TDP-43 and FUS, which account for ~4–6% and ~5% of fALS cases, respectively, are involved in c-NHEJ and in PARP1-dependent alternative NHEJ (a-NHEJ), in repair stabilization in response to oxidative DNA injury and R-loop homeostasis (Mitra et al. 2019; Wang et al. 2013; Giannini et al. 2020). The *C9ORF72*-related G_4C_2 hexanucleotide repeat expansion, which accounts for ~40% of fALS cases, propagates the subcellular relocation of TDP-43 and compromises the efficiency of several DNA repair cascades relevant to neurons, such as c-NHEJ, SSA, and MMEJ/aEJ (Andrade et al. 2020). Strikingly, TDP-43 pathology is common to ~97% of all ALS cases, regardless of a hereditary or sporadic provenance. Such knowledge strongly suggests that ALS pathomechanisms may converge on mono- or oligogenic background characteristics that severely compromise DNA integrity.

Considering a role of the main ALS-associated genes in genome maintenance, coupled with the endogenous propensity of neurons to accumulate DNA injury due to their long lifespan, their inability to perform error-free HR for DSB repair, their endogenous exposure to oxidative stress due to high oxygen consumption, and high transcription of large neuronal genes, we hypothesized to encounter amplified levels of eccDNA in ALS neurodegeneration. In parallel, we asked whether the genomic coordinates of eccDNA shedding might follow a particular pattern relevant to the ALS pathophysiology and imply functionally critical proteome changes.

To address these questions, we investigated eccDNA biogenesis in the murine *hSOD1*^{G93A} model that mimics key features of human ALS (Gurney et al. 1994). Cellularly and molecularly, *SOD1* variants implicate a toxic loss- and gain-of-function that interferes with the neutralization of cytosolic superoxide anion radicals, nuclear DDR signaling, and cell survival (Wang et al. 2017; Sau et al. 2007). When analyzing nine to ten biological replicates per group, we found that the overall abundance of eccDNA was significantly increased in the spinal cord tissue of symptomatic mutants as compared to healthy controls (Gerovska et al. 2023b). Genome-wide analyses revealed 225 DPpGCs that shed several-fold more eccDNA in ALS than from the control genome in a profiled pattern (Gerovska et al. 2023b). Notably, the DPpGC reads with the strongest accumulation in the ALS model showed an inter-sample recurrence rate of 89–100%. Functionally, these hotspot eccDNAs originated predominantly from genes involved in neurodevelopment, neuronal functions, and neuromuscular cell biology, suggesting that eccDNA could qualify for a biomarker role in ALS. Up-production of DPpGC along with altered, primarily repressed protein products from corresponding genes, as we confirmed it for ~40 out of the 225 eccDNA-shedding loci by mass spectrometry, might be caused by DNA injuries that inflict gene expression. In summary, we hypothesize a model, in which DPpGC shedding is linked to transcription stress in failing cells, such as motor neurons, at gene loci that may compensate for a functional loss consecutive to the disease. DNA injuries caused by transcription stress however, such as strand breaks through unscheduled R-loops, will remain unrepaired due to the DNA repair deficits elicited by the index mutation, ultimately interfering with protein translation. Additionally, eccDNAs might release si-like RNAs that interfere with corresponding exons on the linear genome (Paulsen et al. 2019). Theoretically, this scenario can propagate a loss-of-function vicious circle that contributes to disease progression. In support of an ALS-specific functional role, 40% of the DPpGC-related genes that yielded altered protein amounts were detected to be annotated in GWAS repositories and precipitate an increased risk for ALS (Buniello et al. 2019; Rouillard et al. 2016). Moreover, 51 out of the 225 hotspot loci represented on eccDNAs were found to localize to conserved CFS (Araúzo-Bravo et al. 2025; Smith et al. 2006), corroborating the role of genomic instability in ALS. The findings suggest a so far unconsidered link between eccDNA accumulation and ALS neurodegeneration.

To explore a putative contribution of aging to the profile of eccDNAs in ALS, we re-assessed the brain-derived DNA circulomics data of Hong et al. (2024) for a putative implementation of the 225 hotspot eccDNAs we discovered from spinal cord in the murine *hSOD1*^{G93A} model. The results are given in the heatmaps of Fig. 6, showing how the 225 up-DPpGCs identified in the ALS spinal cord are retrievable in the dataset of young and aged brains, according to the six brain regions examined (Fig. 6).

For the majority of the genic eccDNAs corresponding to the 225 up-DPpGC identified in ALS, we did not find any enrichment in young or aged brains. In aged Cb, the set of 225 genes produced a significantly lower amount of eccDNA than in young Cb (Fig. 7). A similar trend was seen in Hip, NAc, and OB regions (Fig. 7). These observations support the notion that the 225 up-DPpGCs represent eccDNA hotspots that are specific for the ALS pathophysiology and not sufficiently explained by aging.

Further studies in other animal models are required to corroborate that eccDNAs differentially produced in ALS-affected tissue can inform about the underlying pathophysiology, e.g., by discovering new variant loci or disclosing those whose transcription is upregulated to counterbalance a loss of function resulting from established hereditary mutations. Interrogation of cf-eccDNA will reveal whether disease stages and progression rates are captured and predicted by an eccDNA biomarker function.

3.4 eccDNA in Multiple Sclerosis

MS, a chronic inflammatory autoimmune disorder of the CNS, remains the most common non-traumatic cause of persistent disability in younger adults. Beyond structural demyelination in the acute attack, progressive neurodegeneration constitutes a relevant cause of functional decline over time in MS. Although the

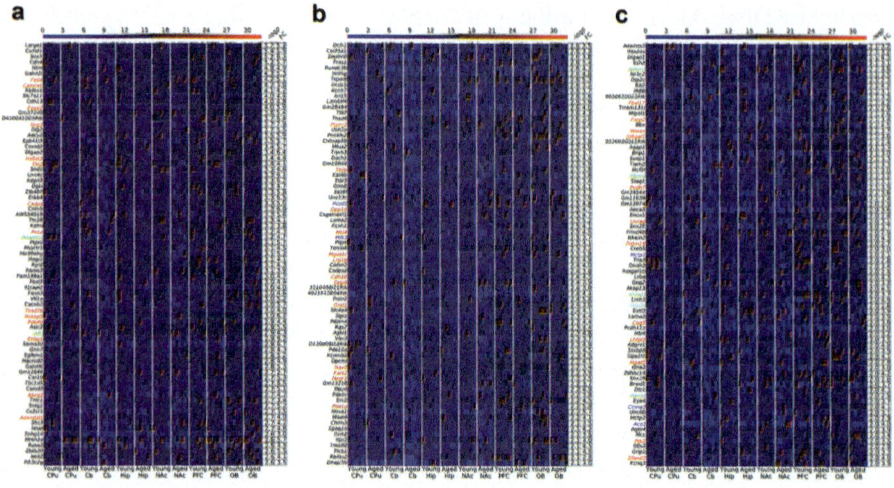

Fig. 6 PpGCs in six regions of young and aged murine brains ($n = 7$ per group) for the 225 genes excising more eccDNA in the spinal cord tissue of the *hSOD1*^{G93A} model of fALS as compared to controls. Gene names colored in red, green, blue, cyan, and magenta highlight the genes associated with CFS induced by aphidicolin, folic acid, BrdU, 5-azacytidine, and distamycin A, respectively. The color bars indicate the PpGC values on a $\log_2$ scale, where higher values correspond to redder colors. The $-\log_{10}(p\text{-value})$ and the absolute FC (in $\log_2$) of the DPpGCs identified in the mutant model are presented in a table to the right of the heatmaps

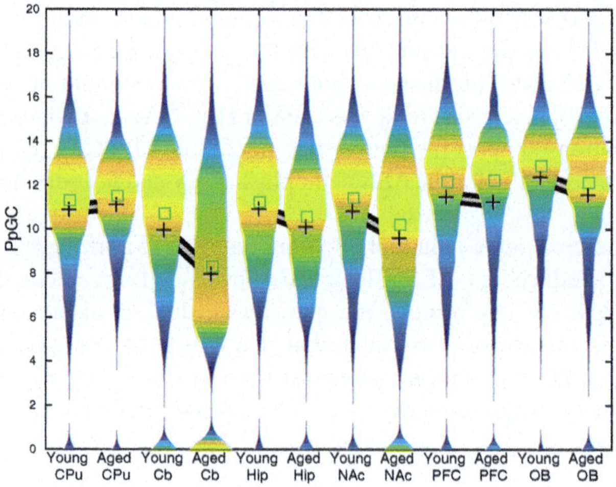

Fig. 7 Violin plots illustrating the distribution of murine brain PpGCs corresponding to the 225 DPpGCs identified in the spinal cord of the *hSOD1*^{G93A} model of fALS compared to controls. The black "plus" and the green "square" signs represent the calculated mean and median of each distribution, respectively. *OB* olfactory bulb; *PFC* prefrontal cortex; *Hip* hippocampus; *CPu* caudate putamen; *NAc* nucleus accumbens; *Cb* cerebellum

pathophysiology and antigen targets of the auto-inflammatory bursts remain incompletely understood, the range of approved pharmacological DMTs has expanded enormously in the recent years, allowing treatment options to be better tailored to each patient. In the course of such advances, refined patient stratification according to the individual disease type, activity, and risk of new attacks has become a paramount task in clinical practice. Approaches of precision medicine have propagated the search for biomarkers that can reliably predict disease progression and prognosis. To date, binary biomarkers that can be used to monitor and prognosticate the MS pathology while remaining unremarkable in healthy or otherwise diseased individuals are unavailable. One promising surrogate marker currently under investigation for monitoring MS relapse activity, disease progression, and therapy responses is the analysis of neurofilament light chains (NfL) in CSF and blood (Khalil et al. 2018). Belonging to a class of high-abundance structural proteins in axons, with an endogenous turnover after 2–3 months, basal levels are detectable also under normal conditions. Increased concentrations emerge in reaction to neuro-axonal damage of different causality; hence NfL are not specific for MS.

As an alternative approach, with the perspective of achieving improved specificity and resistance toward comorbidities, the establishment, validation, and implementation of molecular chromosomal markers appear promising. Unlike recently, genetic markers are meanwhile subtly amenable with modern, less costly molecular genetic and analytical methods.

When interrogating fragmentomics as putative chromosomal markers in MS, it is to realize that eccDNA in non-tumor conditions can originate not only from genomic stress factors, DNA damage events, and subsequent repair processes (Shibata et al. 2012; Paulsen et al. 2021) but also from programmed chromosomal reorganization and targeted genomic plasticity. The latter includes the looping out and excision of circular DNA along with the rearrangement processes of V(D)J recombination (Fujimoto and Yamagishi 1987), complemented by somatic hypermutation (McCormack et al. 1989), and by class switch recombination (CSR) (Von Schwedler et al. 1990). V(D)J recombination involves the genomic editing of polymorphic gene segments encoding the variable regions of TCR and of immunoglobulin heavy (IgH) and light (IgL) chain regions synthesized by members of the B cell lineage. Under physiological conditions, these mechanisms of somatic recombination are essential for the generation of a virtually unlimited repertoire of TCR and BCR and high-affinity immunoglobulins that cannot be attained by germline inheritance alone. The resulting diversity of cell surface TCR, as well as membrane-bound and soluble BCR, ensures the highest precision in antigen-specific, adaptive immune responses. Disruptions in these tightly controlled processes can impair the differentiation between self and foreign antigens, break down self-tolerance, and cause autoimmunity as in MS. Thus, apart from spontaneous somatic events, programmed somatic mutations play a key role in shaping our genome.

At the molecular level, somatic recombination occurs in the precursor cell populations of the T and B cell lineages. It involves the mono-allelic rearrangement of separate gene segments encoding the variable regions of the TCR and BCR, assembling them into a new, complete exon. The light chains of the immunoglobulins are encoded by two gene segments (V, J) and the heavy chains by three gene segments (V, D, J), which exist in different copies on different chromosomes (light chain κ and λ gene loci on Chr 2 and 22, respectively; heavy chain gene locus on Chr 14). Correspondingly, diversity of TCR α-chains involves the recombination of two (V, J) gene segments (on Chr 14), generation of TCR β-chains three (V, D, J) gene segments (on Chr 7). Central to this recombination and reassembly process is the excision of a circular DNA structure at conserved recombination signal sequences (RSS). The released structure has long been known as the TCR excision circle (TREC) and is generated analogously during maturation of the B cell series, thereby forming κ-deleting recombination excision circles, KREC. According to current classifications, these excision circles should belong to the eccDNA type. The excised DNA loops are a direct marker of the recombination events and an indirect marker of the activity of the recombination enzymes involved, encoded by the *recombination activating gene 1/2* (*RAG1/2*) complex. Deregulation of genome-editing RAG enzymes can lead to immunodeficiency syndromes like SCID, CID, and Omenn syndrome (Al-Herz et al. 2014) and, in the case of off-target effects by cryptogenic RSS with aberrant recombination in oncogenes, to genomic instability, cancerogenesis, and leukemias (Papaemmanuil et al. 2014). Hypomorphic *RAG* mutations and the resulting reduced enzyme activity are associated with a restricted B cell and T cell repertoire and the risk of auto-antibody generation due to imprecise elimination and selection of functional premature B cell clones. These auto-antibodies become activated by inflammation and are known to prune humoral autoimmunity (Chen et al. 2014; Csomos et al. 2022). Analogously, restriction of the TCR repertoire and its proper filtering can promote cellular autoimmunity and induce autoimmune pathogenic processes (Walter et al. 2015). Moreover, reduced absolute TREC levels have been suggested to reflect early immune senescence. Accordingly, renewal of the TCR repertoire and reconstitution of a broader clonal diversity and specificity of TCRs, achieved, e.g., by autologous hematopoietic stem cell transplantation (aHSCT), can be therapeutically beneficial in MS and serologically be reflected by the recovery of TREC levels (Muraro et al. 2005). By which mechanism partial RAG deficiency affects immune tolerance checkpoints is still poorly understood (Csomos et al. 2022) and requires further attention. Taken together, these findings suggest that excision of eccDNA from V(D)J recombination loci mediated by the RAG1/2 complex may have a relevant pathophysiological role and biomarker function in MS.

In a recent longitudinal study, mitochondrial eccDNA (mtDNA) as a special form of cell-free circular DNA (cf-mtDNA) was analyzed in the CSF of 48 relapsing-remitting RR-MS patients who had previously received basic or escalation DMT or were therapy-naïve (Pavlovic et al. 2024). The cohort was evaluated before and after aHSCT over a follow-up of 3–5 years and compared with 32 matched controls. The authors demonstrated that median cf-mtDNA levels in the CSF of RR-MS patients were significantly higher than in healthy controls. Patients with actively inflammatory, gadolinium-positive MR-imaging lesions evidenced the strongest increase in cf-mtDNA levels. Pharmaco-treatment with first- and second-line DMT regimens caused a reduction in cf-mtDNA concentrations in all MS treatment groups. Patients receiving high-potency second-line DMT (fingolimod, rituximab, natalizumab) had lower cf-mtDNA concentrations than patients receiving first-line DMT (dimethyl fumarate, glatiramer acetate, interferon, teriflunomide) or therapy-naïve patients.

Intervention with aHSCT reduced CSF cf-mtDNA concentrations independently of prior DMT. In addition, there was some correlation of cf-mtDNA with clinical EDSS scoring, moderate correlation with other CSF parameters (e.g., IgG index), and strong correlation with CSF NfL concentration.

In terms of etiopathogenesis, the authors postulated that cf-mtDNA derives primarily from activated lymphocytes proliferating extensively after antigen exposure. Replication stress in these lymphocytes will activate the DDR that, on the other side, operates as a control factor of the adaptive immune response and a direct contributor to eccDNA synthesis. Accordingly, the authors proposed cf-mtDNA as a potential treatment-sensitive biomarker of inflammatory activity in RR-MS (Pavlovic et al. 2024).

Analogous MS studies of non-mitochondrial eccDNA released from the linear genome are currently not available. However, such a first analysis is interesting and promising to opening new avenues for sensitive and specific biomarker research in inflammatory autoimmune disorders of the CNS.

3.5 eccDNA in Systemic Lupus Erythematosus

A second area of interest for the evaluation of eccDNA is systemic lupus erythematosus (SLE), an autoimmune multisystemic collagenosis with frequent CNS manifestations (Gerovska et al. 2023a; Gerovska and Araúzo-Bravo 2023b). SLE is characterized by the formation of auto-antibodies against deep nuclear and dsDNA structures, histones or complete nucleosomes, RNA-binding proteins, and phospholipids, followed by the initiation of pro-inflammatory cascades, the formation of immune complexes, and the subsequent activation of auto-reactive T cell and auto-antibody-producing B cell clones. Based on circulomics data published by Sin and colleagues (Sin et al. 2022), we re-assessed the eccDNA profile in plasma of patients with a monogenic *DNASE1L3* mutation. DNASE1L3 is a secreted nuclease operating in the degradation of extracellular DNA released, e.g., from apoptotic cells into the blood, and in preventing autoimmune responses against self-DNA, including the generation of DNA- and chromatin-directed auto-antibodies (Sisirak et al. 2016). Accordingly, DNASE1L3 deficiency is associated with autoimmune SLE, while biologicals that replenish enzyme activity can prevent the disease-like phenotype in murine knockout animals (Stabach et al. 2024).

Using DifCir, we found a set of genes that released cf-eccDNA in SLE patients carrying this mutation, while such a panel was absent in healthy controls (Gerovska and Araúzo-Bravo 2023b). Among the top up-produced cf-eccDNAs in SLE were those carrying gene fragments of *LY86* and *IQ-GAP2*. Gene products of these loci are markers for renal dysfunction caused by SLE and Sjögren's syndrome, another collagenosis with kidney manifestation. When performing GSEA for systematic functional assessments of the source genes that up-produced cf-eccDNA in SLE, we found an enrichment for the term "DNA methylation or demethylation." Epigenetic regulation in terms of conveying methylation changes was represented by the genes *FTO*, *MORC1*, *PIWIL4*, and *TET3* (Fig. 8).

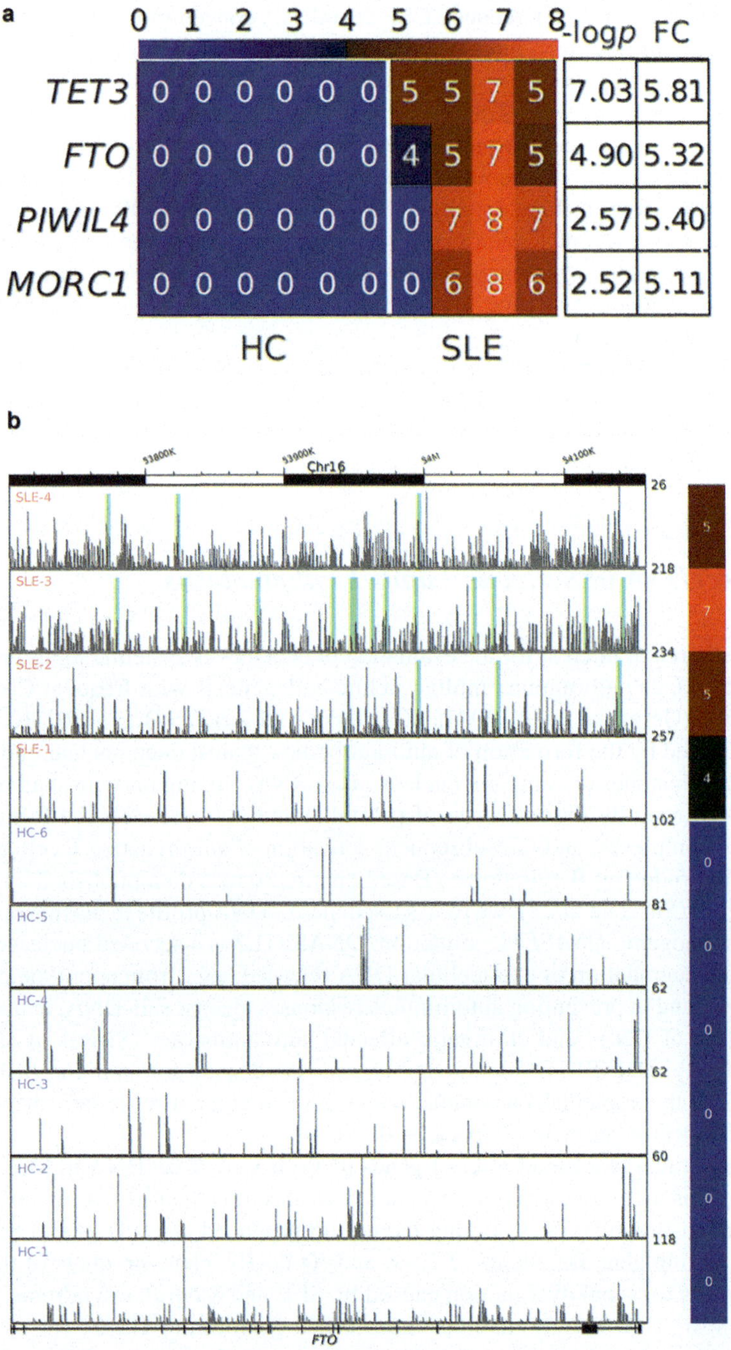

Fig. 8 Genes involved in epigenetic regulation that are related to the cf-eccDNAs discovered in the serum of SLE patients. (**a**) Heatmap of the genes whose split read sequences were found

Multiple variants near and in *TET3* have been associated with SLE so far (Yang et al. 2013), supporting a role of epigenomic regulation in SLE (Scharer et al. 2019). Mechanistically, animal studies suggested that Tet2- and Tet3-mediated chromatin modifications are involved in the repression of the T cell co-stimulatory surface protein CD86 (B7-2) on auto-reactive B cells, an event that may contribute to the protection against autoimmunity. Consistent with this, SLE patients show an increased proportion of $CD86^+$ B cells within the $CD19^+$ B cell population, and this increase has been implicated in autoimmune dysregulation, B cell overstimulation, and hypergammaglobulinemia (Folzenlogen et al. 1997). Consistently, Tet2- and Tet3-deficient B cells contributed to a lupus-like disease phenotype in the affected mice (Tanaka et al. 2020).

In conclusion, cf-eccDNA profiling may reveal SLE-relevant epigenomic regulators, and the related epigenetic fingerprints may be targets for novel approaches to counteract SLE pathophysiology. Whether eccDNA itself may contribute to DNA autoimmunity in SLE, e.g., by ectopic location when extruded from the nucleus, is a question for future research.

4 Open Questions

4.1 eccDNA and Apoptosis

Apoptosis has been proposed as the major cause of functional decline in age-related neurological disorders that progress over time (Mattson 2000). Similarly, in ALS, 50% of the disease-affected motor neurons are expected to have died by the time when functional deficits occur, with an inferred additional 50% loss every 6 months of disease progression (Gooch and Shefner 2004). Accordingly, neuroprotective strategies are being investigated in several disease entities, and the molecular inductive cascades and cellular processes of apoptosis have been extensively studied (Mattson 2000). At the chromatin level, apoptosis involves DNA fragmentation by caspase-activated DNases (CAD) that cleave genomic DNA at inter-nucleosomal linker regions. The resulting products, detectable with TUNEL assay and visualized as "DNA laddering," consist of DNA fragments that are multiples of 180–185 bp in size. Cleavage events that produce serial increments of (oligo-)nucleosomal length are commonly assumed to occur non-hierarchically and randomly in any gene (Winter et al. 1998), leaving a comprehensive breakpoint analysis of the DNA

Fig. 8 (continued) overrepresented on cf-eccDNA isolated from the serum of SLE patients (SLE) relative to healthy controls (HC) and that are functionally engaged in epigenetic regulation, according to GSEA (Gerovska and Araúzo-Bravo 2023b). (**b**) Track plots illustrating the source loci of eccDNAs in the respective gene (in green), with the total read coverage across the gene, and the PpGC values for the four SLE and six HC samples

fragments so far unaddressed. However, several arguments have meanwhile been proposed that support a directed action of CAD and other cleavage factors: First, CAD is found aberrantly expressed in tumors and can induce DNA breaks independently of apoptotic cascades (Larsen et al. 2022). Moreover, histone dephosphorylation or other epigenetic modulations that eventuate during apoptosis are proposed to facilitate the site-specific access of CAD to DNA (Fullgrabe et al. 2010). Finally, when assembled with co-factors, apoptosis-inducing factor, AIF, is endowed with nuclease activity and can propagate the fragmentation of DNA into much larger fragments for degradation (Novo et al. 2022).

Scrutinizing a random topology of apoptotic DNA fragmentation, Fullwood and colleagues set out to create a genome-wide breakpoint map of apoptosis. Using an "Apoptoseq" algorithm, they demonstrated that the distribution of apoptotic cleavage sites is not random across the genome, but is significantly associated with active genes, open chromatin, promoter regions, and transcription factor binding sites. Association with accessible chromatin and sites of active gene transcription was verified using a DNase I hypersensitivity map, H3K4me3 ChIP-Seq map, and CTCF ChIP-Seq map. Moreover, cleavage sites were enriched for CFS as shown, e.g., for *FRA3B*, and for genes that are frequently translocated in cancer cells (Fullwood et al. 2011). The authors also proposed that catastrophic DNA shattering by chromothripsis may be based on DNA hypercleavage during apoptosis and be involved in horizontal gene transfer by integrating DNA fragments into the genome of neighboring cells (Fullwood et al. 2011).

Given these characteristics, and the rate of programmed cell death in tumors, DNA fragments cut early in apoptosis at open chromatin and genic regions may circularize or otherwise contribute to the properties of ecDNA described in recent publications. Considering the periodic enrichment of eccDNA with several times the nucleosomal length in non-tumor tissues, apoptosis could also be considered in benign pathologies and in aging. Accordingly, the pronounced size periodicity of eccDNA in healthy murine tissues, including brain samples, was 378 bp and, although less pronounced, 189 bp, which correspond well to the size of nucleosomal units. Such size distributions were found in both young and aged tissues (Liang et al. 2025) and in a similar bimodal form with peaks at 200 and 350 bp in the liver tissue from murine wild-type and *Dnase 1*$^{-/-}$ or *Dnase 1l3*$^{-/-}$ models of SLE (Sin et al. 2022). Concrete experimental evidence that apoptosis may serve as a source of eccDNA was provided in a study by Wang and colleagues. Using mESC cultures, the authors demonstrated that exposure to toxic, apoptosis-inducing agents increases the amount of DNA fragments produced by enzymatic DNase γ cleavage, which subsequently become sealed by the activity of DNA ligase 3, eventually forming circular eccDNA structures (Wang et al. 2021b).

Although nucleosomal units produced by apoptosis are likely to contribute to the total amount of eccDNA, it is still an open question whether apoptosis can sufficiently explain the enrichment of circular DNA from certain structural regions, such as of microDNA from genic and exon regions (Shibata et al. 2012; Dillon et al. 2015), the high frequency of genic or pseudogenic regions incorporated by 40–52% of eccDNA in healthy tissues including brain tissue (Møller et al. 2018; Hong et al.

2024), and, e.g., the detection of genic regions derived from functionally clustered genes on eccDNA in neurodegeneration (Gerovska et al. 2023b). Similarly unexplored is the role of apoptosis in the positive correlation of eccDNA formation with gene length and transcription, as supported in different models and tissues (Møller et al. 2018; Liang et al. 2025). The extent to which apoptosis contributes to the overall abundance of eccDNA under different conditions requires further investigation. The observation that the total eccDNA amount may not be increased in healthy brain aging (Hong et al. 2024; Liang et al. 2025), but in age-related neurodegenerative pathologies (Gerovska et al. 2023b) could indicate a role of apoptosis.

Notably, fragmentomics studies paralleling the recent bulk of eccDNA research indeed link linear cf-DNA release with cell death mechanisms (Rostami et al. 2020; Schütz et al. 2017). At the same time, and without disaccord, it is widely accepted that linear cf-DNA is released in a non-random pattern, shaped by nucleosomal position, DNA accessibility, and epigenetic footprints that can disclose the tissue type of origin and indicate pathological states such as cancer (Sun et al. 2019; Lo et al. 2021). Incorporating these aspects, profiled origin of eccDNA from apoptotic cells might not per se be a consideration conflicting with biomarker functions.

4.2 Intercellular Distribution, Stability, and Elimination of eccDNA

According to the findings in tumor cells, the abundance of ecDNA circles per cell shows considerable variability. As ecDNAs are devoid of centromeres, and spindle complexes are not formed, the circles segregate asymmetrically into daughter cells after mitosis.

The further fate of ecDNA amplicons is not well understood, but may partly involve the reintegration into the nuclear chromatin of daughter cells. Apart from association with the normal chromosomes and with nucleoli, ecDNA has been proposed to be eliminated either by cell extrusion, e.g., through the cytoplasmic membrane, or by intracellular degradation (Shimizu et al. 1998; Hintzsche et al. 2017). The dominant moiety is expected to undergo spatial separation to specific territories in the nuclear periphery, from which it is extruded through micronucleation prior to karyokinesis (Shimizu et al. 1998). Micronucleus formation occurs through blebbing and budding of the nuclear membrane in S-phase as previously evidenced for DMs and fragments of long chromosomes carrying unstably amplified regions (Toledo et al. 1992; Shimizu et al. 1998) and observed for inactive, condensed X chromosomal Barr bodies (Tucker et al. 1996), particularly under aging conditions (Hando et al. 1994). Micronuclei appearance is a well-established marker of genomic instability, and methods to purify micronuclei were described several decades ago (Shimizu et al. 1996).

Understanding the modes of eccDNA retention and removal is relevant because elimination of extrachromosomally amplified genes, including (proto-)oncogenes and drug resistance genes, may impact the sensitivity to antineoplastic agents, influence tumorigenicity by induction of apoptosis, or prompt differentiation (Nathanson et al. 2014; Von Hoff et al. 1991; Eckhardt et al. 1994). Therefore, elucidation of the mechanisms of ecDNA clearance from the cell may also support the development of new anti-tumor therapies. Likewise, ecDNAs disappear spontaneously in the absence of selection pressure and are reduced in abundance by exposure to low-dose hydroxyurea treatment (Von Hoff et al. 1991; Eckhardt et al. 1994).

Micronucleation in S-phase infrequently eventuates also in normal, non-neoplastic cells and is increased if the cell's response mechanisms to DNA damage are impaired (Shimizu et al. 1998). Elucidating the maintenance and elimination of extrachromosomal genome modifications might therefore also be relevant for non-tumor disorders, in which pathogenetically relevant loci present as or are associated with unstable chromatin. Likewise, mutations in *SETX*, or hexanucleotide repeat expansions (G_4C_2) in the transcribed *C9ORF72* locus, both of which are causal in ALS, give rise to sites of extended DNA unpairing, of difficult-to-resolve secondary structures between the RNA and non-template DNA strand, and of R-loop formation (Abu Diab et al. 2018; Skourti-Stathaki et al. 2011). R-loops are potentially relevant in several other neurodegenerative repeat expansion disorders, such as spinocerebellar ataxia, Huntington's disease, Friedreich's ataxia, and myotonic dystrophy. Though R-loops are now proposed to support DSB repair as well (Brickner et al. 2022), these DNA-RNA hybrids are demonstrated to contribute to repeat instability by propagating repeat expansion through break-induced replication in response to replication fork stalling and collapse (Neil et al. 2018). In opposite, R-loops can promote repeat contraction by conveying mechanisms that lead to a netto deletion of repeats in the process of base excision repair-mediated R-loop resolution (Laverde et al. 2020). When stabilized in repeat regions, R-loops are evidenced to alter the epigenetic landscape, leading to transcriptional silencing of genes whose expression is required to prevent neurodegeneration (Brickner et al. 2022). Moreover, G-quadruplexes (G_4) localized in the single-stranded non-template DNA stabilize the R-loop and enhance transcription, while G_4 in the template strand inhibits transcription (Lee et al. 2020).

Currently, the contribution of R-loops to the progression of neurodegenerative repeat expansion disorders is still unclear, as animal models such as murine *Setx*$^{-/-}$ mutants fail to show increased R-loop abundance in brain tissue or to demonstrate phenotypic alterations (Becherel et al. 2013). Due to the postmitotic nature of neurons, R-loop pathologies in these diseases however need molecular considerations independent of replication. Likewise, R-loops induced by transcription-blocking injuries at the sites of DNA damage allow replication-independent induction of the DDR, e.g, in non-cycling cells such as neurons (Tresini et al. 2015). Moreover, oxidative stress and reactive oxygen species, as they are increased in *SOD1*-related ALS, can provoke and stabilize R-loops at transcriptionally active sites (Tan et al. 2020).

Another context of R-loop formation is at switch recombination sequences in B cells. During transcription of these sequences, long DNA-RNA intermediates spanning more than 1 kb are generated, which might be important for efficient CSR (Yu et al. 2003).

Overall, owing to their diverse regulatory roles on gene expression and silencing, their influence on transcription and chromatin remodeling, and their role in DSB induction and repair, R-loops are interesting genome regulators that might be a source of eccDNA production. Further investigations are required that analyze the progeny, turnover, and maintenance of eccDNA, thereby learning to understand the influence of structural characteristics, and means of eccDNA degradation by cell-specific capacities of enzymatic digestion or autophagy.

4.3 eccDNA: Trend or Revolution

Research of the last decades has concentrated on four key mediators of epigenetic regulation—DNA methylation, histone modification, chromatin remodeling, and effects by non-coding ncRNAs, thereby involving R-loops and RNA modifications, all of which participate in development, aging, and disease.

Here, we extend the scope of epigenetic regulators by considering the role of extrachromosomal circular DNA, an episomally located double-stranded DNA molecule of variable size that can, in the case of ecDNA, provide new spatial and topological interaction sites between autonomously amplified host genes and co-amplified regulatory elements on the circle, and further facilitate cargo gene transcription by favorable nucleosome organisation and chromatin compaction. Acting as mobile super-regulators that provide or interact with enhancer regions, or following chromosomal reinsertion at non-naïve sites, ecDNA can alter the expression and transcription of genes at virtually any location in the linear genome. Moreover, small eccDNAs are evidenced to be transcribed into si-like RNAs with silencer function. Lastly, circular DNAs themselves cluster into interactive hubs, forming hotspots of transcriptional modification.

These features render eccDNA interesting novel targets for therapeutic intervention in many disorders. To date, the role of circular DNA has been extensively studied in tumors but only marginally addressed in neurodegenerative disorders. By analyzing eccDNA in an animal model of ALS, we recently identified a several-fold accumulation of functionally profiled eccDNAs in the disease-affected CNS tissue. Although studies in additional models are needed, this first evidence supports a new perspective by linking hereditary and somatic genetic mutations in neurodegenerative risk and phenotype assessment. In such a role, eccDNA may have the potential to qualify as a new class of somatic genetic biomarkers and to indicate new genes for therapeutic intervention.

Beyond these promising new perspectives, eccDNA is now being explored in several other metabolic, cardiovascular, and immune disorders (Sin et al. 2022; Gerovska and Araúzo-Bravo 2023b; Zhao et al. 2022), as well as in aging (Hong et al. 2024; Liang et al. 2025). Expanding on these results, using customized bioinformatics algorithms and innovative technical pipelines, will clarify whether circular DNA remains an area of benefit mainly in oncology or whether it can become a general principle for improving disease monitoring, risk stratification, and intervention. Large-scale clinical trials will be required to verify the presumed disease-specific genetic fingerprints, followed by the invention of signature- and target-specific assays.

5 Conclusions and Future Prospects

Circular DNA and fragmentomics studies in general have the potential to become established as versatile biomarkers in tumor diagnostics and various other diseases. Although currently only marginally explored for CNS pathologies, the exploitation of eccDNA including cf-eccDNA, the identification of disease-specific signatures, and the understanding of other features such as size, breakpoint coordinates, and stability of eccDNA will be particularly beneficial in age-related neurodegenerative disorders, where tissue biopsies are largely impractical. Implementation of predictive assessments could complement an ongoing biomarker research and become a part of a clinical practice through the development of commercialized assays. Prognostic accuracy will benefit from more standardized eccDNA isolation methods and tailored computational tools that process and annotate circular DNA data.

Author Contributions MJA-B and DG established the pipeline for in silico eccDNA assessments and performed the biocomputational analysis, with support from AK. AK, DG, and MJA-B created the figure and table material. AK, MJA-B, and DG conceived the chapter content. AK, MJA-B, DG, and MS wrote and discussed the chapter.

Compliance with Ethical Standards No ethics approval was required.

Funding MJA-B and DG were promoted by the Circular Vision project with funding from the European Union's Horizon 2020 research and innovation program (under grant agreement no. 899417), by Ministerio de Ciencia e Innovacion, Spain (Grant No. PID2023-152752OB-I00/AEI/10.13039/501100011033), by Instituto de Salud Carlos III, Infrastructure of Precision Medicine associated with Science and Technology (IMPaCT) of the Strategic Action in Health (iDATA-MP), and by the IPerGlio project from the European Union's Horizon 2020 research and innovation program (under grant agreement ERAPERMED2022-245) co-financed by the Fundación Científica de la Asociación Española Contra el Cancer (FCAECC) (Cod. PERME224571ARAU) and the Instituto de Salud Carlos III (ISCIII) of Spain (Exp. AC22/00043).

Conflict of Interest The authors have no conflicts of interest to declare that are relevant to the content of this chapter.

References

Abu Diab M, Mor-Shaked H, Cohen E et al (2018) The G-rich Repeats in *FMR1* and *C9orf72* Loci Are Hotspots for Local Unpairing of DNA. Genetics 210:1239–1252

Al-Herz W, Bousfiha A, Casanova JL et al (2014) Primary immunodeficiency diseases: an update on the classification from the international union of immunological societies expert committee for primary immunodeficiency. Front Immunol 5:162

Andrade NS, Ramic M, Esanov R et al (2020) Dipeptide repeat proteins inhibit homology-directed DNA double strand break repair in C9ORF72 ALS/FTD. Mol Neurodegener 15:13

Araúzo-Bravo MJ, Gerovska D, Schwab M et al (2025) Small extrachromosomal circular DNA in amyotrophic lateral sclerosis matter. Neural Regen Res 20:1411–1413

Becherel OJ, Yeo AJ, Stellati A et al (2013) Senataxin plays an essential role with DNA damage response proteins in meiotic recombination and gene silencing. PLoS Genet 9:e1003435

Blokzijl F, de Ligt J, Jager M et al (2016) Tissue-specific mutation accumulation in human adult stem cells during life. Nature 538:260–264

Bohrson CL, Barton AR, Lodato MA et al (2019) Linked-read analysis identifies mutations in single-cell DNA-sequencing data. Nat Genet 51:749–754

Brickner JR, Garzon JL, Cimprich KA (2022) Walking a tightrope: the complex balancing act of R-loops in genome stability. Mol Cell 82:2267–2297

Brooks WH, Renaudineau Y (2015) Epigenetics and autoimmune diseases: the X chromosome-nucleolus nexus. Front Genet 6:22

Buniello A, MacArthur JAL, Cerezo M et al (2019) The NHGRI-EBI GWAS catalog of published genome-wide association studies, targeted arrays and summary statistics 2019. Nucleic Acids Res 47:D1005–D1012

Cai X, Evrony GD, Lehmann HS et al (2014) Single-cell, genome-wide sequencing identifies clonal somatic copy-number variation in the human brain. Cell Rep 8:1280–1289

Carroll SM, Gaudray P, De Rose ML et al (1987) Characterization of an episome produced in hamster cells that amplify a transfected CAD gene at high frequency: functional evidence for a mammalian replication origin. Mol Cell Biol 7:1740–1750

Carroll SM, DeRose ML, Gaudray P et al (1988) Double minute chromosomes can be produced from precursors derived from a chromosomal deletion. Mol Cell Biol 8:1525–1533

Cesare AJ, Griffith JD (2004) Telomeric DNA in ALT cells is characterized by free telomeric circles and heterogeneous t-loops. Mol Cell Biol 24:9948–9957

Chamorro González R, Conrad T, Stöber MC et al (2023) Parallel sequencing of extrachromosomal circular DNAs and transcriptomes in single cancer cells. Nat Genet 55:880–890

Chen K, Wu W, Mathew D et al (2014) Autoimmunity due to RAG deficiency and estimated disease incidence in RAG1/2 mutations. J Allergy Clin Immunol 133:880–2.e10

Chen YA, Shen YL, Hsia HY et al (2017) Extrachromosomal telomere repeat DNA is linked to ALT development via cGAS-STING DNA sensing pathway. Nat Struct Mol Biol 24:1124–1131

Cobb AM, Larrieu D, Warren DT et al (2016) Prelamin A impairs 53BP1 nuclear entry by mislocalizing NUP153 and disrupting the Ran gradient. Aging Cell 15:1039–1050

Cohen S, Agmon N, Sobol O et al (2010) Extrachromosomal circles of satellite repeats and 5S ribosomal DNA in human cells. Mob DNA 1:11

Cox D, Yuncken C, Spriggs AI (1965) Minute chromatin bodies in malignant tumours of childhood. Lancet 1:55–58

Csomos K, Ujhazi B, Blazso P et al (2022) Partial RAG deficiency in humans induces dysregulated peripheral lymphocyte development and humoral tolerance defect with accumulation of T-bet$^+$ B cells. Nat Immunol 23:1256–1272

D'Angelo MA, Raices M, Panowski SH et al (2009) Age-dependent deterioration of nuclear pore complexes causes a loss of nuclear integrity in postmitotic cells. Cell 136:284–295

Defossez PA, Prusty R, Kaeberlein M et al (1999) Elimination of replication block protein Fob1 extends the life span of yeast mother cells. Mol Cell 3:447–455

Denoth-Lippuner A, Krzyzanowski MK, Stober C et al (2014) Role of SAGA in the asymmetric segregation of DNA circles during yeast ageing. elife 3:e03790

Dillon LW, Kumar P, Shibata Y et al (2015) Production of extrachromosomal microDNAs Is linked to mismatch repair pathways and transcriptional activity. Cell Rep 11:1749–1759

Eckhardt SG, Dai A, Davidson KK et al (1994) Induction of differentiation in HL60 cells by the reduction of extrachromosomally amplified c-myc. Proc Natl Acad Sci USA 91:6674–6678

Eugen-Olsen RAB, Hariprakash JM, Oestergaard VH, et al (2025) Molecular mechanisms of extrachromosomal circular DNA formation. Nucleic Acids Res 53:gkaf122

Falcón AA, Aris JP (2003) Plasmid accumulation reduces life span in Saccharomyces cerevisiae. J Biol Chem 278:41607–41617

Folzenlogen D, Hofer MF, Leung DY et al (1997) Analysis of CD80 and CD86 expression on peripheral blood B lymphocytes reveals increased expression of CD86 in lupus patients. Clin Immunol Immunopathol 83:199–204

Fraser HC, Kuan V, Johnen R et al (2022) Biological mechanisms of aging predict age-related disease co-occurrence in patients. Aging Cell 21:e13524

Fujimoto S, Yamagishi H (1987) Isolation of an excision product of T-cell receptor alpha-chain gene rearrangements. Nature 327:242–243

Fullgrabe J, Hajji N, Joseph B (2010) Cracking the death code: apoptosis-related histone modifications. Cell Death Differ 17:1238–1243

Fullwood MJ, Lee J, Lin L et al (2011) Next generation sequencing of apoptotic DNA breakpoints reveals association with actively transcribed genes and gene translocations. PLoS One 6:e26054

Gerovska D, Araúzo-Bravo MJ (2022) The common incidence-age multistep model of neurodegenerative diseases revisited: wider general age range of incidence corresponds to fewer disease steps. Cell Biosci 12:11

Gerovska D, Araúzo-Bravo MJ (2023a) Skeletal muscles of sedentary and physically active aged people have distinctive genic extrachromosomal circular DNA profiles. Int J Mol Sci 24:2736

Gerovska D, Araúzo-Bravo MJ (2023b) Systemic lupus erythematosus patients with DNASE1L3 deficiency have a distinctive and specific genic circular DNA profile in plasma. Cells 12:1061

Gerovska D, Irizar H, Otaegi D et al (2020) Genealogy of the neurodegenerative diseases based on a meta-analysis of age-stratified incidence data. Sci Rep 10:18923

Gerovska D, Fernández Moreno P, Zabala A et al (2023a) Cell-free genic extrachromosomal circular DNA profiles of DNase knockouts associated with systemic lupus erythematosus and relation with common fragile sites. Biomedicines 12:80

Gerovska D, Noer JB, Qin Y et al (2023b) A distinct circular DNA profile intersects with proteome changes in the genotoxic stress-related $hSOD1^{G93A}$ model of ALS. Cell Biosci 13:170

Giannini M, Bayona-Feliu A, Sproviero D et al (2020) TDP-43 mutations link amyotrophic lateral sclerosis with R-loop homeostasis and R loop-mediated DNA damage. PLoS Genet 16:e1009260

Glück S, Guey B, Gulen MF et al (2017) Innate immune sensing of cytosolic chromatin fragments through cGAS promotes senescence. Nat Cell Biol 19:1061–1070

Gooch CL, Shefner JM (2004) ALS surrogate markers. MUNE Amyotroph Lateral Scler Other Motor Neuron Disord 5:104–107

Gordon PH, Rowland LP, Younger DS et al (1997) Lymphoproliferative disorders and motor neuron disease: an update. Neurology 48:1671–1678

Grabarz A, Guirouilh-Barbat J, Barascu A et al (2013) A role for BLM in double-strand break repair pathway choice: prevention of CtIP/Mre11-mediated alternative nonhomologous end-joining. Cell Rep 5:21–28

Gulen MF, Samson N, Keller A et al (2023) cGAS-STING drives ageing-related inflammation and neurodegeneration. Nature 620:374–380

Gurney ME, Pu H, Chiu AY et al (1994) Motor neuron degeneration in mice that express a human Cu, Zn superoxide dismutase mutation. Science 264:1772–1775

Hando JC, Nath J, Tucker JD (1994) Sex chromosomes, micronuclei and aging in women. Chromosoma 103:186–192
Helmsauer K, Valieva ME, Ali S et al (2020) Enhancer hijacking determines extrachromosomal circular MYCN amplicon architecture in neuroblastoma. Nat Commun 11:5823
Henson JD, Cao Y, Huschtscha LI et al (2009) DNA C-circles are specific and quantifiable markers of alternative-lengthening-of-telomeres activity. Nat Biotechnol 27:1181–1185
Hintzsche H, Hemmann U, Poth A et al (2017) Fate of micronuclei and micronucleated cells. Mutat Res Rev Mutat Res 771:85–98
Hong X, Li J, Han P et al (2024) The extrachromosomal circular DNA atlas of aged and young mouse brains. Sci Data 11:318
Horvath S (2013) DNA methylation age of human tissues and cell types. Genome Biol 14:R115
Hotta Y, Bassel A (1965) Molecular size and circularity of DNA in cells of mammals and higher plants. Proc Natl Acad Sci USA 53:356–362
Hung KL, Yost KE, Xie L et al (2021) ecDNA hubs drive cooperative intermolecular oncogene expression. Nature 600:731–736
Ilić M, Zaalberg IC, Raaijmakers JA et al (2022) Life of double minutes: generation, maintenance, and elimination. Chromosoma 131:107–125
Khalil M, Teunissen CE, Otto M et al (2018) Neurofilaments as biomarkers in neurological disorders. Nat Rev Neurol 14:577–589
Kim H, Nguyen NP, Turner K et al (2020) Extrachromosomal DNA is associated with oncogene amplification and poor outcome across multiple cancers. Nat Genet 52:891–897
Kim J, Huang AY, Johnson SL et al (2022) Prevalence and mechanisms of somatic deletions in single human neurons during normal aging and in DNA repair disorders. Nat Commun 13:5918
Kinzina ED, Podolskiy DI, Dmitriev SE et al (2019) Patterns of aging biomarkers, mortality, and damaging mutations illuminate the beginning of aging and causes of early-life mortality. Cell Rep 29:4276–4284.e3
Kirkwood TB, Austad SN (2000) Why do we age? Nature 408:233–238
Koche RP, Rodriguez-Fos E, Helmsauer K et al (2020) Extrachromosomal circular DNA drives oncogenic genome remodeling in neuroblastoma. Nat Genet 52:29–34
Kumar P, Dillon LW, Shibata Y et al (2017) Normal and cancerous tissues release extrachromosomal circular DNA (eccDNA) into the circulation. Mol Cancer 15:1197–1205
Larsen BD, Benada J, Yung PYK et al (2022) Cancer cells use self-inflicted DNA breaks to evade growth limits imposed by genotoxic stress. Science 376:476–483
Laverde EE, Lai Y, Leng F et al (2020) R-loops promote trinucleotide repeat deletion through DNA base excision repair enzymatic activities. J Biol Chem 295:13902–13913
Lee CY, McNerney C, Ma K et al (2020) R-loop induced G-quadruplex in non-template promotes transcription by successive R-loop formation. Nat Commun 11:3392
Lenormand C, Lipsker D (2021) Somatic mutations in "benign" disease. N Engl J Med 385:e34
Liang X, Arrey G, Qin Y, et al (2025) EccDNA atlas in male mice reveals features protecting genes against transcriptioninduced eccDNA formation. Nat Commun 16:1872
Lin MS, Jo SY, Luebeck J et al (2024) Transcriptional immune suppression and up-regulation of double-stranded DNA damage and repair repertoires in ecDNA-containing tumors. elife 12:RP88895
Ling X, Han Y, Meng J et al (2021) Small extrachromosomal circular DNA (eccDNA): major functions in evolution and cancer. Mol Cancer 20:113
Lo YM, Corbetta N, Chamberlain PF et al (1997) Presence of fetal DNA in maternal plasma and serum. Lancet 350:485–487
Lo YMD, Han DSC, Jiang P et al (2021) Epigenetics, fragmentomics, and topology of cell-free DNA in liquid biopsies. Science 372:eaaw3616
Lodato MA, Walsh CA (2019) Genome aging: somatic mutation in the brain links age-related decline with disease and nominates pathogenic mechanisms. Hum Mol Genet 28:R197–R206
Lodato MA, Rodin RE, Bohrson CL et al (2018) Aging and neurodegeneration are associated with increased mutations in single human neurons. Science 359:555–559

López-Otín C, Blasco MA, Partridge L et al (2013) The hallmarks of aging. Cell 153:1194–1217
Luebeck J, Ng AWT, Galipeau PC et al (2023) Extrachromosomal DNA in the cancerous transformation of Barrett's oesophagus. Nature 616:798–805
Luquette LJ, Miller MB, Zhou Z et al (2022) Single-cell genome sequencing of human neurons identifies somatic point mutation and indel enrichment in regulatory elements. Nat Genet 54:1564–1571
Lv W, Pan X, Han P et al (2022) Circle-Seq reveals genomic and disease-specific hallmarks in urinary cell-free extrachromosomal circular DNAs. Clin Transl Med 12:e817
Mandel P, Metais P (1948) Nuclear acids in human blood plasma. C R Seances Soc Biol Fil 142:241–243
Mattson MP (2000) Apoptosis in neurodegenerative disorders. Nat Rev Mol Cell Biol 1:120–129
Mazzucco G, Huda A, Galli M et al (2020) Telomere damage induces internal loops that generate telomeric circles. Nat Commun 11:5297
McConnell MJ, Moran JV, Abyzov A et al (2017) Intersection of diverse neuronal genomes and neuropsychiatric disease: the brain somatic mosaicism network. Science 356:eaal1641
McCormack WT, Tjoelker LW, Carlson LM et al (1989) Chicken IgL gene rearrangement involves deletion of a circular episome and addition of single nonrandom nucleotides to both coding segments. Cell 56:785–791
Meinema AC, Marzelliusardottir A, Mirkovic M et al (2022) DNA circles promote yeast ageing in part through stimulating the reorganization of nuclear pore complexes. elife 11:e71196
Miller MB, Huang AY, Kim J, et al (2022) Somatic genomic changes in single Alzheimer's disease neurons. Nature 604:714–722
Mitra J, Guerrero EN, Hegde PM et al (2019) Motor neuron disease-associated loss of nuclear TDP-43 is linked to DNA double-strand break repair defects. Proc Natl Acad Sci USA 116:4696–4705
Møller HD, Mohiyuddin M, Prada-Luengo I et al (2018) Circular DNA elements of chromosomal origin are common in healthy human somatic tissue. Nat Commun 9:1069
Morton AR, Dogan-Artun N, Faber ZJ et al (2019) Functional enhancers shape extrachromosomal oncogene amplifications. Cell 179:1330–1341.e13
Muraro PA, Douek DC, Packer A et al (2005) Thymic output generates a new and diverse TCR repertoire after autologous stem cell transplantation in multiple sclerosis patients. J Exp Med 201:805–816
Mustjoki S, Young NS (2021) Somatic mutations in "benign" disease. N Engl J Med 384:2039–2052
Nassiri F, Chakravarthy A, Feng S et al (2020) Detection and discrimination of intracranial tumors using plasma cell-free DNA methylomes. Nat Med 26:1044–1047
Nathanson DA, Gini B, Mottahedeh J et al (2014) Targeted therapy resistance mediated by dynamic regulation of extrachromosomal mutant EGFR DNA. Science 343:72–76
Neil AJ, Liang MU, Khristich AN et al (2018) RNA-DNA hybrids promote the expansion of Friedreich's ataxia (GAA)n repeats via break-induced replication. Nucleic Acids Res 46:3487–3497
Nikolaev S, Santoni F, Garieri M et al (2014) Extrachromosomal driver mutations in glioblastoma and low-grade glioma. Nat Commun 5:5690
Novo N, Romero-Tamayo S, Marcuello C et al (2022) Beyond a platform protein for the degradosome assembly: the apoptosis-inducing factor as an efficient nuclease involved in chromatinolysis. PNAS Nexus 2:pgac312
Olafsson S, Anderson CA (2021) Somatic mutations provide important and unique insights into the biology of complex diseases. Trends Genet 37:872–881
Olecka M, van Bömmel A, Best L et al (2024) Nonlinear DNA methylation trajectories in aging male mice. Nat Commun 15:3074
Papaemmanuil E, Rapado I, Li Y et al (2014) RAG-mediated recombination is the predominant driver of oncogenic rearrangement in ETV6-RUNX1 acute lymphoblastic leukemia. Nat Genet 46:116–125

Paulsen T, Shibata Y, Kumar P (2019) Small extrachromosomal circular DNAs, microDNA, produce short regulatory RNAs that suppress gene expression independent of canonical promoters. Nucleic Acids Res 47:4586–4596

Paulsen T, Malapati P, Shibata Y (2021) MicroDNA levels are dependent on MMEJ, repressed by c-NHEJ pathway, and stimulated by DNA damage. Nucleic Acids Res 49:11787–11799

Pavlovic I, Zjukovskaja C, Nazir FH et al (2024) Cerebrospinal fluid mtDNA concentrations are increased in multiple sclerosis and were normalized after intervention with autologous hematopoietic stem cell transplantation. Mult Scler Relat Disord 84:105482

Poon LL, Leung TN, Lau TK et al (2002) Differential DNA methylation between fetus and mother as a strategy for detecting fetal DNA in maternal plasma. Clin Chem 48:35–41

Redon R, Ishikawa S, Fitch KR et al (2006) Global variation in copy number in the human genome. Nature 444:444–454

Rodin RE, Dou Y, Kwon M et al (2021) The landscape of somatic mutation in cerebral cortex of autistic and neurotypical individuals revealed by ultra-deep whole-genome sequencing. Nat Neurosci 24:176–185

Rose JC, Belk JA, Wong IT et al (2025) Disparate pathways for extrachromosomal DNA biogenesis and genomic DNA repair. Cancer Discov 15:69–82

Rostami A, Lambie M, Yu CW et al (2020) Senescence, necrosis, and apoptosis govern circulating cell-free DNA release kinetics. Cell Rep 31:107830

Rouillard AD, Gundersen GW, Fernandez NF et al (2016) The harmonizome: a collection of processed datasets gathered to serve and mine knowledge about genes and proteins. Database. (Oxford):baw100 2016

Salvi JS, Mekhail K (2015) R-loops highlight the nucleus in ALS. Nucleus 6:23–29

Sato S, Yamamoto K, Matsushita T (2015) Copy number variations in multiple sclerosis and neuromyelitis optica. Ann Neurol 78:762–774

Sau D, De Biasi S, Vitellaro-Zuccarello L et al (2007) Mutation of SOD1 in ALS: a gain of a loss of function. Hum Mol Genet 16:1604–1618

Savas JN, Toyama BH, Xu T et al (2012) Extremely long-lived nuclear pore proteins in the rat brain. Science 335:942

Scharer CD, Blalock EL, Mi T et al (2019) Epigenetic programming underpins B cell dysfunction in human SLE. Nat Immunol 20:1071–1082

Schmauck-Medina T, Molière A, Lautrup S et al (2022) New hallmarks of ageing: a 2022 Copenhagen ageing meeting summary. Aging (Albany NY) 14:6829–6839

Schütz E, Fischer A, Beck J et al (2017) Graft-derived cell-free DNA, a noninvasive early rejection and graft damage marker in liver transplantation: a prospective, observational, multicenter cohort study. PLoS Med 14:e1002286

Seluanov A, Mittelman D, Pereira-Smith OM et al (2004) DNA end joining becomes less efficient and more error-prone during cellular senescence. Proc Natl Acad Sci USA 101:7624–7629

Shamanna RA, Lu H, de Freitas JK et al (2016) WRN regulates pathway choice between classical and alternative non-homologous end joining. Nat Commun 7:13785

Shibata Y, Kumar P, Layer R et al (2012) Extrachromosomal microDNAs and chromosomal microdeletions in normal tissues. Science 336:82–86

Shimizu N, Kanda T, Wahl GM (1996) Selective capture of acentric fragments by micronuclei provides a rapid method for purifying extrachromosomally amplified DNA. Nat Genet 12:65–71

Shimizu N, Itoh N, Utiyama H et al (1998) Selective entrapment of extrachromosomally amplified DNA by nuclear budding and micronucleation during S phase. J Cell Biol 140:1307–1320

Shoshani O, Brunner SF, Yaeger R et al (2021) Chromothripsis drives the evolution of gene amplification in cancer. Nature 591:137–141

Sin STK, Jiang P, Deng J et al (2020) Identification and characterization of extrachromosomal circular DNA in maternal plasma. Proc Natl Acad Sci USA 117:1658–1665

Sin ST, Deng J, Ji L et al (2022) Effects of nucleases on cell-free extrachromosomal circular DNA. JCI Insight 7:e156070

Sinclair DA, Guarente L (1997) Extrachromosomal rDNA circles—a cause of aging in yeast. Cell 91:1033–1042

Sisirak V, Sally B, D'Agati V et al (2016) Digestion of chromatin in apoptotic cell microparticles prevents autoimmunity. Cell 166:88–101

Skourti-Stathaki K, Proudfoot NJ, Gromak N (2011) Human senataxin resolves RNA/DNA hybrids formed at transcriptional pause sites to promote Xrn2-dependent termination. Mol Cell 42:794–805

Smith CA, Vinograd J (1972) Small polydisperse circular DNA of HeLa cells. J Mol Biol 69:163–178

Smith DI, Zhu Y, McAvoy S et al (2006) Common fragile sites, extremely large genes, neural development and cancer. Cancer Lett 232:48–57

Soheili-Nezhad S, Ibáñez-Solé O, Izeta A et al (2024) Time is ticking faster for long genes in aging. Trends Genet 40:299–312

Song K, Minami JK, Huang A et al (2022) Plasticity of extrachromosomal and intrachromosomal BRAF amplifications in overcoming targeted therapy dosage challenges. Cancer Discov 12:1046–1069

Spriggs AI, Boddington MM, Clarke CM (1962) Chromosomes of human cancer cells. Br Med J 2:1431–1435

Stabach PR, Sims D, Gomez-Bañuelos E et al (2024) A dual-acting DNASE1/DNASE1L3 biologic prevents autoimmunity and death in genetic and induced lupus models. JCI Insight 9:e177003

Sun K, Jiang P, Chan KC et al (2015) Plasma DNA tissue mapping by genome-wide methylation sequencing for noninvasive prenatal, cancer, and transplantation assessments. Proc Natl Acad Sci USA 112:E5503–E5512

Sun K, Jiang P, Cheng SH (2019) Orientation-aware plasma cell-free DNA fragmentation analysis in open chromatin regions informs tissue of origin. Genome Res 2:418–427

Tan J, Wang X, Phoon L et al (2020) Resolution of ROS-induced G-quadruplexes and R-loops at transcriptionally active sites is dependent on BLM helicase. FEBS Lett 594:1359–1367

Tanaka S, Ise W, Inoue T et al (2020) Tet2 and Tet3 in B cells are required to repress CD86 and prevent autoimmunity. Nat Immunol 21:950–961

Taylor AMR, Rothblum-Oviatt C, Ellis NA et al (2019) Chromosome instability syndromes. Nat Rev Dis Primers 5:64

Terao C, Suzuki A, Momozawa Y et al (2020) Chromosomal alterations among age-related haematopoietic clones in Japan. Nature 584:130–135

Toledo F, Le Roscouet D, Buttin G et al (1992) Co-amplified markers alternate in megabase long chromosomal inverted repeats and cluster independently in interphase nuclei at early steps of mammalian gene amplification. EMBO J 11:2665–2673

Tomaska L, Nosek J, Makhov AM et al (2000) Extragenomic double-stranded DNA circles in yeast with linear mitochondrial genomes: potential involvement in telomere maintenance. Nucleic Acids Res 28:4479–4487

Tomaska L, Nosek J, Kramara J et al (2009) Telomeric circles: universal players in telomere maintenance? Nat Struct Mol Biol 16:1010–1015

Tresini M, Warmerdam DO, Kolovos P et al (2015) The core spliceosome as target and effector of non-canonical ATM signalling. Nature 523:53–58

Tsang CK, Liu Y, Thomas J et al (2014) Superoxide dismutase 1 acts as a nuclear transcription factor to regulate oxidative stress resistance. Nat Commun 5:3446

Tucker JD, Nath J, Hando JC (1996) Activation status of the X chromosome in human micronucleated lymphocytes. Hum Genet 97:471–475

Turner KM, Deshpande V, Beyter D et al (2017) Extrachromosomal oncogene amplification drives tumour evolution and genetic heterogeneity. Nature 543:122–125

Von Hoff DD, Waddelow T, Forseth B et al (1991) Hydroxyurea accelerates loss of extrachromosomally amplified genes from tumor cells. Cancer Res 51:6273–6279

Von Schwedler U, Jäck HM, Wabl M (1990) Circular DNA is a product of the immunoglobulin class switch rearrangement. Nature 345:452–456

Walter JE, Rosen LB, Csomos K et al (2015) Broad-spectrum antibodies against self-antigens and cytokines in RAG deficiency. J Clin Invest 125:4135–4148
Wang WY, Pan L, Su SC et al (2013) Interaction of FUS and HDAC1 regulates DNA damage response and repair in neurons. Nat Neurosci 16:1383–1391
Wang H, Dharmalingam P, Vasquez V et al (2017) Chronic oxidative damage together with genome repair deficiency in the neurons is a double whammy for neurodegeneration: is damage response signaling a potential therapeutic target? Mech Ageing Dev 161:163–176
Wang H, Kodavati M, Britz GW et al (2021a) DNA damage and repair deficiency in ALS/FTD-associated neurodegeneration: from molecular mechanisms to therapeutic implication. Front Mol Neurosci 14:784361
Wang Y, Wang M, Djekidel MN et al (2021b) eccDNAs are apoptotic products with high innate immunostimulatory activity. Nature 599:308–314
Weideman AM, Tapia-Maltos MA, Johnson K et al (2017) Meta-analysis of the age-dependent efficacy of multiple sclerosis treatments. Front Neurol 8:577
Winter DB, Gearhart PJ, Bohr VA (1998) Homogeneous rate of degradation of nuclear DNA during apoptosis. Nucleic Acids Res 26:4422–4425
Wu S, Turner KM, Nguyen N et al (2019) Circular ecDNA promotes accessible chromatin and high oncogene expression. Nature 575:699–703
Yang W, Tang H, Zhang Y et al (2013) Meta-analysis followed by replication identifies loci in or near CDKN1B, TET3, CD80, DRAM1, and ARID5B as associated with systemic lupus erythematosus in Asians. Am J Hum Genet 92:41–51
Yang L, Jia R, Ge T et al (2022) Extrachromosomal circular DNA: biogenesis, structure, functions and diseases. Signal Transduct Target Ther 7:342
Yang JH, Hayano M, Griffin PT et al (2023) Loss of epigenetic information as a cause of mammalian aging. Cell 186:305–326.e27
Yi E, Chamorro González R, Henssen AG et al (2022a) Extrachromosomal DNA amplifications in cancer. Nat Rev Genet 23:760–771
Yi E, Gujar AD, Guthrie M et al (2022b) Live-cell imaging shows uneven segregation of extrachromosomal DNA elements and transcriptionally active extrachromosomal DNA hubs in cancer. Cancer Discov 12:468–483
Yu K, Chedin F, Hsieh CL et al (2003) R-loops at immunoglobulin class switch regions in the chromosomes of stimulated B cells. Nat Immunol 4:442–451
Yu SC, Chan KC, Zheng YW et al (2014) Size-based molecular diagnostics using plasma DNA for noninvasive prenatal testing. Proc Natl Acad Sci USA 111:8583–8588
Zhao Y, Yu L, Zhang S et al (2022) Extrachromosomal circular DNA: current status and future prospects. elife 11:e81412
Zhu J, Zhang F, Du M et al (2017) Molecular characterization of cell-free eccDNAs in human plasma. Sci Rep 7:10968
Zhu Y, Gujar AD, Wong CH et al (2021) Oncogenic extrachromosomal DNA functions as mobile enhancers to globally amplify chromosomal transcription. Cancer Cell 39:694–707.e7
Zhu Y, Liu Z, Guo Y et al (2022) Whole-genome sequencing of extrachromosomal circular DNA of cerebrospinal fluid of medulloblastoma. Front Oncol 12:934159

Epigenetic Regulation Mediating Immunopathology in Parkinson's Disease

Sabiya Khan and Dharmendra Kumar Khatri

Contents

Abstract The pathogenesis of neurodegenerative disorders such as Parkinson's disease and other similar conditions is significantly influenced by epigenetics. The epigenetic pathway encompasses a wide range of processes, such as the alteration of histones, the methylation of DNA, and the regulation of non-coding RNA (ncRNA). In the central nervous system (CNS), the immune system is an essential component in the regulation of neuroinflammatory processes. In addition, the intersection of the immune system and epigenetic regulation is becoming an intriguing subject for researchers to investigate in order to gain a better knowledge of the pathogenesis of disease. The involvement of such occurrences in the pathogenesis of Parkinson's disease is investigated in detail in this chapter.

Keywords Epigenetic · Parkinson's · Histone modification · MiRNA

S. Khan
Department of Pharmacology and Toxicology, National Institute of Pharmaceutical Education & Research (NIPER), Hyderabad, Telangana, India

D. K. Khatri (✉)
Department of Pharmacology, NIMS Institute of Pharmacy, NIMS University Rajasthan, Jaipur, Rajasthan, India

J. Barciszewski (ed.), *Decoding Aging and Neurodegeneration*,
RNA Technologies 16, https://doi.org/10.1007/978-3-031-91072-2_13

1 Introduction

Parkinson's disease (PD) is a neurodegenerative movement disorder affecting the aged population. α-Synuclein (α-Syn), a presynaptic neuronal protein, aggregates and is found as Lewy bodies (LBs) within the cell body or within the neuronal processes as Lewy neurites in PD cases. It remains an unbeatable hallmark for PD since its discovery of PD correlation. Being a presynaptic protein, it is physiologically involved in synaptic vesicle transportation, neurotransmitter signaling, vesicular recycling, and regulating synaptic plasticity (Sulzer and Edwards 2019). The symptomology evolved from tremors, rigidity, and dyskinesia-like motor symptoms to non-motor symptoms like constipation, rapid eye movement, sleep disorder, anosmia, etc., generally preceding the motor symptoms. PD stages has been delineated based on brain regions and central nervous system (CNS) involvement and motor symptom progression, but the non-motor cardinal symptoms could not be correlated with these hypotheses (Braak et al. 2003). Therefore, we can say that although Lewy neurites and LBs are hallmark features of PD, the symptom progression cannot always be correlated with this pathology, suggesting a stronger participation from other pathological mediators. The complexity of PD pathology increased as the involvement of the gut-brain axis, and the peripheral nervous system evolved through inflammatory pathways and immune system participation in the pathological cascade (Kustrimovic et al. 2024).

The immune system protects the host against invading pathogens, indigenous pathological toxins, and unrecycled waste products. The innate immune response (via microglia, macrophage, monocytes, neutrophils, etc.) helps in the clearance of cellular debris, which, upon failure, activates the adaptive immune system (B- and T-cell), allowing inflammatory cascade activation at the site of action. If not appropriately suppressed, this protective action becomes detrimental. Astrocytes and microglia, in their reactive state, form the key immunogenic mediator in the brain. Coordinating peripheral immune complexes with CNS immunity regulates the overall disease exhibition and progression. Peripheral T-cells infiltrate the CNS and cause microglial activation, amplifying the neuroinflammatory cascade through cytokine and chemokine secretion. This may also be dependent on the blood-brain barrier (BBB) integrity, which is found disrupted in neurodegenerative conditions, allowing peripheral inflammatory cytokines into the CNS and aggravating the inflammation-mediated pathology in the brain. Despite all the studies in this direction for understanding the role of neuroinflammation in PD pathology, the initiation point of this vicious cycle is barely understood considering α-Syn as the main saboteur participating in pathology in coordination with other factors like oxidative damage, mitochondrial and lysosomal dysfunction, endoplasmic reticulum stress, aging, and environmental stressors (Neal and Richardson 2018).

Neurodegenerative conditions like PD have been linked to chronic low-grade inflammation within the CNS. Upon exposure to inflammatory stimuli, the immune cells produce inflammatory mediators like interleukin (IL)-1β, tumor necrosis factor-alpha (TNF-α), IL-18, etc., widely reported to participate in PD pathology through gliosis. Reactive (pro-inflammatory) microglia can induce the conversion of astrocytes to reactive form, which then affects the neurons (Liddelow et al. 2017). Mutation of genes encoding key proteins in PD was initially thought to be a primary cause of the disease, but it was relatable to only 5–10% of the familial PD cases, while the maximum PD cases were of sporadic form. The component regulating the protein expression, except their genetics, was a puzzle until the discovery of the role of epigenetics in such pathological conditions. Epigenetic regulation of these innate and adaptive immunities through trimethylation of histone 3 lysine 27 (H3K27) causes increased pro-inflammatory phenotype of microglia as well as macrophage (M1 response). On the contrary, H3K27 histone demethylase Jmjd3 is required for anti-inflammatory (M2) phenotype promotion in microglia (Tang et al. 2014). PD patients have been reported to have hypomethylation of TNF-α promoter sites, while lipopolysaccharide (LPS) treatment caused increased production of inflammatory factors due to TNF-α promoter region hypomethylation in animal models (Tang et al. 2014). The reported studies suggest the significant role of epigenetics in regulating the immune system in mediating inflammation-mediated PD pathology progression. These epigenetic components have the ability to control the activation and differentiation of microglia and T-cells, directly influencing the ferocity of the pathological progression of the disease. Hence, we can say that epigenetic regulation will control the phenotypic form of the macrophage/microglia expressed upon pathogenic stimuli and, thus, the immune response in PD pathology progression. In the coming sections, we tried to elaboratively discuss these epigenetic regulators and the role of the immune system in PD pathology.

2 Immune System in Parkinson's Disease

The brain is a specialized organ initially thought to be immune-privileged but has evolved as immune-specialized with its specialized immune cells. Aberrant functioning of the immune system has been strongly linked to PD progression. The pathological complexity of PD affects multiple systems with significant participation by neuroinflammation and immune responses, causing early-stage non-motor symptoms like sleep apnea and gastrointestinal dysfunction. Immunosenescence is characterized by age-dependent immunodeficiency and inflammageing. Aging causes competence loss in the innate and adaptive immune responses in PD, affecting the levels of circulatory cytokines and inflammatory markers. Therefore, we can say that neuroinflammation, besides aging, is an essential factor contributing to PD pathogenesis (Calabrese et al. 2018). The BBB protects the integrity of the CNS. Later, the significance of glial cells and monocyte-mediated dysfunctional immune system activation contributing to PD pathology aggravation came to light. Prolonged microglia and astrocyte activation are responsible for BBB integrity loss and act as a connection pathway between the peripheral nervous system and CNS (Tansey et al. 2022). Therefore, understanding the role of immune system activation in neuroinflammation and linked PD pathology becomes crucial.

The first level of physiological defense comes through the innate immune system. In the brain, microglia (brain tissue-resident macrophage) are responsible for providing immune defense to neurons and are found densely populated in substantia nigra pars compacta (SNPc) and striatum regions of the brain (McGeer et al. 1988). Macrophage and dendritic cells are referred to as antigen-presenting cells as they can sense any damage stimuli and express major histocompatibility complex (MHC) molecules on their surfaces, which is detected by other immune cells, and amplification of inflammatory response occurs (Mildner and Jung 2014). T-cells represent the main component of the adaptive immune response in PD pathology reported to participate in chronic inflammation. Other than these cells, natural killer cells, neutrophils, and B-cells, although not well explored, are reported to participate in neuroinflammation mediation in PD.

Microglia, the brain-resident immune cells of the CNS, play a vital role in brain homeostasis sustenance. Glial activation is the most prominent and earliest feature of neuroinflammation in PD. Microglial activation happens through a range of pattern recognition receptors abundantly expressed on glial cells and other immune cells like astrocytes, monocytes, and macrophages (Tansey et al. 2022). The damage stimuli may be categorized as physiological damage signals (damage-associated molecular patterns (DAMPs), e.g., prions, misfolded proteins, damaged organelle, etc.) or pathological external stimuli (pathogen-associated molecular patterns (PAMPs), e.g., lipopolysaccharide, infectious agents, viral or bacterial cell content, etc.). Toll-like receptors (TLR), NOD-like receptors (NLR), and AIM2-like receptors are the major receptors involved in pathological interplay in PD (Marogianni et al. 2020). Activated microglia secretes inflammatory mediators like IL-6, TNF-α, cyclooxygenase 2 (COX-2), nitric oxide synthase 2, reactive oxygen species (ROS), and IL-1β, besides causing excessive ROS generation. They are responsible for phagocytosing the damaged neurons and cellular products, including debris, but a chronic activation in a pathological state causes a vicious inflammatory cycle due to prolonged conversion of the M0 state to M1 pro-inflammatory phenotype leading to neurodegeneration.

The innate immune system coordinates adaptive immunity through antigen-presenting cells expressing MHCs and T-cell activation. At a physiological activation level, a neuroprotective effect is exerted by microglia by secreting these inflammatory markers through its phagocytic mechanism, but over-activation or chronic activation for a prolonged period of time is harmful to the healthy neurons (Choi et al. 2020). Chronic inflammation generally results from disturbed homeostasis between pro-inflammatory and anti-inflammatory responses by macrophages, which makes the protective defense system detrimental to healthy cells. PD patients have an altered expression level of TLR2, TLR4, NLRs, IL-18, IL-1β, TNF-α, and several other inflammatory mediators, which validates the role of innate immune response in PD (Wang et al. 2024a, b). The damage to dopaminergic neurons in PD due to the LB aggregates and other factors like ROS instigate the inflammatory response, leading to pro-inflammatory markers' secretion into the cerebrospinal fluid (CSF) (TGF-β) and peripheral blood (IL-1β, IL-6) (Imamura et al. 2005). Another study reported that soluble TNF-α signaling inhibition rescued dopaminergic neuron death in several animal experiments (Elyaman and Khoury 2017). IL-9 is an essential inflammatory marker that has been found dysregulated in PD patients' serum samples (Deleidi and Gasser 2013). Once the pathogenic stimuli have been identified by macrophage cells, lymphocytes become activated. The CD4 T-cells differentiate into pro-inflammatory (Th1, Th2, Th17) or anti-inflammatory (Treg) subtypes depending on the stimuli, secreting IFN-γ and IL-17 pro-inflammatory cytokines. On the other hand, CD8 cells differentiate into cytotoxic T-cells promoting a pro-inflammatory response, causing an increased IFN-γ, TNF, and other inflammatory cells. This cell can directly attack the affected neuron, and its participation is reported to occur in the later stages of PD (Tan et al. 2020). Treg cells are also reported to be reduced in PD cases, which are majorly responsible for maintaining cellular homeostasis by suppressing excessive immune responses. The B-cells are responsible for producing antibodies against specific antigens, but their role in PD pathology is not widely explored. As α-Syn accumulation is not limited to CNS, inflammatory signaling is also found far and beyond the CNS, specifically in the blood, gastrointestinal tract, enteric nervous system, and the gut microbiome cross-talk (Terkelsen et al. 2022). Like microglia, macrophages get activated peripherally into an M1 pro-inflammatory form due to peripheral protein aggregates and cause intensification of the chronic inflammatory pathological cascade in the CNS. Clinically, PD cases have decreased lymphocyte count and increased neutrophil in blood nearly 10–20 years before diagnosis (Craig et al. 2021). The T-cell and B-cell reactivity, along with anti-α-Syn antibodies, were elevated in prodromal PD cases (Lindestam Arlehamn et al. 2020). Senescence has been reported to cause significant functionality and plasticity loss in microglia cells, leading to an aggravated pathological state. It leads to the generation of dystrophic microglia with increased fragmentation and beading of their branches (Heavener and Bradshaw 2022).

Several genes have been linked to immune dysregulation in PD. In PD cases, human leukocyte antigen (HLA) region and cathepsin B (coding for lysosomal protease) genes are found to be enriched in microglia, altering the antigen presentation to T-cells and, hence, influencing T-cell polarization (Chang et al. 2017). HLA-DRB1 (coding for MHC class II for APC) aids in antigen presentation to CD4+ and CD8+ T-cells, which is important for disease susceptibility and was found to be an important risk factor for PD (Hollenbach et al. 2019). Monocytes from PD patients show higher expression of leucine-rich repeat kinase 2 (LRRK2) (Cook et al. 2017). An animal PD model overexpressing full-length α-Syn showed robust infiltration of C-C chemokine receptor 2 (CCK2) into the substantia nigra region of mice leading to dopaminergic neuronal death (Harms et al. 2018). Another important lysosomal protein-encoding gene, GBA1 (encoding lysosomal acid glucocerebrosidase), is found mutated in PD patients, altering the lysosomal degradation of α-Syn by monocytes (Hughes et al. 2021). Parkin and Pink1 expression have shown suppression of MHC class I molecule-mediated antigen presentation to T-cells (Matheoud et al. 2016). These studies signify the role of mutation of these genes in generating detrimental immune cell responses in pathological conditions like PD, although, besides mutation, epigenetic regulation of these genes has started to evolve as an important pathological factor in PD.

3 Epigenetic Modulations in Parkinson's Disease Pathology

Besides genetic mutations, epigenetic regulation (defined as the processes influencing gene expression without affecting the DNA sequence) has started to evolve as a key mediator for PD pathological progression by directly influencing mRNA expression and thus protein expression. Except for DNA, there are other components in the nucleus which are equally important for genetic regulation. DNA is tightly wrapped by histone octamers (two copies of each H2A, H2B, H3, and H4). This wrapping ensures strict regulation of DNA regions necessary for transcriptional and translational activities while keeping other regions inaccessible (Grunstein 1997). The histone entity along with its post-translational modification, DNA methylation, and microRNA-mediated regulation forms the basis of epigenetic regulation in eukaryotic cells (Miller and Grant 2013). Epigenetics is evolving as an important disease management tool for its diagnostic and therapeutic potency. In the case of PD also, as previously discussed, the genetic component plays a vital pathological role, and, hence, understanding the complex interplay between genetic and epigenetic regulation on disease progression will be helpful. Several key factors like stress, environmental cues like toxicants or heavy metals, trauma, physical exercise, diet, or other underlying conditions are reported to mediate the PD pathology through epigenetic regulation (Miller and Grant 2013). Hence, it is critically important to understand the epigenetic interplay in PD.

3.1 Epigenetic Dysregulation of Astrocyte in PD

Conventionally, PD pathology has been exclusively linked to dopaminergic neuronal degeneration in the substantia nigra region of the midbrain. Lately, researchers have started to realize the significance of the etiology and pathogenesis of PD from the perspective of other important phenomena, often neglected, including the activation and proliferation of brain-resident glial cells. Astrocytes, upon pathological activation, can produce TNF-α, NO, IL-1β, prostaglandin E2 besides other inflammatory markers. The downstream pathways regulated by these inflammatory mediators eventually lead to death and degeneration of dopaminergic neurons. TNF-α binds to TNF-receptor, which causes dopaminergic neuron degeneration by activating the intracellular ceramide C2 pathway. IL-1β is also found to be overexpressed in pathological states and can influence the activity of astrocytes and other glial cells. 1-Methyl-4-phenyl-1,2,3,6-tetrahydropyridine (MPTP) is metabolized to pyridinium ion (MPP^+) majorly at the astrocyte site. S100β is a Ca^{+2}-binding protein found to be highly expressed in injured brain cells, and its expression was found to increase in the astrocytes upon exposure of mice glial cells to MPTP-induced brain injury, leading to dopaminergic neuron injury and aggravating PD pathology. Hence, it can be said that astrocytes play a significant role in PD pathology.

3.1.1 DNA Methylation

Epigenetic regulation of astrocytes, particularly through DNA methylation, has primarily been studied in the context of astrocyte differentiation during development. Although these studies may not directly address neurodegeneration, they offer insights into potential epigenetic mechanisms that could be relevant in neurodegenerative diseases. In the developing brain, DNA methylation plays a critical role in regulating the differentiation of neural precursor cells (NPCs) into astrocytes. For example, the promoter of the astrocyte-specific gene GFAP (glial fibrillary acidic protein) is initially methylated in early-stage NPCs, preventing its expression. However, during later stages of NPC development, this cytosine is demethylated, enabling the activation of STAT3 by leukemia inhibitory factor, which drives GFAP expression and astrocyte differentiation (Fan et al. 2005). GFAP was found elevated in PD plasma samples of patients with tremor and postural instability-gait disturbances depending on the subtype (Lin et al. 2023, Che et al. 2024). Interestingly, it has been found that hypomethylation of the GFAP promoter alone is insufficient to induce gene expression, indicating that histone modifications also regulate GFAP expression. This highlights the importance of chromatin accessibility in controlling gene expression during astrocyte differentiation (Urayama et al. 2013).

Inhibition of DNA methyltransferases (DNMTs) can induce astrocyte differentiation from NPCs (Kunoh et al. 2024). For instance, conditional knockout of DNMT1 in nestin-positive NPCs promotes the expression of astrocyte-specific genes such as GFAP and S100β and activates the Janus kinase-signal transducer and activator of transcription (JAK-STAT) pathway, which is crucial for astrocyte differentiation. Conversely, knockout of DNMT3a in nestin-positive cells impairs re-methylation of the GFAP promoter, suggesting DNA methylation's pivotal role in astrocyte differentiation (Fan et al. 2005). DNA methylation of astrocyte-specific genes may also play a role in neurodegenerative diseases. For example, the JAK-STAT pathway, which regulates reactive astrogliosis, is activated in response to neurodegenerative stress. In PD, activation of STAT3 in astrocytes precedes the upregulation of GFAP and reactive astrogliosis, suggesting that DNA methylation may influence astrocyte reactivity by modulating STAT3 signaling (Neal and Richardson 2018). JAK2/STAT3 pathway is capable of inducing astrocyte reactivity in the mouse model of Alzheimer's disease (Ceyzériat et al. 2018).

Further evidence suggests that DNA methylation in astrocytes influences neurodegeneration and inflammation. For instance, in PD, the peroxisome proliferator-activated receptor pathway has been shown to reduce DNMT expression in astrocytes, attenuating hypermethylation of CpG islands in the miR-181a gene, which protects against corticosterone-induced stress (Ji et al. 2017).

3.1.2 Histone Modification

Histone Methylation: Although it hasn't been thoroughly investigated in relation to CNS dysfunction, histone methylation in astrocytes is essential to astrocyte function. Numerous histone methyltransferases (HMTs) and demethylases, including Ezh2, Jmjd3, and Jmjd5, are expressed by astrocytes and influence H3K27me3 levels. Neural stem cell development into neurons and astrocytes results in downregulated Ezh2, which is responsible for H3K27me3. Ezh2 deletion speeds up astrocyte differentiation, but forced expression of Ezh2 in differentiated astrocytes partially returns them to a stem cell-like state (Sher et al. 2011). A mutation in Jmjd2c increases the number of GFAP-positive cells, which is necessary for astrocyte development. On the other hand, overexpression of the HMT ESET decreases differentiation, whereas its absence increases astrocyte production (Zhang et al. 2017). Additionally, CNS dysfunctions such as ischemia, viral infections, and malignancy are significantly influenced by histone methylation. For example, adult cortical astrocytes have lower H3K9me3 and higher H3K4me3 levels than middle-aged astrocytes following ischemia injury, suggesting altered gene transcription (Kumar et al. 2005). Increased H3K27me3 is associated with higher tumor grades, and histone methylation influences human immunodeficiency virus latency in astrocytes and is connected to the development of astroglial tumors. Through SET7/9-induced H3K4 methylation, the astrocyte protein S100β increases NF-κB inflammatory factors in inflammation (Li et al. 2008). A similar process in astrocytes may contribute to the increased inflammatory response in neurodegenerative disorders (NDDs), as inhibiting Ezh2 in microglia lowers LPS-induced inflammation (Arifuzzaman et al. 2017). Histone methylation in astrocytes is therefore essential for CNS development and dysfunction, suggesting its possible significance in the pathophysiology and management of neurodegenerative diseases.

Histone Acetylation: Several histone deacetylases (HDACs), including HDAC1-11 and the sirtuin family, are expressed by astrocytes and are essential for their differentiation. By acetylating H3K9 and H3K14, the histone acetyltransferase (HAT) p300 is required for astrocyte development (Cheng et al. 2011). While HDAC3 removal results in robust astrocyte differentiation, trichostatin A (TSA) inhibition of HDACs inhibits both neuronal and astrocyte differentiation (Marin-Husstege et al. 2002). In neurodegenerative disorders and CNS dysfunction, histone acetylation plays a crucial role. Histone acetylation is essential for astrocytes in PD to perform tasks such as filament organization, cell survival, and inflammatory response. The astrocyte filament network is reorganized, and GFAP expression is decreased by HDAC inhibition using sodium butyrate (Yang et al. 2014). By lowering neurotrophic factors and generating inflammatory cytokines, reactive astrocytes in PD contribute to neuroinflammation. LPS-activated microglia-conditioned medium reduces astrocytes' H3 and H4 acetylation, which raises cell death and lowers Nrf2 expression. Valproic acid and other HDAC inhibitors lessen these effects. By preventing STAT3 activation, suberoylanilide hydroxamic acid lowers the IFNγ-induced inflammatory response in astrocytes and shields neurons from inflammation-induced cell death (Hashioka et al. 2012). Neuroprotection is provided by increased histone acetylation caused by therapies such as dimethyl fumarate (DMF), which also increases Nrf2 expression and reduces cytokine-induced HDAC gene expression and activity. A byproduct of acetyl-CoA metabolism, sodium acetate, increases anti-inflammatory IL-4 and IL-10 and decreases pro-inflammatory TNF-α while hyperacetylating H3K9 (Soliman et al. 2012). In summary, histone acetylation modulates astrocyte differentiation and function, with increased acetylation reducing inflammatory responses and providing neuroprotection in neurodegenerative diseases like PD. Novel therapeutics targeting histone acetylation in astrocytes, such as DMF, could offer beneficial effects in these conditions.

3.1.3 MicroRNA

MicroRNAs (miRNAs) play a significant role in post-transcriptional regulation of gene expression by target mRNA base-pairing. They are capable of cutting off RNA of target protein and inhibiting gene translation and recombination by silencing or suppression of genetic translational processes. Numerous studies demonstrate how important miRNAs are in astrocytes, impacting inflammatory responses, differentiation, and other cellular processes. For example, Ezh2 (histone-lysine N-methyltransferase) is downregulated by miRNA-124 (miR-124), which favors neuronal over astrocyte development. Furthermore, oligodendrocyte differentiation is promoted by miR-16, miR-184, and miR-118 through their inhibition of the important astrocyte gene (GFAP) (Hamill et al. 2007). MiR-31 was reported to promote development and specification of astrocytes by Lin28-level reduction (stem cell factor implicated in renewal of neural precursor cell). Age- and region-specific miRNA expression is seen in astrocytes. Differentially expressed miRNAs in the SNPc region of the brain include miR-153, miR-223, miR-21, and miR-125b. White matter astrocytes contain more significant amounts of miRNAs linked to immune response and glutamate transport than fetal astrocytes, and pro-inflammatory miRNAs are more common in white matter than grey matter. Aging aids in dysregulated inflammatory miRNAs in astrocytes. MiR-335-3p was reported to be increased in primary murine astrocytes compared to the younger astrocytes. Hence, decreased levels of target HMCGS1 and SFR2 were observed along with decreased cholesterol (in astrocytes) and, therefore, decreased PSD95 protein (in neurons) levels.

MiRNAs have a key role in neurodegenerative illnesses such as PD. For instance, astrocytes exposed to cellular insult have decreased levels of miR-146a and 147b (regulating IL-6 and COX-2 in astrocytes), which results in a surge in inflammatory factors. Exosomes generated from astrocytes also alter miRNA levels, which impacts astrocyte activity and neural plasticity (Guo et al. 2020). MiR-7 is widely explored for slowing PD pathology by regulating neurogenesis, α-Syn production, and other PD-related genes. Overexpression of heme oxygenase 1 in astrocytes contributes to the pathophysiology of PD by decreasing miR-7 and increasing α-synuclein in the ventral tegmental region and substantia nigra (Song et al. 2012). These findings underscore the importance of miRNAs in regulating astrocyte function and their potential roles in neurodegenerative diseases. Overall, miRNAs are crucial regulators of astrocyte function in both health and disease. Further research on miRNA networks in specific neurodegenerative diseases could uncover novel therapeutic approaches.

3.2 *Epigenetic Alteration in Microglial PD Pathology*

Studies reveal a robust genetic association between PD and microglia genes. Based on their phenotypes, microglia display a variety of transcriptional patterns, highlighting their complexity and adaptability. In Parkinson's disease, microglia activity and phenotypes are strongly regulated by epigenetic processes. Different transcriptome signatures in microglia that sustain immunological responses and intracerebral homeostasis have been discovered by RNA sequencing. Although the epigenetic functions of microglia have been shown, the precise molecular pathways underlying Parkinson's disease are still unknown. Developing new treatments for Parkinson's disease requires an understanding of these mechanisms, which include non-coding RNAs, histone changes, and DNA methylation.

3.2.1 DNA Methylation

DNA methyltransferases (DNMT1, DNMT3A, and DNMT3B) are the main enzymes responsible for DNA methylation, an epigenetic alteration in which a methyl group is added to the C5 position of cytosine in CpG dinucleotides. In order to facilitate gene expression, this mechanism usually keeps CpG islands close to gene promoters unmethylated (Moore et al. 2013). By blocking the binding of TFs or polymerase, methylation can suppress the expression of genes. The pathophysiology of PD is influenced by aberrant DNA methylation in microglia, which is connected to alterations in gene expression, microglial activation, and pro-inflammatory cytokine secretion. For instance, in elderly mice, elevated IL-1β expression and persistent illness behavior are linked to hypomethylation of the IL-1β promoter in microglia (Matt et al. 2016). DNA methylation also affects TREM2 variations, which are risk factors for Parkinson's disease (Jay et al. 2017). These alterations affect microglial activity and imply that DNA methylation is a major factor in the neuroinflammatory processes of Parkinson's disease. To completely comprehend the precise genes and mechanisms involved, more in-depth research is required.

3.2.2 Histone Modifications

Histones are essential chromatin components that change chromatin shape, which in turn controls gene expression. Gene activation or repression is impacted by modifications including acetylation, methylation, and phosphorylation, which impact the transition between euchromatin and heterochromatin (Brown 2001). These changes are important in PD, especially histone acetylation and methylation, which are linked to microglial activation and neurodegeneration.

Histone Acetylation: The process of adding and removing acetyl groups from histone tails, known as histone acetylation, is essential for controlling the expression of certain genes. The connection between DNA and histones is weakened when lysine residues on histone tails are acetylated, neutralizing their positive charge. This structural change promotes gene transcription by loosening chromatin, which makes it easier for transcription factors (TFs) to bind (Gujral et al. 2020). On the other hand, the elimination of acetyl groups, or histone deacetylation, results in a more compact chromatin structure and suppresses transcription. HATs and HDACs are two enzyme groups that control the dynamic balance between acetylation and deacetylation. Acetylation homeostasis depends on these enzymes. Recent research has demonstrated the important function that histone acetylation plays in NDDs, namely, in controlling the activity of microglia. For example, in PD models, it has been demonstrated that blocking the HDAC enzyme SIRT2 with AGK2 lowers microglial activation, restores the acetylation-deacetylation balance, and protects dopaminergic neurons (Luthi-Carter et al. 2010). These results highlight the therapeutic potential of reestablishing histone acetylation equilibrium in NDDs and imply that disruption of histone acetylation fuels microglia-mediated pathogenesis in Parkinson's disease. Despite the focus on pan-HDAC inhibitors, the targeting of specific HDAC isoforms or TFs associated with NDDs remains underexplored. Further research is necessary to identify selective modulators of histone acetylation, which may offer new therapeutic approaches for treating NDDs.

Histone Methylation: HMTs are responsible for the important epigenetic alteration known as histone methylation, which controls gene expression by altering arginine and lysine residues. H3K4me3, or trimethylation of H3 at lysine 4, stimulates transcription, whereas H3K9me3 inhibits it. Microglial activation and inflammation are impacted by EZH2-mediated H3K27me3's inhibition of gene expression, including Socs3 (Sun et al. 2023). NeuroD1 may decrease activation by increasing H3K27me3 and decreasing H3K4me3 at microglial promoters. Dextromethorphan has anti-inflammatory properties by lowering H3K4me3 at TNF-α promoters. Jmjd3 demethylates histones, eliminating H3K27me3, which inhibits inflammation and microglial activation. Jmjd3 overexpression may encourage M2 polarization and guard against neurodegeneration, whereas its inhibition worsens neuronal loss in Parkinson's disease (Xu et al. 2023). In summary, histone methylation and demethylation are crucial for regulating microglial activity, influencing inflammatory responses and offering therapeutic potential in neurodegenerative diseases.

3.2.3 Non-coding RNA

Important modulators of gene expression and cellular processes are ncRNAs, which include circulating RNAs, long ncRNAs, and miRNAs. By breaking down target mRNAs, miRNAs, in particular, prevent gene transcription. Their dysregulation is linked to neurodegenerative diseases (NDDs), where it contributes to microglial activation and persistent neuroinflammation (Ling et al. 2013). MiR-124 is downregulated in PD, and restoring it has become a viable treatment strategy. Research indicates that in PD models, miR-124 restoration lowers inflammatory cytokine release, attenuates microglial activation via the MEKK3/NF-κB pathway, and lowers pro-inflammatory indicators (Periyasamy et al. 2018). Furthermore, introduction of miR-124 in microglial cells changes microglial activity and decreases inflammation, indicating that it may be able to reverse the pathophysiology of Parkinson's disease (Wang et al. 2024a, b). These findings highlight the therapeutic potential of modulating miRNA expression to regulate microglial activation and inflammation in NDDs. Accumulating evidence highlights the pivotal role of ncRNAs in modulating microglial function and their significant involvement in the pathophysiology of NDDs. These molecules participate in intricate molecular processes, influencing a broad array of gene networks. To develop effective therapeutic strategies for NDDs, further research is necessary to explore the specific targets and molecular mechanisms of miRNAs, particularly miR-124.

4 Clinical Advances in Epigenetic Avenue Relating to PD

Although genetic and environmental variables have been recognized as major contributors to the etiology of NDDs by continuous research, there are currently few effective treatments available. Instead of curing the condition, the main goal of current treatments is to delay its progression and symptomatic management. The importance of epigenetics in PD has been brought to light by recent research, raising the possibility of genetic and pharmacological treatments.

DNA methylase-targeting epigenetic medications have become attractive options for treating Parkinson's disease. They target the gene silencing mechanism, i.e., methylation at CpG islands of DNA molecule. DNMT inhibitors have the potential to counter abnormal DNA methylation and reprogram gene expression and cellular functions. Green tea's natural polyphenol, epigallocatechin-3-gallate (EGCG), has demonstrated neuroprotective properties in PD animals and finished a phase II clinical trial in 2009 (Singh et al. 2016). By reducing copper-induced ROS, EGCG mitigates α-synuclein toxicity, inhibits DNMT1, stops cytosine methylation, and lessens oxidative stress. When paired with particular lipids, it can penetrate the blood-brain barrier, increasing its potential for therapeutic use. Other DNMT inhibitors have demonstrated a promise, including quercetin, catechins, and derivatives of curcumin (Reuter et al. 2011). The existing medications such as procaine, procainamide, and hydralazine have been found to have a potential for repurposing in NDDs using high-throughput testing. As a potential future treatment option for Parkinson's disease, RG108 has shown DNA demethylation without causing harm in vitro (Zhang et al. 2023a, b). Nevertheless, the majority of DNMT inhibitors have not yet been tested in PD clinical studies.

Inhibitors of HDAC also exhibit notable neuroprotective benefits. Histone deacetylation through the HDAC enzyme leads to chromatin condensation and gene silencing, which is reported to affect the expression of several neuroprotective proteins in PD. Therefore, the evolution of HDAC inhibitors as a potential PD management alternative has gained attention recently. They have been reported to cause a wide spectrum of actions ranging from decreased neuroinflammation, mitochondrial dysfunction, and dopaminergic pathologies in the PD brain. An FDA-approved medication called sodium butyrate inhibits different kinds of HDAC, which causes hyperacetylation and promoter activation. In addition to enhancing the expression of neurotrophic factors like glial cell-derived neurotrophic factor and brain-derived neurotrophic factor, it also crosses the blood-brain barrier and shields dopaminergic neurons from oxidative stress and α-synuclein toxicity (Renani et al. 2019). By lowering α-synuclein toxicity and encouraging neurite outgrowth, TSA and LMK235, two other HDAC inhibitors, have demonstrated comparable neuroprotective benefits (Mazzocchi et al. 2021). Another possible strategy is to target HATs. By increasing HAT activity, CTPB, a small molecule p300/CBP activator, shields neuronal cells and stimulates neurite development. It has been demonstrated that HAT inhibitors like anacardic acid and garcinol can alleviate PD symptoms, including L-DOPA-induced dyskinesia (Hsu et al. 2020).

Therapeutic approaches based on miRNA are also being investigated. MiRNAs regulate the post-transcriptional gene expression, directly impacting the immune response, inflammation, and neuronal survival. Restoring or opposing particular miRNA targets is the goal of miRNA mimics (agomirs) or anti-miRNAs (antagomirs) (Zhang et al. 2023a, b). MiR-7 and miR-155 agomirs, for example, protect dopaminergic neurons by preventing the activation of the NLRP3 inflammasome (Zhou et al. 2016). MiR-425 and miR-29a mimics, on the other hand, decrease oxidative stress and increase protective enzyme activity. The significance of miRNAs in PD can be understood by widespread participation of miRNA regulating neuronal, astrocyte, and microglia function ultimately affecting PD pathology, as depicted in Fig. 1. In PD models, inhibition of miR-155 has been demonstrated to shield neurons from neurotoxicity (Dong et al. 2021). These developments in epigenetic medicines, such as HAT regulators, DNMT and HDAC inhibitors, and miRNA-based approaches, present encouraging paths toward the creation of novel treatments for Parkinson's disease and other neurodegenerative illnesses. More investigation is required to improve these strategies and confirm their clinical effectiveness. The pathological and therapeutic advancements are depicted in Fig. 2.

Besides therapeutic targeting of epigenetic regulators through inhibitors or agomir and antagomirs, advanced genome editing tools like clustered regularly interspaced short palindromic repeats/Cas9-based systems are now evolving as a direct modifier of the epigenome. Direct targeting of the α-Syn gene may help in abnormal protein expression and might aid in mitigating the toxic aggregate formation. This kind of technology might assist in developing tailor-made or targeted therapeutic alternatives for patients. Simultaneously, the use of epigenetic regulators as diagnostic tools has started to flourish immensely due to their promising results in the pathological state. The epigenetic markers are being studied in various biological fluids like blood, CSF, saliva, and sweat glands and even in peripheral blood mononuclear cells and brain tissue. Early diagnosis of PD might promise

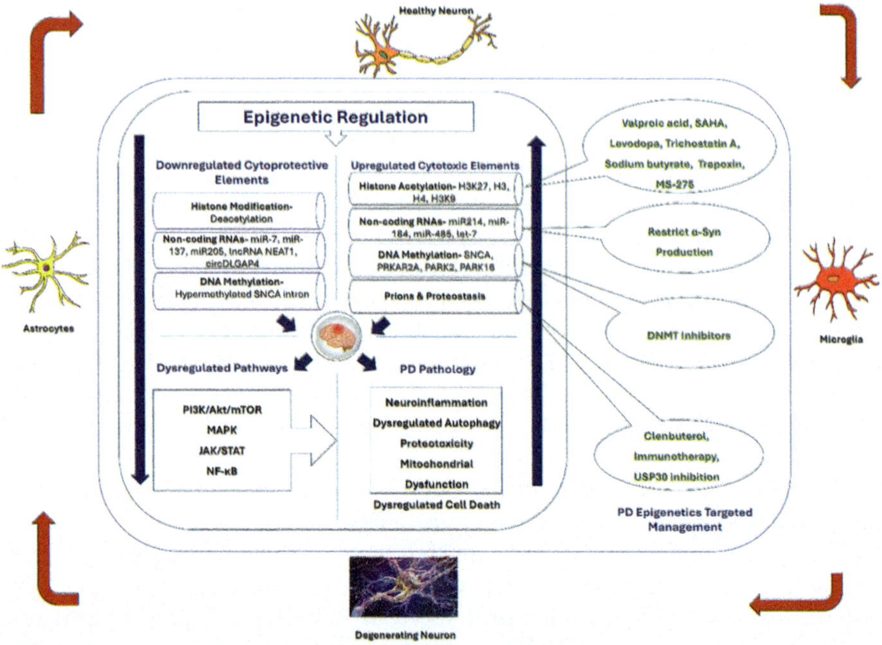

Fig. 1 Epigenetic regulation of Parkinson's disease: Epigenetic components regulate the function of astrocytes, microglia, and neurons in healthy and diseased conditions. In the PD pathological state, the neuroprotective elements like miR-7, NEAT1 regulators, SNCA regulators, etc. are downregulated. At the same time, the neurotoxic components are upregulated, causing aggravated disease conditions through the activation of several pathways leading to pathological processes activation. Several studies have reported different approaches to regulate these epigenetic regulators, like DNMT inhibitors, suppressors of α-Syn production, inhibiting histone acetylation, or directly acting on proteotoxic aggregates

effective disease management by ease in identifying the root underlying cause through the utilization of these advanced technologies. This will help prevent the progression of the disease to advanced stages, preventing further neuronal damage or severed motor impairment causing multiple pathologies substantially.

5 Future Prospectives and Conclusion

Since James Parkinson first identified PD in the early nineteenth century, researchers have continuously explored its etiology and pathogenesis. A substantial body of research has established a strong link between inflammatory responses, oxidative stress, and PD. Neuroinflammation, in particular, plays a crucial role in the initiation and progression of PD. Inflammation within the central nervous system activates glial cells and cytokine release, which can lead to the progressive degeneration of dopaminergic neurons in the substantia nigra and striatum. This pathological

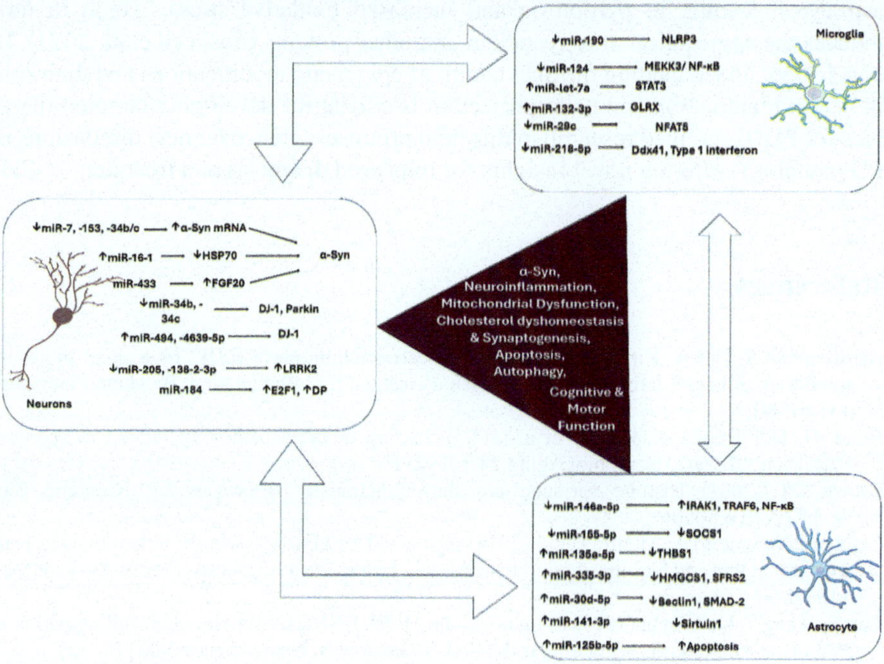

Fig. 2 MicroRNA's regulating PD pathology

activity perpetuates a cycle of neuroinflammation, further advancing PD and underscoring the complexity of the cellular and molecular events involved, many of which remain insufficiently investigated, complicating PD treatment. Recent research has increasingly focused on the role of epigenetics in the central nervous system's development and its regulation of inflammatory responses, glial cell activation, and cytokine production. These epigenetic mechanisms, which can exhibit both neurotoxic and neuroprotective effects, are key to understanding PD pathogenesis. Neurotoxic epigenetic modifications, such as interactions between miR-7, lnc-SNHG1, miR-223-3p, and lncGAS5, promote chronic neuroinflammation in PD models through NLRP3 activation and could serve as early diagnostic markers for PD (Cao et al., 2018, Yang et al. 2023). Conversely, neuroprotective modifications can safeguard dopaminergic neurons against oxidative stress and neuroinflammation. For instance, antagonism of CDR1AS and circSNCA with miR-7 has been proposed to prevent neuronal damage and associated inflammation, highlighting potential therapeutic strategies (Titze-de-Almeida and Titze-de-Almeida 2023).

The interplay between various epigenetic modifications and their coordinated targeting of neuroinflammatory pathways is crucial in PD progression. Methylation of miR-155, miR-124-3p, lncUCA1, and the TNF-α promoter regulates neuroinflammation via the JAK/STAT pathway, while hypomethylation of miR-124 and IL-1β activates the NF-κB pathway. This synergy between epigenetic mechanisms and neuroinflammatory pathways results in the activation of microglia and astrocytes, leading to pyroptosis and increased oxidative stress, which in turn induces the aggregation of α-synuclein and other proteins (Rasheed et al. 2021). In conclusion, understanding the dual nature of epigenetic modifications and their role in neuroinflammation and neuroprotection is crucial for developing targeted therapies for PD. Continued research in this field promises to uncover new dimensions in PD pathology, offering novel insights for improved diagnosis and treatment of PD.

References

Arifuzzaman S, Das A, Kim SH et al (2017) Selective inhibition of EZH2 by a small molecule inhibitor regulates microglial gene expression essential for inflammation. Biochem Pharmacol 137:61–80

Braak H, Del Tredici K, Rüb U et al (2003) Staging of brain pathology related to sporadic Parkinson's disease. Neurobiol Aging 24:197–211

Brown DT (2001) Histone variants: are they functionally heterogeneous? Genome Biol 2(7):REVIEWS0006

Calabrese V, Santoro A, Monti D et al (2018) Aging and Parkinson's disease: inflammaging, neuroinflammation and biological remodeling as key factors in pathogenesis. Free Radic Biol Med 115:80–91

Cao B, Wang T, Qu Q et al (2018) Long noncoding RNA SNHG1 promotes neuroinflammation in Parkinson's disease via regulating miR-7/NLRP3 pathway. Neuroscience 388:118–127

Ceyzériat K, Ben Haim L, Denizot A et al (2018) Modulation of astrocyte reactivity improves functional deficits in mouse models of Alzheimer's disease. Acta Neuropathol Commun 6:104

Chang D, Nalls MA, Hallgrímsdóttir IB et al (2017) A meta-analysis of genome-wide association studies identifies 17 new Parkinson's disease risk loci. Nat Genet 49:1511–1516

Che N, Ou R, Li C et al (2024) Plasma GFAP as a prognostic biomarker of motor subtype in early Parkinson's disease. NPJ Parkinson Dis 10:48

Cheng PY, Lin YP, Chen YL et al (2011) Interplay between SIN3A and STAT3 mediates chromatin conformational changes and GFAP expression during cellular differentiation. PLoS One 6:e22018

Choi I, Zhang Y, Seegobin SP et al (2020) Microglia clear neuron-released α-synuclein via selective autophagy and prevent neurodegeneration. Nat Commun 11:1386

Cook DA, Kannarkat GT, Cintron AF et al (2017) LRRK2 levels in immune cells are increased in Parkinson's disease. NPJ Parkinson Dis 3:11

Craig DW, Hutchins E, Violich I et al (2021) RNA sequencing of whole blood reveals early alterations in immune cells and gene expression in Parkinson's disease. Nat Aging 1:734–747

Deleidi M, Gasser T (2013) The role of inflammation in sporadic and familial Parkinson's disease. Cell Mol Life Sci 70:4259–4273

Dong Y, Xiong J, Ji L et al (2021) MiR-421 aggravates neurotoxicity and promotes cell death in Parkinson's disease models by directly targeting MEF2D. Neurochem Res 46:299–308

Elyaman W, Khoury SJ (2017) Th9 cells in the pathogenesis of EAE and multiple sclerosis. Semin Immunopathol 39:79–87

Fan G, Martinowich K, Chin MH (2005) DNA methylation controls the timing of astrogliogenesis through regulation of JAK-STAT signaling. Development 132:3345–3356

Grunstein M (1997) Histone acetylation in chromatin structure and transcription. Nature 389:349–352

Gujral P, Mahajan V, Lissaman AC et al (2020) Histone acetylation and the role of histone deacetylases in normal cyclic endometrium. Reprod Biol Endocrinol 18:84

Guo M, Wang J, Zhao Y et al (2020) Microglial exosomes facilitate α-synuclein transmission in Parkinson's disease. Brain 143:1476–1497

Hamill CE, Caudle WM, Richardson JR (2007) Exacerbation of dopaminergic terminal damage in a mouse model of Parkinson's disease by the G-protein-coupled receptor protease-activated receptor 1. Mol Pharmacol 72:653–664

Harms AS, Thome AD, Yan Z et al (2018) Peripheral monocyte entry is required for alpha-Synuclein induced inflammation and neurodegeneration in a model of Parkinson disease. Exp Neurol 300:179–187

Hashioka S, Klegeris A, McGeer PL (2012) The histone deacetylase inhibitor suberoylanilide hydroxamic acid attenuates human astrocyte neurotoxicity induced by interferon-γ. J Neuroinflammation 9:113

Heavener KS, Bradshaw EM (2022) The aging immune system in Alzheimer's and Parkinson's diseases. Semin Immunopathol 44:649–657

Hollenbach JA, Norman PJ, Creary LE et al (2019) A specific amino acid motif of HLA-DRB1 mediates risk and interacts with smoking history in Parkinson's disease. Proc Natl Acad Sci USA 116:7419–7424

Hsu SW, Hsu PC, Chang WS et al (2020) Protective effects of valproic acid on 6-hydroxydopamine-induced neuroinjury. Environ Toxicol 35:840–848

Hughes LP, Pereira MMM, Hammond DA et al (2021) Glucocerebrosidase activity is reduced in cryopreserved Parkinson's disease patient monocytes and inversely correlates with motor severity. J Parkinsons Dis 11:1157

Imamura K, Hishikawa N, Ono K et al (2005) Cytokine production of activated microglia and decrease in neurotrophic factors of neurons in the hippocampus of Lewy body disease brains. Acta Neuropathol 109:141–150

Jay TR, Von Saucken VE, Landreth GE (2017) TREM2 in neurodegenerative diseases. Mol Neurodegener 12:56

Ji J, Zeng XN, Cao LL et al (2017) PPARβ/δ activation protects against corticosterone-induced ER stress in astrocytes by inhibiting the CpG hypermethylation of microRNA-181a. Neuropharmacology 113:396–406

Kumar A, Choi KH, Renthal W et al (2005) Chromatin remodeling is a key mechanism underlying cocaine-induced plasticity in striatum. Neuron 48:303–314

Kunoh S, Nakashima H, Nakashima K (2024) Epigenetic regulation of neural stem cells in developmental and adult stages. Epigenomes 8:22

Kustrimovic N, Balkhi S, Bilato G et al (2024) Gut microbiota and immune system dynamics in Parkinson's and Alzheimer's diseases. Int J Mol Sci 25:12164

Li Y, Reddy MA, Miao F et al (2008) Role of the histone H3 lysine 4 methyltransferase, SET7/9, in the regulation of NF-kappaB-dependent inflammatory genes. Relevance to diabetes and inflammation. J Biol Chem 283:26771–26781

Liddelow SA, Guttenplan KA, Clarke LE et al (2017) Neurotoxic reactive astrocytes are induced by activated microglia. Nature 541:481–487

Lin J, Ou R, Li C et al (2023) Plasma glial fibrillary acidic protein as a biomarker of disease progression in Parkinson's disease: a prospective cohort study. BMC Med 21:420

Lindestam Arlehamn CS, Dhanwani R, Pham J et al (2020) α-Synuclein-specific T cell reactivity is associated with preclinical and early Parkinson's disease. Nat Commun 11:875

Ling H, Fabbri M, Calin GA (2013) MicroRNAs and other non-coding RNAs as targets for anticancer drug development. Nat Rev Drug Discov 12:847–865

Luthi-Carter R, Taylor DM, Pallos J et al (2010) SIRT2 inhibition achieves neuroprotection by decreasing sterol biosynthesis. Proc Natl Acad Sci USA 107:7927–7932

Marin-Husstege M, Muggironi M, Liu A et al (2002) Histone deacetylase activity is necessary for oligodendrocyte lineage progression. J Neurosci 22:10333–10345

Marogianni C, Sokratous M, Dardiotis E et al (2020) Neurodegeneration and inflammation-an interesting interplay in Parkinson's disease. Int J Mol Sci 21:8421

Matheoud D, Sugiura A, Bellemare-Pelletier A et al (2016) Parkinson's disease-related proteins PINK1 and Parkin repress mitochondrial antigen presentation. Cell 166:314–327
Matt SM, Lawson MA, Johnson W (2016) Aging and peripheral lipopolysaccharide can modulate epigenetic regulators and decrease IL-1β promoter DNA methylation in microglia. Neurobiol Aging 47:1–9
Mazzocchi M, Goulding SR, Wyatt SL et al (2021) LMK235, a small molecule inhibitor of HDAC4/5, protects dopaminergic neurons against neurotoxin- and α-synuclein-induced degeneration in cellular models of Parkinson's disease. Mol Cell Neurosci 115:103642
McGeer PL, Itagaki S, Boyes BE et al (1988) Reactive microglia are positive for HLA-DR in the substantia nigra of Parkinson's and Alzheimer's disease brains. Neurology 38:1285–1291
Mildner A, Jung S (2014) Development and function of dendritic cell subsets. Immunity 40:642–656
Miller JL, Grant PA (2013) The role of DNA methylation and histone modifications in transcriptional regulation in humans. Subcell Biochem 61:289–317
Moore LD, Le T, Fan G (2013) DNA methylation and its basic function. Neuropsychopharmacology 38:23–38
Neal M, Richardson JR (2018) Epigenetic regulation of astrocyte function in neuroinflammation and neurodegeneration. Biochim Biophys Acta Mol basis Dis 1864:432–443
Periyasamy P, Thangaraj A, Guo ML et al (2018) Epigenetic promoter DNA methylation of miR-124 promotes HIV-1 tat-mediated microglial activation via MECP2-STAT3 Axis. J Neurosci 38:5367–5383
Rasheed M, Liang J, Wang C et al (2021) Epigenetic regulation of Neuroinflammation in Parkinson's disease. Int J Mol Sci 22:4956
Renani PG, Taheri F, Rostami D et al (2019) Involvement of aberrant regulation of epigenetic mechanisms in the pathogenesis of Parkinson's disease and epigenetic-based therapies. J Cell Physiol 234:19307–19319
Reuter S, Gupta SC, Park B et al (2011) Epigenetic changes induced by curcumin and other natural compounds. Genes Nutr 6:93–108
Sher F, Boddeke E, Copray S (2011) Ezh2 expression in astrocytes induces their dedifferentiation toward neural stem cells. Cell Reprogram 13:1–6
Singh NA, Mandal AKA, Khan ZA (2016) Potential neuroprotective properties of epigallocatechin-3-gallate (EGCG). Nutr J 15:60
Soliman ML, Combs CK, Rosenberger TA (2012) Modulation of inflammatory cytokines and mitogen-activated protein kinases by acetate in primary astrocytes. J Neuroimmune Pharmacol 8:287
Song W, Zukor H, Lin SH et al (2012) Schizophrenia-like features in transgenic mice overexpressing human HO-1 in the astrocytic compartment. J Neurosci 32:10841–10853
Sulzer D, Edwards RH (2019) The physiological role of α-synuclein and its relationship to Parkinson's disease. J Neurochem 150:475–486
Sun L, Li X, Luo H et al (2023) EZH2 can be used as a therapeutic agent for inhibiting endothelial dysfunction. Biochem Pharmacol 213:115594
Tan EK, Chao YX, West A et al (2020) Parkinson disease and the immune system—associations, mechanisms and therapeutics. Nat Rev Neurol 16:303–318
Tang Y, Li T, Li J et al (2014) Jmjd3 is essential for the epigenetic modulation of microglia phenotypes in the immune pathogenesis of Parkinson's disease. Cell Death Differ 21:369–380
Tansey MG, Wallings RL, Houser MC et al (2022) Inflammation and immune dysfunction in Parkinson disease. Nat Rev Immunol 22:657–673
Terkelsen MH, Klaestrup IH, Hvingelby V et al (2022) Neuroinflammation and immune changes in prodromal Parkinson's disease and other Synucleinopathies. J Parkinsons Dis 12:S149–S163
Titze-de-Almeida SS, Titze-de-Almeida R (2023) Progress in circRNA-targeted therapy in experimental Parkinson's disease. Pharmaceutics 15:2035
Urayama S, Semi K, Sanosaka T et al (2013) Chromatin accessibility at a STAT3 target site is altered prior to astrocyte differentiation. Cell Struct Funct 38:55–66

Wang R, Lan C, Benlagha K et al (2024a) The interaction of innate immune and adaptive immune system. Med Comm 5:e714

Wang Y, Li D, Zhang et al (2024b) Exosomes derived from microglia overexpressing miR-124-3p alleviate neuronal endoplasmic reticulum stress damage after repetitive mild traumatic brain injury. Neural Regen Res 19:2010–2018

Xu Y, Li Y, Wang C et al (2023) The reciprocal interactions between microglia and T cells in Parkinson's disease: a double-edged sword. J Neuroinflammation 20:33

Yang J, Tang Y, Liu H et al (2014) Suppression of histone deacetylation promotes the differentiation of human pluripotent stem cells towards neural progenitor cells. BMC Med 12:95

Yang R, Yang B, Liu W et al (2023) Emerging role of non-coding RNAs in neuroinflammation mediated by microglia and astrocytes. J Neuroinflammation 20:173

Zhang C, Wang Z, Ji Q et al (2017) Histone demethylase JMJD2C: epigenetic regulators in tumors. Oncotarget 8:91723

Zhang X, Tang B, Guo J (2023a) Parkinson's disease and gut microbiota: from clinical to mechanistic and therapeutic studies. Transl Neurodegener 12:59

Zhang D, Zhang J, Wang Y et al (2023b) Targeting epigenetic modifications in Parkinson's disease therapy. Med Res Rev 43:1748–1777

Zhou Y, Lu M, Du RH et al (2016) MicroRNA-7 targets nod-like receptor protein 3 inflammasome to modulate neuroinflammation in the pathogenesis of Parkinson's disease. Mol Neurodegener 11:28

Decoding the Involvement of miRNAs in Alzheimer's Disease: Focus on Neuroinflammation

Simran Yadav and Urszula Wojda

Contents

Abstract Alzheimer's disease (AD) is a neurodegenerative disorder characterized by amyloid-β aggregation, tau hyperphosphorylation, and neuroinflammation. MicroRNAs (miRNAs) play a critical regulatory role in key mechanisms of AD pathogenesis and show altered expression in AD brain and blood. This chapter explores the multifaceted roles of miRNAs in AD, with a focus on their involvement in neuroinflammation, and highlights their potential as diagnostic biomarkers and therapeutic targets.

S. Yadav · U. Wojda (✉)
Laboratory of Preclinical Testing of Higher Standard, Neurobiology Center, Nencki Institute of Experimental Biology of Polish Academy of Sciences, Warsaw, Poland
e-mail: u.wojda@nencki.edu.pl

J. Barciszewski (ed.), *Decoding Aging and Neurodegeneration*,
RNA Technologies 16, https://doi.org/10.1007/978-3-031-91072-2_14

Keywords Alzheimer's disease · MicroRNA · Non-coding RNA · Neurodegeneration · Amyloid-β · Tau hyperphosphorylation · Neuroinflammation · Inflammatory signaling · Gene regulation · Post-transcriptional regulation · Therapeutic targets · Biomarkers

1 Introduction to Alzheimer's Disease and Its Pathology

1.1 The Global Impact of Alzheimer's Disease

Alzheimer's disease (AD) is a progressive, irreversible, age-related neurodegenerative disorder and the main cause of dementia in the elderly. According to the World Health Organization, the number of people with dementia globally is expected to reach 78 million by 2030 (https://www.who.int/news-room/fact-sheets/detail/dementia). AD is ranked as the seventh leading cause of death (https://www.who.int/news-room/fact-sheets/detail/the-top-10-causes-of-death). Due to its high prevalence, the estimated global economic burden of AD and related dementias is projected to reach $16.9 trillion by 2050 (Nandi et al. 2022). This growing burden is driven by the increasing global life expectancy, which contributes to a larger population at risk for developing AD. Without effective therapeutic interventions, the societal, healthcare, and caregiving costs of AD are expected to rise exponentially in the coming decades.

1.2 Clinical Progression and Molecular Hallmarks of Alzheimer's Disease

AD includes a prolonged preclinical period, often lasting for decades, identified by a silent but progressing pathology. During this asymptomatic stage, small molecular changes occur that initiate and indicate later cognitive and functional impairment. For some, this phase progresses to mild cognitive impairment (MCI), which is commonly seen as a precursor to AD; however, not all MCI cases progress to full-blown dementia. Clinically, AD progresses from mild to moderate to severe stages, with gradual and worsening impairments in memory, cognitive abilities, and self-sufficiency. The neuropathology of AD is defined by two major molecular hallmarks—extracellular amyloid-beta (Aβ) plaques and intraneuronal neurofibrillary tangles (NFTs), composed of hyperphosphorylated tau protein. These pathological changes begin in specific brain regions, such as the entorhinal cortex and hippocampus, which are critical for memory, and progressively spread throughout the brain. This progression mirrors the clinical transition from the asymptomatic or preclinical stage to mild cognitive impairment (MCI) and ultimately to severe dementia. The accumulation of Aβ plaques disrupts synaptic function and promotes neuroinflammation, while NFTs impair neuronal transport and lead to cell death, contributing to widespread neurodegeneration. Understanding these processes is crucial for identifying therapeutic targets and improving early diagnostic approaches.

1.2.1 Amyloid-Beta Plaques

Aβ plaques are composed of aggregated Aβ peptides, which are derived from the proteolytic cleavage of the amyloid precursor protein (APP) (Zott et al. 2024). APP is a transmembrane protein that undergoes sequential cleavage by two enzymes: β-secretase 1 (BACE1) and γ-secretase. The initial cleavage by BACE1 produces a soluble APP fragment (sAPPβ) and a membrane-bound C-terminal fragment, which is subsequently cleaved by γ-secretase to release Aβ peptides of varying lengths, predominantly Aβ40 and Aβ42. Aβ42, being more hydrophobic and prone to aggregation, plays a central role in plaque formation.

Once generated, Aβ peptides can aggregate into soluble oligomers, protofibrils, and ultimately insoluble fibrils, which coalesce into extracellular plaques. These oligomers are particularly toxic, disrupting synaptic communication by impairing long-term potentiation and altering synaptic plasticity. Additionally, Aβ plaques activate microglia and astrocytes, triggering a cascade of neuroinflammatory responses characterized by the release of pro-inflammatory cytokines, such as tumor necrosis factor-alpha (TNF-α), interleukin-1β (IL-1β), and interleukin-6 (IL-6) (Tsai et al. 2022).

The homeostasis of Aβ levels in the brain is tightly regulated by its production, aggregation, and clearance. Degradation and removal of Aβ are facilitated by several pathways, including enzymatic degradation by neprilysin and insulin-degrading enzyme, receptor-mediated transport across the blood-brain barrier (BBB) (e.g., via low-density lipoprotein receptor-related protein 1, LRP1), and clearance through the glymphatic system. Impairments in these clearance mechanisms contribute to the accumulation of Aβ and the progression of Alzheimer's disease. Understanding the balance between Aβ generation and degradation is critical for developing effective therapeutic strategies to mitigate its toxic effects.

1.2.2 Tau Hyperphosphorylation and Neurofibrillary Tangles

Tau, encoded by the *MAPT* gene (17q21.31), is a microtubule-associated protein that stabilizes microtubules in neurons. Alternative splicing of *MAPT* generates six isoforms of tau, containing up to four microtubule-binding domains. Tau's function is regulated through post-translational modifications, particularly phosphorylation. Abnormal hyperphosphorylation of tau leads to its aggregation and formation of NFTs, a key feature of AD pathology (Moore et al. 2023; Ariafar et al. 2023).

The regulation of tau phosphorylation is influenced by various kinases (e.g., extracellular signal-regulated kinase, cyclin-dependent kinase 5, and glycogen synthase kinase 3 beta (GSK3β)), and phosphatases, e.g., protein phosphatase 2A and protein phosphatase 1A. Dysregulation of these processes compromises microtubule stability, contributing to neuronal dysfunction and degeneration in AD (Singh et al. 2025).

1.2.3 Other Proteins Implicated in Complex AD Pathology

AD is broadly categorized into two forms: familial Alzheimer's disease (FAD) and sporadic Alzheimer's disease (SAD). FAD is a rare, autosomal dominant form of the disease, typically with early onset, caused by mutations in key genes such as presenilin 1, presenilin 2, and APP (Cash et al. 2021). These mutations often result in increased production or altered processing of Aβ, favoring the accumulation of the more aggregation-prone Aβ42 peptide. In contrast, SAD, the most common form of AD, occurs sporadically and typically presents with a later onset. Genetic risk factors identified through genome-wide association studies have revealed a variety of genes linked to SAD, including apolipoprotein E4 (APOE4), triggering receptor expressed in myeloid/microglial cells 2 (TREM2), clusterin (CLU), sortilin-related receptor 1 (SORL1), phosphatidylinositol-binding clathrin assembly protein (PICALM), and Myc box-dependent-interacting protein 1/bridging integrator 1 (BIN1) (Bagyinszky and An 2024).

The proteins encoded by these genes have diverse physiological roles that underscore the multifactorial nature of AD pathogenesis. For example, APOE4 is involved in lipid transport and cholesterol metabolism and is a strong genetic risk factor for SAD, with its isoform significantly affecting Aβ aggregation and clearance. TREM2, expressed in microglia, modulates phagocytic activity and neuroinflammation, while SORL1 and PICALM play roles in endocytosis and intracellular trafficking, processes critical for Aβ clearance, and synaptic function. Additionally, BIN1 and CLU are implicated in vesicle-mediated transport and lipid metabolism, both of which are disrupted in AD.

These genetic findings align with pathological hallmarks of AD, including the early disruption of synaptic integrity and progressive accumulation of Aβ plaques and NFTs. Aberrant processing of APP, driven by FAD mutations or SAD-associated risk factors, results in an imbalance between Aβ production and clearance, leading to its aggregation. Lipid metabolism disruptions, influenced by APOE4, further exacerbate Aβ deposition and impair neuronal membrane repair. Microglial dysfunction, driven by genetic factors like TREM2, compromises the clearance of Aβ and contributes to chronic neuroinflammation. Concurrently, tau hyperphosphorylation and aggregation into NFTs impair neuronal transport and promote cytoskeletal instability.

Together, these interconnected mechanisms illustrate the complexity of AD pathogenesis, highlighting the interplay between genetic predispositions, cellular dysfunction, and molecular alterations. Understanding the roles of these genes and pathways offers crucial insights into AD development and potential avenues for therapeutic intervention.

2 MicroRNAs: An Overview and Their Relevance to Alzheimer's Disease

2.1 MicroRNAs as a Key to Decoding Cellular Regulation

MicroRNAs (miRNAs) are short, non-coding RNAs, typically ~22 nucleotides in length, that play an important role in controlling gene expression. The field of miRNA research began with the discovery of lin-4, a key regulator of larval development in *Caenorhabditis elegans* (Lee et al. 1993). Over the past three decades, the field has witnessed an exponential growth. Initially, only one miRNA was identified in a single species. As of the latest release of miRBase (www.mirbase.org, v22; Kozomara et al. 2019), the database now catalogs 48,860 mature miRNA gene sequences across 271 organisms. In humans around 2000 mature miRNAs have been identified, and it is predicted to regulate over 60% of gene targets. Alterations in miRNA profiles can significantly affect cell physiology and such vital processes as cell proliferation, differentiation, DNA repair, apoptosis, and metabolism (Nikolac Perkovic et al. 2021).

In the context of AD, several studies have shown alterations in the levels of a number of miRNAs in the brains, cerebrospinal fluid (CSF), and blood of AD patients compared to healthy controls (reviewed in Nagaraj et al. 2019; Li et al. 2024).

2.2 Biogenesis of miRNAs and RNA Interference

Mature miRNAs are generated through either the canonical or non-canonical biogenesis pathways from their coding sequences in the human genome (Bartel 2018). In the human genome, miRNA genes are distributed across all chromosomes except for the Y chromosome. These genes are found in various genomic contexts, including intergenic, intronic, and, less commonly, exonic regions. Intergenic miRNAs are transcribed as independent units, often controlled by their own promoters. Intronic miRNAs, on the other hand, are embedded within host protein-coding or non-coding genes and are frequently co-transcribed with their host transcripts, utilizing the host gene's regulatory elements.

Human miRNA genes are often organized as monocistronic or polycistronic clusters. Monocistronic miRNAs are encoded individually, while polycistronic miRNA clusters are transcribed as a single primary transcript, yielding multiple mature miRNAs after processing. These clusters often target related pathways, reflecting their coordinated expression and functional significance. Some miRNAs are also derived from non-coding RNAs, such as long non-coding RNAs, further diversifying their regulatory potential.

The organization and regulation of miRNA genes within the human genome highlight their critical roles in maintaining cellular homeostasis and their involvement in various diseases, including cancer, neurodegenerative disorders, and cardiovascular conditions. Understanding the genomic context of miRNA genes provides valuable insights into their biogenesis and functional roles in humans.

2.2.1 Canonical miRNA Biogenesis

The canonical miRNA biogenesis is shown schematically in Fig. 1a. In the canonical pathway, primary miRNA (pri-miRNA) hairpin-like transcripts (~1000 nucleotides) are synthesized from DNA templates by RNA polymerase II/III in the nucleus. These pri-miRNAs often have a 5′ cap and a 3′ poly-A tail, similar to protein-coding mRNAs, facilitating their processing and stability. The microprocessor complex,

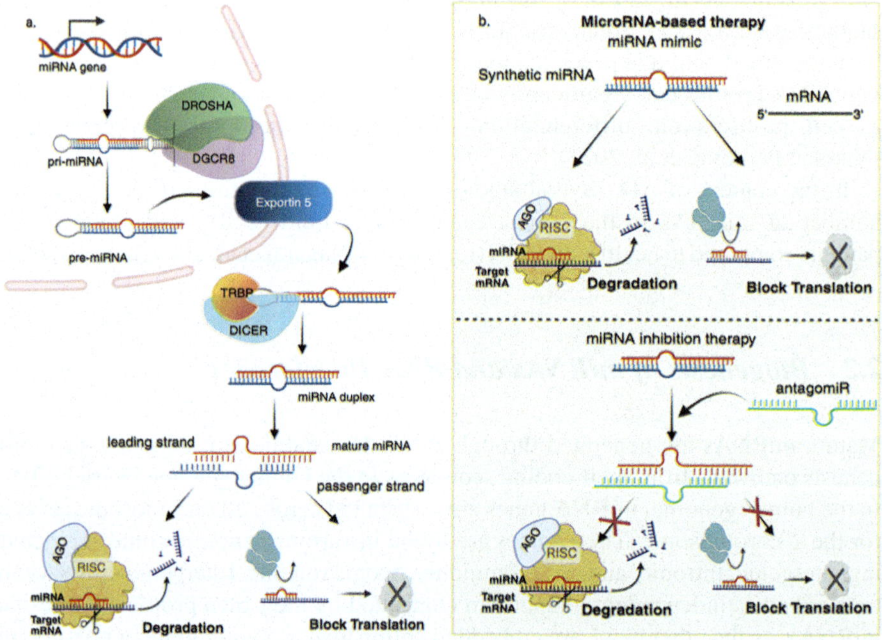

Fig. 1 MicroRNA biogenesis and therapeutic interventions. (**a**) Canonical biogenesis of miRNA and RNA interference: The miRNA biogenesis pathway begins with the transcription of primary miRNAs (pri-miRNAs) in the nucleus. Pri-miRNAs are processed by the Drosha-DGCR8 complex to form precursor miRNAs (pre-miRNAs), which are exported to the cytoplasm via exportin-5. In the cytoplasm, Dicer processes pre-miRNAs into mature miRNA duplexes. Leading strand of the duplex is incorporated into the RNA-induced silencing complex (RISC), which guides the miRNA to its target mRNA for epigenetic regulation through mRNA degradation or translational inhibition, depending on the complementarity between the miRNA and the mRNA. (**b**) Therapeutic application of miRNA mimics and antagomirs: MiRNA mimics are synthetic RNA molecules designed to restore the function of downregulated miRNAs by mimicking their activity and re-establishing gene silencing. Antagomirs, on the other hand, are chemically modified oligonucleotides that specifically bind and inhibit overexpressed miRNAs, blocking their interaction with mRNA targets. These therapeutic strategies allow modulation of RNA interference (RNAi) to either suppress or enhance specific gene expression pathways

consisting of the RNase III enzyme Drosha and DGCR8 protein, processes pri-miRNAs into precursor miRNAs (pre-miRNAs), which are ~70-nucleotide hairpin structures. This cleavage removes the stem-loop structure from the surrounding primary transcript, producing the pre-miRNA. Pre-miRNAs are then exported to the cytoplasm via the exportin-5 and Ran-GTP complex, a process that ensures the precursor molecules reach the cytoplasm for further processing.

Once in the cytoplasm, the Dicer complex, composed of Dicer (another RNase III enzyme) and the double-stranded RNA-binding protein TRBP, cleaves the pre-miRNA into a mature miRNA duplex (~22 nucleotides) with 3′ overhangs. The cleavage step is critical for the generation of the miRNA duplex and determines the final size of the mature miRNA. One strand, termed the functional or guide strand, is incorporated into the RNA-induced silencing complex (RISC) along with Argonaute (Ago2) proteins. The guide strand directs the RISC complex to its target mRNA, where it binds to complementary sequences, usually in the 3′ untranslated region (UTR) of the target mRNA. This interaction typically results in translational repression or mRNA degradation. The passenger strand, which is less stable, is typically degraded in the cytoplasm.

This highly regulated process ensures the precise generation and activity of miRNAs, which are crucial for the regulation of gene expression at the post-transcriptional level. The efficiency and fidelity of miRNA processing, as well as the specificity of target recognition by the RISC complex, contribute to the fine-tuning of cellular functions, such as development, differentiation, and stress responses. The regulation of this pathway is essential for maintaining cellular homeostasis and preventing diseases associated with dysregulated miRNA expression.

2.2.2 RNA Interference

RNAi is a cellular mechanism in which miRNAs in the RISC complex regulate gene expression post-transcriptionally. This process is crucial for maintaining cellular homeostasis and regulating diverse biological pathways. Within the RISC complex, miRNAs bind to the 3′ untranslated region (3′ UTR) of target mRNA transcripts based on sequence complementarity, regulating gene expression either through mRNA degradation or translational inhibition (Bartel 2018). The specific mechanism of regulation depends on the degree of complementarity between the miRNA seed region and the target mRNA. Perfect complementarity typically leads to mRNA degradation, while partial complementarity generally results in translational repression.

In addition to their role in endogenous gene regulation, miRNAs can also influence cellular responses to external signals, such as viral infections, environmental stress, and changes in cell signaling pathways. Their ability to fine-tune gene expression makes miRNAs powerful regulators in diverse physiological processes, including apoptosis, cell proliferation, differentiation, and immune responses. Moreover, alterations in miRNA expression have been linked to a range of diseases, including cancer, neurological disorders, and cardiovascular diseases, further highlighting their critical role in cellular function.

2.2.3 Non-canonical miRNA Biogenesis

Non-canonical miRNA biogenesis pathways involve miRNA maturation processes that bypass specific steps or components of the canonical pathway. For example, mirtrons, short hairpin RNAs, and chimeric hairpins bypass the Drosha cleavage step but still require Dicer for further processing. Mirtrons are derived from introns that contain small hairpin-like structures, which are directly excised from the primary transcript by the spliceosome, bypassing Drosha's usual role. After this excision, they are processed by Dicer to generate mature miRNAs. In contrast, miR-451 is another example of a non-canonical miRNA, which bypasses Dicer for maturation despite requiring Drosha for the initial cleavage step (Bartel 2018).

These non-canonical pathways are thought to play specific roles under certain cellular conditions, providing an additional layer of complexity and flexibility to miRNA biogenesis. They often respond to distinct cellular signals, such as stress or viral infection, and can help fine-tune miRNA regulation under these conditions. This flexibility in miRNA processing allows cells to adapt to dynamic environments and respond to changes in gene expression networks more efficiently. The study of these non-canonical pathways is still ongoing, as they reveal new insights into the diversity of RNA processing and regulation mechanisms in eukaryotic cells.

2.3 *MiRNA Regulation in Alzheimer's Disease: Biomarker and Therapeutic Potential*

MiRNAs are abundant in the central nervous system (CNS) and exhibit both spatial and temporal specificities. They play critical roles in brain development and maintaining homeostasis by regulating processes such as neurogenesis, neuronal differentiation, synaptic plasticity, gliogenesis, and myelin repair (Li et al. 2024). Proper miRNA function is essential for CNS health, as dysregulated expression is associated with neurological disorders, developmental abnormalities, autoimmune conditions, and neurodegenerative diseases including AD. In AD, miRNAs play a critical role in regulating key molecular processes that are heavily impacted in brain neurons, including Aβ production, aggregation, and clearance, hyperphosphorylation of tau, neuronal synaptic functions, and neuroinflammation.

Specific miRNAs have been shown to directly influence Aβ metabolism. For instance, miR-29a and miR-107 are known to regulate the expression of β-secretase 1 (BACE1), an enzyme critical for the cleavage of APP to produce Aβ peptides. Elevated levels of miR-29a have been linked to reduced BACE1 expression, potentially mitigating Aβ production, while decreased expression of miR-107 has been associated with increased BACE1 activity and amyloid accumulation. Similarly, miR-214 has been found to regulate the aggregation of Aβ, potentially influencing plaque formation and neurotoxicity.

In addition to their role in Aβ pathophysiology, miRNAs also regulate tau pathology, particularly through the modulation of tau phosphorylation. MiR-132 and miR-124 have been shown to target tau kinases such as GSK3β, which is responsible for hyperphosphorylating tau in AD. Decreased expression of miR-132 is correlated with increased GSK3β activity, leading to tau hyperphosphorylation and the formation of NFTs.

Furthermore, several studies have shown alterations in the levels of a number of miRNAs in the brain, cerebrospinal fluid, and blood of AD patients compared to healthy controls (Nagaraj et al. 2019; Varma-Doyle et al. 2021). These findings underscore the remarkable potential of miRNAs as biomarkers for AD, owing to their stability, ease of detection in biological fluids, and distinct disease-specific expression profiles. Some of them have shown a promise as diagnostic biomarkers, correlating with both disease presence and severity. Their non-invasive detection in blood and other biofluids offers a promising avenue for early diagnosis and monitoring of disease progression. Early detection of AD using miRNA biomarkers could facilitate timely interventions, potentially slowing disease progression before irreversible neuronal damage occurs.

In addition to diagnostic potential, miRNAs offer a promising approach for distinguishing AD subtypes and endophenotypes, based on the specific signaling pathways they report (Nagaraj et al. 2019). For example, certain miRNA profiles have been linked to different forms of AD pathology, such as the amyloidogenic or tau-centric pathologies, which could help in identifying disease subtypes. By correlating specific miRNA signatures with underlying molecular processes (such as Aβ production versus tau hyperphosphorylation), it may be possible to stratify patients into subgroups with distinct pathophysiological mechanisms. This could pave the way for personalized therapeutic approaches, where miRNA-based therapies can be tailored to the specific signaling pathways driving an individual's AD subtype.

Moreover, miRNAs are emerging as therapeutic targets in AD. Strategies such as synthetic oligonucleotides, miRNA mimics, and antagomiRs (miRNA inhibitors) have shown a promise in modulating disease-related molecular pathways (Fig. 1b). MiRNA mimics are synthetic RNA molecules designed to replicate the function of downregulated miRNAs, restoring their activity to modulate gene expression and restore normal cellular processes. Conversely, antagomiRs are oligonucleotides that specifically inhibit overexpressed miRNAs, blocking their interaction with target mRNAs to prevent disease-associated effects. These approaches aim to either restore the expression of beneficial miRNAs or inhibit the activity of harmful ones, thereby potentially reversing or slowing disease progression. By targeting miRNAs that regulate Aβ production, tau phosphorylation, or neuroinflammation, it may be possible to mitigate key pathological features of AD and provide new therapeutic options for patients.

3 Exploring miRNAs in Neuroinflammation: Implications for Therapeutic Strategies in AD

3.1 *Mechanisms of Neuroinflammation*

The term "neuroinflammation" refers to the inflammatory response within the CNS, which aims to maintain cellular homeostasis under both physiological and pathological conditions. Inflammation significantly contributes to the pathogenesis of CNS diseases, with neuroinflammation playing a pivotal role in AD pathology. Glial cells, including microglia and astrocytes, are key players in the inflammatory response to Aβ toxicity (Kwon and Koh 2020). Under normal conditions, the CNS initiates a controlled inflammatory response through glial cell activation, promoting tissue repair and neuroprotection. However, in pathological states, uncontrolled neuroinflammation can lead to excessive glial activation, resulting in the production of pro-inflammatory cytokines (e.g., IL-1β, IL-6, TNF-α), reactive oxygen species (ROS), nitric oxide, and chemokines, which can compromise the BBB and lead to immune cell infiltration.

Microglia, the CNS's primary defense cells, transition from a resting to an activated state in response to injury. Depending on the stimulus, microglia adopt either an M1 (pro-inflammatory, releasing ROS and TNF-α) or M2 (anti-inflammatory, releasing IL-10 and TGF-β) phenotype. Prolonged M1 activation contributes to neurodegeneration by releasing toxic mediators.

Astrocytes also play a crucial role in maintaining homeostasis by supporting neuronal signaling and synaptic plasticity. Under stress, they influence neuroinflammation by releasing pro- and anti-inflammatory molecules and forming functional barriers. Glial receptors like TLR4 trigger inflammatory cascades via cytokines such as TNF-α and IL-1β. The neurovascular unit, composed of neurons, astrocytes, endothelial cells, and the extracellular matrix, regulates neuroinflammatory signals, blood flow, and BBB permeability (Farhy-Tselnicker and Allen 2018).

3.2 *The Neuroinflammation Hypothesis in AD*

This hypothesis suggests that chronic inflammation in the brain plays a central role in the onset and progression of AD. The most common version of the neuroinflammation hypothesis assumes that activated microglia and astrocytes, the primary immune cells of the central nervous system, respond to pathological features such as Aβ plaques and tau tangles by releasing pro-inflammatory cytokines, chemokines, and ROS. While initially protective, this immune response becomes dysregulated, leading to sustained inflammation that exacerbates neuronal damage, synaptic dysfunction, and cognitive decline.

Another sequence of events in the neuroinflammatory hypothesis, particularly referring to SAD, begins with chronic systemic inflammation, often linked to metabolic syndrome and other age-related diseases (Więckowska-Gacek et al. 2021). These conditions promote a state of low-grade, systemic inflammation, characterized by elevated levels of pro-inflammatory cytokines such as IL-6. Over time, this chronic inflammation can compromise the integrity of the BBB, making it more permeable and allowing peripheral immune cells, including T cells and monocytes, to infiltrate the brain. These infiltrating immune cells, along with circulating cytokines, initiate and sustain neuroinflammation. In turn, neuroinflammation triggers further activation of microglia and astrocytes, amplifying the inflammatory response and contributing to the development of amyloid plaques and tau tangles.

This cascade of immune activation and neuroinflammation is thought to exacerbate the hallmark pathological features of AD. As neuroinflammation intensifies, it accelerates the formation of Aβ plaques and tau phosphorylation, both of which further fuel the inflammatory cycle. Inflammatory mediators, including interleukins and TNF-α, directly affect the processing of APP, leading to an increased Aβ production. Additionally, the persistent activation of microglia and astrocytes can promote tau hyperphosphorylation and aggregation, further impairing neuronal function.

Emerging evidence highlights the involvement of dysregulated microRNAs, Toll-like receptor (TLR) signaling, and peripheral immune cell infiltration in perpetuating this neuroinflammatory cycle, positioning inflammation as a critical therapeutic target in AD. Specific miRNAs have been found to regulate inflammatory pathways, including the TLR pathway, which is activated by Aβ plaques and contributes to the chronic neuroinflammation observed in AD. By modulating these miRNAs, it may be possible to intervene early in the disease process, reducing inflammation and potentially halting or slowing the progression of amyloid and tau pathologies.

3.3 Involvement of miRNAs in Neuroinflammation

The inflammatory response in AD is closely linked to changes in miRNA expression. Identifying the specific miRNAs involved in both pro-inflammatory and anti-inflammatory cellular signaling and processes with precision in AD pathology is essential. Tables 1 and 2 summarize roles of key miRNAs and their therapeutic potential in modulating neuroinflammation in AD.

Among the miRNAs acting as pro-inflammatory agents are miR-155, miR-27b, and miR-326, while others, such as miR-124, miR-146a, miR-21, and miR-223, function as anti-inflammatory agents. Additionally, some miRNAs, like the let-7 family, serve as mixed modulators, playing a dual role in both pro- and anti-inflammatory responses depending on the cellular context and disease stage. These miRNAs regulate the activation of microglia and astrocytes, as well as peripheral immune cells such as neutrophils, macrophages, and T and B cells (Khan et al. 2022).

Table 1 Anti-inflammatory miRNAs

miRNA	Role in neuroinflammation	Target	Potential therapeutic implications	References
miR-146a	Suppresses microglial activation and reduces levels of pro-inflammatory cytokines TNF-α and IL-1β	IRAK1, TRAF6, NF-κB	Mimics may reduce neuroinflammation and help treat depression	Liu et al. (2021)
miR-124	Promotes anti-inflammatory M2-like polarization of microglia/macrophages and protects neurons in ischemic stroke; inhibits pro-inflammatory cytokines	MEKK3/NF-κB; Arg-1 (M2 marker)	Mimics may enhance neuroprotection and promote functional recovery after a stroke	Yao et al. (2018)
miR-21-5p	Reduces levels of pro-inflammatory factors and protects against Aβ-induced damage	SOCS6	Mimics may enhance neuroprotection against Aβ-induced neurodegeneration	Wang et al. (2023)
miR-223, miR-223-3p	Decreases NLRP3 inflammasome activation, alleviates neuroinflammation, reduces nerve damage in AD models; can be released in exosomes from M2 microglia	NLRP3	Mimics may reduce neuroinflammation and increase neuroprotection	Wu et al. (2023); Wei et al. (2024)
miR-486-3p	Attenuates neuroinflammation by modulating microglial phenotype switching, inhibiting SIRT2, and enhancing mitophagy	SIRT2	Targeted delivery of miR-486-3p via exosomes (Exo/miR) may reduce brain edema and neurodegeneration	Sheng et al. (2024)

Abbreviations: *IRAK1* interleukin 1 receptor-associated kinase 1; *TRAF6* TNF receptor-associated factor 6; *NF-κB* nuclear factor of kappa B; *MEKK3* mitogen-activated protein kinase kinase kinase 3; *Arg-1* Arginase-1; *SOCS6* suppressor of cytokine signaling 6; *NLRP3* NLR family pyrin domain containing 3; *SIRT2* sirtuin 2

For instance, studies have demonstrated that miR-146a overexpression in microglia can mitigate cognitive deficits, attenuate neuroinflammation, reduce Aβ levels, and prevent neuronal loss. This is achieved by promoting a shift from the M1 pro-inflammatory phenotype to the M2 anti-inflammatory phenotype, which is associated with tissue repair and neuroprotection (Huang et al. 2023). MiR-146a suppresses microglial activation and reduces levels of pro-inflammatory cytokines such as TNF-α and IL-1β, targeting key proteins such as IRAK1, TRAF6, and NF-κB. The use of miR-146a mimics has shown a promise in reducing neuroinflammation and potentially treating depression, which often coexists with AD (Liu et al. 2021). On the other hand, upregulation of miR-155 leads to enhanced activation of microglia and astrocytes, triggering the release of inflammatory mediators and potentially contributing to AD progression. This occurs through the modulation of T-cell functions during inflammation, as miR-155 regulates genes involved in T-cell differentiation and activation, exacerbating neuroinflammatory responses (Rastegar-Moghaddam et al. 2023).

Table 2 Pro-inflammatory miRNAs

miRNA	Role in neuroinflammation	Target	Potential therapeutic implications	References
miR-155	Pro-inflammatory, modulates immune responses and microglial activation	FADD, SOCS1, IKK, IL13Rα1, and SMAD2	AntagomiRs may reduce neuroinflammation, neurodegeneration, and microgliosis	Zingale et al. (2021)
miR-27b-3p	Pro-inflammatory, modulates microglial inflammation response and apoptosis	A20/TNFAIP3	AntagomiRs may inhibit neuroinflammation and provide neuroprotection	Li et al. (2021)
miR-326	Pro-inflammatory, involved in Th1/Th17 lymphocyte differentiation	TET2	AntagomiRs may restore the Th17/Treg balance	Zhang et al. (2024)
miR-200a-3p	Promotes inflammation in the brain by ROS-induced NLRP3	NLRP3	AntagomiRs may mitigate neuroinflammation	Yu et al. (2019)
miR-34a	Pro-inflammatory, contributes to neuroinflammation, and lowers synaptic plasticity; induces AD-like brain pathology and cognitive impairment	ADAM10, SIRT1, AMPA, and NMDA receptors (miR-34a-5p inhibits NMDAR, miR-34a-3p inhibits AMPAR)	AntagomiRs may increase synaptic plasticity, reduce amyloid β and tau pathology, and improve cognition	Sarkar et al. (2019)

Abbreviations: *FADD* fas-associated death domain; *SOCS1* suppressor of cytokine signaling 1; *IKK* IκB kinase; *IL13Rα1* interleukin-13 receptor alpha 1; *SMAD2* mothers against decapentaplegic homolog 2; *A20/TNFAIP3* tumor necrosis factor alpha-induced protein 3; *TET2* tet methylcytosine dioxygenase 2; *NLRP3* NLR family pyrin domain containing 3; *ADAM10* a disintegrin and metalloproteinase domain-containing protein 10; *NMDAR N*-methyl-d-aspartate receptor; *AMPAR* α-amino-3-hydroxy-5-methyl-4-isoxazolepropionic acid receptor

Similarly, miR-124 promotes anti-inflammatory M2-like polarization of microglia and macrophages, inhibiting pro-inflammatory cytokines and protecting neurons in conditions like ischemic stroke. Its action through the MEKK3/NF-κB pathway and the Arg-1 M2 marker suggests potential neuroprotective benefits for AD, with mimics showing a promise in enhancing neuroprotection and promoting functional recovery after stroke (Yao et al. 2018). Another key miRNA, miR-21-5p, reduces pro-inflammatory factors and protects against Aβ-induced damage by targeting SOCS6. MiR-21-5p mimics may enhance neuroprotection against Aβ-induced neurodegeneration, providing further therapeutic potential for modulating inflammatory responses in AD (Wang et al. 2023).

For example, miR-27b-3p has been shown to modulate microglial inflammation response and apoptosis by targeting A20/TNFAIP3. Thus miR-27b-3p mimics may inhibit neuroinflammation and provide neuroprotection in AD (Li et al. 2021). MiR-223, specifically miR-223-3p, has been shown to alleviate neuroinflammation by decreasing NLRP3 inflammasome activation, reducing nerve damage in AD models, and being released in exosomes from M2 microglia. MiR-223 mimics may reduce neuroinflammation and increase neuroprotection, highlighting their potential in AD treatment (Wu et al. 2023; Wei et al. 2024).

Among the main molecular signaling pathways regulated by miRNAs is the TLR and cytokine signaling pathways. For example, miR-146a, miR-155, and miR-132 respond to stimuli such as lipopolysaccharide, TNF-α, and IL-1β. MiR-146a mitigates excessive inflammation by targeting key cytokine receptors and signaling molecules, preventing an overactive immune response. In contrast, miR-155 exacerbates inflammation by regulating cytokines like interferon (IFN)-λ and IFN-β, promoting a more robust pro-inflammatory response. Additionally, miR-132 has been shown to play a pivotal role in regulating neuroinflammation by modulating the expression of inflammatory cytokines in both microglia and astrocytes, thereby influencing the progression of AD pathology.

MiR-486-3p is another miRNA that plays a crucial role in neuroinflammation by modulating microglial phenotype switching, inhibiting SIRT2, and enhancing mitophagy. Targeted delivery of miR-486-3p via exosomes may reduce brain edema and neurodegeneration, presenting a novel strategy for treating AD (Sheng et al. 2024).

These insights into the roles of specific miRNAs in neuroinflammation offer potential therapeutic targets for modulating inflammatory responses in AD. By selectively targeting the expression or activity of these miRNAs, it may be possible to reduce neuroinflammation, slow disease progression, and ultimately improve cognitive outcomes in AD patients. Further research into the precise mechanisms of miRNA-mediated regulation of immune cells and signaling pathways is crucial to developing effective miRNA-based therapies for AD.

3.4 Perspective of Therapeutic Strategies Targeting miRNAs in AD

Antisense therapeutic strategies targeting miRNAs, including miRNA mimics and antagomiRs as shown in Fig. 1b, hold a significant promise for treating various diseases, including neurodegenerative disorders such as AD (Grabowska-Pyrzewicz et al. 2021). Both strategies aim to correct miRNA dysregulation, offering potential therapeutic interventions for conditions such as AD, where altered miRNA profiles contribute to disease progression.

To influence neuroinflammation and other pathological processes contributing to AD, miRNA-based therapeutics might emerge by targeting dysregulated expression of certain miRNAs. As discussed earlier, aberrant expression of miRNAs can lead to uncontrolled immune responses and the activation of inflammatory pathways, driving pathological processes in AD and other neurodegenerative disorders (Zingale et al. 2021). However, dysregulation of miRNAs can either contribute to the progression of the diseases or act as a homeostatic mechanism in the CNS attempting to restore normal conditions. Therefore understanding the specific roles of these miRNAs is critical for developing therapeutic strategies targeting neuroinflammation in AD.

Moreover, such therapeutic approaches will require not only knowledge on brain functions of a particular miRNA but also on changes in their levels in AD in a particular brain region, with relation to AD stage. Eventually, knowing miRNA profiles from blood, CSF, and brain tissue might lead to the development of specific biomarkers and personalized treatment solutions.

Another necessary tool for miRNA-based therapies is development of advanced delivery techniques, such as nanoparticle-based vehicles or exosome-mediated procedures, which are needed to ensure that miRNA drugs reach the CNS with precision.

4 Conclusions and Future Prospects

Antisense therapeutic strategies targeting miRNAs, such as miRNA mimics and antagomiRs, offer promising approaches for treating diseases like AD by correcting miRNA dysregulation. MiRNA mimics restore the activity of downregulated miRNAs, while antagomiRs inhibit overexpressed miRNAs to prevent disease progression. Targeting dysregulated miRNAs involved in neuroinflammation could help modulate inflammatory pathways contributing to AD. Neuroinflammation plays a critical role in AD pathology by exacerbating neuronal damage through prolonged activation of glial cells and the release of pro-inflammatory cytokines. Multiple miRNAs regulate neuroinflammation, acting as either pro-inflammatory or anti-inflammatory agents, highlighting their potential as therapeutic targets. However, therapeutic success depends on understanding the specific roles of miRNAs in different brain regions and disease stages and developing effective delivery methods to ensure precise targeting in the CNS.

Acknowledgments This research was supported by the Polish National Science Centre Grant OPUS 2022/47/B/NZ7/03005.

References

Ariafar S, Makhdoomi S, Mohammadi M (2023) Arsenic and tau phosphorylation: a mechanistic review. Biol Trace Elem Res 201:5708–5720

Bagyinszky E, An SSA (2024) Haploinsufficiency and Alzheimer's disease: the possible pathogenic and protective genetic factors. IJMS 25:11959

Bartel DP (2018) Metazoan MicroRNAs. Cell 173:20–51

Cash MK, Rockwood K, Fisk JD et al (2021) Clinicopathological correlations and cholinesterase expression in early-onset familial Alzheimer's disease with the presenilin 1 mutation, Leu235Pro. Neurobiol Aging 103:31–41

Farhy-Tselnicker I, Allen NJ (2018) Astrocytes, neurons, synapses: a tripartite view on cortical circuit development. Neural Dev 13:7

Grabowska-Pyrzewicz W, Want A, Leszek J et al (2021) Antisense oligonucleotides for Alzheimer's disease therapy: from the mRNA to miRNA paradigm. EBioMedicine 74:103691

Huang Y, Lin X, Lin X (2023) MiR-146a-5p contributes to microglial polarization transitions associated with AGEs. Mol Neurobiol 60:3020–3033

Khan I, Preeti K, Fernandes V et al (2022) Role of MicroRNAs, aptamers in Neuroinflammation and neurodegenerative disorders. Cell Mol Neurobiol 42:2075–2095

Kozomara A, Birgaoanu M, Griffiths-Jones S (2019) miRBase: from microRNA sequences to function. Nucleic Acids Res 47:D155–D162

Kwon HS, Koh S-H (2020) Neuroinflammation in neurodegenerative disorders: the roles of microglia and astrocytes. Transl Neurodegener 9:42

Lee RC, Feinbaum RL, Ambros V (1993) The C. elegans heterochronic gene lin-4 encodes small RNAs with antisense complementarity to lin-14. Cell 75:843–854

Li L, Qi C, Liu Y et al (2021) MicroRNA miR-27b-3p regulate microglial inflammation response and cell apoptosis by inhibiting A20 (TNF-α-induced protein 3). Bioengineered 12:9902–9913

Li Y-B, Fu Q, Guo M et al (2024) MicroRNAs: pioneering regulators in Alzheimer's disease pathogenesis, diagnosis, and therapy. Transl Psychiatry 14:367

Liu C-P, Zhong M, Sun J-X et al (2021) miR-146a reduces depressive behavior by inhibiting microglial activation. Mol Med Rep 23:463

Moore KBE, Hung T-J, Fortin JS (2023) Hyperphosphorylated tau (p-tau) and drug discovery in the context of Alzheimer's disease and related tauopathies. Drug Discov Today 28:103487

Nagaraj S, Zoltowska KM, Laskowska-Kaszub K et al (2019) microRNA diagnostic panel for Alzheimer's disease and epigenetic trade-off between neurodegeneration and cancer. Ageing Res Rev 49:125–143

Nandi A, Counts N, Chen S et al (2022) Global and regional projections of the economic burden of Alzheimer's disease and related dementias from 2019 to 2050: a value of statistical life approach. eClinicalMedicine 51:101580

Nikolac Perkovic M, Videtic Paska A, Konjevod M et al (2021) Epigenetics of Alzheimer's disease. Biomolecules 11:195

Rastegar-Moghaddam SH, Ebrahimzadeh-Bideskan A, Shahba S et al (2023) Roles of the miR-155 in neuroinflammation and neurological disorders: a potent biological and therapeutic target. Cell Mol Neurobiol 43:455–467

Sarkar S, Engler-Chiurazzi EB, Cavendish JZ et al (2019) Over-expression of miR-34a induces rapid cognitive impairment and Alzheimer's disease-like pathology. Brain Res 1721:146327

Sheng B, Gao S, Chen X et al (2024) Exosomes-mediated delivery of miR-486-3p alleviates neuroinflammation via SIRT2-mediated inhibition of mitophagy after subarachnoid hemorrhage. Stroke Vasc Neurol:svn-2024-003509

Singh S, Khan S, Shahid M et al (2025) Targeting tau in Alzheimer's and beyond: insights into pathology and therapeutic strategies. Ageing Res Rev 104:102639

Tsai AP, Dong C, Lin PB-C et al (2022) PLCG2 is associated with the inflammatory response and is induced by amyloid plaques in Alzheimer's disease. Genome Med 14:17

Varma-Doyle AV, Lukiw WJ, Zhao Y et al (2021) A hypothesis-generating scoping review of miRs identified in both multiple sclerosis and dementia, their protein targets, and miR signaling pathways. J Neurol Sci 420:117202

Wang G, Shen X, Song X et al (2023) Protective mechanism of gold nanoparticles on human neural stem cells injured by β-amyloid protein through miR-21–5p/SOCS6 pathway. Neurotoxicology 95:12–22

Wei H, Zhu Z, Xu Y et al (2024) Microglia-derived exosomes selective sorted by YB-1 alleviate nerve damage and cognitive outcome in Alzheimer's disease. J Transl Med 22:466

Więckowska-Gacek A, Mietelska-Porowska A, Wydrych M et al (2021) Western diet as a trigger of Alzheimer's disease: from metabolic syndrome and systemic inflammation to neuroinflammation and neurodegeneration. Ageing Res Rev 70:101397

Wu C, Xing W, Zhang Y et al (2023) NLRP3/miR-223-3p axis attenuates neuroinflammation induced by chronic intermittent hypoxia. Funct Integr Genomics 23:342

Yao L, Ye Y, Mao H et al (2018) MicroRNA-124 regulates the expression of MEKK3 in the inflammatory pathogenesis of Parkinson's disease. J Neuroinflammation 15:13

Yu J, Chen J, Yang H et al (2019) Overexpression of miR-200a-3p promoted inflammation in sepsis-induced brain injury through ROS-induced NLRP3. Int J Mol Med 44:1811–1823

Zhang G, Wu S, Xia G (2024) MiR-326 sponges TET2 triggering imbalance of Th17/Treg differentiation to exacerbate pyroptosis of hepatocytes in concanavalin A-induced autoimmune hepatitis. Ann Hepatol 29:101183

Zingale VD, Gugliandolo A, Mazzon E (2021) MiR-155: an important regulator of neuroinflammation. IJMS 23:90

Zott B, Nästle L, Grienberger C et al (2024) β-amyloid monomer scavenging by an anticalin protein prevents neuronal hyperactivity in mouse models of Alzheimer's disease. Nat Commun 15:581

Alzheimer's Disease: Link to Aging and Epigenetics

Devangi K. Zala, Shuvomoy Banerjee, and Anand Krishna Tiwari

Contents

Abstract Alzheimer's disease (AD) is a neurodegenerative disorder associated with the accumulation of amyloid-beta protein and the formation of neurofibrillary tangles due to hyperphosphorylation of tau protein. Due to the significantly increased number of AD patients across the globe, it is a matter of global concern and needs urgent action. It has been demonstrated that aging and epigenetic changes such as histone modification, chromatin remodeling, DNA methylation, etc. play a crucial role in the onset of neurodegenerative diseases including AD. Further, with increased age, there is a decline in physical/mental activities, increased oxidative stress, altered mitochondrial dynamics, several biochemical changes, and epigenetic

D. K. Zala · S. Banerjee · A. K. Tiwari (✉)
Genetics and Developmental Biology Laboratory, School of Biotechnology and Bioengineering, Institute of Advanced Research, Gandhinagar, India
e-mail: anandk.tiwari@iar.ac.in

J. Barciszewski (ed.), *Decoding Aging and Neurodegeneration*,
RNA Technologies 16, https://doi.org/10.1007/978-3-031-91072-2_15

modifications that add up the AD and related pathologies. Further, it has been shown that understanding the epigenetic regulation in aging and aging-associated neurodegenerative diseases might help us understand the details of disease mechanisms and discover the potential therapeutic targets for AD. In the current review, we have provided an update on Alzheimer's disease and its link to aging and epigenetic modification.

Keywords Aging · Neurodegenerative disease · Alzheimer's disease · Epigenetics · Histone modification

1 Introduction

Epigenetic regulation has a critical role during the developmental process as well as the progression of age-related neurological disorders mainly in Alzheimer's disease (AD). The ample evidence suggested that altered deoxyribonucleic acid (DNA) methylation, histone modifications, and changes in the expressions of non-coding ribonucleic acids (ncRNAs) contribute significantly to the pathological consequences of the diseases. During aging, these epigenetic changes interrupt gene expression patterns associated with neuronal health, promoting intricate mechanisms that lead to neuroinflammation, oxidative stress, and accumulation of amyloid-beta (Aβ) and tau proteins in the brain. Such modifications are often apparent as age-associated *epigenetic drift*, where dysregulated epigenetic marks accumulate in random fashions, compromising the genomic stability and normal cellular functions that are vital for memory and cognition. By comprehending these age-related epigenetic alterations, researchers mainly aim to recognize potential therapeutic targets to modulate gene expression patterns and delay or prevent Alzheimer's disease progression.

2 Concept of Aging and Related Biological Process

Aging is a complex and inevitable process that involves the deterioration of cells and tissues. It is a process that involves multiple changes at many levels. The decrease in cells can lead to changes at biological levels, and almost all body systems are negatively affected by it. Biological aging does not necessarily refer to the end years of adult human life but is rather a series of changes one has to go through. Aging can be divided into primary and secondary categories. Primary aging is inherent and determined by hereditary influences. The rate of aging is determined by functional decline which is different for different humans. External factors like trauma, environmental factors, or acquired diseases cause secondary aging. Aging is majorly linked with cell loss of irreplaceable cells. This particularly occurs in skeletal muscles, the heart, and the brain. Around the age of 80, striated muscle mass can decrease by almost 50% as cells are replaced by fat and fibrous connective tissue. Thus, the body is provided with more fat cells to store certain drugs. The loss of heart cells affects cardiac function while alterations in cells and supportive tissues decrease pulmonary function. In the brain, neuron shrinkage and loss, particularly in the hypothalamus, may disrupt metabolism and circadian rhythms, contributing to mental and emotional issues in the elderly. Neurotransmitter levels also decline, including dopamine, norepinephrine, and serotonin, while monoamine oxidase activity increases (Hung et al. 2010). The aging brain undergoes significant structural and functional remodeling, particularly in the prefrontal cortex and white matter tracts, which affect cognition and memory. Cellular changes include reductions in spine density, axonal transport, and synapse number, as well as shifts in neurotransmitter levels, mitochondrial dysfunction, and oxidative damage. Age plays a major risk factor in neurodegenerative disease as it might be directly linked with neurodegenerative diseases. It also elevates susceptibility to cardiovascular issues, cancer, neurodegenerative disorders, type II diabetes, and various infections (Partridge et al. 2018). Despite being significant in human biology and its profound implications for society and medicine, the mechanisms behind aging remain poorly understood (De Magalhães 2023). Importantly, the neurological disease burden in old age group of people was found to be a bit high, and the mortality rate for Alzheimer's disease and dementia was also found high worldwide till 2021 according to the World Health Organization data (Fig. 1).

2.1 *Aging at the Cellular Level*

2.1.1 Role of Genomic Instability in the Aging Process

Human DNA is continually exposed to a variety of mutagens, both internal and external. Internal mutagens encompass free radicals created during the electron transport chain's energy-generating process. External mutagens include radiation and toxins. Most of this damage can be fixed by effective cellular mechanisms for DNA repair, but some damage remains unrepaired, and mutations build up gradually (Moskalev et al. 2012). Because of these unrepaired mutations, genetic instability, which can be in many forms, from small point mutations to large chromosomal breakage and translocations, is more common in older cells. According to research, base excision repair (BER) is the main process in charge of repairing minor DNA base changes like oxidative base damage and single-strand breaks. A mutator phenotype may result from mutations in genes related to the BER pathway, which may increase the risk of neurodegeneration and aging (Tell and Demple 2014; Leandro et al. 2015). Neurodegenerative illnesses have been associated with many forms of DNA damage, such as insertions, deletions, base mismatches, bulky adducts, abasic sites, single- and double-strand breaks, and so on (Jeppesen et al. 2011). Reactive oxygen species (ROS) from endogenous sources can induce oxidative DNA damage, which can worsen inflammation, accelerate aging, and raise the risk of

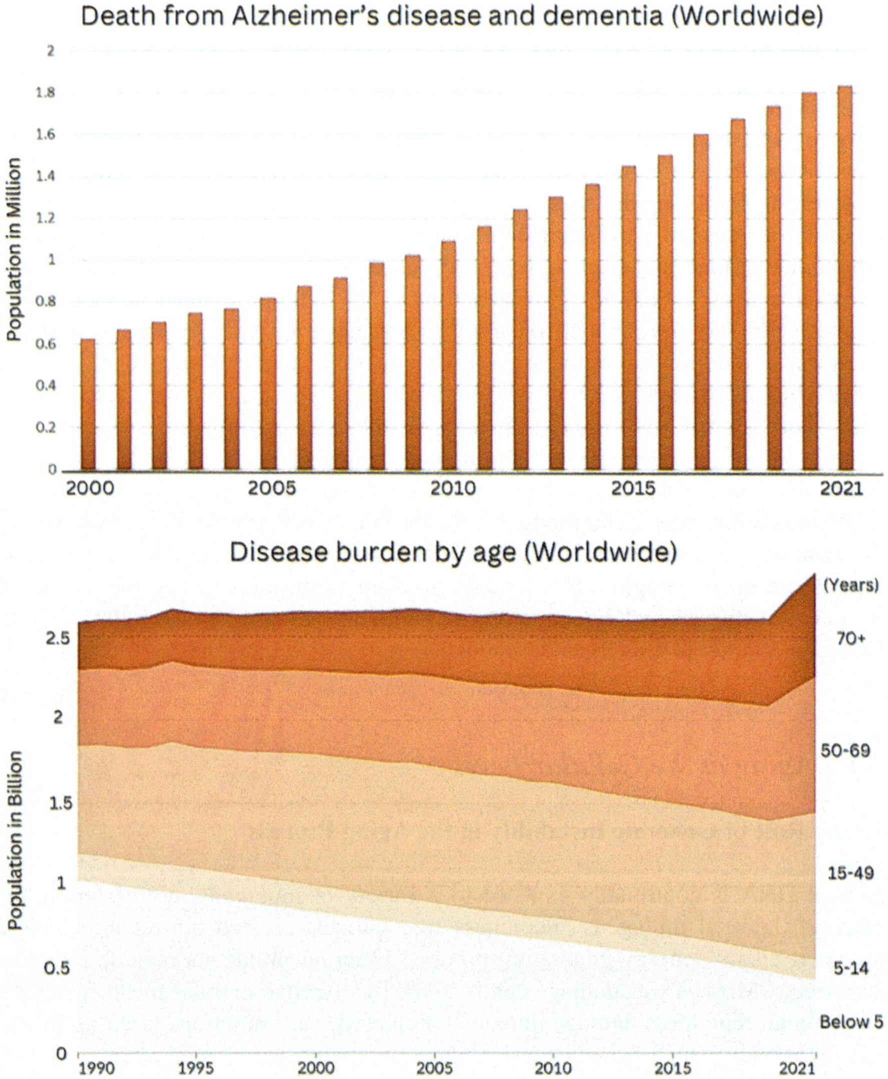

Fig. 1 Estimation of death for neurological disorders and age-related disease burden worldwide. The illustration shows the worldwide estimation of death for Alzheimer's disease and dementia (upper panel) and the total disease burden in different age groups (lower panel) (World Health Organization 2024)

neurological diseases and cancer (Thanan et al. 2014). DNA damage not only creates genomic instability but also triggers signaling pathways that proliferate across the cell (Fang et al. 2017). It encourages inflammation and cellular senescence, both of which are factors in age-related neurodegeneration.

2.1.2 Loss of Proteostasis: Hallmark of Aging

According to the report, proteostasis is the collective term for the set of biological mechanisms that support the stability of the proteome, including human protein synthesis and degradation (Eisenstein 2014). Proteostasis is the steady state achieved through the equilibrium between the synthesis and breakdown of proteins. The maintenance of proteostasis in eukaryotic cells is contingent upon the appropriate control of the proteasome, autophagy, ubiquitination machinery, and the lysosomal system. While autophagy is an intracellular process that breaks down misfolded or unfolded proteins and damaged organelles in lysosomes, it also helps limit the release of inflammatory cytokines. Ubiquitinated proteins are normally targeted for degradation (Tanaka and Matsuda 2013). Protease-mediated folding of proteins, proteasomal degradation, and autophagy are all part of the proteostasis network. Proteotoxic effects and protein aggregation may arise from the failure of any one of these components. Studies have demonstrated that as organisms age, their capacity to preserve proteome stability decreases, resulting in the development of protein inclusions and aggregates in almost all of the older creatures' tissues. Misfolded proteins can accumulate and clump together, which is one of the main causes of many age-related illnesses. These involve a range of illnesses, including non-neurodegenerative ones like cataracts, although they are mostly neurodegenerative disorders like Parkinson's and Alzheimer's disease (Labbadia and Morimoto 2015).

2.1.3 Role of Telomere Attrition in Aging

At the ends of linear chromosomes are DNA-protein sections called telomeres. The telomerase hypothesis of aging and the theory of the Hayflick limit were initially put forward in 1961 when Hayflick and Moorhead noticed that human diploid cell lines could only be cultivated for a limited number of passes. Unless the parental cell expresses telomerase or has another mechanism to avoid this shortening, telomeres shorten with each cell division. Shortening of the telomere causes a cell to age. Cellular aging is believed to be mostly caused by telomere dysfunction and genomic instability (Xie et al. 2015; Pusceddu et al. 2015). Cellular senescence is caused by telomere shortening, which is probably related to organismal aging because telomere maintenance issues can hasten aging in both humans and mice.

2.1.4 The Effect of Dysregulated Nutrient Sensing in Aging

The effect of food and calorie restriction on aging has been the subject of numerous research. An essential part of using substrates for energy production is the function of mitochondria. Consequently, the body as a whole may be impacted by any reduction in mitochondrial DNA, biogenesis, dynamics, or integrity, which could result in the disturbed metabolism that is frequently linked to human aging (Riera and Dillin 2015). The mammalian target of rapamycin (mTOR), which is involved in controlling protein translation and cell growth in response to growth cues and is frequently reported to be increased in cancer, is a crucial component of nutrition sensing and metabolic regulation (Dobashi et al. 2011). AMP-dependent protein kinase, which blocks mTOR signaling, is yet another significant enzyme in food sensing. AMP-dependent protein kinase is activated by factors such as calorie restriction, exercise, and food deprivation, leading to AMP accumulation and ATP depletion. This kinase regulates stress tolerance and prolongs longevity in a variety of species. It is associated with the production of autophagy in many cancer cells, which supports cell survival during environmental stress. Nicotinamide adenine dinucleotide (NAD), another nucleotide linked to starvation, is frequently accumulated in high AMP levels. Though their precise function is yet unknown, sirtuins—deacetylases that are triggered by high NAD levels, which are indicative of nutritional scarcity or growth inhibitory signals (Michan 2014)—might possess an impact on lifespan. Interestingly, sirtuins have been demonstrated to control the insulin signaling pathway and may function as regulators of longevity. Thus, while nutrient excess tends to drive age-related disorders, there are substantial linkages between systems that limit energy and nutrients that promote lifespan extension. A greater amount of evidence suggests that calorie or diet restriction, as well as intermittent fasting, may trigger similar metabolic processes. Enhanced autophagy response and improved mitochondrial efficiency are two of these processes. These result in a host of advantageous biological effects, including enhanced circulation, defense against heart disease, and control over reactive oxygen species and inflammatory cytokines. Furthermore, studies have demonstrated the antimutagenic, antimicrobial, and anticarcinogenic properties of intermittent fasting (Lee and Longo 2011).

2.1.5 Exhaustion of Stem Cells in the Aging Process

In adult life, functional stem cells are crucial for preserving optimum health. However, as organisms age, stem cell activity and proliferation capacity decline. According to the findings of Oh et al. (2014), there are several possible causes for this reduction, including high levels of aging-related DNA damage, diminished DNA repair ability, compromised proteostasis, epigenetic dysregulation, mitochondrial malfunction, telomerase inactivation, and/or cellular senescence. Importantly, rapamycin therapy and upregulating the expression of the heat shock protein HSP70 or the forkhead transcription factor FOXO4 can both improve the activity of stem cells. Heterochronic parabiosis exposes aged animals to youthful blood, which counteracts aging and revives cognitive functions while enhancing stem cell function in tissues like the muscle, liver, spinal cord, and brain (Villeda et al. 2014). An important factor in the aging process is stem cells. Anemia arises from the depletion of hematopoietic stem cells, osteoporosis and fractures are caused by the decline of mesenchymal stem cells, and poor intestinal function is caused by the depletion of intestinal epithelial stem cells. These are typical signs of getting older. Moreover, sarcopenia, the age-related decrease of muscle mass, is linked to malfunction in muscle stem cells, commonly referred to as satellite cells (Du et al. 2014).

2.1.6 Altered Intercellular Communications and Neural Function in Aging

When people get older, their immune systems' ability to defend against infections and precancerous cells deteriorates, which increases the likelihood of opportunistic infections and the growth of cancerous cells. In turn, this results in increases in intercellular inflammatory reactions. The main causes of chronic inflammation include this process, called immunosenescence, and the increased cytokine release by adipose tissue (van Szic et al. 2015). The development of early aging, often known as "inflammaging," is closely associated with immunosenescence and chronic low-grade systemic inflammation. Inflammaging also plays a role in other age-related conditions such as diabetes, arthritis, metabolic syndrome, and cardiovascular disease. Neuronal injury and neurodegeneration are regulated by changes in hormone levels, including those of leptin, ghrelin, insulin, adiponectin, and insulin-like growth factor 1 (IGF1). Major molecular biomarkers and their role in age-related disorders are summarized in Table 1.

Since the immune system and the neurological system both vary with age, the loss of immune response control in the brain is probably a factor in dementia. The immune system is essential for brain development. Recent research indicates that interactions between microglia, the CNS's innate immune cells, and other glial support cells like astrocytes and oligodendrocytes may play a role in the pathogenic processes of neurodegenerative disorders. Regulatory T cells play a critical role in immunological tolerance and privilege by preserving immune homeostasis both peripherally and within the central nervous system (CNS) (He et al. 2013). The immune system's ability to protect against neurodegenerative illnesses depends on the interaction of mast cells, microglia, T cells, oligodendrocytes, and their cross-talk (Cao and Zheng 2018). Aging causes inflammation, a necessary defensive reaction to stressors, to become more noticeable. Reduced inflammation can be advantageous, but excessive or unchecked inflammation can be detrimental and cause long-term age-related neurodegenerative illnesses such as Parkinson's disease (PD) and AD. The important function that inflammation plays in neurodegenerative illnesses is highlighted by studies of gene expression in aging human brains. Age-related neurodegenerative disorders are linked to chronic inflammation, extended elevation of pro-inflammatory mediators, increased oxidative stress, and persistent microglial activation (Hickman et al. 2018). Age-related deficiencies in sex hormones (Uchoa et al. 2016), compromised autophagy and proteasome degradation systems (Nakahira et al. 2010), cellular senescence (Baker and Petersen 2018), increased oxidative DNA damage and inadequate DNA repair (Pan et al. 2016), reduced innate and adaptive immune responses (Frasca and Blomberg 2015), and environmental stressors (Bektas et al. 2017) are some of the factors that contribute to the inflammatory response and pathogenesis of neurodegenerative disorders. Age-related neurodegenerative illnesses may benefit from treatment that targets these neuroinflammatory processes.

Table 1 Major biomarkers related to clinical and molecular features of age-associated neurological disorders

Disease	Molecular characteristics	Clinical symptoms	Genes responsible	References
Alzheimer's disease	Aβ plaques formation, neurofibrillary tangles, neuronal loss, neuroinflammation	Alzheimer's dementia, cognitive difficulties	*Presenilin 1* (*PS1*) on chromosome 14 *Presenilin 2* (*PS2*) on chromosome 1 *Amyloid* β*-protein precursor* (*APP*) on chromosome 21 *Apolipoprotein E* (*APOE*) gene on chromosome 19	Lyman et al. (2013), Peters et al. (2013)
Parkinson's disease	Lewy bodies containing α-synuclein Gray matter atrophy	Visual hallucinations and illusions Cognitive dysfunction, dementia and psychosis, dysregulated sleep-wake cycle, depression, and sensory symptoms	SNCA 22 PRKN	Aarsland Dag et al. (2010), Mak et al. (2013)
Amyotrophic lateral sclerosis	TAR DNA-binding protein 43 aggregation	Bulbar: dysarthria, dysphagia, facial weakness Cervical: weakness in upper limbs Lumbar: weakness in lower limbs	Superoxide dismutase 1 (SOD1), TAR DNA-binding protein (TARDBP), fused in sarcoma (FUS), and chromosome 9 open reading frame 72	Akçimen et al. (2023), Mitchell and Borasio (2007)
Huntington's disease	Formation of mHTT (mutant HTT) protein aggregates and inclusion bodies in neurons	Chorea, dystonia Loss of coordination, involuntary movements, epileptic seizures Cognitive decline, emotional disturbance, and behavioral difficulties	Huntingtin (HTT) gene chromosome 4, polyglutamine aggregation	Penney et al. (1990), Davies et al. (1997)

(continued)

Table 1 (continued)

Disease	Molecular characteristics	Clinical symptoms	Genes responsible	References
Ataxia telangiectasia	DNA damage accumulation, elevated ROS, and cellular senescence	In children: nystagmus, problem walking, motor disabilities In adults: involuntary movements leading to athetosis, cholera, rhythmic or non-rhythmic tremors that affect voluntary movements	Ataxia telangiectasia mutated (ATM) gene on chromosome 11q22–23	Riboldi et al. (2024), Pearson (2016), Perlman et al. (2011)
Multiple sclerosis	Inflammatory demyelination, astrogliosis, neuronal and axonal loss involving the brain and spinal cord	Motor dysfunction, fatigue, visual impairment, sensory changes, and cognitive impairment, resulting from central nervous system demyelination	HLA-DRB1*15:01	Barrie et al. (2024), Papadopoulos et al. (2020)

2.1.7 Role of Mitochondrial Dysfunction in Aging

Damaged mitochondria produce more ROS, which are linked to normal aging and most known neurological disorders. Apart from ATP generation, mitochondria have significance for multiple intercellular processes like apoptosis, calcium signaling, and lipid biosynthesis, all of which are vital for the onset of neurodegenerative diseases (Keogh and Chinnery 2015). Age-related declines in brain mitochondrial function are considered to be a major and early cause of aging (Fig. 2).

Protease degradation, the export of damaged proteins via mitochondria-derived vesicles, mitochondrial fusion and fission, the removal of damaged proteins by proteases, and mitophagy are a few quality control mechanisms that support the maintenance of optimal mitochondrial function. In the beginning, autophagy acts as a rescue mechanism, but if it gets out of control, it can transition to a route that encourages cell death. Mitophagy, the autophagic clearance of damaged mitochondria, is the major reaction to mitochondrial malfunction and is essential for cellular health. The accumulation of misfolded proteins in the mitochondria can be addressed by the mitochondrial unfolded protein response, which activates the protein-folding machinery comprising chaperones and proteases (Haynes et al. 2013; Pellegrino et al. 2012). This response eventually triggers apoptotic mechanisms. Generally

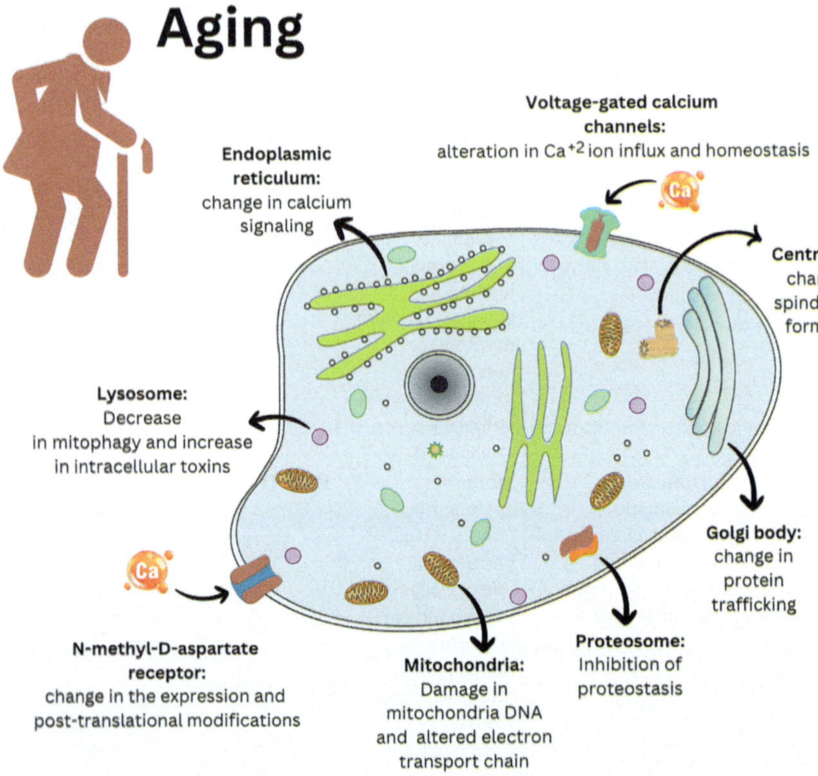

Fig. 2 Deregulations of cellular organelles' function in aging. The figure represents the cellular organelles' functions that are altered during aging. Major cellular receptors, ion channels, and organelles which are associated with energy production, protein trafficking, processing, and degradation, may show deregulated activities through the different stages of the aging process

speaking, autophagy has been shown to decrease with aging; therefore, improving this system is believed to extend life by getting rid of damaged organelles and molecules that may otherwise build up and cause diseases like Alzheimer's (Lane et al. 2015). Mitophagy is a particular kind of autophagy that plays a key role in avoiding age-related illnesses by specifically destroying defective mitochondria. The PTEN-induced putative kinase protein 1 (PINK1)-Parkin route is one of the mitophagy pathways. By recruiting the E3 ubiquitin ligase Parkin, which ubiquitinates mitochondrial proteins and speeds up the breakdown of damaged mitochondria, PINK1 accumulates on the outer membrane of depolarized mitochondria in this route. After autophagosomes combine with lysosomes to generate autolysosomes, damaged organelles and/or the proteins inside are broken down. The locomotor deficiencies in *Drosophila melanogaster* and early-onset recessive Parkinson's disease in humans (Pickrell and Youle 2015) have been linked to mutations in the genes encoding PINK1 or Parkin.

2.1.8 Role of Cellular Senescence in Aging

Cellular senescence is a state of sustained cell cycle arrest brought on by stress and is characterized by the development of a senescence-associated secretory phenotype (SASP) with aging. It was initially identified in fibroblasts by Hayflick (1965). Under stressful circumstances, this procedure might be seen as an attempt to protect the survival of healthy cells while eradicating damaged ones. Multiple mechanisms of senescence have been found, such as replicative senescence, senescence triggered by oncogenes, stress-driven premature senescence, and senescence associated with mitochondrial dysfunction (Kuilman et al. 2010). Although cellular senescence suppresses tumors at first, aging may cause it to change into a mechanism that encourages the beginning of tumors (Loaiza and Demaria 2016). Significant amounts of DNA damage cause cells to go senescent and stop reproducing. The brain tissue can collect considerable quantities of DNA damage resulting from the high energy requirements and increased production of ROS in neurons. This damage tends to grow with age and the corresponding loss in DNA repair capacity (Madabhushi et al. 2014). Because they are postmitotic cells, neurons are especially prone to the accumulation of DNA lesions and mainly rely on the error-prone repair process known as non-homologous end joining instead of the more reliable technique known as homologous recombination for repairing DNA double-strand breaks. Pathological circumstances like Alzheimer's disease often exhibit neuronal re-entry into the cell cycle, which results in neuronal hyperploidy and synaptic failure (Barrio-Alonso et al. 2018). *Senescence-like neurons* are defined as neurons exhibiting continuous activation of the DNA damage response (DDR) pathway, which has several features of cellular senescence (Fielder et al. 2017). Similar to its role in proliferating cells, the researchers found that p21 functions as a critical signal transducer connecting the DDR to the senescence-like phenotype in neurons (Jurk et al. 2012). According to a different study, aging not only reduces the effectiveness of DNA repair but also activates more complex DNA repair pathways, which leads to an increase in the number of mutations. Of note, these less efficient DNA repair pathways are connected to cancer, aging, and possibly even dementia.

2.1.9 Role of Autophagy and Senescence in Aging

Another important mechanism linked to cellular senescence is autophagy, yet it is still unclear if autophagy encourages or inhibits senescence. A possible way that macro-autophagy could contribute to the establishment of the SASP is through increasing secretory protein synthesis. On the other hand, in some circumstances, when p53 and ROS are present, autophagy suppression may worsen senescence (Gewirtz 2013). Age-related decline in autophagy has been observed in the brain. However, more research is needed to determine the precise mechanism relating the decline in autophagy to brain senescence. Telomere shortening, inflammation, autophagy insufficiency, chronic DDR, oxidative stress, and mitochondrial dysfunction are some of the factors linked to cellular senescence that may be involved (Bhatia-Dey et al. 2016).

2.1.10 Epigenetic Modifications: Role in Aging

Conrad Waddington coined the term "epigenetics," which refers to the inheritance of phenotypic changes that happen independently of changed DNA sequences. According to current studies, both pathological and physiological memory formations depend heavily on epigenetic mechanisms (Fischer et al. 2010). Nowadays, this word is frequently used to refer to a variety of activities, such as those that take place in neurons following mitosis and entail the expression of several genes controlled by epigenetic mechanisms (Stilling and Fischer 2011). Three main types of epigenetic processes are (1) DNA hydroxymethylation and methylation, (2) histone modification, and (3) non-coding RNAs. Long-term adaptive alterations in gene expression are the consequence of these processes' substantial contribution to the interaction of genetic and environmental variables. DNA methylation and gene silencing are closely related processes in the field of epigenetics. The methylation of DNA's cytosine residues at the C-5 position is among the most thoroughly researched examples of this. This resulting change commonly takes place in regions of the genome called CpG islands, where guanine succeeds cytosine in the DNA sequence and where cytosine is prevalent. These CpG islands are crucial for controlling gene expression and are generally found in the promoter regions of genes. DNA methyltransferases (DNMTs) are a class of enzymes that assist DNA methylation by moving a methyl group from S-adenosyl-L-methionine (SAM) to the cytosine base. DNMT1, DNMT2, DNMT3a, and DNMT3b are the four different forms of DNMTs that have been found. Of these, DNMT1 is primarily in charge of preserving the pattern of DNA methylation during the process of cell division. It ensures that during replication, these methylation patterns are precisely copied onto newly synthesized DNA strands, maintaining the gene-silencing status for successive generations of cells. 5-Methylcytosine is produced when DNMT1 distinctly moves a methyl group from SAM to the cytosine ring's 5′ position. DNMT2 plays two roles: it is a 5-cytosine DNA methyltransferase and a ribonucleic acid (RNA) methyltransferase. On the other hand, de novo methylation, which entails creating novel methylation patterns during development, particularly during gametogenesis and early embryonic stages, is primarily the function of DNMT3a and DNMT3b (Jalili 2013). Normal development and cellular differentiation depend on the creation of tissue-specific gene expression patterns, which are made possible by these enzymes. According to studies conducted on mice, during these early stages of development, repeated regions that have previously been demethylated are regularly remethylated, which guarantees that particular genes are either activated or silenced in the proper tissues. Such regulation is essential to numerous biological processes. Moreover, DNA methylation plays a role in the development of cancer as well as processes including long-term memory formation, X chromosome inactivation, germline-specific gene imprinting, gene silencing, and more. There are two primary methods by which gene silence happens once a DNA sequence is methylated (Wu et al. 2012). First, the start of gene transcription can be hindered by methylation in the promoter region, which can inhibit transcription factors from binding to the DNA. Second, these methylation regions can connect to methyl-CpG-binding proteins (MBDPs), which in turn can attract other proteins that modify chromatin and cause it to condense into an inactive, heterochromatic form. Further supporting gene silence, this heterochromatin structure prevents transcriptional machinery from accessing the DNA. When MBDPs and other epigenetic factors come together to generate inactive and condensed heterochromatin, gene transcription is reduced and silenced. Furthermore, hydroxylation of 5-methylcytosine on methylated CpGs results in even lower transcription levels than in the methylated state, representing another layer of alteration. Ten-eleven-translocation 1 (TET1) is an enzyme that specifically catalyzes DNA hydroxymethylation in neurons. This oxidation can also occur as a result of oxidative stress (Tahiliani et al. 2009). It has been demonstrated that TET proteins are engaged in both transcriptional repression and activation; however, the precise role of these proteins depends on the other proteins with which they interact (Xu and Wong 2015). As a result, controlling gene expression and preserving cellular identity depend on DNA methylation and hydroxymethylation. According to a recent study, TET proteins aid in the hydroxymethylation of cytosine, which is involved in controlling gene expression in brain tissues. While DNA methylation is known to be involved in the suppression of gene activity, DNA hydroxymethylation's purpose is less understood. However, hydroxymethylation is believed to be a step on the way to DNA demethylation (Tan and Shi 2012).

Histones function as the fundamental proteins that make up the nucleosomes, which are the fundamental components of chromatin. A closer look at the structure of nucleosomes shows that they are an octamer made up of two molecules of each core histone, H2A, H2B, H3, and H4. This structure has a 147 base pair DNA segment wrapped around it. The term *histone tail* refers to the flexible N-terminal portions of histones that extend from the surface of the nucleosome. Many post-translational changes occur in these tails, with lysine acetylation and methylation being the most common (Fischer et al. 2010). Histone modifications include a range of covalent, reversible chemical events that add or remove functional groups from the amino acid residues of histone protein octamers. These reactions include phosphorylations; mono-, di-, and tri-methylations; and acetylations. These modifications predominantly occur on the eight N-terminal histone tails which extend from the histone-DNA complexes (nucleosomes), particularly at lysine and serine residues. The histone-d-DNA complex may become looser or tighter, depending on the type of alteration. This might help attract gene regulatory proteins to the impacted DNA segments, which can then change gene expression in a pattern that repeats throughout the genome. Global depletion is one histone-related process that affects aging.

2.1.11 Role of Non-coding RNA in Aging

Gene regulation and silencing are important functions of non-coding RNAs. With over 1800 identified in humans, microRNAs (miRNAs) are the most studied class of short, single-stranded RNAs (Kozomara and Griffiths-Jones 2010). MicroRNAs are typically about 22 nucleotides in length. As much as 80% of all expressed genes may be regulated by these miRNAs (Lu and Clark 2012). MiRNAs are thought to affect the expression of histone methyltransferases, which leads to the suggestion of feedback loops between miRNA gene regulation and other epigenetic mechanisms. Furthermore, miRNAs control genes and pathways linked to senescence. Their significance in aging is well-established and conserved. They mediate the modulation of the target of rapamycin (TOR) and insulin/insulin-like growth factor (IGF) 1 signaling pathways (Barzilai et al. 2012). Furthermore, immunosenescence—the immune system's deterioration, another sign of aging in humans—is believed to be significantly influenced by miRNAs (Aalaei-Andabili and Rezaei 2015). Alzheimer's disease patients' brains, serum, and cerebrospinal fluids (CSF) have been found to have altered populations and expression levels of miRNAs (Murray et al. 2015). Multiple investigations have demonstrated the critical role that long non-coding RNAs (lncRNAs) play in aging and age-related diseases through their ability to control gene expression (He et al. 2018) and impact several important biological processes, such as cell division, proliferation, and death. According to several studies (Halliday and McCann 2009; He et al. 2018), antisense mRNA strands and pseudogenes derived from promoter regions can produce lncRNAs.

3 Aging and Neurodegenerative Diseases

Specific misfolded proteins accumulate in the central nervous system of many age-related neurodegenerative disorders. The diseases AD, PD, amyloid peptides and tau/phosphorylated tau (p-tau) proteins, amyotrophic lateral sclerosis (ALS), Huntington's disease, and superoxide dismutase (PD) are a few examples. Although the exact connection between misfolded proteins and aging is unknown, it could have something to do with cellular alterations brought on by aging. For example, the quality control mechanisms controlling protein synthesis decrease, and cells tend to shrink with age. This reduction could have an impact on the formation of misfolded protein aggregates, which could cause illness.

3.1 Alzheimer's Disease

According to report, the incidence of AD rose from 2.8 per 1000 person-years in those 65 to 69 to 56.1 per 1000 person-years in those 90 years of age and above (Kukull et al. 2002). The main pathogenic characteristics of AD are tau/

phosphorylated tau proteins and amyloid peptides, which are combined to form neurofibrillary tangles in the cortex and senile plaques, respectively. The accumulation of these misfolded proteins causes synaptic disruption and neuronal death; yet, the mechanisms behind protein misfolding are still not well understood. Age-related cellular alterations are the main risk factor for AD, and they may be connected to the misfolding and aggregation of protein processes. Furthermore, oxidative stress and mitochondrial dysfunction are linked to aging. The brain is particularly susceptible to dysfunctional mitochondria. Although both AD and normal aging of the brain are associated with memory decline, their pathophysiology and neurophysiology are different. Reduced metabolic activity and neuronal loss initially occur in the entorhinal cortex and the CA1 region of the hippocampus, according to research on AD using functional magnetic resonance imaging and histopathological analysis. AD is the neurodegenerative ailment that is most common. Disorientation, mood swings, behavioral issues, and memory and learning difficulties in later life are the main clinical symptoms of AD. AD risk factors include deficiencies in at least three proteins: presenilin 1 and 2, APP, and presenilin 2 (Bekris et al. 2010). Early-onset AD in families is linked to mutations in the genes encoding these proteins. On the other hand, sporadic (late-onset) AD is significantly more common and is thought to be the consequence of intricate interplay between environmental and genetic factors. One major factor contributing to AD is the neuroinflammatory response, which is the brain's immunological response to internal or external stressors, excessive oxidation, the accumulation of Aβ plaques, or p-tau (Fu et al. 2018). The primary cause of this reaction is the aberrant activation of microglia and astrocytes, which results in the production of cytokines that promote inflammation. Additional risk factors for AD could include DNA damage and mitochondrial malfunction. To create viable therapies for AD, new targets and/or different techniques are required, as evidenced by the clinical failures of therapeutic strategies based on the Aβ hypothesis (Hardy 2009). BER protein expression varies depending on the tissue type, cell type, and age of the cells; as the disease progresses, its expression may decrease. According to research, the post-mortem brain tissue from AD patients has reduced levels of uracil DNA glycosylase, β-OGG1 glycosylase, and Polβ in comparison to age-matched controls without AD. Polβ null alleles are fatal during embryonic development and have been connected to neurodevelopmental abnormalities in mice. The etiology of AD appears to be greatly influenced by epigenetic modifications, as evidenced by aberrant DNA methylation patterns in multiple AD-affected brain areas and changes in total methylation levels. Additionally, it has been noted that the hippocampal and lymphocytes of AD patients exhibit higher levels of phosphorylated histone H2AX (H2A histone family member X), suggesting an active response to DNA damage (Silva et al. 2014; Siddiqui et al. 2018). An additional factor in AD may be telomere instability or shortening. In humans and mice, telomere maintenance deficiencies can hasten the aging process. In AD patients, these deficiencies are linked to inflammation and oxidative stress-induced tau hyperphosphorylation, amyloid pathology, and cognitive decline (Herrmann et al. 2018).

As previously mentioned, human age-related disorders are influenced by epigenetic pathways. Epigenetic mechanisms control a variety of phenotypes and biological functions, such as learning, memory, and behavior. It is thought that environmental and epigenetic factors affect a person's susceptibility to sporadic AD. The three types of epigenetic changes that are relevant to AD are methylation, PARylation, and acetylation of DNA and histones. For example, distinct methylation patterns have been found in diverse tissues and brain regions, corresponding with amyloid precursor protein (APP) expression, for the human and monkey APP promoters. Features of AD are associated with specific patterns of DNA methylation and hydroxymethylation (5-methylcytosine and 5-hydroxymethylcytosine) (Bradley-Whitman and Lovell 2013). PARylation, which can result from the overactivation of PARP1, may exacerbate AD pathogenesis (Fang et al. 2016). It is believed that alterations in histone levels and their patterns of modification have a role in the etiology of AD. According to reports, the hippocampal regions of AD patients had decreased amounts of acetylated histone H4 and phosphorylated histone H3 and H3 hyperphosphorylation (Kwon et al. 2016). Furthermore, it has been demonstrated that patients with AD also have considerably higher levels of histone deacetylase 2 (HDAC2) and HDAC6; however, as evidenced by animal models of AD and other neurodegenerative illnesses, these effects may be offset by histone deacetylase (HDAC) inhibitors (Kwon et al. 2016). Recent data suggests that AD is associated with reduced mitophagy, which results in the build-up of defective mitochondria (Kerr et al. 2017). Research has shown that mitophagy lowers inflammation, which lowers the risk of AD, PD, and various other neurodegenerative disorders (Sliter et al. 2018).

Cellular senescence, another characteristic of aging, makes people more susceptible to neurodegenerative illnesses including PD, AD, and others (Chinta et al. 2018; Turnquist et al. 2016). According to research, there is a correlation between the build-up of Aβ fibrils in AD and a rise in the expression of p16INK4a and SASP components. Brain tissues from AD patients have been shown to have increased numbers of senescent astrocytes, microglia, and neurons, together with enhanced expression of senescence-associated β-galactosidase (SA-β-gal) activity (He et al. 2013). Aβ42 oligomers create a senescent phenotype, which increases in SA-β-gal-positive cells, according to in vitro experiments utilizing mouse neural stem cells. These findings are corroborated by in vivo research using AD animal models. Moreover, increased SA-β-gal levels have been found in AD patient plasma samples when compared to control samples (Magini et al. 2014). On the other hand, monocytes and lymphocytes from AD patients showed a decrease in SA-β-gal, which was probably brought on by the overexpression of miR-128 (Tiribuzi et al. 2013). There have been reports of increased p38MAPK activity in the cortex of AD mouse models (Sun 2003) and the brains and lymphocytes of AD patients (Ohyagi et al. 2004). Several SASP components, such as IL-6, IL-1β, TGF-β, and TNF-α, are elevated in AD, which is not surprising considering that p38MAPK is a major regulator of SASP (Wang et al. 2014). Cerebrospinal fluid, serum, and AD brain tissues have all been shown to have elevated levels of these pro-inflammatory cytokines (Swardfager et al. 2010; Gezen-Ak et al. 2013). AD has also been linked to elevated levels of matrix metalloproteinases (MMP-1, MMP-3, and MMP-10) (Horstmann et al. 2010). Senescent neurons and senescent glial cells, such as progenitors of oligodendrocytes, microglia, and astrocytes, are sources of oxidative and inflammatory stress that can cause a similar senescent phenotype in adjacent cells, which can be deleterious (Kirkland and Tchkonia 2017; Bussian et al. 2018). These results suggest that removing senescent cells from the brain could have beneficial therapeutic benefits. Research on AD has indicated that this strategy has a potential.

The chance of acquiring AD may also be influenced by nutritional and metabolic variables. For example, calorie restriction has been demonstrated to offer neuroprotective effects (Fontana et al. 2010), and one prevalent early feature of AD is brain resistance to insulin and IGF1. mTOR signaling is elevated in AD, and rapamycin and other mTOR inhibitors have been shown to ameliorate AD pathology, including cognitive deficits (Oddo 2012). Evidence suggests that increased DNA damage and decreased DNA repair play a role in the development of AD. BER protein expression varies depending on the tissue type, cell type, and age of the cells; as the disease progresses, its expression may decrease. The pathogenesis of AD appears to be significantly influenced by epigenetic modifications, as evidenced by abnormal DNA methylation patterns in several AD-related genes and changes in overall methylation levels in brain regions affected by AD (Qazi et al. 2017; Watson et al. 2016). Furthermore, the hippocampal and lymphocytes of AD patients have been shown to have higher levels of phosphorylated histone H2AX, suggesting an active DNA damage response (Silva et al. 2014; Siddiqui et al. 2018) is connected to abnormalities in mouse neurodevelopment (Sykora et al. 2014).

3.2 Parkinson's Disease

Another prevalent age-related neurodegenerative condition requiring protein misfolding is PD. α-Synuclein, the misfolded protein in Parkinson's disease, builds up to create Lewy bodies. According to studies, Lewy bodies initially develop in the lower brainstem and olfactory regions before gradually spreading to the midbrain and cortex (Braak et al. 2006). Lewy-body deposits have been tracked using a staging approach, and the emergence of clinical symptoms is correlated with this (Halliday and McCann 2009). The principal pathology associated with PD is the degradation of dopamine neurons in the substantia nigra pars compacta, which project to the striatum. About 50% of nigral neurons and 80% of striatal dopamine are destroyed before clinical signs including bradykinesia, stiffness, and resting tremors appear. α-Synuclein aggregates found in intracellular inclusions and substantia nigra neuronal loss, which results in striatal dopamine insufficiency, are characteristics of CliPD. Numerous routes and mechanisms, including α-synuclein proteostasis, oxidative stress, mitochondrial function, calcium regulation, axonal transport, and neuroinflammation, are involved in the molecular pathogenesis of PD (Poewe et al. 2017). Neurons in every PD patient have aggregation of α-synuclein. Large insoluble fibrils of α-synuclein are formed when oligomers made of soluble α-synuclein monomers combine. These fibrils are neurotoxic and have a role in the development of Lewy bodies, which are cytoplasmic inclusions (Melki 2015). In

PD, α-synuclein aggregation and mitochondrial dysfunction seem to be related; in degenerating neurons, mitochondrial dysfunction probably plays a role in facilitating α-synuclein aggregation (Rocha et al. 2017). Senescence is a major factor in PD pathogenesis. Brain samples from Parkinson's disease patients showed higher levels of several cell cycle genes, including p16INK4a mRNA, as compared to control brain samples (Chinta et al. 2018). In another research study, it was discovered that PD patients' substantia nigra neurons had larger quantities of pRB mRNA in their cytoplasm rather than controls (Ihara et al. 2012). The inactive, serine 795 phosphorylated form of pRb has been observed to have a unique distribution pattern in PD cases. In PD, numerous SASP-related factors are elevated. When comparing PD patients with controls, higher levels of IL-1 were found in their cerebrospinal fluid (CSF), serum, and dopaminergic areas of the striatum. According to multiple research (Lindqvist et al. 2012), PD patients' serum levels of IL-6 are elevated, and these levels have been connected to the severity of the disease (Hofmann et al. 2009). Additionally, increased levels of IL-6 were discovered in the CSF of Parkinson's disease patients and the striatal dopaminergic region. When compared to controls, PD patients' striatum and CSF had higher levels of TNF, and their Lewy bodies in the brains of PD patients showed MMP-3 co-localizing with α-synuclein (Choi et al. 2011). It is yet unknown, though, if these increased cytokine levels are a consequence of SASP in PD or if they are a component of a different neuroinflammatory process, which is already recognized as a fundamental aspect of the illness (Kaur et al. 2017). Neuroinflammation plays a major role in the brains of PD patients, according to research on biomarkers and post-mortem brain imaging. According to research evidence, telomere shortening has also been linked to the onset of Parkinson's disease (Eitan et al. 2014). Histone modifications and higher promoter methylation are two epigenetic alterations linked to PD that have been reported. In comparison to control cells, PD patients' cells exhibit a greater rate of protein ubiquitination (Tanaka and Matsuda 2013). Furthermore, PD is associated with mutations in the genes that encode Parkin, PINK1, or PRKN (Pickrell and Youle 2015). In animal models, mitophagy—a process that is partially regulated by the PINK1-Parkin pathway—has been demonstrated to lessen the severity of PD (Sliter et al. 2018). The larger quantity of senescent cells and increased expression of SA-β-gal in PD brain tissue indicate that the deposition of α-synuclein in PD is also linked to increased cellular senescence (Chinta et al. 2018). Nutrient sensing targeting the mTOR and AMP-activated protein kinase (AMPK) pathways also has protective effects in PD (Curry et al. 2018). Additionally, brain cells from models of AD, PD, and HD show decreased rates of oxygen and glucose metabolism (Camandola and Mattson 2017), and brain cells from patients with PD also demonstrate signs of dysglycemia (Cai et al. 2012). According to studies, abnormalities in autophagic and lysosomal pathways are associated with numerous PD-related mutations, suggesting that these pathways play a critical role in the development of PD (Pitcairn et al. 2018). Moreover, many crucial mutations are connected to both mitochondrial dysfunction and autophagic/lysosomal dysfunction, indicating a crucial interplay between both pathways that are essential to the etiology of Parkinson's disease (Plotegher and Duchen 2017).

3.3 Multiple Sclerosis

Inflammatory demyelination, astrogliosis, and the loss of neurons and axons are the hallmarks of multiple sclerosis (MS) which is a chronic immune-mediated disease that affects the brain and spinal cord. Environmental and genetic variables are thought to play a role in the development of MS.

The two main components of MS pathophysiology are oxidative stress and mitochondrial dysfunction (Mahad et al. 2009). Age-dependent cellular senescence is accelerated by chronic inflammation and oxidative stress (Howcroft et al. 2013). ROS and mitochondrial dysfunction are powerful initiators of cellular senescence because they can induce random, irreversible damage to cellular DNA. It is postulated that the rapid accumulation of senescent cells above a threshold could trigger the shift to the secondary progressive phase of MS when cellular function loss associated with senescence drives neurodegeneration. Furthermore, the mild inflammation linked to the SASP may be represented by the "compartmentalized within the blood-brain barrier (BBB)" inflammation that is resistant to immunomodulatory therapies (Lassmann et al. 2007). A greater quantity of senescence-associated β-galactosidase-positive senescent glial cells was discovered in the chronically demyelinated corpus callosum in a cuprizone-induced demyelination model of multiple sclerosis. Histochemistry with GL13 lipofuscin was used to confirm this further. A correlation analysis showed a significant relationship between the amount of demyelination and motor function as well as the number of senescent cells, indicating a relationship between chronic demyelination and the accumulation of senescent glial cells and the subsequent loss of function (Karamita et al. 2018).

3.4 Stroke

The aging process impairs cellular energy metabolism, and one of the main mechanisms that increases tissue susceptibility to ischemia is decreased energy metabolism. Growing older was linked to a higher risk of ischemic tissue turning into infarction in prospective research. Furthermore, energy restriction's protective benefits against cerebral infarction in mice are diminished with age. Intermittent fasting reduced brain injury and functional impairment after middle cerebral artery blockage in young and middle-aged mice, but this effect was not seen in older mice (Arumugam et al. 2009). When compared to reducing inflammatory cytokines in the cerebral cortex and striatum, intermittent fasting increases the levels of neurotrophic factors like brain-derived neurotrophic factor and basic fibroblast growth factor, protein chaperones like heat shock protein 70 and glucose-regulated protein 78, and the antioxidant enzyme hemeoxygenase-1 (Arumugam et al. 2009).

4 Epigenetic Regulation of Neuroinflammatory Pathways in Aging and Alzheimer's Disease

Several research groups proved the significance of epigenetic regulations in modulating neuroinflammatory signaling pathways in both aging conditions and AD progression. Of note, chronic and low-grade inflammations are increasingly known as a key contributor to this complex neurological disease progression. In the aging process, epigenetic changes including DNA methylation, changes in histone modification marks, and alterations of non-coding RNA expressions lead to the upregulation of pro-inflammatory gene sets, and these molecular events persistently activate microglia (the resident immune cells in the brain). Importantly, these age-related modifications prime microglia toward a "pro-inflammatory" phenotype which results in an upregulation of major cytokines including IL-1β, TNF-α, and IL-6 for damaging neural tissue and disrupting the synaptic integrity to inhibit neuron signals. In the context of Alzheimer's disease, similar types of epigenetic changes impair the major neuroinflammatory responsive pathways and promote disease-specific pathological consequences (Fig. 3). For example, increased histone acetylation in promoter regions IL-1B and TNF-alpha has been observed in autopsy samples of AD brains, linking with a raised expression of these pro-inflammatory cytokines and their functions in the disease. This abnormal activation not only quickens amyloid-beta plaque formation at the neural junctions but also promotes tau hyperphosphorylation event which has a great impact on propagating neurodegeneration. Moreover, non-coding RNAs, including specific microRNAs (e.g., miR-146a, miR-155), have been shown to regulate immune-related gene expressions in the aging brain which are distinctly dysregulated by AD. These reports suggested the important role of non-coding RNAs in influencing both amyloidogenic functions and the induction of inflammatory pathways. These epigenetic alterations are being targeted from the therapeutical perspective which offers the potential for modulating neuroinflammation. Recent studies with histone deacetylase inhibitors (HDAC inhibitors) have confirmed the ability to decrease neuroinflammatory responses in AD models. Therefore, in-depth comprehension of the therapeutic promise of epigenetic modulation in improving disease prognosis could be beneficial for the modulation of inflammatory signaling which accelerates both aging and Alzheimer's pathology.

5 Role of Microglia and Astrocytes in Alzheimer's Disease, Influenced by Epigenetic Changes

Both microglia and astrocytes which are primary immune cells of the CNS play crucial roles in AD pathogenesis. Recent research demonstrated how epigenetic changes influence their function and contribute to disease pathology. In AD, microglia and astrocytes undergo special "phenotypic shifts" toward a pro-inflammatory

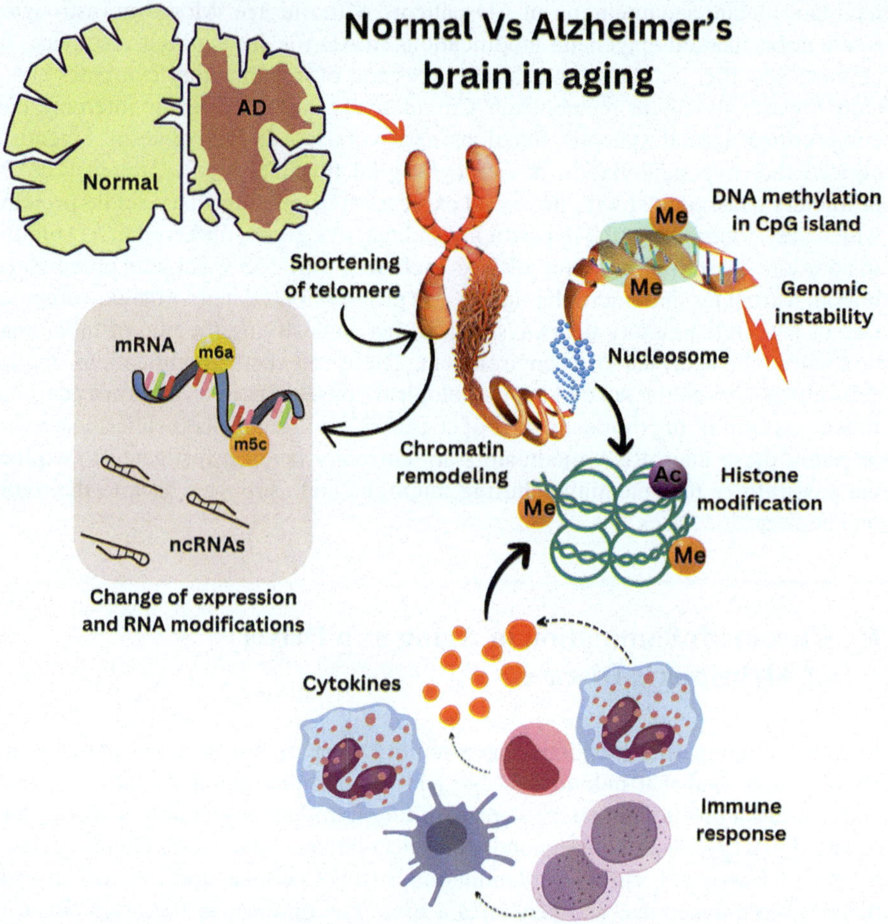

Fig. 3 Epigenetic modifications and its functions in aging-associated Alzheimer's disease. The figure shows the causal factors and consequences of epigenetic modifications in age-related Alzheimer's disease pathology. Genomic instability, epigenetic changes in gene promoters and histones, and telomere shortening may influence the gene expressions of ncRNAs. Also, induction of immune response and cytokine production may affect the histone modifications

state which is largely driven by the major epigenetic modifications and ultimately amplifies neuroinflammation and damage of neurons. Epigenetic changes including histone acetylation and DNA methylation in genes (related to inflammation and phagocytosis) regulate microglia to adopt the feature of an activated, pro-inflammatory phenotype with an elevated level of cytokine production and impaired clearance of Aβ plaques. In particular, hypermethylation of CX3CR1 in AD microglia has been related to reduced expression of this crucial receptor which plays an important role in disrupting the communications between the neuron and microglia impairing their potentialities to phagocytose Aβ. Therefore, these molecular events facilitate plaque accumulation in AD patients with old age. Moreover, astrocytes also exhibit marked epigenetic modifications in AD for driving their reactivity. To be more specific, it allowed increased expression of inflammatory cytokines and a shift toward glycolytic metabolism. Cumulatively, such molecular interplay can compromise normal synaptic functions and worsen AD pathogenesis. Scientists noticed that hypomethylation in astrocytic glial fibrillary acidic protein (GFAP) promoters is associated with increased expression of glial fibrillary acidic proteins which may contribute to the hypertrophy and reactive gliosis observed in AD brains. In addition, a specific set of microRNAs including miR-155 was found upregulated in both astrocytes and microglia during AD progression related to the downregulation of anti-inflammatory pathways and further intensifying the rate of inflammation within the neuronal microenvironment. These epigenetic modifications in glial cells either worsen the Aβ and tau pathological consequences or interrupt neuronal functions, amplifying the declination of cognitive functions. Therapeutic approaches targeting these microRNA modulators are currently being investigated to explore the possibilities for potentially shifting microglia and astrocytes back to their normal homeostatic states.

6 Chronic Inflammation in Aging as a Driver of Alzheimer's Disease

Chronic inflammation was found as a major driver of AD in aging populations. Reports indicated that individual's age and the immune functions face a gradual shift toward a pro-inflammatory state, resulting in higher production of IL-6, TNF-α, and IL-1β cytokines. Consequently, they accumulate in the CNS. Therefore, such a state of low-grade, chronic inflammation disturbs cellular homeostasis, impairs neural communications, and generates a toxic environment in the brain tissue to prime neuronal cells for degenerative processes. In this process, microglia become sensitized in aged brains and lead to Aβ accumulation at a high rate which triggers other pathological conditions of AD. Epigenetic modifications are often associated with changes in microglial and astrocytic gene expressions which further impair inflammatory responses. Of note, reinforcing a cycle of neuroinflammation was observed to be related to neuronal damage. Moreover, different research groups also identified that chronic inflammation contributed to the breakdown of the BBB and allowed peripheral immune cells and pro-inflammatory cytokines to target CNS. As a result, it intensified the local inflammation and Aβ deposition in the brain tissue. Recent studies suggested that targeting systemic inflammation with anti-inflammatory dietary components or the use of senolytic drugs may alleviate these higher inflammatory responses and reduce the risk of the progression of AD. Future AD research may address the significance of epigenetically regulated inflammatory pathways related to aging for developing therapeutic interventions to break the cycle of chronic inflammation in AD.

7 Therapeutics of Neurodegenerative Diseases Linked with Aging

In recent times, there have been no effective, definitive medicines or preventive strategies against neurodegenerative diseases, especially among older people. Traditional treatment options for AD that aim to target amyloid-beta have been found unsuccessful despite multiple attempts. Amyloid plaques have been identified in a large number of older persons who are cognitively healthy, therefore casting doubt on the amyloid hypothesis (Anderson et al. 2017; Cummings 2017). Similarly, PD treatments try to increase dopamine levels but do not stop neuronal death or delay the course of the disorder. The report demonstrated that the use of siponimod (a modulator of S1P1/S1P5 receptors) could be a bit effective in delaying the course of impairment in the case of MS, although it has some limitations (Karabiyik et al. 2021). Some neuroprotective treatments also become important as recent data suggested targeting cellular senescence and related signaling molecules as a possible therapeutic option. As per research, both pro- and anti-senescent strategies may show some beneficial effects, but they are dependent on the physiological conditions and disease progression of patients (Dickson et al. 2013). Some anti-senescent therapies were found effective for eliminating or altering senescent cells to delay brain tissue damage. Researchers also adopted some specific techniques like senomorphism (modification of the activity of senescent cells) or senolysis (elimination of senescent cells), which may reduce the negative consequences of damaged cells and delay the pathological consequences. Experimental pieces of evidence suggested some interesting results with the BubR1 progeroid mice model where researchers were able to eradicate senescent cells throughout life which prevents age-related disorders and illness (Baker et al. 2011). Apart from these, several drugs including dasatinib and metformin have recently drawn research attention. Interestingly, their actions were observed as different based on the cell types and pathological conditions. Therefore, their therapeutic applications become more difficult (Myrianthopoulos et al. 2018). Research evidence shows that metformin may improve cognitive responses, but other studies also suggest its role in enhancing the risk of AD and PD; therefore the therapeutic applications of this drug are controversial (Ng et al. 2014; Kuan et al. 2017). Scientists are trying to investigate the complex role of cellular senescence and the potential of senolytic vaccines (for the clearance of immune-mediated senescent cells). Currently, two primary groups of compounds are being employed such as mTOR-dependent autophagy enhancers' molecules. From the research viewpoints, autophagy could be negatively regulated by mTOR signaling, and researchers used specific pathway inhibitors such as rapamycin and rapalogs. These inhibitors have demonstrated promising results in lowering hyperphosphorylated tau and Aβ and increasing cognitive functions in P301S mutant tau and PDAPP transgenic mice AD models. In addition, other autophagy enhancers were identified which are dependent on AMPK signaling (Karabiyik et al. 2021). Therefore, AMPK modulators, including nilotinib, bosutinib, and metformin, showed some promising results. After the induction of autophagy, AMPK phosphorylates ULK1 directly (Hansen et al. 2018; Menzies et al. 2017), and this second group of inhibitors has demonstrated hopeful therapeutic interventions with enhanced memory function and the long prevention of neuronal degeneration in different AD models. Moreover, trehalose, lithium, and other compounds such as imidazoline receptor agonists (e.g., clonidine, rilmenidine) and L-type Ca^{2+} channel antagonists (e.g., felodipine, verapamil) activate AMPK signaling through decreasing ATP production or Ca^{2+} levels which further promote autophagy in the cells (Siddiqi et al. 2019). Researchers also tried to comprehend the clearance of Aβ and tau aggregates including BECN1 to impair its association with anti-apoptotic Bcl-2 by overexpressing BECN1 at same time delaying the effect of calpains. These innovative approaches have shown promising results in a wide range of AD models (Rocchi et al. 2017).

8 The Use of Epigenetic Drugs in Alzheimer's Disease Treatment

Epigenetic drugs, particularly histone deacetylase (HDAC) inhibitors and DNA methylation inhibitors, are being used as emerging and promising therapeutic candidates in the treatment of AD. HDAC inhibitors were known for inhibiting enzymes responsible for removing acetyl groups from histone proteins, and they generate a more open chromatin structure that enables gene expression. During AD pathogenesis, HDAC inhibition has shown the potential to increase the expression of genes that are involved in synaptic plasticity and neuronal survival, potentially opposing the cognitive decline features of AD. Recent studies also demonstrated that HDAC inhibitors like vorinostat and sodium butyrate may reduce Aβ plaque formation and hyperphosphorylation of tau by variable neuroprotective pathways with the reduction of neuroinflammation in preclinical models. Additionally, these inhibitors seem to restore the balance between microglial and astrocytic phenotypes. DNA methylation inhibitors block the addition of methyl groups to DNA (particularly in promoter regions of genes). This demethylation effect has been initiated to reactivate already silenced genes which is essential for neuronal function and reduction of the production of pro-inflammatory cytokines. As a consequence, AD progression may be delayed. Of note, DNMT inhibitors (such as 5-azacytidine) have shown effectiveness in preclinical models by increasing memory and cognitive functions while reducing neurotoxic inflammatory responses. Although these epigenetic drugs may hold substantial promises, their application in clinical settings was found very complex with challenges for target-specific applications in regions of the aging brain and systemic side effects. However, ongoing clinical trials and research are trying to optimize epigenetic drug delivery mechanisms involving nanoparticle-based carriers and intranasal administrations to improve target specificity which may offer a novel, mechanistic approach to treat Alzheimer's disease.

9 Gene Therapy Approaches to Correct Epigenetic Limitations

Several lines of research showed that gene therapy is a rapidly advancing strategy to correct epigenetic abnormalities in different pathological conditions, including AD and human malignancies. This approach includes direct targeting and modifying the epigenetic landscape including DNA methylation patterns, histone modifications, and non-coding RNA dysregulation. Recent techniques, like CRISPR-based epigenome editing, have empowered precise modulation of epigenetic marks at specific gene loci. Research demonstrated that CRISPR-dCas9 (deactivated Cas9 fused with epigenetic modifiers) showed a good potential in targeting DNA methylation and histone acetylation to silence or activate specific genes that are important for inflammation, neuronal function, and tumor suppression. The accuracy of this technique enables gene-specific epigenetic reprogramming without changing the underlying DNA sequence, thus offering a safer and more target-specific involvement compared to traditional gene-editing methods. One promising application was explored in neurodegenerative disease pathology where gene therapy was successfully used to downregulate pro-inflammatory signaling pathways and enhance the expression of neuroprotective genes. For example, targeting the epigenetic regulation of genes like BACE1 can potentially reduce plaque formation and slow down disease progression. Some studies demonstrated that targeting TNF-α and IL-1β cytokine promoters with dCas9-HDAC complexes can reduce neuroinflammation, therefore lessening neuronal damage. Moreover, the adjustment of methylation patterns at the BACE1 gene may have a critical impact on Aβ production. Experimental pieces of evidence have shown the potential of CRISPRs to reduce Aβ levels and plaque formation in AD. Though these applications are still under clinical investigation, CRISPR-based epigenome editing could be optimized for delivery into the brain tissue using viral vectors or nanoparticle-based systems to expand targeting accuracy and decrease off-target effects.

10 Discussion

The process of aging was identified as the single most significant risk factor for the progression of neurodegenerative diseases, including AD, PD, and frontotemporal dementia. With the increase of individuals' age, their brains sense and directly experience different molecular and cellular changes that contribute to cognitive impairment and vulnerability to neurodegeneration. Several age-related processes including oxidative stress, chronic inflammation, and mitochondrial dysfunction gradually impair neuronal functions, compromising the brain's flexibility and rapidly inducing degenerative pathways. In addition, aging also involves epigenetic modifications including DNA methylation drift, histone modifications, and changes in the expression of non-coding RNAs. These epigenetic changes disrupt gene expressions in critical neural networks and affect their functionalities. Such epigenetic changes collectively create an environment that was found conducive to protein misfolding, neuroinflammation, etc. in AD. It is essential to understand these aging-driven signaling pathways which are essential for developing interventions for the aging brain. The strategy could be important for potentially staving off neurodegenerative diseases before their clinical manifestations. Despite extensive observations of epigenetic changes in AD patients, aberrant DNA methylations in genes including APP and PSEN1, and histone modifications affecting inflammatory pathways, the exact impact of these changes remains poorly unstated. Therefore, the current situation is facing the basic question of whether they initiate, impair, or simply accompany the neurodegeneration process during AD. Extensive research is needed to clarify these crucial molecular mechanisms as they could reveal and identify specific epigenetic marks for driving amyloid-beta accumulation, tau pathology, or neuroinflammation in the brain tissue. Moreover, age-related epigenetic drift where stochastic variations in DNA methylation, histone modification patterns, and non-coding RNA expression were observed over time ultimately presents both a risk factor and potential intervention point in AD pathogenesis. Recent experiments have demonstrated that these modifications contribute to an inflammatory environment and a decline in proteostasis in the aging brain, generating pathological conditions helpful for the onset of AD. However, some studies suggested that targeting these age-induced deviations could offer novel and preventive or therapeutic strategies that counteract the feature of cognitive decline. For example, researchers showed age-related histone deacetylation patterns selectively reversing the specific neural pathways in animal models. Their results demonstrated an effective reduction of amyloid plaque formation with improved cognitive outcomes. In light of modern medical advancements, the potential for personalized epigenetic therapies in AD aims to tailor interventions based on an individual's unique epigenetic landscape. Such an innovative approach was found particularly promising with the significant inter-individual variability in the epigenetic profiles involving the proper shaping of genetic background, lifestyle factors, and environmental exposures of AD patients. By connecting personalized data to design target-specific epigenetic modulators including the inhibitors of specific histone deacetylases (HDACs) or methyltransferases, some therapies could be customized to alleviate individual-specific risk factors. Therefore, it could be beneficial for reducing the probability of the onset and progression of neurodegenerative diseases. Such therapeutic interventions not only show the potential for improved efficacy but also signify a new avenue toward precision medicine in AD with old people. However, the practical challenges and limitations of applying these personalized epigenetic therapies remain significant, necessitating advancements in diagnostic precision, therapeutic specificity, and longitudinal monitoring of the epigenomic landscape of age-related neurodegenerative disorders.

Acknowledgments The authors are thankful to the Puri Foundation for Education in India and the management of IAR, Gandhinagar, for supporting this work. The IAR Reference No. IAR/2024-25/RO/Research/237 is duly acknowledged.

Conflict of Interest The authors declare that there is no conflict of interest.

References

Aalaei-Andabili SH, Rezaei N (2015) MicroRNAs (MiRs) precisely regulate immune system development and function in immunosenescence process. Int Rev Immunol 35:57–66

Aarsland Dag K, Bronnick C, Williams-Gray D et al (2010) Mild cognitive impairment in Parkinson disease: a multicenter pooled analysis. Neurology 75:1062–1069

Akçimen F, Lopez ER, Landers JE et al (2023) Amyotrophic lateral sclerosis: translating genetic discoveries into therapies. Nat Rev Genet 24:642–658

Anderson RM, Hadjichrysanthou C, Evans S et al (2017) Why do so many clinical trials of therapies for Alzheimer's disease fail? Lancet 390:2327–2329

Arumugam TV, Phillips TM, Cheng A et al (2009) Age and energy intake interact to modify cell stress pathways and stroke outcome. Ann Neurol 67:41–52

Baker DJ, Petersen RC (2018) Cellular senescence in brain aging and neurodegenerative diseases: evidence and perspectives. J Clin Invest 128:1208–1216

Baker DJ, Wijshake T, Tchkonia T et al (2011) Clearance of p16Ink4a-positive senescent cells delays ageing-associated disorders. Nature 479:232–236

Barrie W, Yang Y, Irving-Pease EK et al (2024) Elevated genetic risk for multiple sclerosis emerged in steppe pastoralist populations. Nature 625:321–328

Barrio-Alonso E, Hernández-Vivanco A, Walton CC et al (2018) Cell cycle reentry triggers hyperploidization and synaptic dysfunction followed by delayed cell death in differentiated cortical neurons. Sci Rep 8:14316

Barzilai N, Huffman DM, Muzumdar RH et al (2012) The critical role of metabolic pathways in aging. Diabetes 61:1315–1322

Bekris LM, Yu CE, Bird TD et al (2010) Review article: genetics of Alzheimer disease. J Geriatr Psychiatry Neurol 23:213–227

Bektas A, Schurman SH, Sen R et al (2017) Aging, inflammation and the environment. Exp Gerontol 105:10–18

Bhatia-Dey N, Kanherkar RR, Stair SE et al (2016) Cellular senescence as the causal nexus of aging. Front Genet 7:13

Braak H, Bohl JR, Müller CM et al (2006) Stanley Fahn lecture 2005: the staging procedure for the inclusion body pathology associated with sporadic Parkinson's disease reconsidered. Mov Disord 21:2042–2051

Bradley-Whitman MA, Lovell MA (2013) Epigenetic changes in the progression of Alzheimer's disease. Mech Ageing Dev 134:486–495

Bussian TJ, Aziz A, Meyer CF et al (2018) Clearance of senescent glial cells prevents tau-dependent pathology and cognitive decline. Nature 562:578–582

Cai H, Cong W-N, Ji S et al (2012) Metabolic dysfunction in Alzheimers disease and related neurodegenerative disorders. Curr Alzheimer Res 9:5–17

Camandola S, Mattson MP (2017) Brain metabolism in health, aging, and neurodegeneration. EMBO J 36:1474–1492

Cao W, Zheng H (2018) Peripheral immune system in aging and Alzheimer's disease. Mol Neurodegener 13:51

Chinta SJ, Woods G, Demaria M et al (2018) Cellular senescence is induced by the environmental neurotoxin Paraquat and contributes to neuropathology linked to Parkinson's disease. Cell Rep 22:930–940

Choi DH, Kim YJ, Kim YG et al (2011) Role of matrix metalloproteinase 3-mediated α-synuclein cleavage in dopaminergic cell death. J Biol Chem 286:14168–14177

Cummings J (2017) Lessons learned from Alzheimer disease: clinical trials with negative outcomes. Clin Transl Sci 11:147–152

Curry DW, Stutz B, Andrews ZB et al (2018) Targeting AMPK signaling as a neuroprotective strategy in Parkinson's disease. J Parkinsons Dis 8:161–181

Davies SW, Turmaine M, Cozens BA et al (1997) Formation of neuronal intranuclear inclusions underlies the neurological dysfunction in mice transgenic for the HD mutation. Cell 90:537–548

De Magalhães JP (2023) Ageing as a software design flaw. Genome Biol 24:51
Dickson MA, Tap WD, Keohan ML et al (2013) Phase II trial of the CDK4 inhibitor PD0332991 in patients with advanced CDK4-amplified well-differentiated or dedifferentiated liposarcoma. J Clin Oncol 31:2024–2028
Dobashi Y, Watanabe Y, Miwa C et al (2011) Mammalian target of rapamycin: a central node of complex signaling cascades. Int J Clin Exp Pathol 4:476–495
Du Y, Karvellas CJ, Baracos V et al (2014) Sarcopenia is a predictor of outcomes in very elderly patients undergoing emergency surgery. Surgery 156:521–527
Eisenstein M (2014) Molecular biology: remove, reuse, recycle. Nature 514:S2–S4
Eitan E, Hutchison ER, Mattson MP (2014) Telomere shortening in neurological disorders: an abundance of unanswered questions. Trends Neurosci 37:256–263
Fang EF, Kassahun H, Croteau DL et al (2016) NAD+ replenishment improves lifespan and healthspan in ataxia telangiectasia models via mitophagy and DNA repair. Cell Metab 24:566–581
Fang EF, Lautrup S, Hou Y et al (2017) NAD+ in aging: molecular mechanisms and translational implications. Trends Mol Med 23:899–916
Fielder E, Von Zglinicki T, Jurk D (2017) The DNA damage response in neurons: die by apoptosis or survive in a senescence-like state? J Alzheimers Dis 60:S107–S131
Fischer A, Sananbenesi F, Mungenast A et al (2010) Targeting the correct HDAC(s) to treat cognitive disorders. Trends Pharmacol Sci 31:605–617
Fontana L, Partridge L, Longo VD (2010) Extending healthy life span—from yeast to humans. Science 328:321–326
Frasca D, Blomberg BB (2015) Inflammaging decreases adaptive and innate immune responses in mice and humans. Biogerontology 17:7–19
Fu WY, Wang X, Ip NY (2018) Targeting neuroinflammation as a therapeutic strategy for Alzheimer's disease: mechanisms, drug candidates, and new opportunities. ACS Chem Neurosci 10:872–879
Gewirtz DA (2013) Autophagy and senescence. Autophagy 9:808–812
Gezen-Ak D, Dursun E, Hanağası H et al (2013) BDNF, TNFα, HSP90, CFH, and IL-10 serum levels in patients with early or late onset Alzheimer's disease or mild cognitive impairment. J Alzheimers Dis 37:185–195
Halliday GM, McCann H (2009) The progression of pathology in Parkinson's disease. Ann N Y Acad Sci 1184:188–195
Hansen M, Rubinsztein DC, Walker DW (2018) Autophagy as a promoter of longevity: insights from model organisms. Nat Rev Mol Cell Biol 19:579–593
Hardy J (2009) The amyloid hypothesis for Alzheimer's disease: a critical reappraisal. J Neurochem 110:1129–1134
Hayflick L (1965) The limited in vitro lifetime of human diploid cell strains. Exp Cell Res 37:614–636
Haynes CM, Fiorese CJ, Lin YF (2013) Evaluating and responding to mitochondrial dysfunction: the mitochondrial unfolded-protein response and beyond. Trends Cell Biol 23:311–318
He N, Jin W, Lok K et al (2013) Amyloid-β1–42 oligomer accelerates senescence in adult hippocampal neural stem/progenitor cells via formylpeptide receptor 2. Cell Death Dis 4:e924
He J, Tu C, Liu Y (2018) Role of lncRNAs in aging and age-related diseases. Aging Med 1:158–175
Herrmann M, Pusceddu I, März W et al (2018) Telomere biology and age-related diseases. Clin Chem Lab Med 56:1210–1222
Hickman S, Izzy S, Sen P et al (2018) Microglia in neurodegeneration. Nat Neurosci 21:1359–1369
Hofmann KW, Schuh AFS, Saute J et al (2009) Interleukin-6 serum levels in patients with Parkinson's disease. Neurochem Res 34:1401–1404
Horstmann S, Budig L, Gardner H et al (2010) Matrix metalloproteinases in peripheral blood and cerebrospinal fluid in patients with Alzheimer's disease. Int Psychogeriatr 22:966–972
Howcroft TK, Campisi J, Louis GB et al (2013) The role of inflammation in age-related disease. Aging 5:84–93

Hung CW, Chen YC, Hsieh WL et al (2010) Ageing and neurodegenerative diseases. Ageing Res Rev 9:S36–S46

Ihara Y, Morishima-Kawashima M, Nixon R (2012) The ubiquitin-proteasome system and the Autophagic-lysosomal system in Alzheimer disease. Cold Spring Harb Perspect Med 2:a006361

Jalili M (2013) Effect of diet and nutrients on molecular mechanism of gene expression mediated by nuclear receptor and epigenetic modulation. Open Nutraceutic J 6:27–34

Jeppesen DK, Bohr VA, Stevnsner T (2011) DNA repair deficiency in neurodegeneration. Prog Neurobiol 94:166–200

Jurk D, Wang C, Miwa S et al (2012) Postmitotic neurons develop a p21-dependent senescence-like phenotype driven by a DNA damage response. Aging Cell 11:996–1004

Karabiyik C, Vicinanza M, Rubinsztein DC (2021) AMPK. In: Elsevier eBooks. p 9–39

Karamita M, Nicholas R, Kokoti L et al (2018) Cellular senescence correlates with demyelination, brain atrophy and motor impairment in a model of multiple sclerosis (P2.405). Neurology 90

Kaur K, Gill JS, Bansal PK et al (2017) Neuroinflammation—a major cause for striatal dopaminergic degeneration in Parkinson's disease. J Neurol Sci 381:308–314

Keogh MJ, Chinnery PF (2015) Mitochondrial DNA mutations in neurodegeneration. Biochim Biophys Acta Bioenerg 1847:1401–1411

Kerr JS, Adriaanse BA, Greig NH et al (2017) Mitophagy and Alzheimer's disease: cellular and molecular mechanisms. Trends Neurosci 40:151–166

Kirkland JL, Tchkonia T (2017) Cellular senescence: a translational perspective. EBioMedicine 21:21–28

Kozomara A, Griffiths-Jones S (2010) miRBase: integrating microRNA annotation and deep-sequencing data. Nucleic Acids Res 39:152–157

Kuan YC, Huang KW, Lin CL et al (2017) Effects of metformin exposure on neurodegenerative diseases in elderly patients with type 2 diabetes mellitus. Prog Neuro-Psychopharmacol Biol Psychiatry 79:77–83

Kuilman T, Michaloglou C, Mooi WJ et al (2010) The essence of senescence: figure 1. Genes Dev 24:2463–2479

Kukull WA, Higdon R, Bowen JD et al (2002) Dementia and Alzheimer disease incidence. Arch Neurol 59:1737–1746

Kwon MJ, Kim S, Han MH et al (2016) Epigenetic changes in neurodegenerative diseases. Mol Cell 39:783–789

Labbadia J, Morimoto RI (2015) The biology of proteostasis in aging and disease. Annu Rev Biochem 84:435–464

Lane RK, Hilsabeck T, Rea SL (2015) The role of mitochondrial dysfunction in age-related diseases. Biochim Biophys Acta Bioenerg 1847:1387–1400

Lassmann H, Brück W, Lucchinetti CF (2007) The immunopathology of multiple sclerosis: an overview. Brain Pathol 17:210–218

Leandro GS, Sykora P, Bohr VA (2015) The impact of base excision DNA repair in age-related neurodegenerative diseases. Mutation Res 776:31–39

Lee C, Longo VD (2011) Fasting vs dietary restriction in cellular protection and cancer treatment: from model organisms to patients. Oncogene 30:3305–3316

Lindqvist D, Kaufman E, Brundin L et al (2012) Non-motor symptoms in patients with Parkinson's disease—correlations with inflammatory cytokines in serum. PLoS One 7:e47387

Loaiza N, Demaria M (2016) Cellular senescence and tumor promotion: is aging the key? Biochim Biophys Acta Rev Cancer 1865:155–167

Lu J, Clark AG (2012) Impact of microRNA regulation on variation in human gene expression. Genome Res 22:1243–1254

Lyman M, Lloyd DG, Ji X et al (2013) Neuroinflammation: the role and consequences. Neurosci Res 79:1–12

Madabhushi R, Pan L, Tsai L-H (2014) DNA damage and its links to neurodegeneration. Neuron 83:266–282

Magini A, Polchi A, Tozzi A et al (2014) Abnormal cortical lysosomal β-hexosaminidase and β-galactosidase activity at post-synaptic sites during Alzheimer's disease progression. Int J Biochem Cell Biol 58:62–70

Mahad DJ, Ziabreva I, Campbell G et al (2009) Mitochondrial changes within axons in multiple sclerosis. Brain 132:1161–1174

Mak E, Zhou J, Tan LCS et al (2013) Cognitive deficits in mild Parkinson's disease are associated with distinct areas of grey matter atrophy. J Neurol Neurosurg Psychiatry 85:576–580

Melki R (2015) Role of different alpha-synuclein strains in synucleinopathies, similarities with other neurodegenerative diseases. J Parkinsons Dis 5:217–227

Menzies FM, Fleming A, Caricasole A et al (2017) Autophagy and neurodegeneration: pathogenic mechanisms and therapeutic opportunities. Neuron 93:1015–1034

Michan S (2014) Calorie restriction and NAD+/sirtuin counteract the hallmarks of aging. Front Biosci 19:1300

Mitchell J, Borasio G (2007) Amyotrophic lateral sclerosis. Lancet 369:2031–2041

Moskalev AA, Shaposhnikov MV, Plyusnina EN et al (2012) The role of DNA damage and repair in aging through the prism of Koch-like criteria. Ageing Res Rev 12:661–684

Murray CJL, Barber RM, Foreman KJ et al (2015) Global, regional, and national disability-adjusted life years (DALYs) for 306 diseases and injuries and healthy life expectancy (HALE) for 188 countries, 1990–2013: quantifying the epidemiological transition. Lancet 386:2145–2191

Myrianthopoulos V, Evangelou K, Vasileiou PVS et al (2018) Senescence and senotherapeutics: a new field in cancer therapy. Pharmacol Ther 193:31–49

Nakahira K, Haspel JA, Rathinam VK et al (2010) Autophagy proteins regulate innate immune responses by inhibiting the release of mitochondrial DNA mediated by the NALP3 inflammasome. Nat Immunol 12:222–230

Ng TP, Feng L, Yap KB et al (2014) Long-term metformin usage and cognitive function among older adults with diabetes. J Alzheimers Dis 41:61–68

Oddo S (2012) The role of mTOR signaling in Alzheimer disease. Front Biosci (Schol Ed) S4:941–952

Oh J, Lee YD, Wagers AJ (2014) Stem cell aging: mechanisms, regulators and therapeutic opportunities. Nat Med 20:870–880

Ohyagi Y, Asahara H, Chui D-H et al (2004) Intracellular Aβ42 activates p53 promoter: a pathway to neurodegeneration in Alzheimer's disease. FASEB J 19:1–29

Pan M-R, Li K, Lin SY et al (2016) Connecting the dots: from DNA damage and repair to aging. Int J Mol Sci 17:685

Papadopoulos D, Magliozzi R, Mitsikostas DD et al (2020) Aging, cellular senescence, and progressive multiple sclerosis. Front Cell Neurosci 14:178

Partridge L, Deelen J, Slagboom PE (2018) Facing up to the global challenges of ageing. Nature 561:45–56

Pearson TS (2016) More than ataxia: hyperkinetic movement disorders in childhood autosomal recessive ataxia syndromes. PubMed 6:368

Pellegrino MW, Nargund AM, Haynes CM (2012) Signaling the mitochondrial unfolded protein response. Biochim Biophys Acta, Mol Cell Res 1833:410–416

Penney JB, Young AB, Shoulson I et al (1990) Huntington's disease in Venezuela: 7 years of follow-up on symptomatic and asymptomatic individuals. Mov Disord 5:93–99

Perlman SL, Boder E, Sedgewick RP et al (2011) Ataxia–telangiectasia. Handb Clin Neurol 103:307–332

Peters ME, Rosenberg PB, Steinberg M et al (2013) Neuropsychiatric symptoms as risk factors for progression from CIND to dementia: the Cache County study. Am J Geriatr Psychiatry 21:1116–1124

Pickrell AM, Youle RJ (2015) The roles of PINK1, Parkin, and mitochondrial fidelity in Parkinson's disease. Neuron 85:257–273

Pitcairn C, Wani WY, Mazzulli JR (2018) Dysregulation of the autophagic-lysosomal pathway in Gaucher and Parkinson's disease. Neurobiol Dis 122:72–82

Plotegher N, Duchen MR (2017) Crosstalk between lysosomes and mitochondria in Parkinson's disease. Front Cell Dev Biol 5:110

Poewe W, Seppi K, Tanner CM et al (2017) Parkinson disease. Nat Rev Dis Primers 3:17013

Pusceddu I, Farrell C-JL, Di Pierro AM et al (2015) The role of telomeres and vitamin D in cellular aging and age-related diseases. Clin Chem Lab Med 53:1661–1678

Qazi TJ, Quan Z, Mir A et al (2017) Epigenetics in Alzheimer's disease: perspective of DNA methylation. Mol Neurobiol 55:1026–1044

Riboldi GM, Samanta D, Asuncion RMD et al (2024) Ataxia-Telangiectasia. In: StatPearls—NCBI Bookshelf. https://www.ncbi.nlm.nih.gov/books/NBK519542/

Riera CE, Dillin A (2015) Tipping the metabolic scales towards increased longevity in mammals. Nat Cell Biol 17:196–203

Rocchi A, Yamamoto S, Ting T et al (2017) A Becn1 mutation mediates hyperactive autophagic sequestration of amyloid oligomers and improved cognition in Alzheimer's disease. PLoS Genet 13:e1006962

Rocha EM, De Miranda B, Sanders LH (2017) Alpha-synuclein: pathology, mitochondrial dysfunction and neuroinflammation in Parkinson's disease. Neurobiol Dis 109:249–257

Siddiqi FH, Menzies FM, Lopez A et al (2019) Felodipine induces autophagy in mouse brains with pharmacokinetics amenable to repurposing. Nat Commun 10:1817

Siddiqui MS, Francois M, Hecker J et al (2018) γH2AX is increased in peripheral blood lymphocytes of Alzheimer's disease patients in the South Australian Neurodegeneration, Nutrition and DNA Damage (SAND) study of aging. Mutat Res Genet Toxicol Environ Mutagen 829–830:6–18

Silva ART, Santos ACF, Farfel JM et al (2014) Repair of oxidative DNA damage, cell-cycle regulation and neuronal death may influence the clinical manifestation of Alzheimer's disease. PLoS One 9:e99897

Sliter DA, Martinez J, Hao L et al (2018) Parkin and PINK1 mitigate STING-induced inflammation. Nature 561:258–262

Stilling RM, Fischer A (2011) The role of histone acetylation in age-associated memory impairment and Alzheimer's disease. Neurobiol Learn Mem 96:19–26

Sun A (2003) P38 MAP kinase is activated at early stages in Alzheimer's disease brain. Exp Neurol 183:394–405

Swardfager W, Lanctôt K, Rothenburg L et al (2010) A meta-analysis of cytokines in Alzheimer's disease. Biol Psychiatry 68:930–941

Sykora P, Misiak M, Wang Y et al (2014) DNA polymerase β deficiency leads to neurodegeneration and exacerbates Alzheimer disease phenotypes. Nucleic Acids Res 43:943–959

Szic KSV, Declerck K, Vidaković M et al (2015) From inflammaging to healthy aging by dietary lifestyle choices: is epigenetics the key to personalized nutrition? Clin Epigenetics 7:33

Tahiliani M, Koh KP, Shen Y et al (2009) Conversion of 5-methylcytosine to 5-hydroxymethylcytosine in mammalian DNA by MLL partner TET1. Science 324:930–935

Tan L, Shi YG (2012) Tet family proteins and 5-hydroxymethylcytosine in development and disease. Development 139:1895–1902

Tanaka K, Matsuda N (2013) Proteostasis and neurodegeneration: the roles of proteasomal degradation and autophagy. Biochim Biophys Acta, Mol Cell Res 1843:197–204

Tell G, Demple B (2014) Base excision DNA repair and cancer. Oncotarget 6:584–585

Thanan R, Oikawa S, Hiraku Y et al (2014) Oxidative stress and its significant roles in neurodegenerative diseases and cancer. Int J Mol Sci 16:193–217

Tiribuzi R, Crispoltoni L, Porcellati S et al (2013) miR128 up-regulation correlates with impaired amyloid β (1-42) degradation in monocytes from patients with sporadic Alzheimer's disease. Neurobiol Aging 35:345–356

Turnquist C, Horikawa I, Foran E et al (2016) p53 isoforms regulate astrocyte-mediated neuroprotection and neurodegeneration. Cell Death Differ 23:1515–1528

Uchoa MF, Moser VA, Pike CJ (2016) Interactions between inflammation, sex steroids, and Alzheimer's disease risk factors. Front Neuroendocrinol 43:60–82

Villeda SA, Plambeck KE, Middeldorp J et al (2014) Young blood reverses age-related impairments in cognitive function and synaptic plasticity in mice. Nat Med 20:659–663

Wang HZ, Bi R, Hu QX et al (2014) Validating GWAS-identified risk loci for Alzheimer's disease in Han Chinese populations. Mol Neurobiol 53:379–390

Watson CT, Roussos P, Garg P et al (2016) Genome-wide DNA methylation profiling in the superior temporal gyrus reveals epigenetic signatures associated with Alzheimer's disease. Genome Med 8:5

World Health Organization (2024) With major processing by Our World in Data. "Deaths from Alzheimer's". World Health Organization, "Global Health Estimates". https://ourworldindata.org/grapher/deaths-from-alzheimers-other-dementias. Accessed 11 Dec 2024

Wu H, Tao J, Sun YE (2012) Regulation and function of mammalian DNA methylation patterns: a genomic perspective. Brief Funct Genomics 11:240–250

Xu GL, Wong J (2015) Oxidative DNA demethylation mediated by Tet enzymes. Natl Sci Rev 2:318–328

Toward a Systematic Understanding of Epigenetic Contribution to Normal Aging and Alzheimer's Disease

Zhou Zhang, Xiao-Long Cui, and Wei Zhang

Contents

Abstract Normal aging and aging-associated diseases, such as Alzheimer's disease (AD), are complex processes influenced by various genetic and non-genetic factors, including diet, lifestyle, and socioeconomic status. While previous genome-wide association studies have implicated multiple genetic variants in aging and AD, there is a significant individual heterogeneity that cannot be explained by genetics alone. Epigenetic modifications, such as cytosine modifications (DNA methylation and hydroxymethylation) and RNA modifications, are significant contributors to gene regulation underlying normal biological processes and disease pathobiology. Technical advances have enabled the interrogation of epigenome-wide features using clinical biospecimens, including tissues, cells, and liquid biopsies. To date the most investigated epigenetic modifications for aging and aging-associated diseases, including AD, are 5-methylcytosines (5mC), which are known to primarily suppress gene expression. More recently, other important cytosine modifications, particularly 5-hydroxymethylcytosines (5hmC), have been implicated in various biological processes and diseases as well. Besides epigenetic modifications of cytosines on DNA, RNA modifications, specifically N^6-Methyladenosine (m^{6}A) modifications on mRNA molecules, have emerged as critical gene regulators with importance in

Z. Zhang · X.-L. Cui · W. Zhang (✉)
Department of Preventive Medicine, Northwestern University Feinberg School of Medicine, Chicago, IL, USA
e-mail: wei.zhang1@northwestern.edu

J. Barciszewski (ed.), *Decoding Aging and Neurodegeneration*,
RNA Technologies 16, https://doi.org/10.1007/978-3-031-91072-2_16

determining disease development. In this chapter, we review current advances in the epigenetic contributions to normative aging and neurodegenerative diseases, focusing primarily on AD and related technological advances. Although the majority of the current research in this area has been on 5mC, the expansion of more recent research to such epigenetic modifications as 5hmC and m^6A will greatly enhance our understanding of normative aging and aging-associated diseases, improve biomarker discovery, and lay the foundation for individualized prevention and treatment.

Keywords Aging · Alzheimer's disease · Complex disease · Epigenetics · Cytosine modification · RNA modification · Biomarker

1 Introduction

Aging is a complex process that characterized by the deterioration of the human body's physical functions accompanied by the flow of time (Clarke and Korotchenko 2011). Normal aging, experienced by every individual is different from age-related disease, and their relationship and interaction are complex (Guo et al. 2022). Of particular interest are the epigenetic mechanisms underlying normal aging as well as Alzheimer's disease (AD) and AD-related dementia, which affects approximately five to seven million people and nearly 40% of the population of 85 years or older in the United States alone (Alzheimer's Association 2024). Therefore, AD represents one of the most common diseases associated with aging, and its prevalence is rapidly increasing. In this chapter, we discussed the difference between normal aging and a disease of aging, intending to use AD as the main example. In the meantime, we briefly introduced related technological advances, as the enabling techniques play a key role in enhancing our understanding of the systems biology of aging and aging-associated diseases. We emphasized why studying epigenetic modifications can both improve our understanding of the biology and lay the foundation for developing powerful clinical tools (e.g., diagnosis, prognosis). We also summarized the current epigenetic association studies of aging and AD. Importantly, because of the complex nature of aging and aging-related diseases (e.g., AD) as well as the advancement in enabling profiling technologies (e.g., suitable for precious, limited clinical biospecimens that have been difficult to access) during the past decades, we look forward to a systematic understanding of epigenetic contributions to normal aging and AD through, for example, integrating novel epigenetic modifications (e.g., DNA methylation and hydroxymethylation, RNA modification) using innovative techniques into related research. A comprehensive understanding of the epigenetic mechanisms underlying normal aging and AD will ultimately improve related biomarker discovery studies, for example, those using more convenient clinical biospecimens (e.g., blood) for disease prediction, detection, diagnosis, surveillance, and prognosis.

2 Normal Aging and Aging-Related Diseases, e.g., AD

Normal aging is a natural and inevitable biological process that involves the gradual decline of physiological and cognitive functions over time. These changes manifest in various ways, including a reduction in physical capacity, hearing and vision impairments, decreased metabolic efficiency, and cognitive slowing (Clarke and Korotchenko 2011; Guo et al. 2022). While normal aging is a universal experience, its progression and severity can vary significantly among individuals due to genetic predisposition, environmental exposure, and lifestyle choices. Key genetic factors, such as the presence of longevity-associated alleles in genes like *APOE* (encoding apolipoprotein E) (Deelen et al. 2019), *FOXO3* (encoding forkhead box O3) (Broer et al. 2015), and *CETP* (encoding cholesteryl ester transfer protein) (Sanders et al. 2010), have been implicated in determining the rate of biological aging (Melzer et al. 2020). Beyond genetics, external influences such as diet, physical activity, sleep quality, and exposure to pollutants can modulate aging trajectories, highlighting the interplay between intrinsic and extrinsic factors in the aging process (Nianogo et al. 2022).

Although normal aging involves a slow and progressive decline in cognitive and physical function, it is distinct from pathological aging and age-related diseases such as AD. Aging is the single most significant risk factor for many chronic diseases, including cardiovascular diseases, type 2 diabetes, cancer, and neurodegenerative disorders (Guo et al. 2022). These conditions often arise due to accumulated cellular damage, impaired metabolic homeostasis, and dysregulated inflammatory responses, which can exacerbate the effects of aging on the body (Nianogo et al. 2022). For example, in the case of cardiovascular diseases, endothelial dysfunction and chronic inflammation contribute to increased arterial stiffness and heightened susceptibility to atherosclerosis (Xu et al. 2021). Similarly, in cancer, aging-associated genomic instability and epigenetic alterations can promote oncogenic transformation and tumor progression (Fenoglio et al. 2018).

A particularly devastating age-related disease is AD, which is the most common form of dementia and a major cause of morbidity and mortality in the aging population. While normal aging may lead to mild cognitive decline, such as occasional forgetfulness or slower information processing, AD is characterized by profound neurodegeneration and significant disruption of brain function (Gonzales et al. 2022). Pathologically, AD is distinguished by the accumulation of β-amyloid (Aβ) plaques and neurofibrillary tangles (NFTs) composed of hyperphosphorylated tau protein, which contribute to synaptic dysfunction, neuronal loss, and ultimately brain atrophy (Alzheimer's Association 2024). The prevalence of AD is staggering, with approximately five to seven million individuals affected in the United States and over 55 million cases worldwide (Alzheimer's Association 2024). This number is projected to rise dramatically, reaching an estimated 13 million cases in the United States and 153 million globally by 2050 due to increasing life expectancy (GBD 2019). The economic burden of AD is also a growing concern, with global costs for care projected to range between $11.3 trillion and $27.3 trillion by 2050

(Nandi et al. 2022). These figures emphasize the urgency of developing early diagnostic tools and effective therapeutic strategies to mitigate the impact of AD on both individuals and healthcare systems.

Several risk factors contribute to AD beyond chronological aging, including genetic susceptibility, metabolic disorders, and lifestyle influences. The APOE4 allele is the most well-established genetic risk factor for sporadic AD, significantly increasing an individual's likelihood of developing the disease (Fenoglio et al. 2018). However, genetic predisposition alone does not fully account for AD risk, as environmental and modifiable lifestyle factors also play a crucial role. Diabetes, obesity, hypertension, physical inactivity, smoking, poor diet, and social isolation have all been associated with an increased risk of developing AD (Hou et al. 2019). Importantly, evidence suggests that preventive strategies such as maintaining a Mediterranean diet, engaging in regular physical activity, ensuring high-quality sleep, and managing cardiovascular risk factors can help delay or reduce the severity of AD symptoms (Alzheimer's Association 2024).

At the molecular level, AD pathogenesis is influenced by a combination of genetic, epigenetic, and inflammatory mechanisms. Beyond the hallmark amyloid and tau pathologies, increasing attention has been given to epigenetic dysregulation in AD, including altered DNA methylation, hydroxymethylation, histone modifications, and RNA modifications (Fenoglio et al. 2018; Andrews et al. 2023). These modifications can affect gene expression in key neurobiological pathways involved in synaptic plasticity, oxidative stress response, and neuroinflammation. For instance, recent studies have identified aberrant 5-hydroxymethylcytosine (5hmC) patterns in the brain tissue and cell-free DNA from AD patients, highlighting potential epigenetic biomarkers for early diagnosis and disease progression monitoring (Beadell et al. 2023).

Given the multifactorial nature of AD, future research must integrate genomic, transcriptomic, proteomic, and epigenomic data to achieve a more comprehensive understanding of its pathogenesis. Advancements in liquid biopsy techniques and machine learning-based biomarker discovery hold a significant promise for non-invasive detection, personalized risk assessment, and early interventions (Beadell et al. 2023). As our knowledge of the biological mechanisms underlying both normal aging and age-related diseases expands, targeting epigenetic pathways may offer novel therapeutic opportunities for delaying cognitive decline and improving brain health in aging populations.

However, despite these advancements, significant challenges remain in distinguishing normal aging from pathological aging, particularly in the context of neurodegenerative diseases such as AD. Aging is an inherently complex and multifaceted process that influences disease susceptibility, yet not all aging trajectories lead to pathology. A deeper understanding of the intricate interactions between genetic predisposition, environmental exposures, and epigenetic modifications is crucial for identifying individuals at higher risk and developing targeted interventions (Badhwar et al. 2020). Future studies should aim to refine integrative models that encompass molecular, cellular, and systemic factors, advancing our understanding of neurodegeneration and aging-related diseases while facilitating the development of clinically relevant diagnostic and therapeutic tools.

3 DNA Methylation Underlying Normal Aging and Aging-Related Diseases

Since the launch of the Human Genome Project (Collins and Fink 1995; Gibbs 2020), genome-wide association studies (GWAS) have been performed to investigate genetic alleles associated with various complex traits and diseases, including normal aging, for which genetics has been appreciated as one of many factors that affect its process (Rosoff et al. 2023; Sollis et al. 2023). For example, genetic association studies have implicated several genes in longevity, such as *APOE*, *FOX3*, and *CETP*. Notably, many of these genes are critical players in metabolism of important nutrients such as cholesterol and fatty acids and have been associated with aging-related cardiovascular diseases, indicating further the complex relationship between aging and aging-related diseases. Since the beginning of the GWAS era, it has long been recognized that while genetics is an important factor in the normal aging process, it is not the only determinant (Rosoff et al. 2023). A variety of non-genetic and environmental factors, such as lifestyle (Ng et al. 2020) (personal choices regarding diet, physical activity, smoking, alcohol intake, etc.), environment (Misra 2020) (exposure to pollution, chemicals, toxins, etc.), and life circumstances (Steptoe and Zaninotto 2020) (social support, socioeconomic status, education, access to quality healthcare, etc.), may have a stronger impact on the aging process than genetic factors alone (Fig. 1).

While these non-genetic factors (i.e., lifestyle, environment, and life circumstances) do not change an individual's germ-line alleles in the human genome, their effects could be mediated through epigenetic modifications, e.g., cytosine modifications on the DNA in gene dysregulation, and in turn reflect in cellular and whole body-level biological processes, including the process of aging (Soto-Palma et al. 2022). Of particular interest during the past 15–20 years are the investigations of the relationships between various epigenetic modifications and complex traits and diseases (Cortijo et al. 2014). To date, DNA methylation, i.e., 5-methylcytosine (5mC), at CpG dinucleotides in the human genome has been the most investigated cytosine modification type (Greenberg 2019).

The human genome contains an estimated 30 million CpG sites, which are potential targets for methylation (Rivera and Ren 2014). Throughout life, age-related changes in DNA methylation occur at specific genomic loci, including both hyper- and hypomethylation patterns. Global hypomethylation is commonly observed in aging cells, leading to genomic instability, activation of transposable elements, and increased mutation rates. Conversely, hypermethylation is frequently detected in promoter regions of genes associated with cellular senescence, inflammation, and cancer, suggesting a role in age-related functional decline (Fraga et al. 2005).

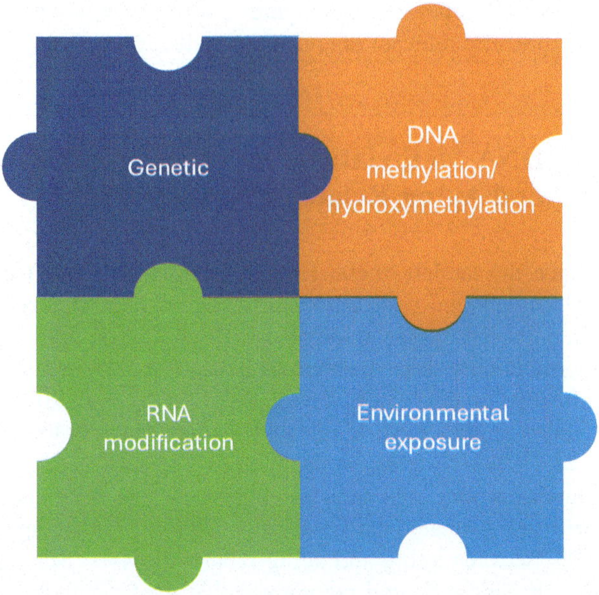

Fig. 1 A view of systems biology needed for understanding aging and aging-associated diseases. Development of complex diseases and traits is contributed by various genetic and non-genetic factors

Briefly, altered 5mC levels, whether increased or decreased compared to baseline, can result from lifestyle factors such as smoking and alcohol use. These changes may lead to gene dysregulation, affecting biological processes and pathways involved in aging (Johnson et al. 2012; Unnikrishnan et al. 2020). For example, aberrant DNA methylation has been implicated in numerous aging-related diseases, including cardiovascular diseases, neurodegenerative disorders, diabetes, and cancer (Unnikrishnan et al. 2020). Many of these diseases exhibit methylation changes in key regulatory genes, disrupting cellular homeostasis, and contribute to pathogenesis. Neurodegenerative diseases such as AD have been linked to epigenetic changes in genes like ANK1, B1N1, and BHBDF2, highlighting a role of methylation in neurodegeneration (De Jager et al. 2014). Figure 2 provides an overview of the tissue and cell sources used in epigenetic aging research, highlighting the brain tissue, cerebrospinal fluid (CSF), and blood as critical biological sources for studying epigenetic markers. This figure contextualizes how 5hmC and 5mC modifications on genomic DNA (gDNA) or circulating cell-free DNA (cfDNA), along with N^6-methyladenosine (m^6A) on RNA, contribute to our understanding of aging-related processes.

Using the whole-genome DNA methylation profiling techniques, such as the Illumina Infinium DNA methylation arrays and the "gold standard" bisulfite sequencing or its related techniques (e.g., RRBS (reduced representation bisulfite sequencing)), the research community has been actively investigating the relationships between DNA methylation and aging (Unnikrishnan et al. 2020). A milestone in this area of research has been the development of so-called epigenetic clock, which translates DNA methylation profiles into an "epigenetic age" that can estimate biological age and help predict the risk for aging-related diseases and

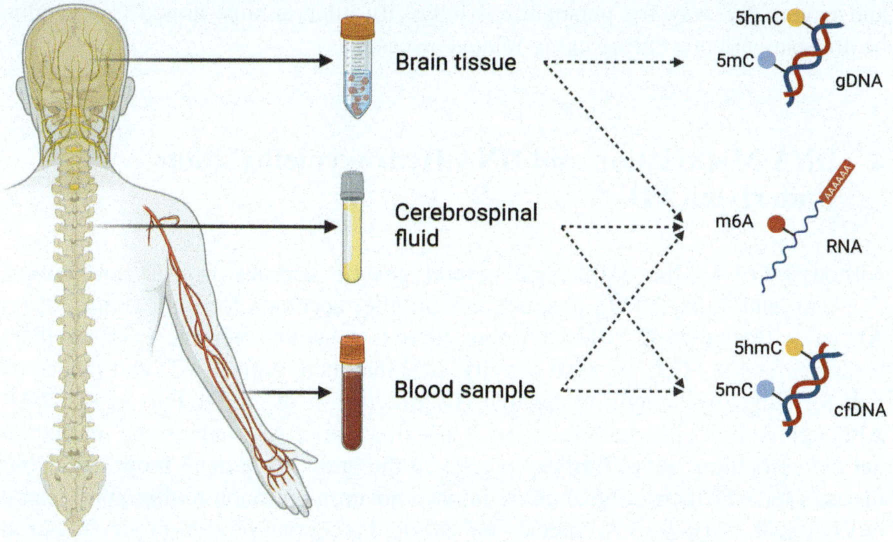

Fig. 2 Overview of tissue and cell sources for epigenetic aging research. The figure highlights the brain tissue, CSF, and blood as sources for studying epigenetic markers, including 5hmC and 5mC on gDNA or cfDNA and m6A on RNA, to explore aging-related processes

mortality (Horvath 2013). These clocks provide insights into age acceleration and serve as biomarkers for age-related diseases, mortality risk, and overall health status (Lu et al. 2019).

In a recent study of exceptionally long-lived species, Wilkinson et al. established that longevity-related DNA methylation changes are associated with genes involved in innate immunity or tumorigenesis (Wilkinson et al. 2021). This finding suggests that the enhanced immune response and cancer suppression observed in these species may contribute to normal aging and longevity, which offers a compelling explanation for the relationship between DNA methylation and aging-related diseases, considering that immune response is critical not only for cancers (Pardoll 2017) but also for cardiovascular diseases (Mann 2012) and AD (Van Eldik et al. 2016). In particular, immune system dysregulation has been identified as a key factor in AD, for example, through the recognition of misfolded Aβ and tau protein, which, if not cleared from brain cells, will trigger a complex immune response that ultimately leads to neuroinflammation and neurodegeneration (Chen and Yu 2023). Persistent low-grade inflammation plays a crucial role in age-related diseases, with DNA methylation changes in key inflammatory pathway genes, including interleukin-6, tumor necrosis factor alpha, and nuclear factor kappa B, being well-documented in aging and disease processes (Franceschi et al. 2018).

However, as the field advances, challenges remain, particularly in understanding causality versus correlation in DNA methylation changes and identifying modifiable epigenetic targets for intervention. By expanding our understanding of DNA methylation dynamics, we move closer to unlocking the molecular basis of aging and paving the way for personalized longevity interventions aimed at extending health span and preventing aging-related diseases.

4 DNA Methylation and DNA Hydroxymethylation Underlying AD

Although GWAS has implicated several genetic variants in AD development (Bertram and Tanzi 2009), genetic susceptibility accounts for only about 30% of AD cases (Ridge et al. 2013). Of note, there is growing scientific evidence that a healthy lifestyle (diet, physical activity, etc.) may help prevent AD and reduce the risk for subjective cognitive decline (Dominguez et al. 2021; Puri et al. 2023). Although pathological hallmarks, such as extracellular Aβ plaques and intraneuronal tau-containing neurofibrillary tangles in the brain, have been proposed as biomarkers for AD, their clinical use requires a positron emission tomography scan or biochemical analyses of Aβ species and amyloid precursor protein-processing products in CSF, and importantly their underlying molecular mechanisms are not fully understood (Guo et al. 2020; Dubois et al. 2023). Therefore, investigation of the relationships between epigenetic modifications and AD will not only provide novel insights into its pathogenesis but also provide biomarkers potentially useful in its clinical management, such as monitoring cognitive decline.

Like in other complex diseases, the most investigated epigenetic modification for AD is 5mC, although more recent studies have also started to investigate 5hmC (Branco et al. 2011), another major cytosine modification in the human genome, which is biochemically stable, reflects an active demethylation process from 5mC, and exhibits distinct genomic distribution and gene regulatory roles. Majority of earlier studies between cytosine modifications and AD have largely focused on a single pathology measure (i.e., Braak NFT staging) (Braak et al. 2006) and targeted the brain tissue samples. For example, Smith et al. utilized oxidative bisulfite conversion, in conjunction with the Illumina Infinium Human Methylation 450K microarray, to identify neuropathology-associated differential DNA methylation and DNA hydroxymethylation in the entorhinal cortex (Smith et al. 2019). In comparison, Lardenoije et al. also utilized the bisulfite conversion version and oxidative bisulfite conversion version of the Illumina arrays to identify AD-associated 5mC and 5hmC in the middle temporal gyrus (Lardenoije et al. 2019). Notably, in the Lardenoije study, the authors also showed the possibility of detecting altered 5mC and 5hmC between converters to AD dementia and non-converters at "a pre-clinical stage," indicating the feasibility of future development of blood-based epigenetic biomarkers for AD, although this study itself was limited in sample size. In contrast, in a recent study by Shireby et al., DNA methylation signatures of AD neurobiology in the cortex have been found to be primarily driven by variation in non-neuronal cell types (Shireby et al. 2022). Specifically, their team profiled DNA methylation in two regions of the cortex from 631 donors, performing an epigenome-wide association study of multiple measures of AD neuropathology. Their findings suggested that the majority of differentially methylated loci identified in "bulk" cortex tissue reflected DNA methylation differences occurring in non-neuronal cells, thus highlighting the importance of characterizing disease-associated variation in purified cell types and the simultaneous analysis of multiple measures of pathology.

More recently, with the aim of developing highly sensitive epigenetic biomarkers for AD, especially to evaluate the possibility of developing non-invasive biomarkers in liquid biopsies—circulating cfDNA, a highly sensitive chemical labeling technique, the 5hmC-Seal was invented (Song et al. 2011; Li et al. 2017). The 5hmC-Seal technique has been utilized by our team and the research community in biomarker discovery for various human diseases, including cancers (Cai et al. 2019, 2021; Ren et al. 2023; Zeng et al. 2023), diabetic complications (Han et al. 2021; Yang et al. 2022), and AD (Beadell et al. 2023). Given that diabetes is a recognizable risk factor for dementia (Kodl and Seaquist 2008; McCrimmon et al. 2012), we reasoned that deeper understanding of the mechanisms linking diabetes to the clinical expression of AD dementia could provide insights into much-needed dementia therapeutics. Specifically, in a set of 40 age- and sex-matched AD patients (brain- and serum-derived cfDNA) and 40 non-AD controls, with and without type 2 diabetes background from the Rush University Memory and Aging Project cohort (Bennett et al. 2019), we investigated diabetes as a dementia risk factor by examining genome-wide distributions of 5hmC and compared their distributions between blood and brain samples. After obtaining genome-wide 5hmC profiles in the 80 samples across four groups (AD, diabetes clinically defined, AD with diabetes, and individuals with neither disease—controls), we performed differential analysis and demonstrated the potential diagnostic value of 5hmC profiling in circulating cfDNA. Specifically, an 11-gene weighted model was found to distinguish AD from controls with an area under the curve (AUC) of 91.8%, while a four-gene model could distinguish diabetes-associated AD from AD alone with an AUC of 87.9%. Our findings echoed a previous study by Zhao et al., who used the 5hmC-Seal technique in brain samples to show the possibility of identifying differential 5hmC features associated with neuritic plaques and neurofibrillary tangles (Zhao et al. 2017), although Zhao et al. did not show associated 5hmC with AD directly, possibly due to some analytical challenges that our team tried to overcome later in our recent study. Therefore, we demonstrated in this pilot study, although relatively small in sample size, the feasibility of detecting and characterizing 5hmC in diabetes-associated AD and of using 5hmC information contained in circulating cfDNA to detect AD in high-risk individuals, such as those with diabetes.

Table 1 and Fig. 3 provide a summary of current techniques for profiling DNA modifications. Technically, due to the limited quantity of cfDNA, traditional bisulfite sequencing methods typically yield sparse genomic coverage. To address this limitation, we have developed the LABS technique (Cui et al. 2024), which employs linear amplification of bisulfite-converted DNA fragments. This approach enhances genome-wide coverage by reducing amplification bias. Additionally, the mixed composition of cfDNA benefits from linear amplification, as underrepresented DNA fragments can be efficiently captured and amplified. As a result, the LABS

Table 1 Sequencing techniques for profiling of DNA and RNA modification

Methylation type	Sequencing technique	Resolution	References
5mC	MeDIP-seq	Limited	Weber et al. (2005)
	WGBS	Single-base	Lister et al. (2009)
	TAPS	Single-base	Liu et al. (2019)
	EM-seq	Single-base	Vaisvila et al. (2021)
	LABS	Single-base	Cui et al. (2024)
	UBS-seq	Single-base	Dai et al. (2024)
5hmC	hMeDIP-seq	Limited	Ficz et al. (2011)
	5hmC-seal	Limited	Song et al. (2011)
	CMS-seq	Limited	Pastor et al. (2011)
	TAB-seq	Single-base	Yu et al. (2012)
	oxBS-seq	Single-base	Booth et al. (2012)
	ACE-seq	Single-base	Schutsky et al. (2018)
	CAPS	Single-base	Liu et al. (2021)
m^6A	MeRIP-seq	Limited	Dominissini et al. (2013)
	miCLIP	Single-base	Linder et al. (2015)
	m6A-LAIC-seq	Limited	Molinie et al. (2016)
	MAZTER-seq	Single-base	Garcia-Campos et al. (2019)
	m6A-REF-seq	Single-base	Zhang et al. (2019)
	DART-seq	Single-base	Meyer (2019)
	m6A-label-seq	Single-base	Shu et al. (2020)
	m6A-SAC-seq	Single-base	Hu et al. (2022)
	GLORI	Single-base	Liu et al. (2023)
	eTAM-seq	Single-base	Xiao et al. (2023)

could offer increased sensitivity, making it a valuable method for liquid biopsy applications. We anticipate that such technical improvements will play a significant role in advancing the discovery of epigenetic biomarkers for AD in future studies.

5 Epigenetic Clock and Aging

The concept of the "epigenetic clock" has revolutionized aging research, offering a molecular biomarker capable of estimating biological age based on specific DNA methylation patterns at CpG sites across the genome (Horvath 2013) (Table 2). As previously discussed, DNA methylation serves as a critical epigenetic modification influencing gene regulation and aging processes. The epigenetic clock builds on this principle, leveraging age-associated methylation signatures to develop predictive models of aging. Prominent examples, such as the Horvath clock (Horvath 2013) and Hannum clock (Hannum et al. 2013), have demonstrated remarkable accuracy in correlating with chronological age while also providing insights into biological age disparities driven by environmental and genetic factors.

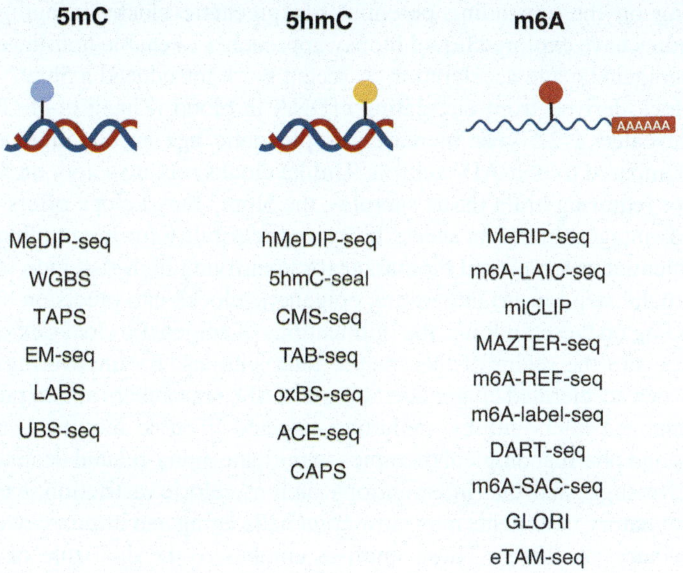

Fig. 3 Sequencing techniques for profiling DNA and RNA methylation

Table 2 Overview of epigenetic clocks

Name	Number of CpGs	Tissue/cell source	References
Horvath clock	353	Multi-tissue (51 tissues and cell types)	Horvath (2013)
Hannum clock	71	Whole blood	Hannum et al. (2013)
PhenoAge	513	Whole blood	Levine et al. (2018)
GrimAge	1030	Whole blood	Lu et al. (2019)
Cortical clock	~347	Prefrontal cortex	Shireby et al. (2020)

One of the most compelling applications of epigenetic clocks is their ability to detect accelerated epigenetic aging, where biological age surpasses chronological age. This phenomenon has been linked to an increased risk of mortality, heightened vulnerability to age-related diseases (Bell et al. 2019), including neurodegenerative conditions such as AD (Shadyab et al. 2022). For instance, Grodstein et al. analyzed dorsolateral prefrontal cortex specimens from 721 older participants in the Religious Orders Study and the Rush Memory and Aging Project (Bennett et al. 2019) to evaluate five epigenetic clocks: Hannum, Horvath, PhenoAge (Levine et al. 2018), GrimAge (Lu et al. 2019), and the recently developed cortical clock (Shireby et al. 2020), which is uniquely trained in the postmortem cortical tissue. Among these, the cortical clock exhibited the strongest association with AD pathology and was significantly linked to clinical phenotypes such as dementia, cognitive decline across multiple domains, and Parkinsonian signs. These findings highlight the importance of brain-specific epigenetic clocks in advancing our understanding of neurodegenerative diseases and developing strategies for early diagnosis and intervention.

Building on the expanding potential of epigenetic clocks in aging research, recent studies have explored liquid biopsy approaches to enable non-invasive monitoring of epigenetic age acceleration. A recent work introduced a 5hmC-based epigenetic clock derived from circulating cfDNA in blood (Zhang et al. 2023). This method revealed a 2.5-year increase in epigenetic age acceleration among AD patients compared to non-AD controls. Unlike clocks relying solely on 5mC modifications or requiring brain tissue samples, the 5hmC-based clock offers significant clinical advantages due to its accessibility and feasibility for longitudinal monitoring. Its potential as a practical biomarker for identifying high-risk individuals highlights a crucial step toward integrating epigenetic clocks into precision medicine.

Expanding on these findings, the implications of epigenetic clocks extend beyond diagnostics into the realm of therapeutic interventions. By uncovering molecular links between accelerated epigenetic aging and AD, researchers aim to pave the way for personalized interventions, including targeted lifestyle modifications, dietary strategies, and pharmacological therapies to mitigate aging-related decline. Notably, studies have suggested that interventions such as caloric restriction, exercise, and anti-inflammatory treatments may slow epigenetic aging and improve overall health span (Levine et al. 2018). These findings emphasize the dual role of epigenetic clocks as both aging biomarkers and predictors of therapeutic efficacy, bridging the gap between molecular diagnostics and precision gerontology. The ongoing advancements in epigenetic clock methodologies hold a significant promise for transforming our understanding of aging, enabling early interventions, and ultimately fostering longer and healthier lifespans.

6 Future Directions: RNA Modifications and Challenges

Beyond cytosine modifications such as 5mC and 5hmC, recent advancements in sequencing technologies have enabled robust profiling of emerging epigenetic modifications, shedding new light on the intricate regulatory mechanisms underlying aging and AD. As discussed in previous sections, DNA methylation-based epigenetic clocks have provided powerful tools for aging research. However, these clocks primarily rely on DNA modifications, whereas RNA modifications—a rapidly growing field—are gaining recognition for their critical roles in gene regulation and disease pathogenesis. Given the fast-accumulating evidence supporting their significance in normal biological processes and complex disorders, further investigation into RNA epigenetics is imperative for understanding age-associated diseases such as AD.

Among RNA modifications, m^6A stands out as the most abundant internal mRNA modification in mammals (Zaccara et al. 2019). It plays a pivotal role in numerous biological processes, including stem cell differentiation, neurodevelopment, and immune responses (Jiang et al. 2021). Dysregulated m^6A modifications and their associated regulatory machinery—including writers (METTL3, METTL14), erasers (FTO), and readers (YTHDC1, IGFBP)—have been strongly implicated in complex diseases such as diabetes, cancer, and neurodegenerative disorders (Wang et al. 2022; Li et al. 2023). Given the well-established connection between diabetes and an increased risk of AD, it is crucial to delineate the role of m^6A modifications in both metabolic dysfunction and neurodegeneration. However, despite their potential, the precise contributions of RNA modifications in diabetes-associated AD remain poorly understood, largely due to technical limitations in detecting and quantifying these modifications across the transcriptome.

Sequencing techniques of m^6A have significantly advanced our understanding of RNA modifications, yet they face key challenges in resolution, sensitivity, and clinical applicability. Current methodologies for m^6A profiling face significant challenges (Table 1, Fig. 3). The commonly used antibody-based approaches (e.g., MeRIP-seq) suffer from low resolution, limited stoichiometric information, and challenges in cross-condition comparisons (McIntyre et al. 2020). Although emerging antibody-independent techniques such as MAZTER-seq and m^6A-REF-seq offer improvements in specificity, they still cover only ~16% of m^6A sites and exhibit low detection sensitivity (Zhong et al. 2023). Furthermore, more recent approaches like m^6A-SEAL improve selectivity but still lack quantification capabilities (Hu et al. 2022). Additionally, cross-linking-based methods, such as miCLIP (Linder et al. 2015), provide single-nucleotide resolution but can introduce RNA degradation artifacts and require complex bioinformatics analysis. Enzyme- and chemical-based approaches, including DART-seq (Meyer 2019) and m^6A-label-seq (Shu et al. 2020), improve site specificity but are often restricted to sequence motifs, limiting transcriptome-wide coverage. Most existing techniques also fail to provide reliable stoichiometric data, making it difficult to quantify the functional impact of m^6A modifications. These challenges, coupled with computational complexity and limited suitability for low-input clinical samples, hinder our ability to precisely map m^6A dynamics and assess their functional relevance in aging and neurodegeneration.

To overcome these limitations, the development of next-generation epitranscriptomic tools is paramount. Novel quantitative and high-resolution profiling methods such as m^6A-SAC-seq provide enhanced detection sensitivity and quantitative resolution, making them particularly valuable for studying rare or limited clinical biospecimens (Hu et al. 2022; Ge et al. 2023). As discussed in previous sections, liquid biopsy-based epigenetic markers have revolutionized non-invasive diagnostics. Expanding these methods to include RNA modifications could significantly enhance our ability to track epigenetic aging signatures and develop novel biomarkers for age-related diseases, including AD.

Future advancements should focus on hybrid methods that combine high resolution with broader transcriptomic coverage, improved antibody specificity, enhanced quantification, and machine learning-driven analysis to refine m^6A site detection and expand its clinical applications in neurodegenerative disease research. Integrating RNA modifications into multi-omics frameworks alongside DNA methylation, histone modifications, and proteomic data will be essential for a comprehensive understanding of the epigenetic landscape of aging. Additionally, prioritizing longitudinal and tissue-specific RNA modification profiling will help uncover their temporal dynamics in aging and disease progression. By advancing both methodological innovations and biological insights, RNA epigenetics has the potential to reshape aging research, offering new diagnostic and therapeutic strategies to mitigate age-related diseases and promote healthy longevity.

7 Conclusions

While GWAS has identified multiple genetic variants associated with normal aging and AD, genetics alone cannot fully explain the observed individual heterogeneity in aging and disease susceptibility. As explored in previous sections, epigenetic modifications—including 5mC, 5hmC, and m^6A—serve as crucial regulatory layers that mediate gene expression and influence biological processes central to aging and AD pathogenesis. These modifications bridge the gap between genetic predisposition and environmental influences, reinforcing the importance of a multi-dimensional approach in studying aging biology.

Recent advancements in high-resolution epigenomic technologies have revolutionized the ability to profile 5mC, 5hmC, and m^6A using minimal clinical biospecimens, particularly cfDNA and RNA from liquid biopsies. These breakthroughs provide promising avenues for developing non-invasive, clinically relevant biomarkers for aging and AD. As discussed in Fig. 2, diverse tissue sources—including brain tissue, CSF, and blood—can be leveraged to study these modifications, enabling researchers to establish epigenetic aging signatures that may guide diagnostic and therapeutic strategies.

While significant progress has been made in deciphering the roles of 5hmC in normal aging and m^6A in AD, research remains in its early stages. Future studies must systematically integrate these epigenetic players to construct a comprehensive model of aging and neurodegeneration. This effort will require multi-omics approaches, combining genomics, epigenomics, transcriptomics, and proteomics, to uncover complex regulatory networks underpinning disease progression. Additionally, as highlighted in previous sections, the development of more cost-effective, accurate, and high-throughput technologies will be crucial in making large-scale epigenetic studies feasible and accessible for clinical application.

Despite the existing challenges, emerging AI-driven analytical frameworks and access to expansive scientific consortia, such as the Epigenomics Roadmap Project (Bernstein et al. 2013), provide unprecedented opportunities for decoding epigenetic mechanisms in aging and AD. AI and machine learning techniques can facilitate pattern recognition across massive datasets, enabling researchers to identify novel epigenetic biomarkers and therapeutic targets with greater precision. As these methodologies evolve, they hold the potential to transform how we understand, predict, and manage aging-related diseases.

In conclusion, epigenetics has emerged as a pivotal factor in shaping aging trajectories and modulating AD risk. The integration of 5mC, 5hmC, and m^6A into aging research not only deepens our mechanistic understanding but also propels the field toward personalized medicine and early intervention strategies. The convergence of technological advancements, computational innovations, and interdisciplinary collaborations will ultimately drive forward the goal of unraveling the epigenetic foundations of normal aging and aging-related diseases. Future studies leveraging these cutting-edge tools and approaches will bring us closer to realizing precision aging medicine, with the potential to enhance longevity and improve quality of life in aging populations.

Acknowledgments This work was supported in part by grants (R21AG091553, RF1AG074549) from the National Institutes of Health.

References

Alzheimer's Association (2024) 2024 Alzheimer's disease facts and figures. Alzheimers Dement 20:3708–3821

Andrews SJ, Renton AE, Fulton-Howard B et al (2023) The complex genetic architecture of Alzheimer's disease: novel insights and future directions. EBioMedicine 90:104511

Badhwar A, McFall GP, Sapkota S et al (2020) A multiomics approach to heterogeneity in Alzheimer's disease: focused review and roadmap. Brain 143:1315–1331

Beadell AV, Zhang Z, Capuano AW et al (2023) Genome-wide mapping implicates 5-Hydroxymethylcytosines in diabetes mellitus and Alzheimer's disease. J Alzheimers Dis 93:1135–1151

Bell CG, Lowe R, Adams PD et al (2019) DNA methylation aging clocks: challenges and recommendations. Genome Biol 20:249

Bennett DA, Buchman AS, Boyle PA et al (2019) Religious orders study and rush memory and aging project. J Alzheimers Dis 64:S161–S189

Bernstein BE, Stamatoyannopoulos JA, Costello JF et al (2013) The NIH roadmap epigenomics mapping consortium. Nat Biotechnol 28:1045–1048

Bertram L, Tanzi RE (2009) Genome-wide association studies in Alzheimer's disease. Hum Mol Genet 18:R137–R145

Booth MJ, Branco MR, Ficz G et al (2012) Quantitative sequencing of 5-methylcytosine and 5-hydroxymethylcytosine at single-base resolution. Science 336:934–937

Braak H, Alafuzoff I, Arzberger T et al (2006) Staging of Alzheimer disease-associated neurofibrillary pathology using paraffin sections and immunocytochemistry. Acta Neuropathol 112:389–404

Branco MR, Ficz G, Reik W (2011) Uncovering the role of 5-hydroxymethylcytosine in the epigenome. Nat Rev Genet 13:7–13

Broer L, Buchman AS, Deelen J et al (2015) GWAS of longevity in CHARGE consortium confirms APOE and FOXO3 candidacy. J Gerontol A Biol Sci Med Sci 70:110–118

Cai J, Chen L, Zhang Z et al (2019) Genome-wide mapping of 5-hydroxymethylcytosines in circulating cell-free DNA as a non-invasive approach for early detection of hepatocellular carcinoma. Gut 68:2195–2205

Cai J, Zeng C, Hua W et al (2021) An integrative analysis of genome-wide 5-hydroxymethylcytosines in circulating cell-free DNA detects noninvasive diagnostic markers for gliomas. Neurooncol Adv 3:vdab049

Chen Y, Yu Y (2023) Tau and neuroinflammation in Alzheimer's disease: interplay mechanisms and clinical translation. J Neuroinflammation 20:165

Clarke LH, Korotchenko A (2011) Aging and the body: a review. Can J Aging 30:495–510

Collins FS, Fink L (1995) The human genome project. Alcohol Health Res World 19:190–195

Cortijo S, Wardenaar R, Colome-Tatche M et al (2014) Mapping the epigenetic basis of complex traits. Science 343:1145–1148
Cui XL, Nie J, Zhu H et al (2024) LABS: linear amplification-based bisulfite sequencing for ultrasensitive cancer detection from cell-free DNA. Genome Biol 25:157
Dai Q, Ye C, Irkliyenko I et al (2024) Ultrafast bisulfite sequencing detection of 5-methylcytosine in DNA and RNA. Nat Biotechnol 42:1559–1570
De Jager PL, Srivastava G, Lunnon K et al (2014) Alzheimer's disease: early alterations in brain DNA methylation at ANK1, BIN1, RHBDF2 and other loci. Nat Neurosci 17:1156–1163
Deelen J, Evans DS, Arking DE et al (2019) A meta-analysis of genome-wide association studies identifies multiple longevity genes. Nat Commun 10:3669
Dominguez LJ, Veronese N, Vernuccio L et al (2021) Nutrition, physical activity, and other lifestyle factors in the prevention of cognitive decline and dementia. Nutrients 13:4080
Dominissini D, Moshitch-Moshkovitz S, Salmon-Divon M et al (2013) Transcriptome-wide mapping of N(6)-methyladenosine by m(6)A-seq based on immunocapturing and massively parallel sequencing. Nat Protoc 8:176–189
Dubois B, von Arnim CAF, Burnie N et al (2023) Biomarkers in Alzheimer's disease: role in early and differential diagnosis and recognition of atypical variants. Alzheimers Res Ther 15:175
Fenoglio C, Scarpini E, Serpente M et al (2018) Role of genetics and epigenetics in the pathogenesis of Alzheimer's disease and frontotemporal dementia. J Alzheimers Dis 62:913–932
Ficz G, Branco MR, Seisenberger S et al (2011) Dynamic regulation of 5-hydroxymethylcytosine in mouse ES cells and during differentiation. Nature 473:398–402
Fraga MF, Ballestar E, Paz MF et al (2005) Epigenetic differences arise during the lifetime of monozygotic twins. Proc Natl Acad Sci USA 102:10604–10609
Franceschi C, Garagnani P, Parini P et al (2018) Inflammaging: a new immune-metabolic viewpoint for age-related diseases. Nat Rev Endocrinol 14:576–590
Garcia-Campos MA, Edelheit S, Toth U et al (2019) Deciphering the "m6A Code" via antibody-independent quantitative profiling. Cell 178:731–747.e716
GBD 2019 Dementia forecasting collaborators (2022) Estimation of the global prevalence of dementia in 2019 and forecasted prevalence in 2050: an analysis for the global burden of disease study 2019. Lancet Public Health 7:e105–e125
Ge R, Ye C, Peng Y et al (2023) m(6)A-SAC-seq for quantitative whole transcriptome m(6)A profiling. Nat Protoc 18:626–657
Gibbs RA (2020) The human genome project changed everything. Nat Rev Genet 21:575–576
Gonzales MM, Garbarino VR, Pollet E et al (2022) Biological aging processes underlying cognitive decline and neurodegenerative disease. J Clin Invest 132:e158453
Greenberg MVC, Bourc'his D (2019) The diverse roles of DNA methylation in mammalian development and disease. Nat Rev Mol Cell Biol 20:590–607
Guo T, Zhang D, Zeng Y et al (2020) Molecular and cellular mechanisms underlying the pathogenesis of Alzheimer's disease. Mol Neurodegener 15:40
Guo J, Huang X, Dou L et al (2022) Aging and aging-related diseases: from molecular mechanisms to interventions and treatments. Signal Transduct Target Ther 7:391
Han L, Chen C, Lu X et al (2021) Alterations of 5-hydroxymethylcytosines in circulating cell-free DNA reflect retinopathy in type 2 diabetes. Genomics 113:79–87
Hannum G, Guinney J, Zhao L et al (2013) Genome-wide methylation profiles reveal quantitative views of human aging rates. Mol Cell 49:359–367
Horvath S (2013) DNA methylation age of human tissues and cell types. Genome Biol 14:R115
Hou Y, Dan X, Babbar M et al (2019) Ageing as a risk factor for neurodegenerative disease. Nat Rev Neurol 15(10): 565–581
Hu L, Liu S, Peng Y et al (2022) m(6)A RNA modifications are measured at single-base resolution across the mammalian transcriptome. Nat Biotechnol 40:1210–1219
Jiang X, Liu B, Nie Z et al (2021) The role of m6A modification in the biological functions and diseases. Signal Transduct Target Ther 6:74

Johnson AA, Akman K, Calimport SR et al (2012) The role of DNA methylation in aging, rejuvenation, and age-related disease. Rejuvenation Res 15:483–494
Kodl CT, Seaquist ER (2008) Cognitive dysfunction and diabetes mellitus. Endocr Rev 29:494–511
Lardenoije R, Roubroeks JAY, Pishva E et al (2019) Alzheimer's disease-associated (hydroxy) methylomic changes in the brain and blood. Clin Epigenetics 11:164
Levine ME, Lu AT, Quach A et al (2018) An epigenetic biomarker of aging for lifespan and healthspan. Aging (Albany NY) 10:573–591
Li W, Zhang X, Lu X et al (2017) 5-Hydroxymethylcytosine signatures in circulating cell-free DNA as diagnostic biomarkers for human cancers. Cell Res 27:1243–1257
Li X, Yang Y, Chen Z (2023) Downregulation of the m(6)A reader protein YTHDC1 leads to islet beta-cell failure and diabetes. Metabolism 138:155339
Linder B, Grozhik AV, Olarerin-George AO et al (2015) Single-nucleotide-resolution mapping of m6A and m6Am throughout the transcriptome. Nat Methods 12:767–772
Lister R, Pelizzola M, Dowen RH et al (2009) Human DNA methylomes at base resolution show widespread epigenomic differences. Nature 462:315–322
Liu Y, Siejka-Zielińska P, Velikova G et al (2019) Bisulfite-free direct detection of 5-methylcytosine and 5-hydroxymethylcytosine at base resolution. Nat Biotechnol 37:424–429
Liu Y, Hu Z, Cheng J et al (2021) Subtraction-free and bisulfite-free specific sequencing of 5-methylcytosine and its oxidized derivatives at base resolution. Nat Commun 12:618
Liu C, Sun H, Yi Y et al (2023) Absolute quantification of single-base m6A methylation in the mammalian transcriptome using GLORI. Nat Biotechnol 41:355–366
Lu AT, Quach A, Wilson JG et al (2019) DNA methylation GrimAge strongly predicts lifespan and healthspan. Aging (Albany NY) 11:303–327
Mann DL (2012) The emerging role of innate immunity in the heart and vascular system: for whom the cell tolls. Circ Res 108:1133–1145
McCrimmon RJ, Ryan CM, Frier BM (2012) Diabetes and cognitive dysfunction. Lancet 379:2291–2299
McIntyre ABR, Gokhale NS, Cerchietti L et al (2020) Limits in the detection of m(6)A changes using MeRIP/m(6)A-seq. Sci Rep 10:6590
Melzer D, Pilling LC, Ferrucci L (2020) The genetics of human ageing. Nat Rev Genet 21:88–101
Meyer KD (2019) DART-seq: an antibody-free method for global m6A detection. Nat Methods 16:1275–1280
Misra BB (2020) The chemical exposome of human aging. Front Genet 11:574936
Molinie B, Wang J, Lim KS et al (2016) m6A-LAIC-seq reveals the census and complexity of the m6A epitranscriptome. Nat Methods 13:692–698
Nandi A, Counts N, Chen S et al (2022) Global and regional projections of the economic burden of Alzheimer's disease and related dementias from 2019 to 2050: a value of statistical life approach. EClinicalMedicine 51:101580
Ng R, Sutradhar R, Yao Z et al (2020) Smoking, drinking, diet and physical activity-modifiable lifestyle risk factors and their associations with age to first chronic disease. Int J Epidemiol 49:113–130
Nianogo RA, Rosenwohl-Mack A, Yaffe K et al (2022) Risk factors associated with Alzheimer disease and related dementias by sex and race and ethnicity in the US. JAMA Neurol 79:584–591
Pardoll D (2017) Cancer and the immune system: basic concepts and targets for intervention. Semin Oncol 42:523–538
Pastor WA, Pape UJ, Huang Y et al (2011) Genome-wide mapping of 5-hydroxymethylcytosine in embryonic stem cells. Nature 473:394–397
Puri S, Shaheen M, Grover B (2023) Nutrition and cognitive health: a life course approach. Front Public Health 11:1023907
Ren Y, Zhang Z, She Y et al (2023) A highly sensitive and specific non-invasive test through genome-wide 5-hydroxymethylation mapping for early detection of lung cancer. Small Methods:e2300747

Ridge PG, Mukherjee S, Crane PK et al (2013) Alzheimer's disease: analyzing the missing heritability. PLoS One 8:e79771

Rivera CM, Ren B (2014) Mapping human epigenomes. Cell 155:39–55

Rosoff DB, Mavromatis LA, Bell AS et al (2023) Multivariate genome-wide analysis of aging-related traits identifies novel loci and new drug targets for healthy aging. Nat Aging 3:1020–1035

Sanders AE, Wang C, Katz M et al (2010) Association of a functional polymorphism in the cholesteryl ester transfer protein (CETP) gene with memory decline and incidence of dementia. JAMA 303:150–158

Schutsky EK, DeNizio JE, Hu P et al (2018) Nondestructive, base-resolution sequencing of 5-hydroxymethylcytosine using a DNA deaminase. Nat Biotechnol 36:1083–1090

Shadyab AH, McEvoy LK, Horvath S et al (2022) Association of epigenetic age acceleration with incident mild cognitive impairment and dementia among older women. J Gerontol A Biol Sci Med Sci 77:1239–1244

Shireby GL, Davies JP, Francis PT et al (2020) Recalibrating the epigenetic clock: implications for assessing biological age in the human cortex. Brain 143:3763–3775

Shireby G, Dempster EL, Policicchio S et al (2022) DNA methylation signatures of Alzheimer's disease neuropathology in the cortex are primarily driven by variation in non-neuronal cell-types. Nat Commun 13:5620

Shu X, Cao J, Cheng M et al (2020) A metabolic labeling method detects m6A transcriptome-wide at single base resolution. Nat Chem Biol 16:887–895

Smith AR, Smith RG, Pishva E et al (2019) Parallel profiling of DNA methylation and hydroxymethylation highlights neuropathology-associated epigenetic variation in Alzheimer's disease. Clin Epigenetics 11:52

Sollis E, Mosaku A, Abid A et al (2023) The NHGRI-EBI GWAS catalog: knowledgebase and deposition resource. Nucleic Acids Res 51:D977–D985

Song CX, Szulwach KE, Fu Y et al (2011) Selective chemical labeling reveals the genome-wide distribution of 5-hydroxymethylcytosine. Nat Biotechnol 29:68–72

Soto-Palma C, Niedernhofer LJ, Faulk CD et al (2022) Epigenetics, DNA damage, and aging. J Clin Invest 132:e158446

Steptoe A, Zaninotto P (2020) Lower socioeconomic status and the acceleration of aging: an outcome-wide analysis. Proc Natl Acad Sci USA 117:14911–14917

Unnikrishnan A, Freeman WM, Jackson J et al (2020) The role of DNA methylation in epigenetics of aging. Pharmacol Ther 195:172–185

Vaisvila R, Ponnaluri VKC, Sun Z et al (2021) Enzymatic methyl sequencing detects DNA methylation at single-base resolution from picograms of DNA. Genome Res 31:1280–1289

Van Eldik LJ, Carrillo MC, Cole PE et al (2016) The roles of inflammation and immune mechanisms in Alzheimer's disease. Alzheimers Dement (NY) 2:99–109

Wang Y, Wang Y, Gu J et al (2022) The role of RNA m6A methylation in lipid metabolism. Front Endocrinol (Lausanne) 13:866116

Weber M, Davies JJ, Wittig D et al (2005) Chromosome-wide and promoter-specific analyses identify sites of differential DNA methylation in normal and transformed human cells. Nat Genetics 37:853–862

Wilkinson GS, Adams DM, Haghani A et al (2021) DNA methylation predicts age and provides insight into exceptional longevity of bats. Nat Commun 12:1615

Xiao Y-L, Liu S, Ge R et al (2023) Transcriptome-wide profiling and quantification of N6-methyladenosine by enzyme-assisted adenosine deamination. Nat Biotechnol 41:993–1003

Xu S, Ilyas I, Little PJ et al (2021) Endothelial dysfunction in atherosclerotic cardiovascular diseases and beyond: from mechanism to pharmacotherapies. Pharmacol Rev 73:924–967

Yang Y, Zeng C, Yang K et al (2022) Genome-wide analysis reflects novel 5-Hydroxymethylcytosines implicated in diabetic nephropathy and the biomarker potential. Extracell Vesicles Circ Nucl Acids 3:49–60

Yu M, Hon GC, Szulwach KE et al (2012) Tet-assisted bisulfite sequencing of 5-hydroxymethylcytosine. Nat Protocols 7:2159–2170

Zaccara S, Ries RJ, Jaffrey SR (2019) Reading, writing and erasing mRNA methylation. Nat Rev Mol Cell Biol 20:608–624
Zeng C, Song X, Zhang Z et al (2023) Dissection of transcriptomic and epigenetic heterogeneity of grade 4 gliomas: implications for prognosis. Acta Neuropathol Commun 11:133
Zhang Z, Chen LQ, Zhao YL et al (2019) Single-base mapping of m(6)A by an antibody-independent method. Sci Adv 5:eaax0250
Zhang Z, Beadell A, Cui X-l et al (2023) A 5-hydroxymethylation-based clock derived from cell-free DNA in blood reveals accelerated epigenetic aging in patients with Alzheimer's disease. Alzheimers Dement 19:e076673
Zhao J, Zhu Y, Yang J et al (2017) A genome-wide profiling of brain DNA hydroxymethylation in Alzheimer's disease. Alzheimers Dement 13:674–688
Zhong ZD, Xie YY, Chen HX et al (2023) Systematic comparison of tools used for m(6)A mapping from nanopore direct RNA sequencing. Nat Commun 14:1906

The manufacturer's authorised representative in the EU is Springer Nature Customer Service Centre GmbH, Europaplatz 3, 69115 Heidelberg, Germany. If you have any concerns regarding our products, please contact ProductSafety@springernature.com

Printed and bound by CPI Group (UK) Ltd, Croydon, CR0 4YY

30/09/2026

02225454-0010